PHYSICS, CHEMISTRY AND APPLICATIONS OF NANOSTRUCTURES

PROCEEDINGS OF
INTERNATIONAL CONFERENCE NANOMEETING – 2017

PHYSICS, CHEMISTRY AND APPLICATIONS OF NANOSTRUCTURES

REVIEWS AND SHORT NOTES

Minsk, Belarus, 30 May – 2 June 2017

editors

V E Borisenko

Belarusian State University of Informatics and Radioelectronics, Belarus

S V Gaponenko

National Academy of Sciences of Belarus, Belarus

V S Gurin

Belarusian State University, Belarus

C H Kam

Nanyang Technological University, Singapore

World Scientific

NEW JERSEY • LONDON • SINGAPORE • BEIJING • SHANGHAI • HONG KONG • TAIPEI • CHENNAI

Published by

World Scientific Publishing Co. Pte. Ltd.
5 Toh Tuck Link, Singapore 596224
USA office: 27 Warren Street, Suite 401-402, Hackensack, NJ 07601
UK office: 57 Shelton Street, Covent Garden, London WC2H 9HE

British Library Cataloguing-in-Publication Data
A catalogue record for this book is available from the British Library.

PHYSICS, CHEMISTRY AND APPLICATION OF NANOSTRUCTURES
Reviews and Short Notes to Nanomeeting-2017

ISBN 978-981-3224-52-0

Desk Editor: Rhaimie Wahap

Printed in Singapore

INTERNATIONAL CONFERENCE

NANOMEETING-2017

Minsk, Belarus, May 30 - June 2, 2017

ORGANIZERS

Ministry of Education of Belarus

Belarusian State University of Informatics and Radioelectronics

Nanyang Technological University

Centre National de la Recherche Scientifique

SOL Instruments

Belarusian Republican Foundation for Fundamental Research

INTERNATIONAL ORGANIZING COMMITTEE

V. E. Borisenko – Co-chairman	(Belarus)
C. H. Kam – Co-chairman	(Singapore)
F. Arnaud d'Avitaya	(France)
L. T. Canham	(Great Britain)
N. G. Galkin	(Russia)
S. V. Gaponenko	(Belarus)
S. A. Gavrilov	(Russia)
D. Grützmacher	(Germany)
J.-L. Lazzari	(France)
A. Maffucci	(Italy)
S. A. Maksimenko	(Belarus)
A. Nassiopoulou	(Greece)
S. Ossicini	(Italy)
T. Suemasu	(Japan)
B. K. Tay	(Singapore)

BELARUSIAN NATIONAL ORGANIZING COMMITTEE

V. A. Bogush – Chairman
M. P. Batura
H. V. Bandarenka
V. E. Borisenko
N. V. Gaponenko
V. S. Gurin
A. V. Krivosheeva
V. A. Labunov
S. K. Lazarouk
A. A. Leshok
D. B. Migas
V. L. Shaposhnikov

FOREWORD

Nanotechnologies and their resulting nanostructures and nanomaterials are rapidly penetrating into modern science and engineering. Their scientific and applied impact is becoming increasingly valuable. It can be easily traced through the topics discussed at the series of International Conferences on Physics, Chemistry and Applications of Nanostructures, NANOMEETING, being regularly organized in Minsk (Belarus) since 1995. They show that starting research interest in fundamental electronic and optical properties of silicon-based nanostructures has been enormously extended to many other materials, *i.e.* conductors, semiconductors, superconductors and magnetics promising for nanoelectronics, spintronics, renewable energetics and nanomedicine. After experimental demonstration by A. K. Geim and K. S. Novoselov in 2004 of the possibility to divide graphite crystals into graphene sheets two dimensional hexagonal crystals (in particular those made of transition metal dichalcogenides) along with traditionally attractive quantum dots, became objects of enhanced research interest. Nanotechnologies were enriched by advanced techniques based on chemical vapor deposition, colloidal chemistry and electrochemistry.

Current state-of-the-art in nanoscience and nanotechnologies is presented and discussed at NANOMEETING-2017 taking place in Minsk within the period of May 30 – June 2, 2017. The most interesting results have been selected for presentation and discussion at the Conference.

This book includes invited reviews and short notes of the contributions to the Conference. The papers are arranged within the traditional sections of the previous publications: Physics of Nanostructures, Chemistry of Nanostructures, Nanotechnology and Nanostructure-Based Devices. Meanwhile, the papers for the special Sessions "Nanoelectromagnetics" and "Frontiers of Molecular Diagnostics with Nanostructures" are arranged in separate sections. All the papers have been mainly preserved in their original forms.

The camera-ready version of the Proceedings was composed by V. L. Shaposhnikov and A. V. Krivosheeva. The art design of the book cover belongs to V. A. Pushkarchuk.

The Sponsors who kindly provided the financial support for the Conference are deeply acknowledged.

V. E. Borisenko — Minsk
C. H. Kam — Singapore
Co-chairmen of NANOMEETING-2017 — February 2017

CONTENTS

FRONTIERS OF MOLECULAR DIAGNOSTICS WITH NANOSTRUCTURES

CHEMISTRY OF NANOSTRUCTURES

PHYSICS OF NANOSTRUCTURES

PHYSICS, CHEMISTRY AND APPLICATION OF NANOSTRUCTURES, 2017

INVITED

MULTIPLE EXCITON GENERATION IN SILICON NANOCRYSTALS

I. Marri

Centro S3, CNR-Istituto Nanoscienze, via Campi 213/a, 41125 Modena, Italy
marri@unimo.it

S. Ossicini

University of Modena and Reggio Emilia, Department of Science and Methods for Engineering (DISMI), via Amendola 2 Reggio Emilia, 42122, Italy
stefano.ossicini@unimore.it

M. Govoni

Institute for Molecular Engineering, the University of Chicago, 5747 South Ellis Avenue 60637 Chicago, Illinois (USA)
mgovoni@uchicago.edu

In this paper we present density functional theory calculations of carrier multiplication (CM) processes in a system of strongly coupled silicon nanocrystals. By inducing new charge and energy transfer processes, nanocrystals interplay can lead to a reduction of the carrier multiplication energy threshold and can influence the dynamics of the excited states. A detailed analysis of the carrier multiplication processes as a function of the nanocrystal-nanocrystal separation is reported. The validity of the so called Förster approximation for CM based energy transfer processes is discussed.

1. Introduction

The advent of nanoscience and nanotechnologies has opened new perspectives in the development of materials with new properties. At the nanoscale, electronic properties can be tuned by reducing the size of the system, by changing passivation and shape of the nanostructure, by doping or by functionalizing with molecules to fulfill specific requirements deriving from different technological fields [1-9]. Effects induced by the reduced dimensionality of nanocrystals (NCs) can be also exploited to emphasize the relevance of recombination processes which are negligible in the bulk phase. This is the case of the carrier multiplication (CM) effect that is a non-radiative recombination mechanism driven by the Coulomb interaction between carriers. CM is a counterpart of the Auger recombination [10] and can be exploited to minimize the occurrence of loss factors in solar cell devices. CM occurs when a highly excited carrier (excess energy of the excited carrier higher that the energy gap E_g) decays toward the band edges by transferring its excess energy to generate extra e-h pairs. When CM involves states of the same NC the effect is namely one-site CM. When it involves states delocalized onto two NCs it is termed two-site CM. Because of the restrictions imposed by energy and momentum conservation, and by the occurrence of concurrent fast phonon relaxation processes, CM is often inefficient in bulk semiconductors.

At the nanoscale, instead, CM can alter high-energy carrier relaxation dynamics [11–13]. Effects induced by one-site CM on the carriers population of the excited states have been observed in PbSe and PbS [14–15], CdSe and CdTe [16–17], InAs [18] and Si [19] while two-site energy transfer CM effects, termed space separated quantum cutting (SSQC), were observed in a dense array of Si-NCs by Timmerman et al. [20] and Trinh *et al.* [21]. One-site and two-site CM effects have been recently quantified by first principles calculations by M. Govoni *et al.* [22] and I. Marri *et al.* [23–25]. They clarified the existence of two different two-site CM processes (the SSQC, where a carrier decays toward the band edge by transferring its excess energy to a nearby NC where an extra e-h pair is generated, and the coulomb driven charge transfer (CDCT), where an excited carrier moves from a NC to the other one where an extra e-h pair is generated) and the existence of a well defined lifetime hierarchy (one-site CM processes are faster than two-site CM effects and CDCT is faster than SSQC).

In this paper we focus our attention on the study of effects induced by NCs interplay on the detectable CM energy threshold. Moreover, we prove that only a detailed description of the Coulomb interaction by a first-principle approach can lead to a correct evaluation of the CM energy transfer mechanisms when systems of closely packed NCs are considered.

2. Results and discussion

2.1. *CM activation energy threshold*

In this section we report the calculated CM lifetimes for a system formed by two interacting hydrogenated Si-NCs, that is the $Si_{147}H_{100}xSi_{293}H_{172}$. We consider four different configurations where NC-NC separation d moves from 1.0 to 0.4 nm by steps of 0.2 nm (NC-NC separation is calculated as the shortest distance between NCs surface atoms). Calculated one-site (left panel) and two-site (right panel) CM lifetimes are presented in Fig. 1 as a function of d and of the energy of the carrier igniting the process E_i (only mechanisms generated by the relaxation of an electron are considered; processes ignited by relaxation

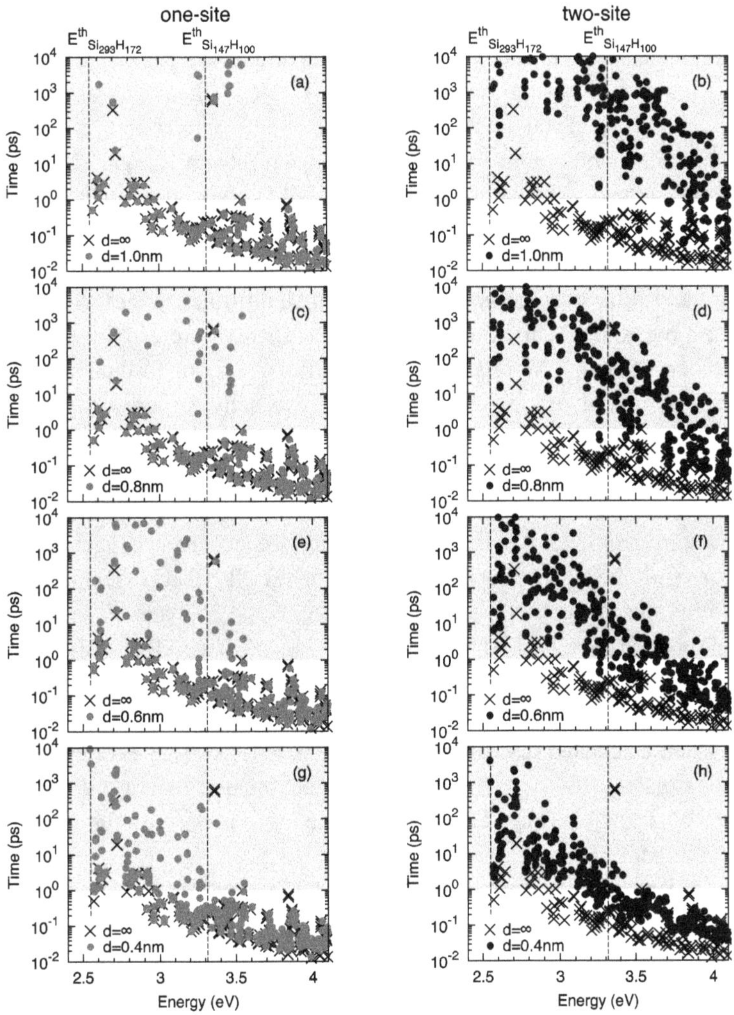

Figure 1. One- and two-site CM lifetimes. The zero of the energy is placed at the half gap. The shared region contains processes with a lifetime greater than 1 ps.

of holes lead to similar trends). Results reported in Fig. 1 are obtained by explicitly calculating both direct and exchange Coulomb matrix elements following the *ab initio* procedure described in [22–25]. Dashed vertical lines denote the ideal CM activation energies for the single isolated NCs, that is $E^{th}_{\mathrm{Si293H172}}$ and $E^{th}_{\mathrm{Si147H100}}$. They are obtained by multiplying by a factor of two the NCs energy gaps (1.7 and 2.0 eV, respectively). $E^{th}_{\mathrm{Si293H172}}$ and $E^{th}_{\mathrm{Si147H100}}$ represent the ideal lower limits for the detectable CM energy threshold for the

$Si_{293}H_{172}$ and the $Si_{147}H_{100}$ NCs. Cross-type points keep tracks of the CM lifetimes calculated for the system of noninteracting NCs ($d \rightarrow \infty$), that is an ideal configuration comprising sparse arrays of Si-NCs where interaction between NCs can be neglected. At d=1 nm, NC-NC interaction is weak and wavefunctions are almost totally confined onto a single NC. In this configuration two-site CM events are slow and the overall efficiency of the CM is determined only by the occurrence of one-site CM decay dynamics (see panels (a) and (b) of Fig. 1). When NC-NC separation is reduced, wavefunctions start to delocalize over the entire system and significant changes in both plots of one- and two-site CM lifetimes appear. Such changes are essentially due to the occurrence of new CM relaxation mechanisms that are forbidden when NCs interplay is turned off. NC-NC interaction can activate new CM decay paths (such processes cannot occur in systems of non-interacting NCs) or can, by adding new contributions, make more complicated CM dynamics that already exist in systems of non-interacting NCs.

Moving from $d = 0.8$ nm to $d = 0.4$ nm, we observe changes in both the number and the distribution of points of Fig. 1, that essentially concern transitions that occur with energies between $E^{th}_{Si293H172}$ and $E^{th}_{Si147H100}$ with a lifetime greater that 1 ps (shared region of Fig. 1). Two effects simultaneously appear when d decreases: new points are induced in the CM lifetimes plot and a large number of points move toward slower CM lifetimes. As a final result, at $d = 0.4$ nm, calculated CM lifetimes for the $Si_{147}H_{100}xSi_{293}H_{172}$ resemble the one of the isolated $Si_{293}H_{172}$ (the NC with the smaller energy gap) although, especially at low energies, CM is now made more intense by the presence of a larger number of CM decay paths.

In order to evaluate effects induced by CM on excited states population, we have to compare the relevance of this process with the one of other competitive relaxation processes, in particular the thermalization mechanisms (in experiments, CM is detectable when it is faster than all the other competitive decay processes). A detailed *ab initio* determination of thermalization dynamics is currently an unattainable task due to the large numerical resources involved. However, for Si-NCs, non-CM relaxation times are expected to be of the order of 1–10 ps [26]. These values will be taken as a reference for thermalization mechanisms and will be compared with the ones calculated for the CM decay processes.

By summing the contributions derived by both one-site and two-site CM processes, we count in the histogram of Fig. 2 the number of transitions that fall in the range of 1–10 ps (left panel) and in the sub-picosecond (subps) region, as a function of E_i and d.

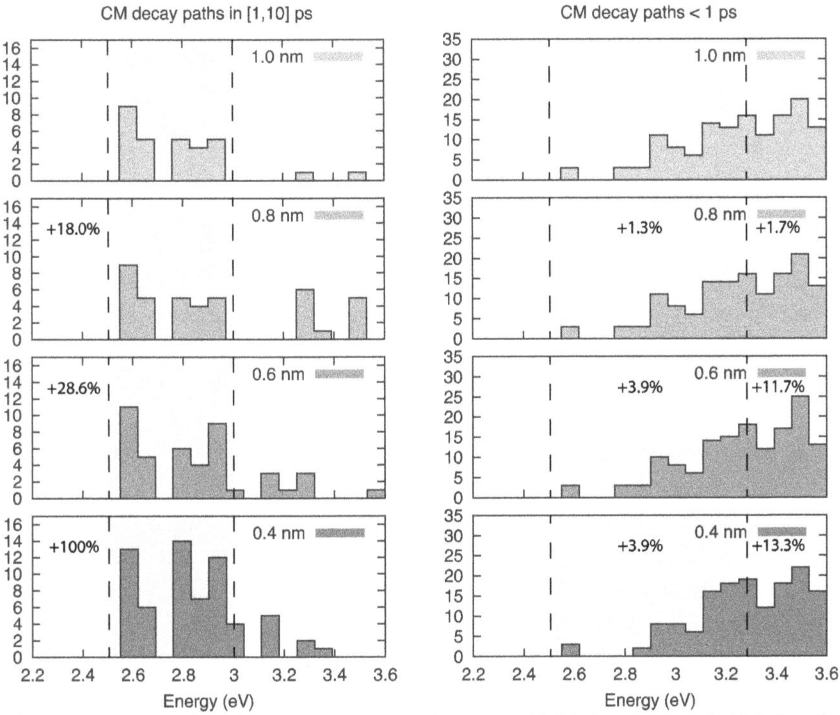

Figure 2. The distribution of the CM decay events that occur with lifetimes between 1–10 (<1) ps is reported in the left (right) panel of the figure for the cases d=1.0 0.8, 0.6 and 0.4 nm.

The counted CM transitions can be purely one-site, purely two-site or can be a combination of one-site and two-site processes. The results of Fig. 2 point out that NCs interplay produces an intensification of CM effects especially for energies between $E^{th}{}_{\mathrm{Si293H172}}$ and $E^{th}{}_{\mathrm{Si147H100}}$ and lifetimes between 1–10 ps. In particular, at d=0.4 nm, in the range of 2.55–3.0 eV, we count a double number of CM decay processes with respect to the ones recorded at d=1.0 nm. When NCs strongly interact, we observe a non-negligible intensification of the CM dynamics at low energies with respect to the one recorded for an equivalent system of non-interacting NCs. This effect is very important and can lead to a red-shift of the detectable CM energy threshold, thus explaining the CM energy threshold reduction observed in different systems of closely packed NCs [27].

A different behavior is instead observed when sub-ps CM decay paths are analyzed (right panel of Fig. 2). In this regime we do not observe a relevant modification of the intensity of the CM activity when d is reduced from 1.0 to 0.4 nm. Our results, therefore, point out that CM benefits from NCs interplay especially at low energies, where NC-NC interaction leads to a significant intensification of the CM dynamics and thus to a possible reduction of the detectable CM energy threshold. Far from the activation threshold the occurrence of new CM processes does not lead to a significant intensification of the CM activity.

2.2. *Förster approximation*

In this section we focus on SSQC processes. Energy transfers have been often interpreted considering a simple expansion of the Coulomb interaction in the formulation of Förster approximation [28] (dipole-dipole approximation) and Dexter approach [29] (multiple interaction). They can be used to describe energy transfer processes between non-delocalized states (as often occurs when band-edge states are involved in the process) but often fail in the description of processes between delocalized states. Förster and Dexter approaches are therefore expected to not well described SSQC processes when the effect occurs in systems of closely packed NCs.

In order to estimate errors induced by the Förster resonance energy transfer (FRET) approximation in the description of SSQC mechanisms, we calculated the ratio between the SSQC rate obtained at $d = 0.8$, 0.6 and 0.4 nm with the one obtained at $d = 1$ nm. The calculations were performed both by explicitly evaluating the Coulomb screened interaction (full *ab initio* approach) and by adopting the Förster approximation. The results obtained are reported in Fig. 3.

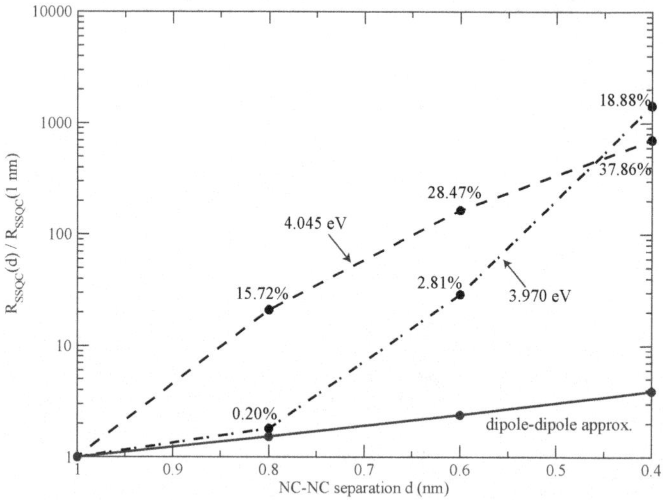

Figure 3. The increment of SSQC rate calculated by selecting two different decay paths ignited by the states with the energies of 4.045 eV and 3.970 eV. For all the considered NC-NC separations we report the localization of the initial state on the smaller NC. Results are compared with the ones obtained by using the dipole-dipole approximation (continuous line).

The dashed and dashed-dot lines were obtained by using a full *ab initio* approach by selecting two different SSQC decay processes ignited by the states with the energy of 4.045 and 3.970 eV, respectively. For each decay path the localization of the wavefunction of the initial state on the smaller NC is reported as a function of the NC-NC separation. The continuous line indicates the results obtained by using the Förster approximation, that is by treating the screened Coulomb potential in term of dipole-dipole interaction.

From the results of Fig. 3 we observe that when a full *ab initio* technique is adopted to calculate the SSQC processes, the ratio $R_{SSQC}(d)/R_{SSQC}(1\text{ nm})$ (here $R_{SSQC}(d)$ denotes the SSQC rate calculated at a NC-NC separation d) increase up to 3 order of magnitude when d moves from 1.0 to 0.4 nm. The Förster approximation instead strongly underestimates the increment of the SSQC rate and cannot correctly describe the effects induced by NCs interplay on energy transfer CM processes. Moreover, SSQC rates depend on the selected decay path so that it is not possible to define an universal rule to describe such effects. Due to the delocalization of wavefunctions that occur at high energies, neither dipole-dipole approximation nor multiple interactions can be exploited to describe SSQC processes.

3. Conclusion

The results presented point out that NCs interplay leads to an intensification of the CM dynamics at low energies and can be responsible for reduction of the detectable CM energy threshold. Moreover, we have proven than CM energy transfer processes can be described only by explicitly calculating the screened Coulomb matrix elements: dipole-dipole approximation fails to describe SSQC processes.

Acknowledgments

The Authors thank the Super-Computing Interuniversity Consortium CINECA. Simulations have been performed thank to the ISCRA and PRACE initiative. I. Marri thanks the European center of excellence MAX, GA 676598.

References

1. I. Marri, S. Ossicini, *Solid State Commun.* **147**, 205 (2008).
2. R. Guerra *et al.*, *Superlattice Microst.* **46**, 246 (2009).
3. S. Ossicini *et al.*, *J. Nanosci. Nanotechnol.* **8**, 479 (2008).
4. E. Degoli *et al.*, *CR Phys.* **10**, 575 (2009).
5. F. Iori *et al.*, *J. Lumin.* **121**, 335 (2006).

6. A. Iacomino *et al.*, *Phys. Rev. B* **78**, 075405 (2008).
7. F. Iori *et al.*, *Phys. Rev. B* **76**, 085302 (2007).
8. R. Guerra *et al.*, *Phys. Rev. B* **79**, 155320 (2009).
9. O. Regan, M. Grätzel, *Nature* **353**, 737 (1991).
10. M. Govoni *et al.*, *Phys. Rev. B* **84**, 075215 (2011).
11. V. I. Klimov *et al.*, *Science* **290**, 314 (2000).
12. M.C. Beard, *J. Phys. Chem. Lett.* **2**, 1282 (2011).
13. A. J. Nozik, *Physica E: Low Dimens. Syst. Nanostruct.* **14**, 115 (2002).
14. R. J. Ellingson *et al.*, *Nano Lett.* 5, 865 (2005).
15. R. D. Schaller, V. I. Klimov, *Phys. Rev. Lett.* **92**, 186601 (2004).
16. R. D. Schaller *et al.*, *Appl. Phys. Lett.* **87**, 253102 (2005).
17. D. Gachet *et al.*, *Nano Lett.* **10**, 164 (2010).
18. R. D Schaller *et al.*, *Nano Lett.* 7, 3469 (2007).
19. M. C. Beard *et al.*, *Nano Lett.* 7, 2506 (2007).
20. D. Timmerman *et al.*, *Nat. Photonics* **2**, 105 (2008).
21. M. T. Trinh *et al.*, *Nature Photonics* **6**, 316 (2012).
22. M. Govoni *et al.*, *Nature Photonics* **6**, 672 (2012).
23. I. Marri *et al.*, *J. Am. Chem. Soc.* **136**, 13257 (2014).
24. I. Marri *et al.*, *Beilstein J. Nanotechnol.* **6**, 343 (2015).
25. I. Marri *et al.*, *Sol. Energ. Mat. Sol. Cells* **145**, 162 (2016).
26. W. D. A. de Boer *et al.*, *Nature Nanotech.* **5**, 878 (2010).
27. C. S. S. Sandeep *et al.*, *Nature Commun.* **5**, 878 (2010).
28. T. Förster, *Annalen der Physik* **1-2**, 55 (1948).
29. D. L. Dexter, *J. Chem. Phys.* **21**, 836 (1953).

INVITED

STRUCTURAL CHARACTERIZATION OF Si-BASED NANOWIRE AND NANOSHEET BUNDLES

P. Yuan[1], K. Sasaki[2], M. Nakayama[2], Y. Kumazawa[2], K. Hikichi[2], H. Tatsuoka[2]

[1]*Graduate School of Science and Technology, Shizuoka University*
Hamamatsu, 432-8011 Shizuoka, Japan
ts049272@ipc.shizuoka.ac.jp

[2]*Graduate School of Integrated Science and Technology, Shizuoka University*
Hamamatsu, 432-8561 Shizuoka, Japan
tatsuoka.hirokazu@shizuoka.ac.jp

A variety of Si-based nanostructured bundles have been prepared. $CrSi_2$, SiO_x, Mg_2Si/MgO composite, and Mg_2Si and $MnSi_{1.7}$ nanowire bundles were synthesized. In addition, the hexagonal $MoSi_2$ nanosheet bundles were synthesized using a MoS_2 layered material as a template. Powders consisting of Si-based nanosheet bundles were synthesized by Ca atom extraction from $CaSi_2$ powders using inositol hexakisphosphate (IP6) in an aqueous solution. The Si-based nanosheet bundles were also synthesized by thermal treatment of $CaSi_2$ microwalls grown on Si substrates under $FeCl_2$ vapors. The surface property of the Si-based nanosheets was evaluated by infrared optical absorption. This study demonstrated a simple method for preparing Si-based nanosheets and their surface modification. The structural properties of the nanostructured bundles were also investigated.

1. Introduction

Semiconducting silicides have attracted much interest as environmentally-friendly semiconductors, consisting of non-toxic and abundant materials [1]. In addition, low-dimensional materials have attracted much interest because of their enhanced or modified optical, electronic, thermoelectric and mechanical properties compared to those of bulk materials [2]. A nanowire or nanosheet bundle is one of the important structures for technological applications, such as thermoelectricity, and use in solar cells and Li-ion batteries.

In this paper, the nanowire bundle is defined as an aggregate of long nanowires that are highly oriented, well-aligned, dense and tied together, and have longer and higher densities than the so-called arrays of wires or rods fabricated on substrates. In this sense, the nanosheet bundle is a collection of nanosheets overlapping each other with a void space. However, silicon or silicide nanowire bundle structures have been rarely synthesized. On the other hand, the formation of silicene, siloxane, polysilane and other Si two-dimensional (2D) layered structures have been reported. In this context, it has been demonstrated that Si-based nanowire and nanosheet bundles can be prepared.

A series of fabrication techniques to synthesize the Si-based nanowire and nanosheet bundles based on Si and silicide materials are reported. $CrSi_2$ [3,4], SiO_x [5,6], Mg_2Si/MgO composite bundles [7], and Mg_2Si and $MnSi_{1.7}$ [8,9] nanowire bundles were successfully synthesized. In addition, a bundle consisting of hexagonal (h-) $MoSi_2$ semiconducting silicide nanosheets formed using MoS_2 as a template is described [10].

Inositol hexakisphosphate (IP6), $C_6H_{18}O_{24}P_6$, is known as a phytate for metal storage found in cereals and grains. The active reactions with the divalent metal ions to form phytate complexes were reported. Thus, it is expected that the metallic atoms can be extracted from the silicides using IP6. The synthesis of Si nanosheet bundles by extraction of the Ca atoms from $CaSi_2$ crystals using IP6 has been reported [11]. Moreover, the free-standing Si nanosheets were synthesized on Si substrates by Ca extraction from $CaSi_2$ microwalls by thermal treatment under $FeCl_2$ vapors [12]. The structural properties of these nanostructure bundles were characterized and compared to each other.

2. Structural property of Si-based nanowire and nanosheet bundles

An appropriate selection of the fabrication technique is required to form various bundle structures for specific applications. For a direct growth, it is important to consider the reaction process during the silicidation. Several chlorides were used to synthesize silicide nanowire bundles. In this case, the pyrolysis ratio of the chloride and formation of the catalyst precursor are important to design the direct growth of the nanostructures. For the template techniques, it is important to consider the diffusion coefficient of the constituents of the silicides, the volume change by the silicidations, and the difference in the thermal expansion coefficients between the silicide and the host crystal. The choice of the starting materials is also important. The crystalline structure, the enthalpy of formation of starting materials and resultant compounds, and in addition, if the reaction takes place under vacuum, the vapor pressure should be discussed.

On the other hand, the structural and morphological properties of the resultant compounds significantly depend on the precise artificial control of the starting materials. In addition, concerning the fabrication procedures for application to energy devices, a simple fabrication technique is required for the formation of a large volume and/or area materials, for example, for applications, in thermoelectric devices, solar cells, Li-ion batteries.

2.1. *Si-based nanowire bundles*

Figs. 1a–e show scanning electron microscopy (SEM) images of the fabricated nanowire bundles. The $CrSi_2$ nanowire bundles were synthesized by an exposure of the Si(111) substrates to $CrCl_2$ vapors with argon carrier gas under atmospheric pressure. The Si substrate and the $CrCl_2$ source were heated to 750–900 °C. The detailed growth temperature profile is shown elsewhere [4].

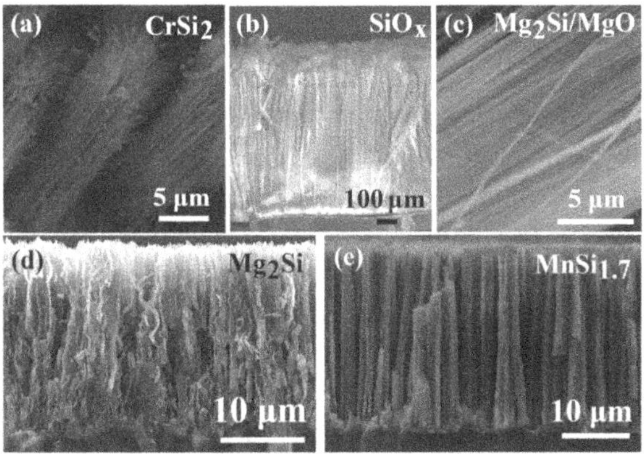

Figure 1. (a) – (e) SEM images of $CrSi_2$ [3], SiO_x (x~2) [5], Mg_2Si/MgO composite [7], and Mg_2Si and $MnSi_{1.7}$ [8,9] nanowire bundles, respectively. Reproduced by permission of The Electrochemical Society.

Fig. 1a demonstrates that well-aligned $CrSi_2$ nanowires were densely grown to form bundles on a microrod [3]. It was also found that SiO_x spheres are observed on top of the nanowires. It is considered that $SiCl_4$ droplets are formed during the reaction and act as catalysts for the growth of the $CrSi_2$ nanowires. The grown wires were exposed to air, and SiO_x spheres were probably formed by the reaction between the $SiCl_4$ and the air [4].

The direct growth of a dense nanowire bundle has also been demonstrated for amorphous SiO_x (x~2) nanowires with the lengths up to nearly 1 mm. It is considered that one of the key factors to obtain a dense nanowire bundle is formation of high density catalytic nanoparticles covering the substrates. The SiO_x nanowire bundle was simply obtained by annealing of FeSi plates with Ga droplets.

Fig. 1b shows a cross-sectional SEM image of the nanowires grown on the FeSi substrate by the heat treatment at 900 °C for 114 h. Highly oriented, well-aligned nanowires grew in an orderly fashion with a high density on the substrate. It was revealed that the nanowires mainly contain Si and O [5].

It is usually difficult to control the growth morphology and the structures of the silicide materials for the direct growth of the nanowire bundle. One alternate preparation method is a template assisted fabrication technique. Si or Si-based nanowire bundle templates are used to obtain the silicide bundles. Fig. 1c shows SEM image of the Mg_2Si/MgO composite nanowire bundles fabricated using SiO_x nanowire bundles as a template material by thermal treatment at 450 to 630 °C under Mg vapors. The bundle consists of Mg_2Si and MgO, and even though Mg_2Si and MgO are randomly distributed, the bundle structure is maintained as the SiO_x nanowire bundle [7].

Recently, Si nanowire arrays have been easily obtained by chemical etching with Ag catalysts. The Mg_2Si nanowire bundles were synthesized using the Si nanowire arrays prepared by that technique. The silicidation processes of the Si structures would be useful for the easy and simple fabrication of silicide nanowire bundles using the nanowire arrays as a template. Figs. 1d and 1e show the cross-sectional SEM images of Mg_2Si and $MnSi_{1.7}$ nanowire bundles, respectively [8,9]. The Mg_2Si and $MnSi_{1.7}$ nanowire bundles were synthesized by the thermal treatment of the Si nanowire arrays under Mg vapors at 500 °C for 5 h and $MnCl_2$ vapors at 600 °C for 5 h, respectively. The silicide nanowires roughly maintain the nanowire array structures. However, the 3.2 times volumetric increase from Si to Mg_2Si increased the diameter of the nanowires. The nearby nanowires became joined together [8].

As the Si nanowire arrays reacted with the $MnCl_2$ vapors, the morphology of the nanowires is shown in Fig. 1e [9]. The volumetric change after the complete conversion of the Si nanowires into the $MnSi_{1.7}$ nanowires is $V_{MnSi1.7}/V_{Si} = 0.96$, which results in a shrinkage of the volume. The length of the nanowires is almost unchanged. The growth strongly depends on the boundary conditions and not simply on the free energy of formation. The $MnSi_{1.7}$ phase is the first nucleation phase in the reaction system of the $MnCl_2$ vapors and Si substrate under this growth condition. It is considered that the thermal treatment at 600 °C provides the appropriate dissociation ratio of $MnCl_2$ molecules, though the enthalpy of formation of $MnCl_2$ (–38.4 kcal/g-atm) is much higher than those of the Mn silicides. It is demonstrated that well-aligned Mg_2Si nanowire bundles and $MnSi_{1.7}$ nanowire bundles were successfully synthesized by the reaction of the artificial Si nanowires with Mg and $MnCl_2$ vapors, respectively.

2.2. *Si-based nanosheet bundles*

Fig. 2a shows SEM, transmission electron microscopy (TEM) and enlarged TEM images of the $MoSi_2$ nanosheet bundle synthesized by a molten salt technique using MoS_2 as a template material [8,10]. The surface with a locally smooth, stepped morphology and thin layered lamellar structure was observed. It is interesting that the h-$MoSi_2$ does not have a layered structure. However, the silicide growth using MoS_2 leads to the formation of nanoscale lamellar structures, which are due to the layered structure of the MoS_2 starting material. The MoS_2 has a layered structure, and the replacement of Si atoms with S sites in the hexagonal MoS_2 leads to the formation of the single phase h-$MoSi_2$ crystals [10].

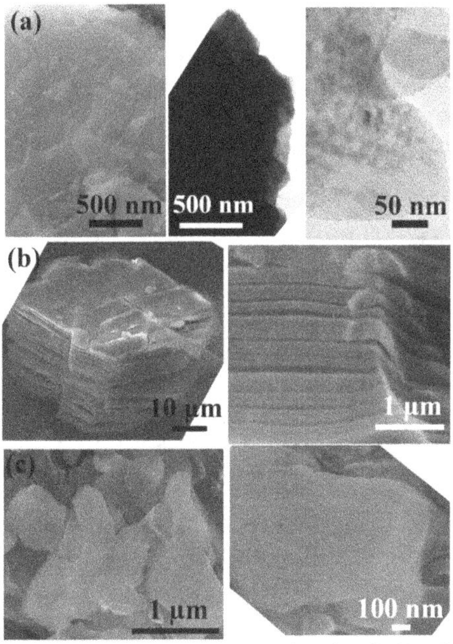

Figure 2. (a) SEM and TEM images of h-$MoSi_2$ nanosheet bundle [8,10], (b) SEM images of a powder consisting of nanosheet bundles synthesized by Ca atom extraction from $CaSi_2$ powder using IP6, (c) SEM images of Si nanosheet bundles rooted on Si substrate synthesized from $CaSi_2$ microwalls by the thermal treatment under $FeCl_2$ vapors. Reproduced by permission of The Electrochemical Society.

Fig. 2b shows a powder consisting of Si-based nanosheet bundles synthesized by Ca atom extraction from the $CaSi_2$ powder. The raw $CaSi_2$ powders were simply immersed in a dilute 0.43 mol/l IP6 solution for 1 h, then dried [11].

The 2D sheets could be clearly seen with an estimated thickness of tens to hundreds of nanometers. The nanosheets are alternately stacked with small void spaces to form a sheet bundle, and the thickness of the bundles could reach tens of micrometers. Fig. 2c shows SEM images of the Si nanosheet bundles rooted on Si (111) substrate synthesized from the $CaSi_2$ microwalls [12]. They were formed by the thermal treatment at 540 °C for 10 min under $FeCl_2$ vapors. The resultant Si-based nanosheets were rinsed with ethanol to remove the residual chloride sources.

Various chlorides were examined to extract Ca atoms from the $CaSi_2$ crystals. It was reported that by exposure of the $CaSi_2$/Si substrates to the $CrCl_2$ vapors, the Si-based nanosheets rooted on the Si substrates were formed [13]. It was also reported that transition metal chlorides, such as $FeCl_2$, $FeCl_3$, and NH_4Cl for the vapor phase growth [12], and $NiCl_2$ $FeCl_2$ and $MnCl_2$ for the solid-state exfoliation reaction [14,15], were also used to extract Ca atoms from the $CaSi_2$ crystals to form Si-based nanosheets.

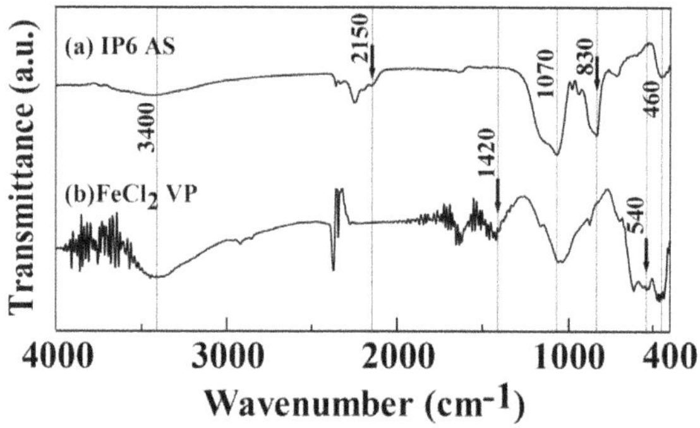

Figure 3. FTIR spectra of the powders consisting of Si-based nanosheets synthesized (a) using IP6 aqueous solution and (b) by thermal treatment under $FeCl_2$ vapors.

Fig. 3a shows the Fourier-transform infrared spectroscopy (FTIR) spectrum of the product treated by the reaction of the $CaSi_2$ powders with the 0.43 mol/L IP6 aqueous solution for 1 h. It indicates a possible presence of the oxygen-terminated surface of the Si sheets or SiO_x species in the products synthesized. It is suggested that the edges and the majority of the surface area of the nanosheets may have been oxidized and hydroxylated [11].

Fig. 3b shows the FTIR spectrum of the Si-based nanosheets thermally treated in the $FeCl_2$ vapor. Most of the peaks observed for the IP6 aqueous solution in Fig. 3a are also observed in the spectrum, except for the absorption peak at 830 cm^{-1} and peak around 2150–2250 cm^{-1}. On the other hand, a new relatively broad absorption peak appeared around 520–620 cm^{-1}. Several vibration modes can possibly be attributed to the absorption. In addition, another new absorption peak appeared around 1420 cm^{-1}. The formation of the H-Si-O (network) (830 cm^{-1}) structural unit for the Si-based nanosheets is considered to be due to the preferential reaction in the aqueous solution. On the other hand, the formation of the structure unit of Si-CH=CH_2 or Si-R(R=CH_3 or C_2H_5) is considered to be due to the ethanol washing after the process for the Si-based nanosheets synthesized in the vapor phase.

3. Conclusion

It has been demonstrated that Si-based nanowire and nanosheet bundles can be prepared. In this study, $CrSi_2$, Mg_2Si/MgO composite, Mg_2Si and $MnSi_{1.7}$ nanowire bundles, h-$MoSi_2$ and Si nanosheet bundles were successfully synthesized. The structural control and morphological modifications of the nanowire and nanosheet bundles have been successfully demonstrated by the appropriate thermal or chemical treatments with specific starting materials. The bundle structure consisting of nanowires or nanosheets can also be considered as one of the highly efficient nanostructured materials for thermoelectric applications. Further characterization of the nanostructured semiconductors will be expected in order to evaluate the enhanced electric, thermal and thermoelectric properties of the bundles. It is also expected that the simple and easy nanostructure fabrication technique will encourage us to develop improved energy devices with a large volume and/or large area.

Acknowledgments

Part of this study was supported by MEXT KAKENHI (grant 26420270).

References

1. Y. Maeda (ed.), *Science And Technology Of Semiconducting Silicides And Related Materials* (Shokabo, 2014) – in Japanese.
2. Y. Huang, K.-N. Tu (eds.), *Silicon and Silicide Nanowires: Applications, Fabrication, and Properties* (Pan Stanford Publishing, Singapore, 2013).
3. H. Tatsuoka *et al.*, *Solid State Phenom.* **213**, 35 (2014).
4. W. Li *et al.*, *J. Cryst. Growth* **365**, 11 (2013).
5. T. Inaba *et al.*, *Thin Solid Films* **515**, 8158 (2007).
6. T. Inaba *et al.*, *Jpn. J. Appl. Phys.* **45**, L1320 (2006).
7. W. Li *et al.*, *E-Journal Surf. Sci. Nanotechnol.* **10**, 297 (2012).
8. H. Tatsuoka, W. Li, E. Meng, D. Ishikawa, *ECS Trans.* **50**, 3 (2012).
9. W. Li, K. Nakane, H. Tatsuoka, *Phys. Status Solidi* **10**, 1800 (2013).
10. D. Ishikawa *et al.*, *E-Journal Surf. Sci. Nanotechnol.* **10**, 55 (2012).
11. X. Meng *et al.*, *Jpn. J. Appl. Phys.*, accepted.
12. X. Meng *et al.*, submitted.
13. X. Meng *et al.*, *J. Ceram. Soc. Japan.* **122**, 618 (2014).
14. S.-Y. Oh, H. Imagawa, H. Itahara, *J. Mater. Chem. A* **2**, 12501 (2014).
15. S.-Y. Oh, H. Imagawa, H. Itahara, *Chem. Asian J.* **9**, 3130 (2014).

POSSIBLE PLASMONIC ENHANCEMENT OF ELECTROLUMINESCENCE

D. V. Guzatov

Physical-Technical Faculty, Yanka Kupala State University of Grodno
230023 Grodno, Belarus
guzatov@gmail.com

S. V. Gaponenko

Laboratory of Nano-Optics, B. I. Stepanov Institute of Physics, NASB
220072 Minsk, Belarus
s.gaponenko@ifanbel.bas-net.by

H. V. Demir

EEE and Physics Departments and UNAM–Institute of Materials Science and Nanotechnology, Bilkent University, 06800 Ankara, Turkey;
LUMINOUS! Center of Excellence, School of EEE, School of Physical and Mathematical Sciences, Nanyang Technological University
Nanyang Ave. 50, 639798 Singapore
volkan@bilkent.edu.tr

Plasmonic effect on electroluminescence (EL) is studied on the basis of the model of modified radiative and nonradiative decay rates for a dipole source near a metal spherical nanoparticle (NP) including finite size effects. Unlike photoluminescence where local incident field factor mainly determines an enhancement possibility, an EL enhancement is only possible by means of quantum yield rise up. Therefore EL enhancements become feasible only for the intrinsic quantum yield $Q_0 < 1$. The plasmonic effect is found to be independent of the intrinsic emitter lifetime defined by Q_0, emission spectrum, NP diameter and emitter-NP spacing. Ag NPs are shown to enhance LED/OLED intensity by several times ($0.1 < Q_0 < 0.25$).

1. Introduction

Plasmonic enhancement of luminescence intensity with metal NPs has become an active field of research because of potential impact on lighting, display and analytical instruments. Till recently only photoluminescence (PL) enhancement by metal NPs has been examined [1–6]. The PL intensity is defined by emitter excitation rate and its quantum yield Q, both being affected by metal NP. For emitters with a high intrinsic quantum yield $Q_0 \rightarrow 1$ in typical experiments, the relation $Q < Q_0$ holds since the enhancement of nonradiative decay component dominates over the radiative one [6,7]. Therefore, a local incident field enhancement becomes the major factor for PL intensity rise-up. For emitters with intrinsically low Q_0, the relative growth of nonradiative transitions rate near a metal NP is low and much higher enhancement factors become affordable for PL.

In the context of extensive progress in LEDs and OLEDs technologies, possible positive effect from metal NPs embedded is important. However, for electroluminescence (EL), the plasmonic effect does reduce to the modified quantum yield, *i.e.* to the modified ratio of radiative versus nonradiative decay components. The desirable EL enhancement becomes possible only for emitters with lower intrinsic quantum yield $Q_0 < 1$. Recently [8–10], plasmonic enhancement of LED and OLED efficiency has been experimentally observed for LEDs based on quantum wells and colloidal quantum dots.

In this paper, calculation of plasmonic effects in EL structures is reported based on the model of light emission by a dipole source near a metal spherical NP with including the finite size effects. We use the approach presented in [11] for calculation of modified decay rates of a dipole emitter near a metal NP using its modified dipole transitions probability. The calculation scheme has been described by us recently [6]. Change in the emission intensity I near metal NP versus intensity I_0 without metal NP equals to the quantum yield change:

$$\frac{I}{I_0} = \frac{Q}{Q_0} = \left(\frac{\gamma_r}{\gamma_0}\right) / \left[1 + Q_0\left(\frac{\gamma_r + \gamma_{nr}}{\gamma_0} - 1\right)\right], \tag{1}$$

where $Q_0 = \gamma_0/(\gamma_0 + \gamma_{int})$ is the emitter quantum yield with no metal NP, γ_0 is the intrinsic radiative decay rate and γ_{int} is the intrinsic nonradiative decay rate describing internal losses of the emitter. $Q = \gamma_r/(\gamma_r + \gamma_{nr} + \gamma_{int})$ is the quantum yield in the presence of metal NP, γ_r and γ_{nr} are radiative and nonradiative decay rates modified by metal NP, respectively.

One can find analytical expressions for γ_r/γ_0 and $(\gamma_r + \gamma_{nr})/\gamma_0$ in the case of spherical NP in [11], and we use them in the most favorable case of the dipole moment oriented along the line connecting emitter with NP center. The expressions in [11] are general and take into account scattering contribution. The other cases of dipole orientations can be considered alike. Notably, Eq. (1) shows that the plasmonic effect on EL does not depend on the absolute values of intrinsic radiative γ_0 and nonradiative γ_{int} decay rates, *i.e.* it is independent on intrinsic emitter lifetime. The ratio enhancement/quenching is entirely defined by Q_0, emission spectrum, metal permittivity, nanoparticle size and emitter-metal spacing.

2. Results and discussion

Calculations were made for Ag NPs of 20–80 nm diameters (Fig. 1). The range of $Q_0 = 0.1–0.5$ was chosen since the higher values do not allow for a plasmonic enhancement in most of the practical cases.

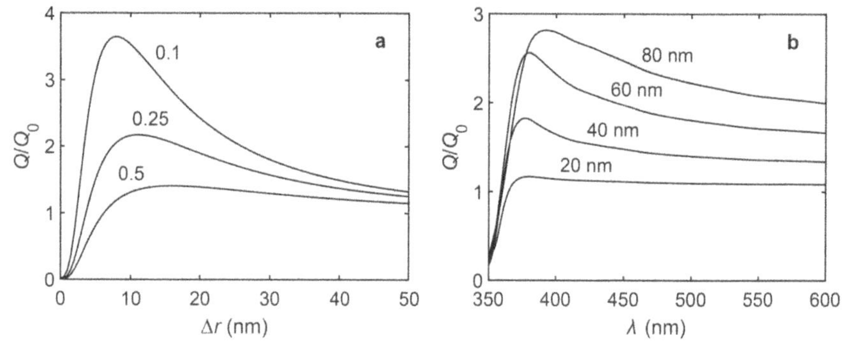

Figure 1. Calculated ratio Q/Q_0 for a dipole emitter in air in the presence of Ag NP: (a) versus distance Δr to NP surface in the case of NP diameter of 50 nm, $\lambda = 450$ nm, and Q_0 equals to 0.1, 0.25, and 0.5; (b) versus wavelength λ in the case of $\Delta r = 25$ nm, $Q_0 = 0.25$, and NP diameters of 20, 40, 60, and 80 nm.

The calculated data in Fig. 1a indicate that the ratio Q/Q_0 may exceed unity for the selected wavelength $\lambda = 450$ nm (*i.e.* the typical wavelength of commercial blue LEDs) for Ag NP. At the optimal distance between the emitter and NP surface Q/Q_0 may reach several times. It is important to note that for the emitters locating far away (20–30 nm) from the NP surface the enhancement occurs as well because Q_0 is sufficiently small. We emphasize that initial smallness of the intrinsic quantum yield and enhancement of luminescence near metal NP correspond to experimental conditions at which the plasmonic EL enhancement was registered [9,10].

From the data presented in Fig. 1b one can see that increasing diameter of Ag NP may additionally lead to an increase of Q/Q_0. This is due to the fact that the extinction maximum spectrum approaches the LED wavelength and leads to higher radiative decay rates. This indicates the importance of correct account for NP size in the theory.

3. Conclusion

The calculations described in the paper predict a noticeable plasmonic enhancement of visibly emitting LED and OLED performance for the internal quantum efficiency below 0.3 by the factor of 1.5–4 near Ag NPs. The improvement is higher for bigger nanoparticles and exists for metal-emitter spacing 5–15 nm. The results obtained are in a reasonable agreement with experiments. They are believed to stimulate further experiments in the field.

Acknowledgments

The work has been supported by the Belarus-Turkish project (BRFFR-TUBITAK) #F16T/A-010.

References

1. V. V. Klimov, *Nanoplasmonics* (Fizmatlit, Moscow, 2009).
2. S. V. Gaponenko, *Introduction to Nanophotonics* (Cambridge University Press, Cambridge, 2010).
3. C. D. Geddes (ed.), *Metal-Enhanced Fluorescence* (Wiley-VHC, New York, 2010).
4. J. R. Lakowicz, *Plasmonics* **1**, 5 (2006).
5. N. Strekal, S. Maskevich, *Plasmonic gold and silver films: selective enhancement of chromophore Raman scattering or plasmon-assisted fluorescence*, in: *Reviews in Plasmonics* 2010, ed. C. D. Geddes (Springer, Berlin, 2010), pp. 283–302.
6. D. V. Guzatov, S. V. Vaschenko, V. V. Stankevich, A. Y. Lunevich, Y. F. Glukhov, S. V. Gaponenko, *J. Phys. Chem. C* **116**, 10723 (2012).
7. V. V. Klimov, D. V. Guzatov, *Quant. Electron.* **37**, 209 (2007).
8. A. Fujiki, T. Uemura, N. Zettsu, M. Akai-Kasaya, A. Saito, Y. Kuwahara, *Appl. Phys. Lett.* **96**, 043307 (2010).
9. C. Y. Cho, S. J. Lee, J. H. Song, S. H. Hong, S. M. Lee, Y. H. Cho, S. J. Park, *Appl. Phys. Lett.* **98**, 051106 (2011).
10. X. Yang, P. L. Hernandez-Martinez, C. Dang, E. Mutlugun, K. Zhang, H. V. Demir, X. W. Sun, *Adv. Opt. Mater.* **3**, 1439 (2015).
11. V. V. Klimov, V. S. Letokhov, *Laser Phys.* **15**, 61 (2005).

CERAMICS STRUCTURES, ENERGY AND FRACTALS

V. V. Mitic

University of Niš, A. Medvedeva 14, 18000 Niš, Serbia
Institute of Technical Sciences of Serbian Academy of Sciences
Knez Mihailova 35, 11000 Belgrade, Serbia
vmitic.d2480@gmail.com

H.-J. Fecht

Institute of Micro and Nanomaterials, University of Ulm, D-89081 Ulm, Germany
hans.fecht@uni-ulm.de

B. Vlahovic

University Research Center for Aerospace Device Research and Education North Carolina Central University
vlahovic@nccu.edu

The goal of this theoretic-experimental note is to stress the connection of ceramics micro- and nanostructures with the issue of energy and fractals as a natural "bridge" between these two entities. Physical properties derived or computed on the unit cell which generates the entire crystal by translation in space can be analyzed taking Euclidian geometric dimensions. However, surfaces of nanostructures are characterized by randomness, which is difficult to be represented by Euclidean geometry while the fractal geometry approach is alternative that is considered by the increasing number of authors. It is favorable to the fact that energy transformations are permitted on a small scale. The surface asperity stays in positive correlation with energy, rough surfaces of higher fractal dimension having a greater potential in energy receiving and emission. Besides, the intergranular space, grain contacts and pore geometry as fractal objects are responsible for energy storage. This makes the firm connection fractality-energy, supported by the property of fractal objects that are scale invariant, *i.e.* is equally applicable on macro-, micro- or nanosizes.

1. Introduction

After Mandelbrot [1], fractal geometry deals with objects having broken, fragmented, wrinkled or amorphous form or being highly irregular in some other way. Fig. 1 shows familiar SEM images of sintered Ho doped barium-titanate ceramics surfaces. Many authors referred ceramics as fractal structures. Professor Mitic and his group [2–5] are the first among them.

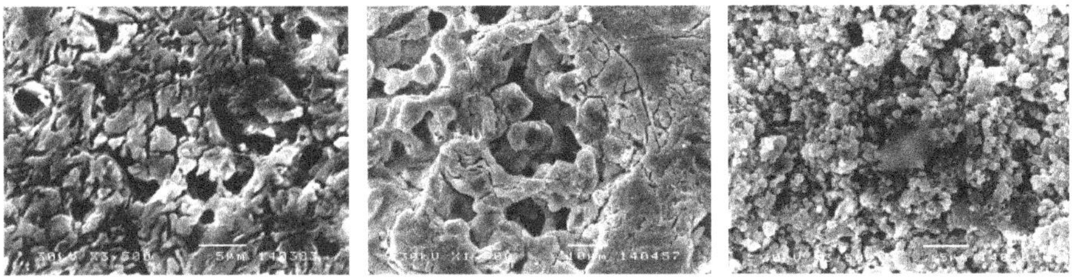

Figure 1. SEM images of sintered $BaTiO_3$ ceramics doped with holmium. From left to right: 0.01 wt.% of Ho, 1350 °C; 0.1 wt.% of Ho, 1380 °C; 0.5 wt.% of Ho, 1350 °C.

The most important numerical parameter of a fractal object is its fractal dimension D_H. Being attached to some fractal object F, the fractal dimension $D_H(F)$ strictly exceeds its ordinary geometric dimension $D_G(F)$, or $D_H(F)>D_G(F)$.

2. Nanoscale and energy

A fractal object has a remarkable property of preserving its "degree of fractalness", expressed by the fractal dimension, in all scales of magnitude. This means that if some feature exists in the scale of millimeters, the same type of feature will be observable down to the nanoscale as shown in Fig. 2.

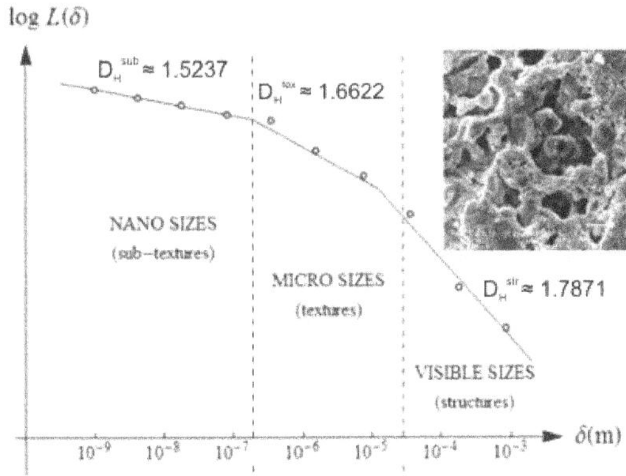

Figure 2. The Richardson-Mandelbrot diagram of the middle specimen from Fig. 1.

Three items are essential regarding energetic questions: i) free energy stocks location, ii) energy harvesting, iii) short and long term energy storage.

All three items have their specific common points with fractals. Also, the concept of energy as a property of physical objects share some features characteristic to fractal objects [6]:

- both fractals and energy are properties, energy is physical objects property, fractality is geometrical;
- as energy, the fractality is omnipresent, unlimited entity concerning space and time;
- both entities, energy and fractality, exist in micro- and macroworld and any fractal is dividable down to nanoscale as well as the energy field;
- energy is closely connected with geometry structures and especially with nanogeometry while some energetic extremal problems as a solution have smooth, symmetric Euclidean objects, others have fractals;
- if an energetic situation can be described by fractals, then potential energy corresponds to fractals constant in time while kinetic energy needs time dependent fractals;
- energy spectrum and energy spectral density of some time series (signals), by the rule, has fractal structure.

From six common features of the two entities, energy-fractal relationship, with some examples provided, one may conclude that the “classic” attitude that extremal solutions always lead to Euclidean and simple forms may lead to solutions that are contrary to fractal conclusions. For example, the soap film problem results in a sphere that represents the smallest surface containing the biggest volume. Contrarily, the smallest volume with the largest surface size question inevitably has fractal solutions. An interesting parallel is found concerning the wind energy equations and the behavior of nanofluids [6].

Deeply connected, the simple material model based on fundamentals of geometry, through physical, planar and volume, constraints was reviewed to link genomic-like properties of coordination and temperature dependent ionic properties, radii and polarizability, with temperature dependent material properties: lattice parameter and volume, (b) crystal structure, (c) dielectric constant, (d) polarization induced phase transition temperature and (e) temperature and frequency dependent conductivity through the complex permittivity as related to Maxwell’s conduction equation. The later three properties are linked to the Clausius-Mossotti relation which has been shown to incorporate both the Curie and Curie-Weiss laws. Most importantly, the model provides a method for *a priori* determining the fabric of space and therefore time and energy, through providing to relatively high precision, the crystal symmetry, starting position(s) of atom(s), ion(s) and anion(s) and their temperature dependent evolution within the structure.

Also, fractals can be used to discuss and describe a wide range of constructs, including but not limited to structures, contact surfaces, energy, flow of energy over large length and time scales. The existence of the fractal nature of ceramic materials has been shown for a wide range of phenomena like ferroelectrics, ferromagnetics, piezo- and optoelectronic properties as well as electrochemical thermodynamic and fluid dynamics parameters. Thin film layers and chemical vapor deposition techniques naturally produce fractal structures which under appropriate growth conditions provide new, desirable properties.

3. Conclusion

In this note we discuss the key property of fractals which is their independence of the scale. This brings the valuable conclusion that properties of fractals are valid on any scale, macro, micro or nano. Fractality, as a scale-independent morphology, provides significant opportunity toward energy storage. Through dilatation symmetry, fractals allow mapping of energy in any scale from sub-nano to distances in light years. On the other hand, time as a quantity conjugated to energy has translation symmetry on all feasible scales. As a consequence, the law of conservation of energy can be applied in the same range. Thus, fractals have a generalized use for the discussion of a wide range of physical phenomena.

References

1. B. B. Mandelbrot, *The Fractal Geometry of Nature* (Freeman, N.Y. 1982).
2. V. V. Mitić, P. Petković, Lj. Kocić, *Proc. Micro Mat. Conf.* (Berlin, 1997), pp. 1103–1105.
3. V. V. Mitić, Lj. M. Kocić, M. Miljković, I. Petković, *Mikrochim. Acta* [Suppl.] 15, (Springer-Verlag, Berlin, 1998), pp. 365–369.
4. V. V. Mitić, V. Paunović, Lj. Kocić, *J. Ceram. Sci. Tech.* 7, 365 (2016).
5. V. Paunović, V. V. Mitić, Lj. Kocić, *Ceramics Int.* **42**, 11692 (2016).
6. V. V. Mitić, Lj. M. Kocić, S. Tidrow, H.-J. Fecht, *Nanotechnology for Energy Sustainability*, Vol. 2, edited by B. Raj, M. Van de Vorde, Y. Mahajan (Wiley-VCH, Weinheim) – to be published.

ELECTRIC FIELD EFFECT ON RADIATIVE RECOMBINATION FROM DEFECT LEVELS IN CdSe NANOPLATELETS

A. O. Muravitskaya, L. I. Gurinovich
B. I. Stepanov Institute of Physics NASB, Nezavisimosti Ave. 68, 220072 Minsk, Belarus
gphms@imaph.bas-net.by

A. V. Prudnikau, M. V. Artemyev
Research Institute for Physico-Chemical Problems of BSU
Leningradskaya Str. 14, 220006 Minsk, Belarus
m_artemyev@yahoo.com

Photoluminescence (PL) of colloidal CdSe nanoplatelets in an external electric field has been studied experimentally. It was shown that the low-energy PL band consists of two peaks of the gaussian form with various amplitudes and half-widths, which can correspond to radiative recombination on defects of a crystal lattice or on surface traps.

1. Introduction

Control of PL properties of semiconductor quantum-sized nanocrystals by external electric field is a promising area of nanophotonics [1]. Study of generation and recombination mechanisms of spatially confined electron-hole pairs in a strong electric field can help to reveal fundamental properties of an electronic subsystem of nanocrystals. This is a motivation for deeper research of quantum confined Stark effect (QCSE) [2] in semiconductor nanomaterials.

Unlike spherical quantum dots, nanoplatelets (NPLs) have much weaker spatial confinement of the charge carriers motion in lateral directions than in the cross-cut direction. This leads to an additional opportunity to control the spatial separation of electrons and holes by the external electric field.

The purpose of the present work is to study the effect of external electric field on PL spectra of quantum-sized CdSe nanoplatelets at room temperature.

2. Experimental

Colloidal quantum-sized CdSe NPLs with the average sizes of $15 \times 25 \times 1.2$ nm (4 monolayers of Cd atoms and 3 monolayers of Se ones) have been synthesized as it is described elsewhere [3].

Photoluminescence of CdSe NPLs embedded in the thin transparent dielectric matrix of poly-(maleic anhydride-alt-1-octadecene) (PMAO) film was excited by a blue LED at 404 nm. The PL spectra were recorded with the cooled CCD detector [4] under an external electric field in the range of 0–200 kV/cm.

3. Results and discussion

The mathematical analysis of the PL spectrum structure of CdSe NPLs has been carried out by the method of multigaussian approximation. It was established that a broad band of CdSe NPLs PL contained the exciton peak and a wide component, consisting of at least two gaussian peaks with equal amplitudes and different half-widths. It is illustrated in Fig. 1.

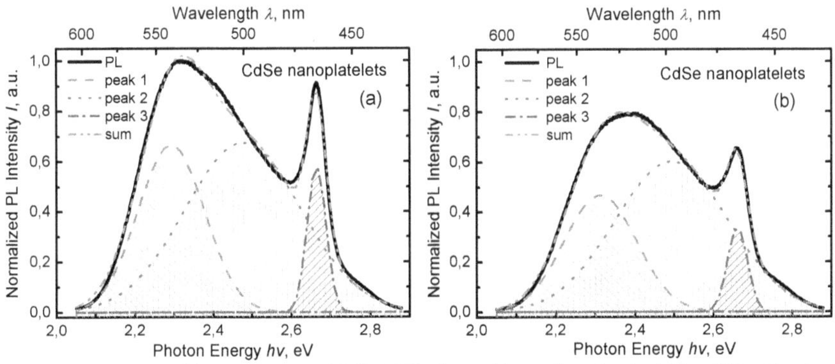

Figure 1. Fitting of the PL spectra of CdSe NPLs by multigaussians (peaks 1, 2 and 3) without electric field (a) and under electric field ~200 kV/cm (b).

The subsequent analysis of the PL structure changes in the external electrical field shows that the spectral shift and the shape distortion of the wide component can be treated as a variation of the component amplitudes.

Effective quenching of the long-wavelength peak can be associated with radiative recombination of electron-hole pairs through surface donor levels (centers with low binding energy), while the less efficient quenching of short-wavelength peak through deep acceptor levels (centers with high binding energy). Thus, the external electric field separates with different efficiency free charges and charges located on surface defects and impurities.

The PL intensity decrease of all three gaussian peaks in the electric field is proportional to the square root of the field strength. Its spectral shift (and energy level shift) is also proportional to the field strength (Fig. 2). The character of the peak amplitude changes agrees with the PL quenching character of CdSe nanorods [5].

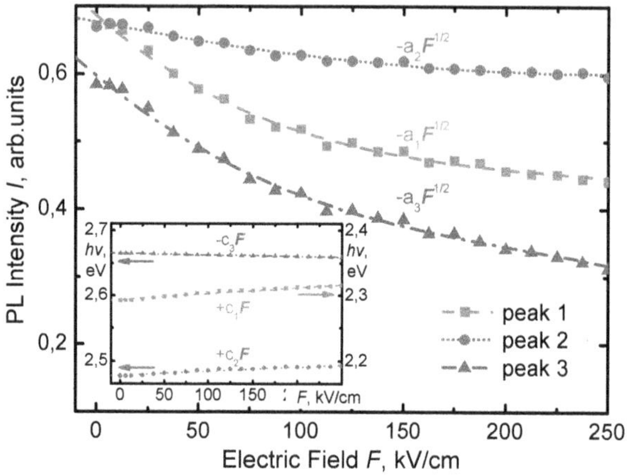

Figure 2. Intensity of PL peaks versus electric field F. Inset: Spectral shift of CdSe NPLs PL peaks on electric field F.

In contrast to the exciton peak, the spectral shifts of the both peaks within the wide band are opposite with respect to the linear Stark shift in quantum-sized NPLs and nanorods. These peaks can correspond to radiative recombination on defects of the crystal lattice (deep acceptor levels) or on surface traps (shallow donor levels), which are appearing due to disconnected crystalline bindings. It is illustrated in Fig. 3.

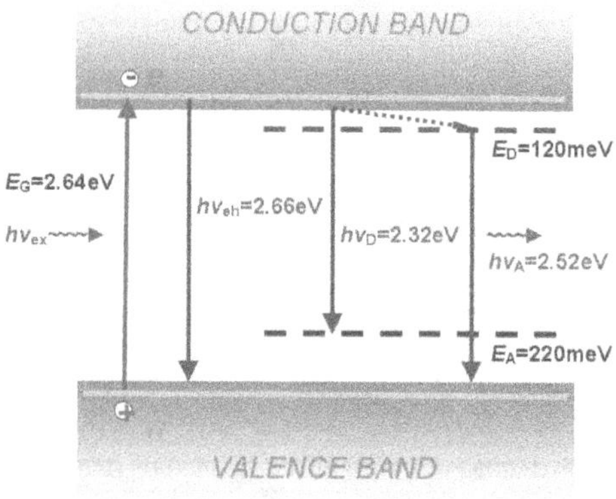

Figure 3. Energy diagram of energy levels in CdSe NPLs.

4. Conclusion

Based on the investigation of the PL spectra of CdSe NPLs in the electric field we can conclude that the radiative recombination from low excited levels occurs through two competing channels: excitons and defects. The recombination through defect states proceeds with a contribution of both donor and acceptor levels, but with the different efficiency. The results can be used for creation of new optoelectronic devices with electrically controlled optical characteristics, including white LEDs with operated color temperature.

Acknowledgments

The authors acknowledge Prof. S. V. Gaponenko for the fruitful discussion of the results obtained. This work was partially supported by Belarusian State Program "Photonics, Opto- and Microelectronics 1.2.02" and CHEMREAGENTS program.

References

1. S. V. Gaponenko, *Introduction to Nanophotonics* (Cambridge University Press, New York, USA, 2010).
2. A. I. Ekimov, Al. L. Efros, T. V. Shubina, A. P. Skvortsov, *J. Luminescence* **46**, 97 (1990).
3. A. W. Achtstein, A. V. Prudnikau, M. V. Ermolenko, L. I. Gurinovich, S. V. Gaponenko, U. Woggon, A. V. Baranov, M. Yu. Leonov, I. D. Rukhlenko, A. V. Fedorov, M. V. Artemyev, *ACS Nano* **8**, 7678 (2014).
4. L. I. Gurinovich, A. A. Lutich, A. P. Stupak, S. Ya. Prislopsky, E. K. Rusakov, M. V. Artemyev, S. V. Gaponenko, H. V. Demir, *Semiconductors* **43**, 1008 (2009).
5. L. I. Gurinovich, M. V. Artemyev, A. P. Stupak, S. Ya. Prislopskii, S. V. Zhukovsky, S. V. Gaponenko, *J. Appl. Spectros.* **79**, 95 (2012). Erratum: *J. Appl. Spectrosc.* **79**(2), 337 (2012).

"DARK" MODES BACKSCATTERING AS POSSIBLE RATIONALE FOR ANOMALOUS RETROREFLECTION FROM STRONGLY ABSORBING POROUS NANOSTRUCTURES

V. V. Sergentu
Institute of Applied Physics, Academy of Sciences of Moldova
Academy Str. 5, MD-2028 Chisinau, Moldova

V. Ursaki, Ed. Monaico
National Center for Materials Study and Testing, Technical University of Moldova
MD-2004 Chisinau, Moldova

I. M. Tiginyanu
Institute of Electronic Engineering and Nanotechnologies "D.Ghitu"
Academy of Sciences of Moldova, MD-2028 Chisinau, Moldova
National Center for Materials Study and Testing, Technical University of Moldova
Chisinau MD-2004, Moldova

S. Ya. Prislopski, S. V. Gaponenko
B.I.Stepanov Institute of Physics NASB, Nezavisimosti Ave. 68, 220072 Minsk, Belarus
s.gaponenko@ifanbel.bas-net.by

The previously discovered anomalous retroreflection in a strongly absorbing nanostructured medium is explained by using "dark" and "bright" modes. The consideration provides rationale not only for the retroreflection itself but explains correlations with absorption and differences for s- and p-polarized radiation retroreflection.

1. Introduction

Anomalous retroreflection has been discovered in 2011 for a fishnet nanoporous semiconductor material strongly absorbing in the visible range [1]. Retroreflection occurs in a narrow solid angle along with a diffusive specular reflection for all angles of incidence. Retroreflection is apparent by the naked eye with a day light illumination and exhibits no selectivity with respect to a wavelength featuring minor depolarization of back reflected light. The phenomenon can be classified neither as coherent backscattering nor as Anderson localization of light. Later on we found that retroreflection does correlate with high absorption and vanishes in the spectral range of low absorbance of the semiconductor material [2]. In this contribution we provide consideration for the observed phenomenon in terms of "dark" and "bright" modes. The consideration provides rationale not only for the retroreflection itself but explains correlations with absorption and differences for s- and p-polarized radiation retroreflection.

2. Results and discussion

We propose to use the phenomenon of coherent backscattering for an explanation of anomalous retroreflection. However, unlike traditional consideration based on transverse modes exclusively, we consider a possibility of longitudinal modes generated in a nanostructured medium. In case of porous absorbing medium, two types of waves exist in such a system: usual transverse waves and longitudinal waves of electrical type, which can exist only in nanostructured media, and each of them contribute to the coherent backscattering. The longitudinal waves represent inhomogeneous in space purely electrical longitudinal fields in the low-frequency limit, they do not carry energy and cannot be excited by an external transverse wave [3]. We will consider the situation when the wave vector is of the order of $k_0 \infty 1/a$ (where a is the lattice constant of nanopores), and such a wave will carry energy in such a case. In a porous medium consisting of a semiconductor material attenuation of the longitudinal waves is two orders of the magnitude weaker than that for transverse waves. It becomes possible since the energy of the wave is distributed non-uniformly in the space being concentrated predominantly inside the pores. This is a key issue to describe the observed anomalous retroreflection in terms of the coherent backscattering with the major contribution from scattering of longitudinal waves in the otherwise strongly absorbing medium.

We use additional boundary conditions in the form of

$$\vec{P}_z \,|_{z=0} \equiv 0\,, \tag{1}$$

where $\vec{P}$ is the electrical polarization of the medium. Its component perpendicular to the surface of the medium has to be equal to zero at the boundary of the porous medium with vacuum by analogy with that in [4]. The Eq. (1) is transformed to a trivial identity 0≡0 in the case of s-polarization, which means that the longitudinal wave is not excited. In the case of p-polarization, the longitudinal wave is excited, but for both longitudinal "dark mode" and transversal mode at the medium/vacuum interface the polarization vector is

$$\vec{P} = (\varepsilon_t - 1)\vec{E}\,, \tag{2}$$

where the transverse dielectric constant $\varepsilon_t = \varepsilon_t(k)\,|_{k\equiv 0}$ is used, which actually is defined by methods described in [5].

Fig. 1 shows calculated scattering indicatrices for s- and p-polarization for the angle of incidence of 30 deg without ($\lambda = 1000$ nm) and with absorption ($\lambda = 500$ nm).

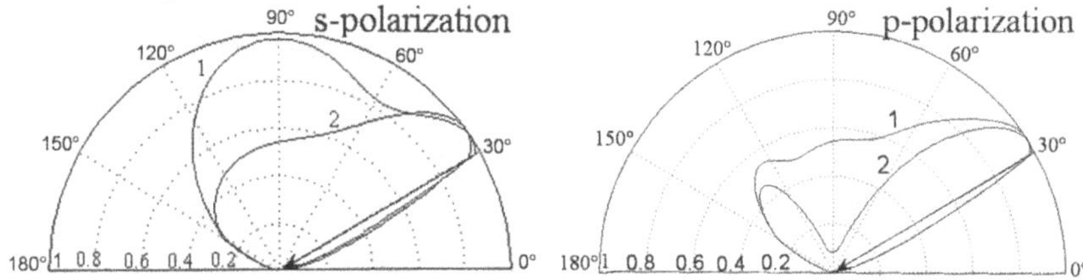

Figure 1. Calculated scattering indicatrices for s- and p-polarization for the angle of incidence of 30 deg without (1) and with (2) absorption.

Scattering indicatrices obtained experimentally for nanoporous InP sample are presented in Fig. 2 to show principal regularities of the observed 'anomalous' backscattering. Left panel demonstrates vanishing of retroreflection for wavelengths corresponding to low intrinsic absorption ($\lambda = 1064$ nm) in comparison with wavelengths corresponding to high intrinsic absorption ($\lambda = 654$ and 531 nm). Right panel shows that the backscattering is more pronounced for p-polarization as compared to s-polarization of incident radiation.

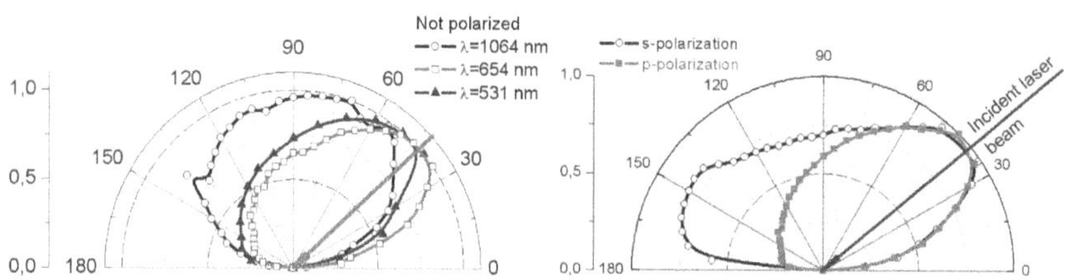

Figure 2. Experimental data for angular dependent scattering of nanoporous InP sample for the angle of incidence of 40 deg, different wavelength corresponding to low (1064 nm) and high (532 and 654 nm) intrinsic interband absorption in InP, s- and p-polarization.

It is obvious that the experimental data are in a reasonable agreement with the principal theory predictions.

The results of this work explain the mechanisms of anomalous retroreflection as well as the main experimental peculiarities observed experimentally in nanoporous InP for the case of strong absorption by using both "dark" and "bright" modes. The observed regularities are:

- Only usual transverse waves ("bright modes") and the associated with them backscattered radiation can be excited with s-polarized incident wave. The absorption strongly changes the scattering indicatrix suppressing the scattered radiation at small incidence/scattering angles.

- Ultrashort modes (“dark modes”) and usual transverse waves (“bright modes”) as well as associated with them backscattered radiation can be excited with p-polarized incident wave. The contribution from “dark modes” which practically do not depend on the absorption is predominant at large incidence/scattering angles. The absorption significantly changes the shape of the scattering indicatrix suppressing the scattering radiation by “bright modes” at small incidence/scattering angles.

Due to these regularities it was possible to observe the phenomenon of anomalous retroreflection in a strongly absorbing nanostructured medium in a wide frequency range. Therefore, the dissipation does not prevent multiple scattering of longitudinal modes, because their energy is distributed irregularly in the space and is being concentrated predominantly inside the pores.

3. Conclusion

Based on the consideration involving the additional boundary condition for electromagnetic waves scattering in a strongly absorbing medium with nanopores we provide the rationale of anomalous retroreflection based on development of the “dark” modes with inhomogeneous space distribution that do not attenuate and therefore experience multiple scattering resulting in coherent backscattered signal. In accordance with the experiment, our theory explains predominating retroreflection for p-polarized versus s-polarized incident light since the latter does not generate the “dark” modes.

References

1. S. Ya. Prislopski, E. K. Naumenko, I. M. Tiginyanu, L. Ghimpu, E. Monaico, L. Sirbu, S. V. Gaponenko, *Optics Letters* **36**, 3227 (2011).
2. S. Y. Prislopski, I. M. Tiginyanu, L. Ghimpu, E. Monaico, L. Sirbu, S. V. Gaponenko, *Applied Physics A* **117**, 467 (2014).
3. V. Sergentu, V. Ursaki, L. Sirbu, *Phys. Stat. Solidi (b)* **250**, 2215 (2013).
4. S. I. Pekar, *Crystal Optics and Additional Light Waves* (Benjamin/Cummings Pub. Co., Menlo Park, Calif. 1983), 469 p.
5. L. A. Golovan, V. Yu. Timoshenko, P. K. Kashkarov, *Phys.-Usp.* **50**, 595 (2007).

ELECTROMAGNETIC INTERACTION OF ELECTRODEPOSITED NICKEL NANOWIRES WITH A NIOBIUM THIN FILM

S. V. Redko, E. B. Chubenko, A. L. Dolgiy, Al. L. Dolgyi, V. A. Petrovich, S. L. Prischepa, A. L. Danilyuk, A. V. Andreyenka, V. P. Bondarenko

Department of Micro- and Nanoelectronics
Belarusian State University of Informatics and Radioelectronics
P. Browka 6, 220013 Minsk, Belarus
eugene.chubenko@gmail.com, sv.redko@bsuir.by

M. Trezza, C. Cirillo, C. Attanasio

CNR-SPIN Salerno and Dipartimento di Fisica "E. R. Caianiello"
Universita degli Studi di Salerno
via Giovanni Paolo II 132, I-84084 Fisciano (Sa), Italy

In this paper we present superconducting properties of a Nb thin film deposited on an array of ferromagnetic (Ni) nanowires embedded in a porous template. By investigating the $T_c(H)$ phase boundary and by measuring $V(I)$ characteristics and critical currents as a function of the applied magnetic field, we find that the Nb film exhibits properties similar to those of a network of one-dimensional superconducting nanowires. We attribute this behavior to the stray fields of the magnetic dipoles, which create an almost regular lattice of normal regions in the superconductor, ultimately changing its topology.

1. Introduction

Arrays of ferromagnetic nanowires (NWs) embedded in porous self-assembled templates received a considerable interest, due to the possibility to prepare in these structures regular magnetic arrays with characteristic dimensions, *i.e.* diameter of the wires and nearest neighbor distance, in the nanometer range. Self-assembled methods are versatile and reliable techniques, which allow one to fabricate low-cost patterns of nanostructures on very large surfaces with a high degree of reproducibility. Recently, these systems were used in conjunction with superconducting films in order to create periodic magnetic pinning centers and study matching effects between the artificial non-superconducting regular structure and the vortex lattice [1]. The system behaves like a superconducting wire network when the superconducting coherence length ξ is of the same order as the width of the channels between the holes w.

Here we present superconducting properties of a 30 nm thick niobium (Nb) film sputtered on a dense array of nickel (Ni) nanowires grown in a nanoporous Si template. Nb is separated from the magnetic dipoles by an insulating Al_2O_3 layer to obtain an F/I/S hybrid.

2. Experimental

Monocrystalline silicon wafers with (100) orientation and 0.01 Ω×cm resistivity were used as the substrates for porous silicon (PS) formation. The electrochemical anodization process provided the formation of an uniform PS layer 10 μm thick with 72 % of porosity and pore diameter of ≈100 nm [2]. After the formation of PS, the sample were washed in deionized water. Ni was electrochemically deposited into the PS matrix from aqueous $NiSO_4$ based electrolyte at the current density of 3.5 mA/cm^2. After that a thin insulating layer of Al_2O_3 (8 nm) was sputtered in vacuum chamber. Then Nb thin film was deposited in the same sputtering system. The sample under investigation is a Nb film of thickness $d = 30$ nm (sample W). Due to the presence of the Al_2O_3 insulating buffer layer between the superconductor (Nb) and the PS filled with the Ni nanopillars, the system under study is a F/I/S hybrid. In this way, the proximity effect and exchange of electrons are completely avoided and the interaction between the superconducting and the ferromagnetic systems is totally electromagnetic. A reference sample (sample F) was also fabricated and analyzed. It was formed by depositing a continuous 1 μm thick Ni layer onto PS template. Subsequently, Al_2O_3 and Nb layers were grown exactly as described previously.

Superconducting electric transport properties of the samples were measured in a ^{4}He cryostat using a standard DC four-probe technique. The superconducting transition temperature in the absence of applied field has been taken as the temperature at which the resistance drops 50 % of the normal state resistance (R_N). It was $T_{c0} = 5.0$ and 5.5 K for the samples W and F, respectively. $V(I)$ characteristics were recorded using a pulsed technique [3]. Before starting the measurements, the Ni NWs were first magnetized at low temperatures in a perpendicular magnetic field of 1 T which is far above the saturation field of Ni NWs.

3. Results and discussion

Fig. 1a shows the $T_c(H)$ phase diagram constructed from the measured resistive transitions in various perpendicular magnetic fields for the sample W. The three curves have been obtained using different resistive transition criteria of the normal state resistance, namely 10 %, 50 %, and 90 % of R_N. The phase boundary shows a pronounced anomaly in the field range of 0.19–0.23 T where the critical temperature apparently increases with respect to what is conventionally expected almost reaching T_{c0}.

The bump is also present at higher fields, in the range of 0.53–0.64 T, though with a lower amplitude. Furthermore, the amplitude of the two bumps decreases when a higher resistance criterion is adopted, consistently with previous observations in similar systems [4]. In particular, the anomaly present at higher fields is smeared out when the criterion of 90 % of R_N is adopted (open circles in Fig. 1a). A selection of the $R(T)$ curves in different fields is shown in Fig. 1b. In Fig. 1c the phase boundary of the reference sample F, constructed with the criterion of 50 % of R_N, is compared with the one of the sample W, obtained using the same resistive criterion. We see that the curve is smooth and does not show any anomalous feature.

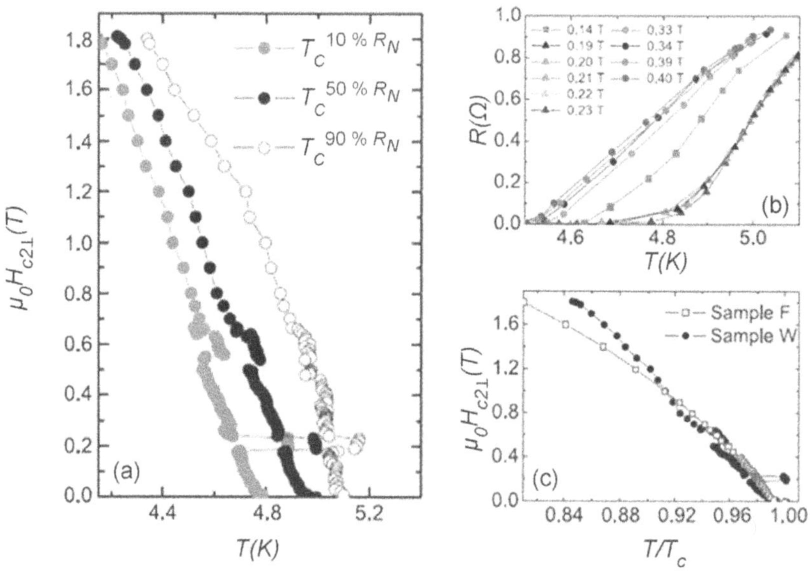

Figure 1. (a) $T_c(H)$ phase diagrams of the sample W obtained with resistive criteria of 10 %, 50 %, and 90 % of R_N. (b) Resistive transitions of the sample W for a selection of applied magnetic fields. The curves from 0.19–0.23 T (at 0.14 T and from 0.33–0.40 T) are representative of the temperature dependence of the resistance at the magnetic fields where the bump is (is not) present in the phase diagram. (c) H-T phase diagrams of the samples W and F, both obtained with the 50 % of R_N criterion.

By combining magnetoresistance and $V(I)$ measurements, we establish that the matching effects observed in the Nb film are caused by the Little-Parks effect more than by the vortex pinning, in agreement with theoretical predictions [5]. Therefore, the system behaves like a one-dimensional (1D) superconducting wire network where the confinement of the Cooper pairs is not caused by geometrical constrictions, but rather by the stray fields of the magnetic dipoles that modify the topology of the plain film creating regions where the superconductivity is strongly depressed. Due to the values of the parameters ξ and w, the wire network behavior is presented at relatively low temperatures and high magnetic fields.

4. Conclusion

Porous Si templates have been used to host a dense array of electrodeposited Ni NWs that electromagnetically interact with a Nb thin film through the intermediate Al_2O_3 layer to form an F/I/S hybrid. The analyzed F/I/S system, even though not very highly ordered, can be conveniently used to obtain a network of 1D superconducting NWs in the presence of a magnetic array, with the advantage of depositing the superconducting film on a smooth buffer layer instead of an irregular porous template.

Acknowledgments

This work is supported by Belarusian Republican Foundation for Fundamental Research (grant T16M-012).

References

1. M. Trezza, C. Cirillo, A. I. Vorobjeva, E. A. Outkina, S. L. Prischepa, C. Attanasio, *Supercond. Sci. Technol.* **26**, 035001 (2013).
2. A. Dolgiy, H. Bandarenka, S. Prischepa, K. Yanushkevich, P. Nenzi, M. Balucani, V. Bondarenko, *ECS Trans.* **41**, 1118 (2012).
3. C. Cirillo, E. A. Ilyina, C. Attanasio, *Supercond. Sci. Technol.* **24**, 024017 (2011).
4. U. Patel, Z. L. Xiao, J. Hua, T. Xu, D. Rosenmann, V. Novosad, J. Pearson, U. Welp, W. K. Kwok, G. W. Crabtree, *Phys. Rev.* B **76**, 020508(R) (2007).
5. A. Yu. Aladyshkin, D. A. Ryzhov, A. V. Samokhvalov, D. A. Savinov, A. S. Mel'nikov, V. V. Moshchalkov, *Phys. Rev. B* **75**, 184519 (2007).

GRAIN EFFECT IN THE CARRIER MOBILITY OF $BaSi_2$ NANOFILMS

D. A. Shohonov[1], I. S. Samusevich[1], A. B. Filonov[1], D. B. Migas[1], K. Morita[2], T. Suemasu[2]

[1]*Belarusian State University of Informatics and Radioelectronics, Minsk, Belarus*
migas@bsuir.by
[2]*Institute of Applied Physics, University of Tsukuba, Tsukuba, Ibaraki 305-8573, Japan*
suemasu@bk.tsukuba.ac.jp

According to the experimental data the electron mobility of $BaSi_2$ nanofilms has rather high values, as compared with other semiconducting silicides, reaching 1230 cm^2/Vs at 218 K and 816 cm^2/Vs at 300 K. We demonstrate that the temperature dependence of the mobility cannot be adequately reproduced by the use of standard carrier scattering mechanisms. The modified approach which accounts for the grained nature of the films was proposed for the correct description of the mobility behavior.

1. Introduction

Semiconducting silicides being well compatible with conventional silicon technology and promising for various applications are also known as environmentally friendly materials. Barium disilicide ($BaSi_2$) with the band gap of 1.15–1.30 eV [1], having rather large values of the absorption coefficient near the absorption edge and a large minority-carrier diffusion length in addition to a long minority-carrier lifetime can be viewed as a potential material for efficient conversion of sun's light to electricity [2]. $BaSi_2$ is also attractive not only for photovoltaic applications but for thermoelectric applications because of the large Seebeck coefficient [3]. There are some investigations devoted to the study of $BaSi_2$ basic electronic and optical properties [*e.g.* 4], but at the same time there is almost lack of information on the transport properties of the material under consideration. Thus the main purpose of our paper is to present experimental and theoretical data on the carrier mobility in $BaSi_2$.

2. Experimental

An ion-pumped molecular beam epitaxy (MBE) system equipped with a standard Knudsen cell for Ba and an electron-beam evaporation source for Si was used for the growth. After cleaning a floating-zone n-Si(111) substrate ($\rho > 1000\ \Omega\cdot cm$) at 900 °C for 30 min in ultrahigh vacuum, a 5 nm-thick $BaSi_2$ template layer was grown on the Si substrate at 500 °C by reactive deposition epitaxy. The $BaSi_2$ template works as seeds for overlayers. Next, Ba and Si were codeposited on the template layer at 600 °C to form a 250 nm-thick $BaSi_2$

epitaxial layer by MBE [5]. It was found from X-ray diffraction measurements that the *a* axis of the $BaSi_2$ layer with grain size of 200 nm is normal to the substrate surface. The main results are shown in Fig. 1 (the left panel). The self-carrier concentration is estimated to be $n = 5.0\times10^{15}$ cm^{-3} by Hall measurements.

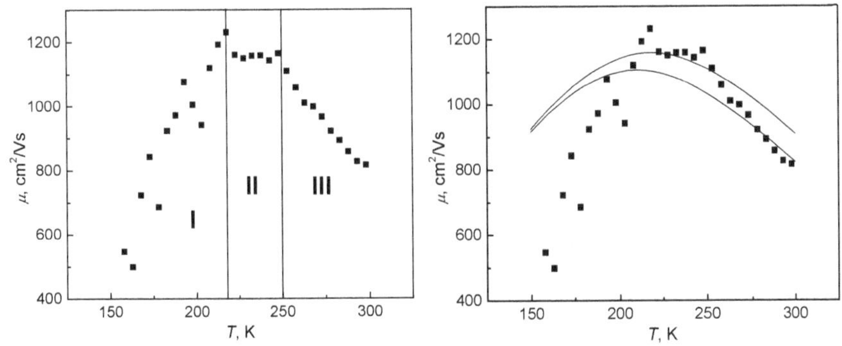

Figure 1. The left panel: the experimentally measured Hall mobility of electrons versus temperature in $BaSi_2$ thin films. The right panel: mobility versus temperature (experiment – black squares, simulation – the upper curve at $T_D = 900$ K and lower curve at $T_D = 850$ K.

It is clearly seen that the dependence of the Hall mobility with temperature can be conventionally subdivided into three main regions: I – $160 \leq T \leq 218$ K displaying a sharp mobility raise with temperature till it reaches the maximum; II – $218 \leq T \leq 243$ K including the region of the mobility maximum (the position of the mobility maximum is shifted to the high temperature region as compared to other semiconducting silicides as well as the mobility has rather high value); III – $243 \leq T \leq 298$ K characterizing the region of the fast mobility drop.

3. Simulation

Generally in semiconductors there are several standard well-known carrier scattering mechanisms [6]. Each mechanism can be described by the complex function of meaningful physical values which can be treated as an appropriate set of parameters. In our case if one accounts for all the possible scattering mechanisms the resulting mobility according to the Mathiessen's formula is $1/\mu = 1/\mu_{AC} + 1/\mu_{NPO} + 1/\mu_{PO} + 1/\mu_{CIM}$, where μ_{AC}, μ_{NPO}, μ_{PO}, μ_{CIM} are partial mobilities due to the carrier scattering on the acoustic phonons, nonpolar and polar optical phonons, and charged impurity centers, respectively. Previously this approach was used for a number of semiconducting silicides [*e.g.* 7]. The electron effective masses were used from [4].

We have found a lot of reasonable sets of realistic parameters to reproduce the region III of the experimental curve in addition to a few reasonable sets of realistic parameters to reproduce regions II and III of the experimental curve simultaneously as shown in Fig. 1 (the right panel). The latter results were obtained for the following set of parameters: $D = 5.0$ eV is the constant defined by the components of the acoustic deformation potential tensor; $\eta = 5$ is a ratio of the nonpolar optical deformation potentials to the acoustic one; T_D is the Debay temperature; $\varepsilon = 13.5$ and $\varepsilon_\infty = 3.0$ are the static and high frequency dielectric constants [4]; $n = 5.0\times10^{15}$ cm^{-3} is the self-carrier concentration measured experimentally; $N_i = 2.0\times10^{16}$ cm^{-3} is the charged impurity center concentration [8]. Also we need to emphasize that the Debay temperature is one of the key parameters to reproduce the experimental data correctly. In our case the Debay temperature is considerably higher than the possible experimental one of 260 K [3].

4. Modified theory

For an adequate description of the observed experimental data the theoretical approach needs to be modified. We make the following assumptions. The first one is the grained nature of the studied samples. Thus the intergrain boundaries can be treated as small potential barriers. And the resulting mobility can be written as $\mu^* = \mu \times exp(-\Delta E / kT)$ where ΔE is the potential barrier height. And especially such the correction is important for the low temperature range since not all the carriers have sufficient energy to overcome such barriers.

The results of the simulation with a use of the modified approach are presented in Fig. 2 (the left panel).

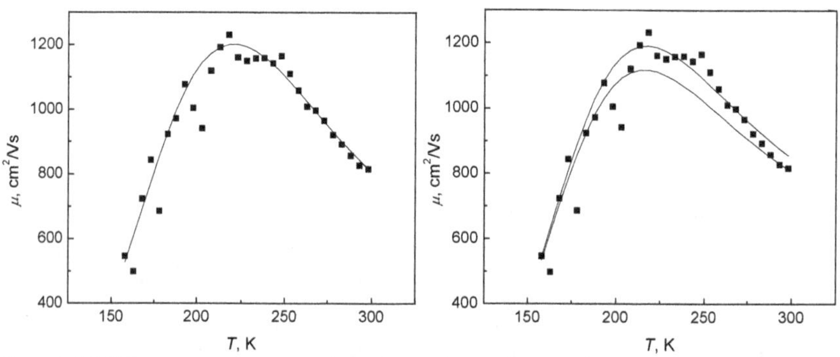

Figure 2. Mobility versus temperature: (the left panel) experiment – black squares, simulation – solid curve at $T_D = 600$ K; (the right panel) experiment – black squares, simulation – the upper curve at $T_D = 450$ K and lower curve at $T_D = 400$ K.

The results in Fig. 2 (the left panel) were obtained for the following set of parameters: $D = 5.0$ eV; $\eta = 5$; $T_D = 600$ K; $\varepsilon = 13.5$; $\varepsilon_\infty = 3.0$; $n = 5.0\times10^{15}$ cm^{-3};

$N_i = 2.0\times10^{16}$ cm^{-3} and ΔE = 73 meV. Here we note that the obtained barrier height correlates well with the experimentally estimated [8]. But the Debay temperature is still higher than the experimental one [3]. Thus we made one more assumption that it is necessary to account for the fact that generally Hall mobility is larger 1.2…1.9 times than the drift one depending on the carrier scattering mechanism [6]. The corresponding dependencies are shown in Fig. 2 (the right panel) and the following set of parameters for the upper (D = 5.0 eV; $\eta = 3$; $\varepsilon = 14$; ε_∞=3.0; $n = 5.0\times10^{15}$ cm^{-3}; $N_i = 2.0\times10^{16}$ cm^{-3} and ΔE = 73 meV) and lower (D = 2.5 eV; $\eta = 1$; $\varepsilon = 14$; $\varepsilon_\infty = 3.0$; $n = 5.0\times10^{15}$ cm^{-3}; $N_i = 2.0\times10^{16}$ cm^{-3} and ΔE = 73 meV) curves have been obtained with the Debay temperature of 400–450 K.

5. Conclusion

The results obtained in our study show that the low temperature part of the experimental data *i.e.* the sharp raise of the mobility is mainly defined by the presence of the intergrain boundaries in the structure under consideration. However, it is also important to account for the carrier scattering on the charged impurity centers at least at relatively low concentration (N_i of the order of 10^{16} cm^{-3}) of such centers. Moreover, it was recently proposed [9] that Si vacancies in $BaSi_2$ are responsible for the presence of such centers and their estimated concentration is in a very good agreement with the results of our study.

References

1. V. E. Borisenko, *Semiconducting Silicides* (Springer Series in Materials Science, 2000).
2. T. Suemasu, N. Usami, *J. Phys. D: Appl. Phys.* **50**, 023001 (2017).
3. K. Hashimoto *et al.*, *J. Appl. Phis.* **102,** 063703 (2007).
4. D. B. Migas *et al.*, *phys. stat. sol. (b)* **244**, 2611 (2007).
5. K. Morita, Y. Inomata, T. Suemasu, *Thin Solid Films* **508**, 363 (2006).
6. B. K. Ridley, *Quantum Processes in Semiconductors* (Clarendon Press, Oxford, 1892); V. L. Bonch-Bruevich, S. G. Kalashnikov, *Semiconductors Physics* (Nauka, Moscow, 1990).
7. D. B. Migas *et al.*, *Phys. Rev. B* **77**, 075205 (2008).
8. M. Baba *et al.*, *J. Appl. Phis.* **103,** 142113 (2013).
9. M. Kumar, N. Umezawa, M. Imai, *2016 Spring Meeting of the Japan Society of Applied Physics* (21a-S223-2).

MECHANISM OF CHARGE TRANSPORT IN $Sr_2FeMoO_{6-\delta}$ CERAMICS WITH INTERGRAIN DIELECTRIC SHEATHS

M. V. Yarmolich, N. A. Kalanda, S. E. Demyanov, A. V. Petrov
SSPA "Scientific-Practical Materials Research Center NASB"
P. Browka 19, 220072 Minsk, Belarus
kalanda@physics.by

S.-C. Yu, S. K. Oh, D.-H. Kim
Chungbuk National University, 361-763 Cheongju, South Korea
donghyun@cbnu.ac.kr

A heterostructure based on $Sr_2FeMoO_{6-\delta}$ with $SrMoO_4$ dielectric sheaths has been fabricated. Temperature ranges of the hopping conduction mechanism were established, and the field dependence of magnetoresistance of the composite material was measured.

1. Introduction

Magnetic and magnetoresistive properties of $Sr_2FeMoO_{6-\delta}$ (SFMO) magnetic half-metal with the double perovskite structure make it a promising material for spintronic applications due to its high Curie temperatures ($T_C \sim 420$ K), large negative magnetoresistance ($MR \sim 35$ % at 4.2 K) and practically 100 % spin-polarisation of conduction electrons [1].

The observed high magnetoresistance of SFMO is caused, in the first line, by the presence of grain boundaries which provide potential barriers for electron tunnelling. Besides, the MR value depends on the spin polarisation degree and superstructural ordering of iron and molybdenum cations. The present paper aims at studying thermal treatment modes which lead to an appearance of potential barriers for the electrons caused by the formation of an insulating oxide phase at the SFMO grain boundaries, and thus to manifestation of the tunnel magnetoresistance in $Sr_2FeMoO_{6-\delta}/SrMoO_4/Sr_2FeMoO_{6-\delta}$ heterostructure.

2. Experimental

For the synthesis of nanosized $Sr_2FeMoO_{6-\delta}$ powders by the citrate-gel route, high-purity strontium nitrate $Sr(NO_3)_2$, iron nitrate $Fe(NO_3)_3 \cdot 9H_2O$, ammonium molybdate $(NH_4)_6Mo_7O_{24}$ and citric acid monohydrate $C_6H_8O_7 \cdot H_2O$ were used. To obtain the colloidal sol, water solutions of strontium and iron nitrates were mixed at molar ratio of 2Sr:Fe. The citric acid was added to the solution at the molar ratio of citric acid:Fe = 6.5:1. After this, $(NH_4)_6Mo_7O_{24}$ water mixture was added to the prepared aqueous solution of strontium and iron nitrates at the molar ratio of 1Mo:1Fe. The resulting solutions were continuously stirred upon adding ethylenediamine until the pH value achieved 4. The resulting gel was placed in a furnace in air at 372 K and heated up to 470 K, followed by holding at this temperature for 18 h. At this step, a solid foam was obtained which was ground up and heat treated at 773 K and $p(O_2) = 0.21\times10^5$ Pa for 10 h. The final annealing was performed in the reducing gas mixture of 5 % H_2/Ar in several steps with a final heat treatment at 1223 K for 4 h. Nanosized single-phase SFMO powders were pressed into pellets with the diameter of 10 mm at various pressures 1–4 GPa and temperatures 300–1100 K.

The degree of superstructural ordering of the iron and molybdenum cations (P) has been calculated by the Rietveld method applying the PowderCell, FullProf software packages fitting to experimental X-ray diffraction (XRD) patterns obtained at room temperature on a "DRON-3" setup with CuK_α radiation. The magnetic and magnetoresistive properties of the samples were studied in a multipurpose measurement system (Cryogenic Ltd.) in the temperature range of 4.2–300 K and magnetic fields up to 8 T. Electrical resistivity and magnetoresistance measurements were performed by the conventional four-probe method.

3. Results and discussion

The minimum resistivity was obtained for the SFMO sample after annealing at the pressure of 4 GPa and the temperature of 800 K. The sample exhibited metallic conductivity in the temperature range of 4.2–300 K (Fig. 1). According to XRD data, the pressed SFMO samples were single-phase compounds of the tetragonal ($I4/m$, $Z = 2$) structure with the superstructural ordering of Fe^{3+}/Mo^{5+} cations $P = 88$ %. The unit cell parameters were: $a = b = 0.556(2)$ nm, $c = 0.789(3)$ nm and $V = 0.244(2)$ nm^3. After additional annealing in Ar at 700 K for 5 h we have detected an appearance of the $SrMoO_4$ phase. Its content did not exceed 5.5 % (Fig. 2), whereas the parameters a, b, c and P of the SFMO samples did not change.

The dependences $\rho(T)$ of the obtained SFMO ceramics with dielectric sheaths exhibited the pronounced semiconductor conductivity with the negative temperature resistance coefficient ($dR/dT < 0$) in the entire range of magnetic fields and temperatures (Fig. 3). The power laws $\ln(\rho/\rho_0) \propto f(T^{-1/4})$ (the Mott mechanism in the range of 300–250 K) and $\ln(\rho/\rho_0) \propto f(T^{-1/2})$ (the Efros-Shklovskii mechanism in the range of 200–130 K) are typical of the variable hopping range over localized states in the dielectric $SrMoO_4$ sheaths [2–3].

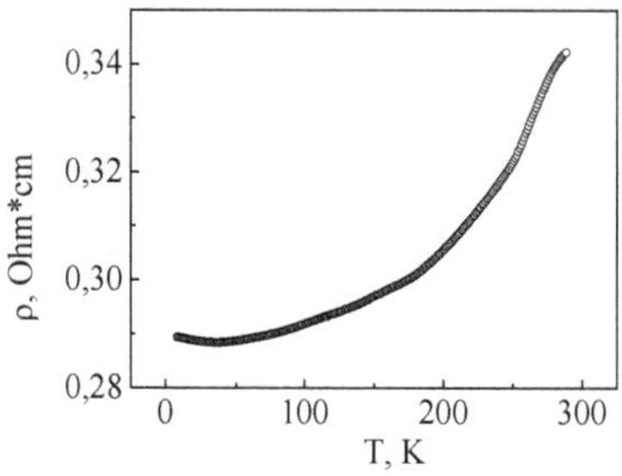

Figure 1. Temperature dependence of the electrical resistance of SFMO powders pressed at 4 GPa and 800 K.

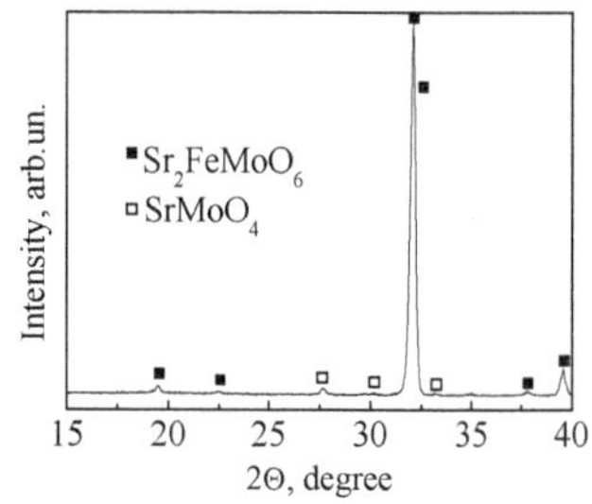

Figure 2. X-ray diffractograms of the pressed SFMO pellets additionally annealed at 700 K and oxygen pressure of 10 Pa for 5 h.

Within the framework of this model, the resistivity can be expressed through the barrier parameters as

$$\rho \propto \exp[2(2\pi/h)(2m\phi)^{1/2}s], \tag{1}$$

where h is the Planck constant, m is the electron effective mass, ϕ is the effective barrier height, s is the barrier width equal to the minimum jump length. In this case, the electrical neutrality is violated at the jump, so that the hopping electron must overcome the Coulomb barrier, E_C. As a result, only the electrons thermally activated above the barrier can be considered as free ones. Taking into account the combination of the two processes, we can write the following expression for electrical resistivity:

$$\rho = \rho_0 \cdot exp[2\sqrt{Å_{à} / k_B T}], \tag{2}$$

where $E_a = (2\pi/h)(2m\phi)^{1/2}E_C$ is the tunnel barrier height. This expression finally leads to the power law

$$\ln(\rho/\rho_0) = 2[(E_a/(k_B T)]^{1/2}. \tag{3}$$

When the magnetic field is applied parallel to the electrical current flow, a negative magnetoresistance is observed in the samples with dielectric sheaths: the electrical resistance measured at the current $I_{meas} = 1$ μA decreases with increasing of the field (Fig. 3). It has a quantum nature, since conduction electrons tunnel between SFMO grains through dielectric sheaths. The charge transfer becomes a spin-dependent, and the largest value of $MR = -41$ % is achieved in the field of 8 T (Fig. 4). The maximum change of the magnetoresistance, ΔMR, is observed in relatively low magnetic fields $B \leq 0.2$ T, and with further increasing of field, MR saturates.

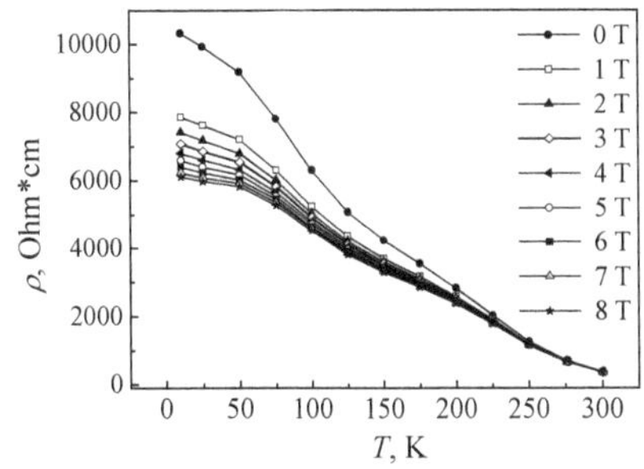

Figure 3. Temperature dependences of the resistivity of the SFMO sample measured at various magnetic fields.

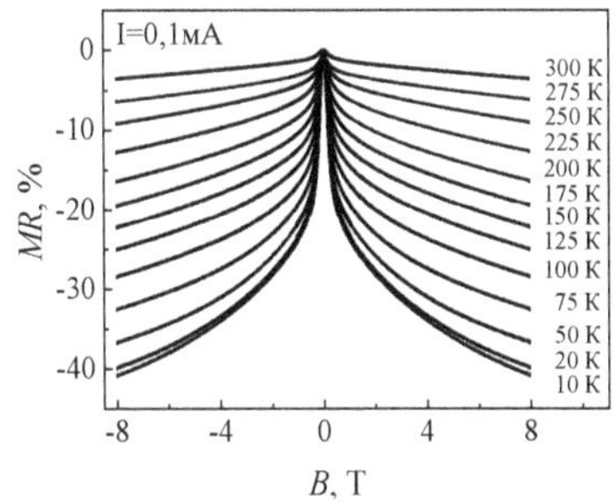

Figure 4. Field dependence of the magnetoresistance of the SFMO sample measured in the magnetic fields parallel to the electrical current.

4. Conclusion

Annealing of pressed nanosized $Sr_2FeMoO_{6-\delta}$ powder at 700 K and oxygen pressure of 10 Pa for 5 h leads to the $SrMoO_4$ phase formation at the grain boundaries. $\rho = f(T)$ dependencies of the resulting material has the pronounced semiconductor conductivity with a negative temperature coefficient of resistance ($dR/dT<0$) in the entire range of the external magnetic field up to 10 T and temperatures 300–4.2 K. In the 300–130 K temperature region, the variable range hopping conductivity has been observed, with $\ln(\rho/\rho_0) \propto f(T^{-1/4})$ and $\ln(\rho/\rho_0) \propto f(T^{-1/2})$ in the temperature ranges of 300–250 K and 200–130 K, respectively.

Acknowledgments

This work has been funded by National Research Foundation of Korea and Belarusian Republican Foundation for Fundamental Research (grant F16KOR-002).

References

1. D. Serrate *et al.*, *J. Phys.: Condens. Matter* **19**, 023201 (2007).
2. S. Mitani *et al.*, *J. Magn. Magn. Mater.* **165**, 141 (1997).
3. J. C. Slonczevski, *Phys. Rev. B* **39**, 6995 (1989).

ELECTRICAL CONDUCTION PROPERTIES OF NANOSTRUCTURED INDIUM OXIDE FILMS

A. E. Pochtenny, V. G. Luhin, S. S. Shikanov, V. S. Volobuev
Department of Physics, Belarusian State Technological University
Sverdlova 13a, 220006 Minsk, Belarus
pae@tut.by

Electrical conduction properties of the nanostructured indium oxide thin films were investigated. Structure and chemical composition of the films were studied by electron diffraction, scanning electron microscopy and X-ray photoelectron spectroscopy. The temperature dependence of conductivity was measured at a constant oxygen concentration by cyclic thermal desorption. The conduction mechanism based on the experimental results is proposed.

1. Introduction

At the present time, thin indium oxide (In_2O_3) films are widely applied in manufacturing of gas sensors, solar cells, electrodes, *etc.* [1,2]. A use of indium oxide as a gas-sensitive material is restrained to high «quasimetallic» conductivity and low selectivity. A modification of chemical composition and structure of In_2O_3 films improves their sensitivity and selectivity and makes actual recent researche in this field.

2. Experimental

Thin In_2O_3 films have been fabricated by thermal oxidation of thin indium films formed by *DC* magnetron sputtering by argon. The pressure in the process of sputtering was 0.1–1 Pa. The indium metal target of "analytical grade" was used as a cathode. The target sputtering was carried out at the accelerating voltage of 0.5 kV. The discharge current was 0.15 A. The indium films were deposited on the single-crystal silicon wafers coated with SiO_2, Al_2O_3 and mica.

After the deposition, indium films were oxidized in a muffle electric furnace in the non-isothermal mode: heating to the temperature of 500–600 °C for 40–60 min and annealed in the isothermal mode at 500 °C and 600 °C for 60 min (temperature and oxidation time were chosen experimentally). Phase structure and surface morphology have been investigated by electron diffraction using TEM H-800 (Hitachi) and SEM S-806 devices.

DC conductivity of In_2O_3 films and the temperature dependence of conductivity were measured in vacuum of 10^{-2} Pa with the aid of a V7E-42 electrometer. The investigation was performed by the cyclic thermal desorption method based on the fact that the conductivity of In_2O_3 films depends on concentration of adsorbed oxygen which can be decreased by heating a sample. Upon heating the sample in vacuum to a certain temperature, the adsorbed oxygen concentration was set to a definite controlled level. Subsequent cooling in vacuum allows to record the temperature dependence of conductivity at a constant oxygen concentration. Thus, using the cycles of heating the sample to a stepwise increased temperature followed by measurement of the temperature variation of conductivity in the course of cooling. A family of the temperature dependences of conductivity corresponding to various concentrations of adsorbed oxygen on the same sample can be obtained.

The specific conductivity σ of films depends on the absolute temperature T as described by the equation $\sigma = \sigma_0 \exp(-E_a/kT)$, where σ_0 is the pre-exponential factor, E_a is the conductivity activation energy, and k is the Boltzmann constant. Using this expression and a series of experimental temperature dependences, it is possible to determine the conductivity σ and the parameters σ_0 and E_a corresponding to various concentrations of adsorbed oxygen.

3. Results and discussion

Analysis of electron diffraction patterns for the films fabricated shows that the only identifiable crystal phase formed in the process of indium oxidation is the polycrystalline cubic phase of In_2O_3. The films are characterized by grain structure with the particle size, the bulk of which ~80 % in the range of 20–55 nm, with the maximum in the range of 25–30 nm (Fig. 1).

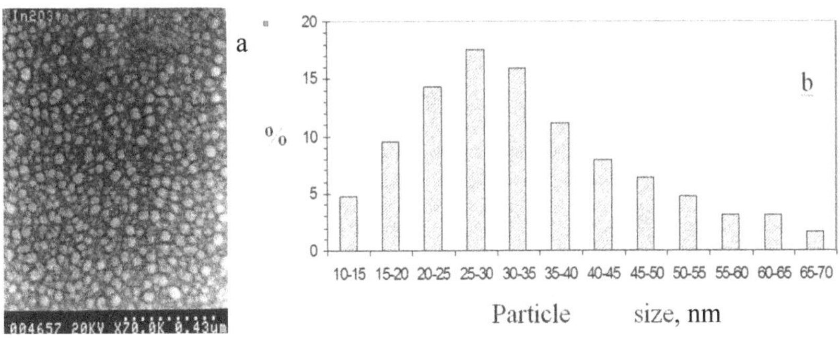

Figure 1. SEM image (a) and particle size distribution (b) of indium oxide films fabricated on mica.

X-ray photoelectron spectrum of indium is characterized by the presence of two spectral lines with binding energies of 444.4 and 452 eV due to the $3d$ level multiplet splitting. The chemical shift of the In $3d5/2$ line in the oxide film is

0.9 eV, which can be attributed to the In^{3+} state. By the reference data, the change in the binding energy due to oxidation of In into In_2O_3 is +0.8–1.2 eV.

The oxygen photoelectron line is shown in Fig. 2. It has a pronounced asymmetry indicating the presence of the overlapping peaks different in energy, shape and intensity, with binding energies of 529.95 and 531.9 eV.

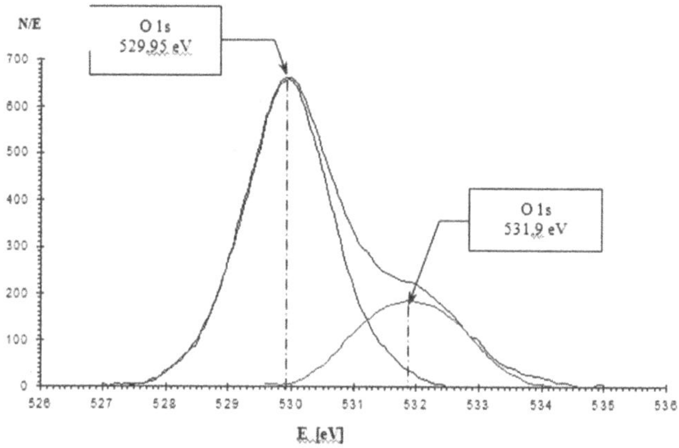

Figure 2. XPS data of oxygen on the surface of In_2O_3 films.

The photoelectron line with $E_b = 529.95$ eV has the highest intensity and corresponds to the oxygen lattice state (reference data for the average O_2-binding energy in In_2O_3 is 530 eV and varies from 529.1 to 530.9 eV). The additional component of O 1s line in the range of 531–534.5 eV can be associated with a number of oxygen forms and its compounds adsorbed on the surface. The binding energy of 531.9 eV may correspond to the surface adsorbed oxygen in various forms and to the oxygen of hydroxyl groups.

The data on temperature dependence of conductivity (Fig. 3) show that the conductivity increases with the desorption of oxygen. The activation energy of conductivity is not changed and its value is set to 10 meV. Therefore, the increase of conductivity, which is associated with a decrease in the concentration of the adsorbed oxygen, due to the fact that oxygen is adsorbed by carrier scattering centers.

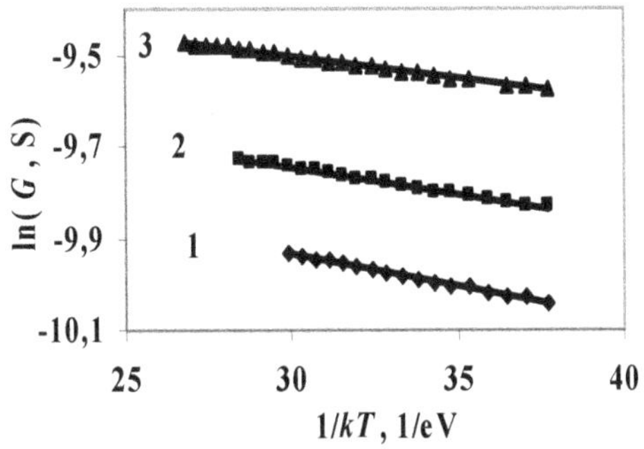

Figure 3. Temperature dependence of the conductivity of In_2O_3 films in vacuum on cooling from 120° (1), 140° (2) and 160 °C.

Desorption of oxygen from the surface of In_2O_3 films results in an increased charge carrier mobility. This effect is to be of practical application in adsorption-resistive oxygen sensors.

4. Conclusion

The structure, composition and morphology of the surface of indium oxide films and their conductivity have been investigated by electron diffraction, scanning electron microscopy and X-ray photoelectron spectroscopy. The temperature dependence of conductivity of these films was measured at constant oxygen concentration using cyclic thermal desorption. The obtained results can be used for design of microelectronic sensors.

Acknowledgments

This study was supported by the «Convergence» Program (GB16-196).

References

1. N. M. Shaalan, M. Rashad, M. A. Abdel-Rahim, *Mat. Sci. Semicond. Proc.* **56**, 260 (2016).
2. A. Sudha, S. L. Sharma, T. K. Maity, *Mat. Lett.* **157**, 19 (2015).

BAND GAP CALCULATION OF BULK AND MONOLAYER TRANSITION METAL DICHALCOGENIDES WITH NEW GVJ-2E APPROACH WITHIN DFT FRAMEWORK

J. Gusakova[1], B. K. Tay[1, 2], X. Wang[1, 2], L. L. Shiau[1]
[1]*Novitas Center, Nanyang Technological University, Singapore*
50 Nanyang Avenue, 639798 Singapore
julia001@e.ntu.edu.sg
[2]*CINTRA UMI CNRS/NTU/THALES, Singapore, 50 Nanyang Drive, 637553 Singapore*

V. E. Gusakov
Scientific-Practical Materials Research Center NASB
P. Browka 19, 220072 Minsk, Belarus
gusakov@ifttp.bas-net.by

V. E. Borisenko
Belarusian State University of Informatics and Radioelectronics
P. Browka 6, 220013 Minsk, Belarus

We have calculated fundamental band gaps for bulk and monolayer transition metal dichalcogenides (TMDs: MoS_2, $MoSe_2$, WS_2, and WSe_2) with recently proposed by us GVJ-2e method. The calculated band gaps are in a good agreement with experimental ones for both bulk and monolayer TMDs, having mean absolute error (MAE) of about 0.03 eV. The errors of GVJ-2e method are significantly smaller than those of other widely used ones such as GW (MAE 0.35 eV) and hybrid functional HSE (MAE 0.17 eV).

1. Introduction

Density functional theory (DFT) is a tenet for electronic-structure calculations across wide variety of fields, from organic chemistry to condensed matter physics. Although for many properties DFT produces results which correlate with good accuracy with experimental ones, for electronic properties it is not always the case due to the existence of well-known band gap problem [1]. As widely used LDA and GGA approximations of the exchange-correlation energy functional lead to the electronic structure and band gap values not accurate enough or underestimated severely.

Various approaches have been developed to overcome this issue and to improve prediction of electronic properties of materials [2]. This includes hybrid functionals such as HSE [3]. Approximate effective potential was introduced by Becke and Johnson which resembles optimized effective potential (OEP) for the energy exchange [4]. Modified BJ potential (MBJ) [5] was later introduced. GW approach represents Green's function based method [6]. Nevertheless, the methods mentioned above still give large errors for band gaps of some materials (~20 %). The problem is complicated by the lack of correlation between calculated and experimental band gaps. Some methods [5] use adjustable parameters, which is not appropriate for *ab initio* methods. Moreover, it should be mentioned that hybrid functionals and GW methods are computationally expensive. This fact complicates significantly theoretical search for new (1D-3D) materials with desired electronic (optical, thermoelectric) properties.

Recently, we offered a new general approach (GVJ-2e) for evaluation of a fundamental band gap of semiconductors [7]. Last decades monolayers of transition metal dichalcogenides (TMDs) [8], as representatives of two dimensional (2D) materials, have attracted attention due to their rich physics and promising applications in nanoelectronics and photonics. In the present paper, we demonstrate the application of the proposed method to the TMDs in their bulk and monolayer 2D forms.

2. Model

The main expression for the band gap calculation has been derived using total energy functionals of charge density only in consideration of neutral and charged states of the system (Z = 0, +2, -2). The fundamental band gap is defined as a difference between the first ionization potential and the first electron affinity in the form $E_g = E_{N+1}^{(0)} + E_{N-1}^{(0)} - 2E_N^{(0)}$. We consider the expression for the band gap through adding or removing two electrons from the system:

$$E_g = 0.5\left(E_{N+2}^{(0)} + E_{N-2}^{(0)} - 2E_N^{(0)}\right). \tag{1}$$

The final expression for the band gap has been derived with representation of the exchange-correlation energy functional as a sum of local (LDA) and nonlocal parts, and has the following form:

$$E_g^{(0)} = E_g^{(LDA)} + \frac{1}{2}\Delta_{XC}^{\infty} - \frac{1}{2}E_{(1,0)}^{(\infty,\mathrm{xc})}\left(1 + \exp\left(-\frac{E_g^{(0)}}{E_0}\right)\right), \tag{2}$$

where $E_g^{(LDA)}$ is the band gap in LDA approximation calculated using (1), E_0 is used for the dimensionless energy in the exponent and in the following calculations is taken equal to 1 eV. The correction terms Δ_{XC}^{∞} and $E_{(1,0)}^{(\infty,\mathrm{xc})}$ are determined by a nonlocal part of the exchange correlation energy. The term $E_{(1,0)}^{(\infty,\mathrm{xc})}$ is the nonlocal part of the exchange-correlation energy of interaction of the HOMO pair of electrons with all electrons in the lower energy bands. It can be represented in the form

$$E_{(1,0)}^{(\infty,\mathrm{xc})} = \left(E_{Z=0}^{(0)} - E_{Z=0}^{(LDA)}\right) - \left(E_{Z=+2}^{(0)} - E_{Z=+2}^{(LDA)}\right). \tag{3}$$

The definition of Δ_{XC}^{∞} is given through the total energies of the system by

$$\Delta_{XC}^{\infty} = \left(E_{Z=+2} + E_{Z=-2} - 2E_{Z=0}\right)^{(0)} - \left(E_{Z=+2} + E_{Z=-2} - 2E_{Z=0}\right)^{(LDA)} \tag{4}$$

Here, we have focused on the application of the proposed method for calculation of band gaps of 2D materials.

3. Results and discussion

We studied monolayers of TMDs with hexagonal symmetry, which included MoS_2, $MoSe_2$, WS_2, WSe_2. First we considered bulk TMDs which are formed by the stacking monolayers together with weak interactions. TMD monolayer typically consists of a metallic plane (Mo or W) sandwiched between two chalcogen planes (S or Se). The calculations of structural and electronic properties of the materials was performed in Quantum ESPRESSO implementation of DFT, using PZ LDA and PBE GGA approximations of the exchange-correlation energy and pseudopotentials. The Brillouin zone was sampled using the Monkhorst-Pack approach, $12\times12\times3$ and $10\times10\times1$ meshes were used for the bulk and the monolayer, respectively.

The following band gaps have been obtained using (3) for bulk TMDs: 1.23 eV (MoS_2), 1.09 eV ($MoSe_2$), 1.32 eV (WS_2) and 1.21 eV (WSe_2). These gaps agree well with experimentally observed ones: 1.23 eV (MoS_2), 1.09 eV ($MoSe_2$), 1.35 eV (WS_2) and 1.20 eV (WSe_2). For the monolayers of TMDs we got: 1.88 eV (MoS_2), 1.57 eV ($MoSe_2$), 2.03 eV (WS_2), 1.67 eV (WSe_2).

There exists a variability of the published experimental band gaps in monolayers. So we also measured optical band gaps for TMD monolayers, which were grown using chemical vapour deposition (CVD). The band gaps retrieved from PL spectra are: 1.87 eV (MoS_2), 1.52 eV ($MoSe_2$), 1.99 eV (WS_2), 1.62 eV (WSe_2). They agree with other published experimentally obtained band gaps as well as those calculated by GVJ-2e method.

4. Conclusion

The application of the proposed GVJ-2e method for the band gap calculations of the TMDs in bulk (3D) and monolayer (2D) forms has shown that the method predicts the band gaps almost with an experimental accuracy. The mean absolute error of GVJ-2e method for TMDs is about 0.03 eV, which overcomes that of other widely used methods such as GW (0.35 eV), G_0W_0 (0.52 eV), HSE (0.17 eV). We have shown the crucial role of the nonlocal part of the exchange-correlation energy in the accurate determination of a fundamental band gap of TMDs, which agrees also with the case of conventional semiconductors. The results obtained also clearly demonstrate applicability of the method for complex materials which have a layered structure with van der Waals interactions between the layers and for 2D layers.

References

1. S. Kümmel, L. Kronik, *Rev. Mod. Phys.* **80**, 3 (2008).
2. J. P. Perdew, K. Schmidt, *AIP Conf. Proc.* **577**, 1 (2001).
3. J. Heyd, G. E. Scuseria, M. Ernzerhof, *J. Chem. Phys.* **118**, 8207 (2003).
4. A. D. Becke, E. R. Johnson, *J. Chem. Phys.* **124**, 221101 (2006).
5. F. Tran, P. Blaha, *Phys. Rev. Lett.* **102**, 226401 (2009).
6. W. G. Aulbur, L. Jonsson, J. W. Wilkins, *Solid State Phys.* **54**, 1 (2000).
7. J. Gusakova, B. K. Tay, V. Gusakov, *Phys. Status Solidi A* **213**, 2834 (2016).
8. A. K. Geim, I. V. Grigorieva, *Nature* **499**, 419 (2013).

MAGNETIC ORDERING IN DOPED TWO-DIMENSIONAL DICHALCOGENIDES

A. V. Krivosheeva, V. L. Shaposhnikov
Belarusian State University of Informatics and Radioelectronics
P. Browka 6, 220013 Minsk, Belarus
anna@nano.bsuir.edu.by

J.-L. Lazzari
Aix Marseille Université, CNRS, Centre Interdisciplinaire de Nanoscience de Marseille (CINaM) UMR 7325, Campus de Luminy, 13288 Marseille, France

Magnetic properties of the individual layers of MoS_2, WS_2, WSe_2, $MoSe_2$ upon substitution of metal atoms by manganese ones are studied. The magnetic moments on the impurity atoms and a possibility of a sustainable magnetic order are evaluated.

1. Introduction

Semiconducting two-dimensional transition metal dichalcogenides (TMDs), including MoS_2, $MoSe_2$, $MoTe_2$, WS_2, and WSe_2, have emerged as promising materials for a range of applications [1,2]. Monolayers of MoS_2 have already found application in low-power field-effect transistors [3], logic circuits [4], and phototransistors [5–9]. Adsorption energy, stable geometry, as well as magnetic and electronic properties of configurations with transition metal atoms adsorbed on the MoS_2 monolayer were investigated in [10] and MoS_2 monolayers with adsorbed TM atoms were found to be excellent candidates for spintronics. In [11] the unexpected strong magnetism of Cu-doped single-layer MoS_2 was discovered. There is a work to be done in order to study the magnetic ordering in TMDs layers doped with elements, possibly giving the magnetic moment to the system. We suppose that manganese might be a promising element and consider the case of substitution of molybdenum and tungsten atoms by manganese ones.

2. Details of calculations

In order to perform the calculation of atomic arrangement of structures doped with manganese atoms, the full structural optimizations within density-functional theory (DFT) with a plane-wave basis set was done first. The PAW-LDA approximation (VASP package [12]) was used for structural optimization and calculation of band gap values. In the calculations $4p(5p)$ semi-core states of Mo(W) were treated as valence. In order not to deal with the van-der-Waals interaction, only one monolayer was considered. The energy cutoff of 380 eV and 8×8×3 Γ-centered grid of k-points were applied. 3×3×1 supercells were considered for representation of TMD structures. The vacuum gap of 21 Å was used to suppress an influence of neighboring layers.

3. Results and discussion

We found that non-doped MoS_2, WS_2, WSe_2, and $MoSe_2$ are non-magnetic materials. The spin-polarized total densities of states (DOS) for TMDs layers considered with substitutional impurity atoms of manganese in place of Mo and W atoms are shown in Fig. 1. Such structures possess high spin polarization which is defined by the states at the Fermi level in spin-up and spin-down channels. In the case of substitution of molybdenum atoms in MoS_2 and $MoSe_2$ by manganese ones the materials become semi-metals with spin polarization of 100 %. The magnetic moments on the impurity atom are shown in Table 1.

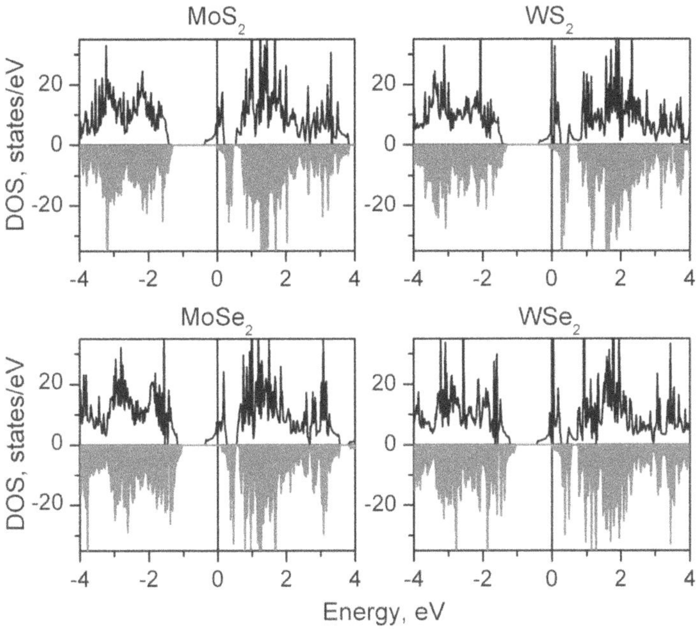

Figure 1. Spin-polarized total DOS for MoS_2, WS_2, WSe_2, $MoSe_2$ layers with substitutional impurity of manganese atoms in place of molybdenum and tungsten atoms of the TDMs 3×3×1 cell. Zero on the scale corresponds to the energy of the Fermi level.

Table 1. Magnetic moments on the impurity Mn atoms in MoS_2, WS_2, WSe_2, and MoSe.

	MoS_2	WS_2	$MoSe_2$	WSe_2
Magnetic moment, μ_B/atom	0.780	0.679	0.940	0.928

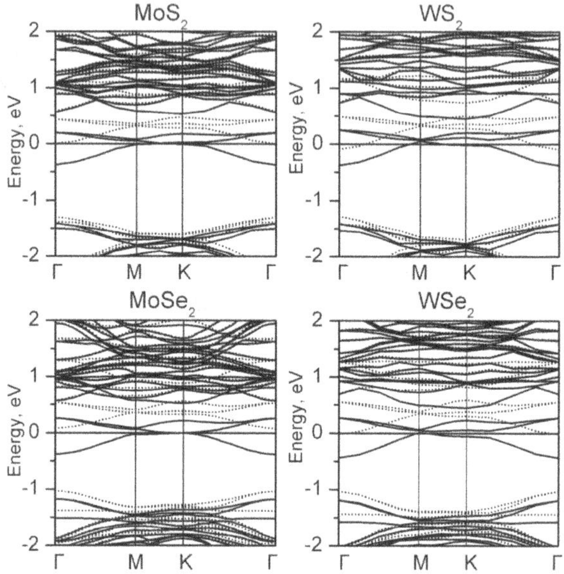

Figure 2. Spin-polarized electronic band structures for MoS_2, WS_2, WSe_2, $MoSe_2$ layers with a manganese substituting molybdenum and tungsten atoms. Solid lines relate to spin-up, dotted lines to spin-down states. Zero on the scale corresponds to the energy of the Fermi level.

Fig. 2 shows the spin-polarized electronic band structures for MoS_2, WS_2, WSe_2, $MoSe_2$ layers, where molybdenum and tungsten atoms were substituted by manganese atoms. One may observe the opening of the gap in the spin-down channel for the case of doped MoS_2 and $MoSe_2$, which show half-metallic behavior, while for WS_2 and WSe_2 the states cross the Fermi level in both channels. The analysis of the partial DOS (Fig. 3) has shown, that *d*-states of Mn hybridized with *d*-states of the host Me atoms and *p*-states of chalcogen atoms make the main contribution to the states at the Fermi level.

4. Conclusion

Two-dimensional transition metal dichalcogenides MoS_2, $MoSe_2$, WS_2, and WSe_2 were studied in order to check their possible application in spintronics. It was found, that pure TMDs are non-magnetic materials. However, they demonstrate an appearance of magnetic ordering when magnetic impurity like Mn substitutes metal atoms. The replacement of one Mo atom with Mn one in 3×3 cell of MoS_2 or $MoSe_2$ leads to a half-metallic behavior with weak magnetic moments of 0.78 and 0.94 μB/atom, respectively.

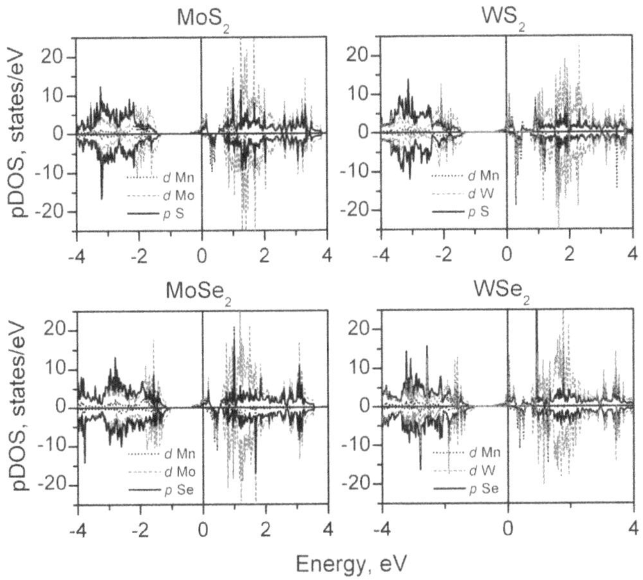

Figure 3. Spin-polarized partial DOS for MoS_2, WS_2, WSe_2, $MoSe_2$ layers with substitutional impurity of manganese atoms in place of molybdenum and tungsten atoms. Zero on the scale corresponds to the energy of the Fermi level.

Acknowledgments

This work was supported by joint BRFFR-CNRS project (grant F15F-003). The authors are grateful to Professor V. E. Borisenko for the critical comments and helpful discussion of the results presented.

References

1. A. V. Krivosheeva *et al.*, *J. Semicond.* **36**, 122002 (2015).
2. A. V. Krivosheeva, *Doklady BSUIR* **3**, 12 (2016) – in Russian.
3. B. Radisavljevic *et al.*, *Nature Nanotechnology* **6**, 147 (2011).
4. B. Radisavljevic, M. B. Whitwick, A. Kis, *ACS Nano* **5**, 9934 (2011).
5. Z. Yin *et al.*, *ACS Nano* **6**, 74 (2012).
6. Y. Zhang, J. Ye, Y. Matsuhashi, Y. Iwasa, *Nano Lett.* **12**, 1136 (2012).
7. H. Wang, L.Yu, Y.-H. Lee *et al.*, *Nano Lett.* **12**, 4674 (2012).
8. I. Popov, G. Seifert, D. Tománek, *Phys. Rev. Lett.* **108**, 156802 (2012).
9. H. Qiu, L. Pan, Z. Yao *et al.*, *Appl. Phys. Lett.* **100**, 123104 (2012).
10. Y. Wang *et al., Physica E* **63**, 276 (2014).
11. W. S. Yunab, J. D. Lee, *Phys. Chem. Chem. Phys.* **16**, 8990 (2014).
12. G. Kresse, J. Furthmüller, *J. Comput. Mater. Sci.* **6**, 15 (1996).

RADIATION DAMAGE IN Si NANOSTRUCTURES: MOLECULAR DYNAMICS SIMULATION

V. I. Belko, N. N. Dorozhkin
Belarusian State University, Nezavisimosti Ave. 4, 220030 Minsk, Belarus
belko@bsu.by, Dorozhkin@bsu.by

V. Gusakov
Scientific-Practical Materials Research Centre NASB
P. Browka 19, 220072 Minsk, Belarus
gusakov@ifttp.bas-net.by

Molecular dynamics simulation of atomic displacement cascades was performed for silicon nanowires. It has been shown that with a decrease of the size of nanowires, the number of primary radiation defects also decreases. The number of vacancies is described by the non-monotonous function of the primary knock-on atom energy.

1. Introduction

Nanoscale elements of integrated circuits provide a significant reduction of active regions of semiconductor elements. So, for a FinFET transistor, the channel width becomes ≤ 7 nm. Increasing the radiation hardness of integrated circuits requires detailed studies of radiation defect formation processes in nanoscale semiconductor devices. Radiation defect formation in nanostructures and bulk materials can be considerably different [1,2]. This paper presents a study of primary radiation defects formation in compositional silicon nanowires.

2. Results and discussion

We performed the molecular dynamics simulation of atomic displacement cascades in silicon nanowires. We used the package XMD [3]. A rectangular fragment of silicon crystal elongated in the direction of z-axis was modeled. Periodic boundary conditions in the z direction and free surfaces in x and y directions were applied. We consider two cases: the silicon nanowire and silicon nanowire composite. In the latter one face (parallel to yOz plane) is covered by four monolayers of gold (Fig. 1). In both cases the crystal was previously put into equilibrium at a predetermined temperature (300 K) for a sufficiently long time.

In the first numerical experiment we examined the model with cross-sectional dimensions of 3.8×3.26 nm and 5.41×4.89 nm of the Fd3m lattice symmetry. It corresponds to the size of the computational domain of

7×6×20 and 10×9×20 unit cells, while the unit cell parameter is 0.5431 nm. An initial speed was imparted to one of the atom near the center of the wire (imitation of the primary knock-on atom (PKA)). The initial kinetic energy of PKA was varied from 30 eV to 1900 eV and from 30 eV to 3300 eV in the larger and smaller wires, respectively. For each initial velocity, PKAs were created at 5 various random directions close to the following vectors in the standard coordinate system: (0, –1, 0), (0, 0, 1), (1, 1, 0), (0, 1, 1), (2, 0, 1).

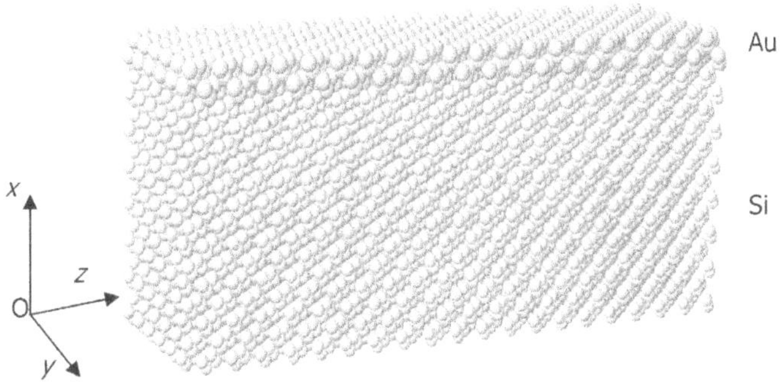

Figure 1. The model for atomic displacement cascades in the composite silicon nanowire.

After the development of a displacement cascade, the system was maintained at 300 K and relax to equilibrium during 5 ns. Then, analysis of intrinsic defects formed in the inner region was performed (departing from the free surfaces for 0.5 unit cell length).

In the second experiment we considered the model with cross-sections of 4.34×3.26 nm and 8.15×6.52 nm (8×6×20 and 15×12×20 in unit cells) of the Fd3m lattice symmetry. One surface parallel to the *yOz* plane was covered with four Au monolayers. Due to limitations in the XMD molecular dynamics package, the interactions between metal and silicon atoms and between silicon atoms in the bulk were described by the same empirical interatomic potentials. The initial kinetic energy for the primary knock-on atoms was varied from 30 eV to 1900 eV in both models. For each initial velocity, PKAs were created in three different directions: along the *z* axis, along the *x* axis of the metal, and along the *x* axis to the metal. After the formation of atomic displacement cascade, the system was allowed to relax torwards equilibrium for 5 ns. Then, we analyzed intrinsic defects formed in the inner region, departing from the free surfaces over 0.5 unit cell length.

Statistical analysis of the results obtained is illustrated in Fig. 2. It shows: (1) Reduction of the nanowire cross-section leads to the decrease of the number of primary radiation defects. (2) The number of radiation-formed vacancies depends on the PKA energy non-monotonously. For nanowire models in the size range considered, the maximum number of vacancies is observed at the PKA energy of 1400 eV, and decrease of the size shifts the maximum of distribution to higher energies. (3) For the composite nanowire (Si + Au), distribution function maximum of the number of vacancies is shifted to higher energies. (4) The distribution function of the number of vacancies depends on the nanowire relative orientation and the velocity vector of the primary knock-on atom.

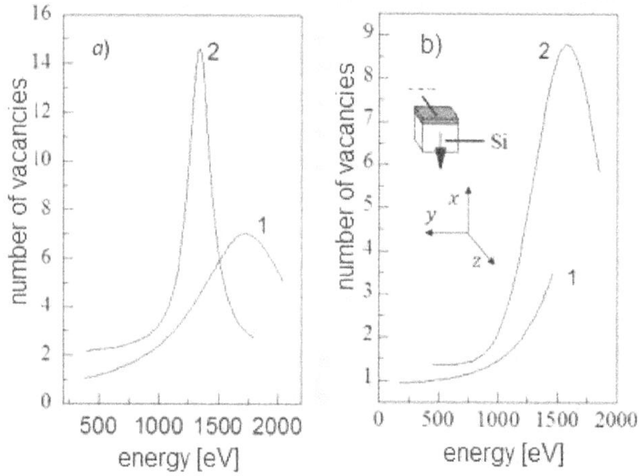

Figure 2. Dependence of the number of vacancies in the nanowire on PKA energy. (a) Si nanowire with the dimensions of 1 - 7×6×20; 2 - 10×9×20; (b) composite Au + Si nanowire with the dimensions of 1 - 8×6×20; 2 - 15×12×20. Arrow indicates the PKA direction.

3. Conclusion

Atomic displacement cascades in silicon nanowires and composite silicon nanowires (Au + Si) were simulated with the molecular dynamics approach (XMD package). The size of nanowires was shown to have a significant effect upon the number of primary radiation defects in both types of nanowires, while the dependence of the number of vacancies on energy of PKA is extremal.

This work was partly supported by the Ministry of Education of the Republic of Belarus, SPSR "Computer science, space and security".

References

1. A. A. Leino, F. Djurabekova, K. Nordlund, *Eur. Phys. J. B* **87**, 242 (2014).
2. K. Nordlund, F. Djurabekova, *J. Comp. Electr.* **13**, 122 (2014).
3. XMD, *Molecular Dynamics for Metals and Ceramics* (Electronic resource, – March, 2010), http://xmd.sourceforge.net/about.html.

SOLUTE-DEPENDENT DFT SIMULATIONS OF ELECTRON AND SPIN PROPERTIES OF $C_{60}(OH)_{24}(Ni(CpCOOH)_2)_2$ CONJUGATES OF DIFFERENT MULTIPLICITY

A. L. Pushkarchuk[1,3], T. V. Bezyazychnaya[1], V. I. Potkin[1], E. A. Dikusar[1],
A. G. Soldatov[1,4], S. Ya. Kilin[2], A. P. Nizovtsev[2], S. A. Kuten[3],
E. M. Shpilevsky[5], V. A. Pushkarchuk[6]

[1]*Institute of Physical Organic Chemistry NASB*
Surganova 13, 220072 Minsk, Belarus
[2]*B. I. Stepanov Institute of Physics NASB, Nezavisimosti Ave.68, 220072 Minsk, Belarus*
[3]*Institute for Nuclear Problems of Belarusian State University*
Bobruiskaya 11, 220030 Minsk, Belarus
[4]*The Scientific and Practical Materials Research Center NASB*
P. Browka 19, 220072 Minsk, Belarus
[5]*A.V. Luikov Heat and Mass Transfer Institute NASB*
P. Browka 15, 220072 Minsk, Belarus
[6]*Belarusian State University Informatics and Radioelectronics*
P. Brovka 6, 220013 Minsk, Belarus

The results of DFT modeling of atomic and spin structure of the water-soluble fullerene derivatives, as well as possible mechanisms for the formation of the ferromagnetic properties of the complexes based on these structures are presented.

1. Introduction

Targeted delivery of water-soluble fullerene derivative (WSFD) into the specified body regions can be implemented in different ways. In particular, it can be done by means of an external magnetic field, if these complexes have ferromagnetic properties. In this context, this paper presents the results of the study of methods of computer quantum chemistry and structural characteristics of the WSFD-based spin-conjugates, as well as possible mechanisms for the formation of the ferromagnetic properties of these complexes. As far as the study of biological activity and possible delivery of WSFD in the experiments (*in vitro* and *in vivo*) take place in an aqueous solution, the modeling of such structures must take into account the presence of water as the solvent. The WSFD object under consideration was fullerenol–nickelocene based conjugate $C_{60}(OH)_{24}(Ni(CpCOOH)_2)_2$ named below FNC conjugate.

2. Results and discussion

The COSMO method [1] was used in our simulations to take into account the presence of the solvent. This method is implemented in the package ORCA programs [2] used for quantum-chemical simulations. The results of calculations of the FNC conjugate structural parameters by DFT, without and with taking into account the solvent are shown in Fig. 1.

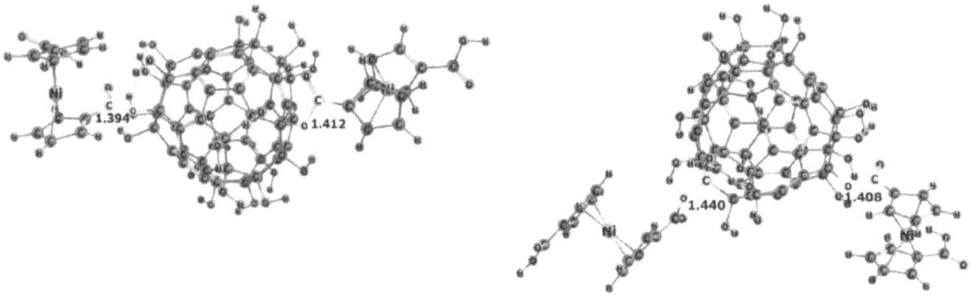

a) b)

Figure 1. The FNC complex after DFT calculations a) in an anhydrous medium, b) in an aqueous solution. Distances are given in Å.

As one can see in Fig. 1, the solvation effect influences the distance "C-O" at the location of the fullerenol and nickelocene junction: the distance increases in an aqueous solution in comparison with anhydrous one. In this regard, we consider the question of how to influence such structural changes on the variations of the spin-conjugate parameters

The main characteristics of the spin structure of the complex of interest were the spin density localization and the exchange integral J.

The geometric structure of the complex was calculated using the DFT/UKS/B3LYP/3-21G/TZVP level of the theory for maximum spin multiplicity following the procedure described in [3] for transition metal complexes. Then, having the FNC complex geometry obtained the calculations of the exchange constant J and the spin density in the complex for its several high-spin states were carried out using DFT/UKS/BP86/RI/SV level of the theory. The constant J was calculated using the broken-symmetry approach (BS) [4]:

$$J = - (E [HS] - E [BS]) / S2max)$$

where E[HS] and E[BS] are total energies of high spin state and broken symmetry state, respectively, and S2max is the maximal value of the spin squared of the system.

The calculation results for the non-solvated and solvated complexes are presented in Tables 1 and 2, respectively. Energies E[HS] and E[BS] in HS and BS states are given in the Tables in a.u.

Table 1. The calculated exchange constants J (cm^{-1}) and the energies of non-solvated conjugate for different total spin S_{max} of the system in its different spin states.

Total spin S_{max}	E[HS]	E[BS]	Exchange integral J
5	-8493.212172	-8493.212131	2.26
11	-8493.212172	-8493.212131	-2009.78

Table 2. The calculated values of the exchange constants J (cm^{-1}) and the energies of solvated conjugate for different total spin S_{max} of the system in its different spin states.

Total spin S_{max}	E[HS]	Total spin S_{max}	E[HS]
5	-8493.308880	-8493.308816	3.52
11	-8492.865493	-8493.333718	-1605.53

As a result, using the DFT/UKS/BP86/RI/SV method we found that covalent bound conjugates $C_{60}(OH)_{24}(Ni(CpCOOH)_2)_2$ in the spin state $S_{max} = 5$ have a positive value of exchange interaction "metal-metal". The sign of the exchange constant J is not dependent on the presence of the solvent (water). As a result, the complexes of interest have ferromagnetic properties, and therefore a control of biologically active nanostructures generated on the FNC complex basis can be possible using an external magnetic field.

At the same time, covalent bound FNC conjugates in spin state $S_{max} = 11$ have a negative value of the exchange constant independently on the presence of the water. FNC conjugates in this state are antiferromagnetics. The presence of a solvent affects the spin properties of the complex in its $S_{max} = 11$ and $S_{max} = 5$ states in specific ways. In general, the presence of a solvent leads to the lengthening of C-O bonds in the complex. The latter has no effect on the spin properties of the complex, but the exchange constant in the presence of a solvent is increased in $S_{max} = 5$ state and reduced in $S_{max} = 11$ state.

The analysis of the charge density distribution has showed that there is a little transfer of the electron density onto the fullerenol molecule. This result corresponds to the data obtained in the experiments for $C_{60}Fe(CpCOOH)_2$ complex [5].

Analysis of the spin structure has showed that the spin density is localized mainly in the *d*-orbitals of Ni atoms belonging to the nickelocene molecules $Ni(CpCOOH)_2$.

3. Conclusion

Quantum-chemical simulations based on the DFT/UKS/BP86/RI/SV method indicate that covalent bound FNC conjugates in anhydrous and aqueous medium in their high-spin state $S_{max} = 5$ have a positive value of exchange interaction "metal-metal". Complexes of interest in this state can have ferromagnetic properties, and therefore a control of biologically active nanostructures generated on such complex basis can be possible using an external magnetic field.

The exchange integral of the conjugate in the spin state $S_{max} = 11$ is negative independently on the presence of the solvent and this higher-spin state is antiferromagnetic.

Acknowledgments

Support by the "Convergence-2020" national scientific programs, Belarus, is gratefully acknowledged.

References

1. A. Klamt, G. Schürmann, *J. Chem. Soc.*, *Perkin Trans.* **2**, 799 (1993). DOI: 10.1039/P29930000799
2. F. Neese, *Computational Molecular Science* **2**, 73 (2012).
3. D. A. Pantzis, V. Krewald, M. Orio, F. Neese, *Dalton Trans.* **39**, 4959 (2010).
4. J. M. Mouesca, *Methods Mol. Biol.* **1122,** 269 (2014).
5. N. Guskos, A. G. Soldatov, G. Zolnierkiewicz, V. Likodimos, S. J. Glenis, *Noncryst Solids* **354**, 4334 (2008).

INVERSION OF THE SnPc MOLECULE ON GRAPHENE

N. A. Poklonski, S. A. Vyrko, S. V. Ratkevich, A. I. Siahlo
Belarusian State University, 220030 Minsk, Belarus
poklonski@bsu.by

Inversion of the funnel-shaped tin phthalocyanine molecule near the graphene plane by moving the tin atom to the graphene is studied by quantum mechanical methods.

1. Introduction

Recently electromechanical generators of electrical oscillations, based on vibrations of funnel-shaped molecule of tin phthalocyanine ($C_{32}H_{16}N_8Sn$ or SnPc) placed between the graphene planes [1] or on the graphene plane under the probe tip of an atomic force microscope [2] (see Fig. 1), have been proposed. As in the case of the lead phthalocyanine [3,4], the molecule funnel-like shape causes an existence of a nonzero dipole moment perpendicular to the macrocycle plane [5–8]. A change in the dipole moment at the inversion (turning inside out) of the SnPc molecule may lead to the emission of a photon that is important to form single-photon sources for quantum cryptography.

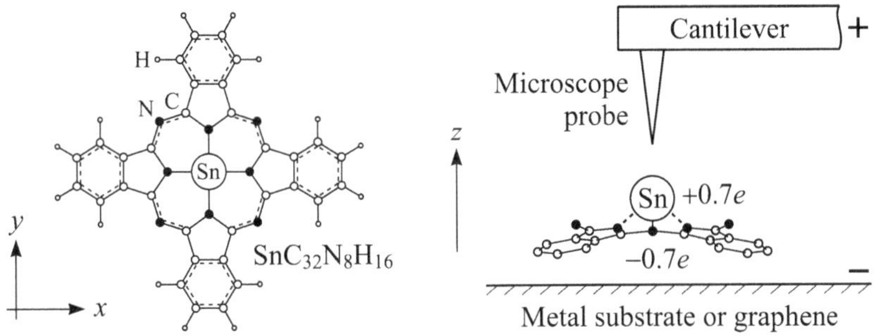

Figure 1. The scheme of the electromechanical generator of electrical oscillations with "movable" field-emission cathode of the funnel-shaped SnPc macromolecule located on the "metal" surface under the probe tip of an atomic force microscope.

The purpose of this paper is to study the inversion of SnPc when moving the tin atom to the graphene plane in arrangement shown in Fig. 1.

2. Methodology

The structural optimization of the SnPc macromolecule near the fragment of the graphene plane was carried out using the software package MOPAC 2012 [9] with the semi-empirical molecular orbital PM3 method [10]. The changes in the geometry of the funnel-shaped molecule caused by its conjugation with the graphene surface were found. The graphene surface is modeled by the graphene fragment with $\boldsymbol{D}_{6h}$ symmetry made of 150 carbon atoms passivated with 30 hydrogen atoms along the perimeter. By optimizing the SnPc molecule its $\boldsymbol{C}_{4v}$ symmetry was kept invariable while the graphene fragment remained "frozen".

3. Results and discussion

The SnPc molecule was placed near the graphene plane in a way that its macrocycle was closer to graphene than the tin atom as shown in Fig. 1. Three points in the graphene plane (the hexagon center, the bond center, and the graphene lattice site) were selected as reference points. The tin atom of the SnPc molecule was placed over one of these reference points. It is found that by turning the SnPc molecule relative to the graphene plane, the energy changes slightly (within a few millielectronvolts) and the minimum energy is achieved when placing the tin atom above the hexagon center. The obtained equilibrium geometry of the SnPc molecule is shown in Fig. 2 (the distance from the tin atom to the graphene plane is $R_{\text{Sn-Gr}} = 0.6818$ nm, the distance from the tin atom to the macrocycle is 0.0986 nm).

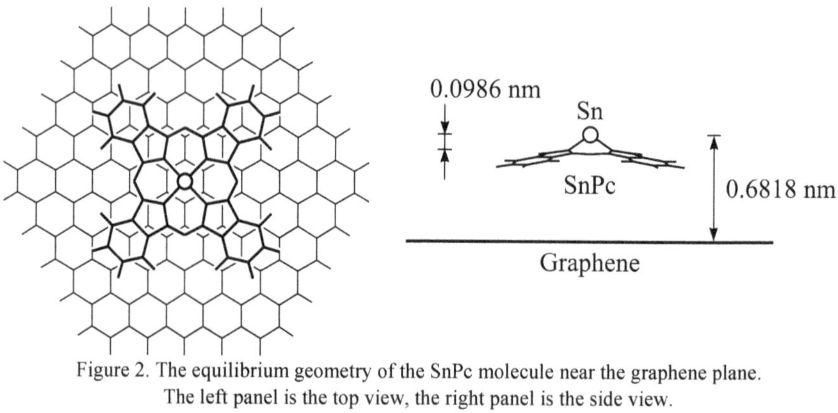

Figure 2. The equilibrium geometry of the SnPc molecule near the graphene plane. The left panel is the top view, the right panel is the side view.

The heat of formation H was calculated as a function of the distance $R_{\text{Sn-Gr}}$ from the tin atom in the molecule SnPc to the graphene for the equilibrium configuration (see Fig. 3). At the distance $R_{\text{Sn-Gr}} = 0.36$–0.35 nm there is an abrupt change in H caused by the inversion of the SnPc molecule. The inversion of the SnPc molecule occurs if the energy barrier of about 2.5 eV is overcome. For comparison, the energy barrier of 4.8 eV is found to invert the lead phthalocyanine molecule (PbPc) with the C_{4v} symmetry [4], and for tunneling of a nitrogen atom through the plane of three hydrogen atoms in ammonia (NH_3), the energy barrier is 0.2 eV [11].

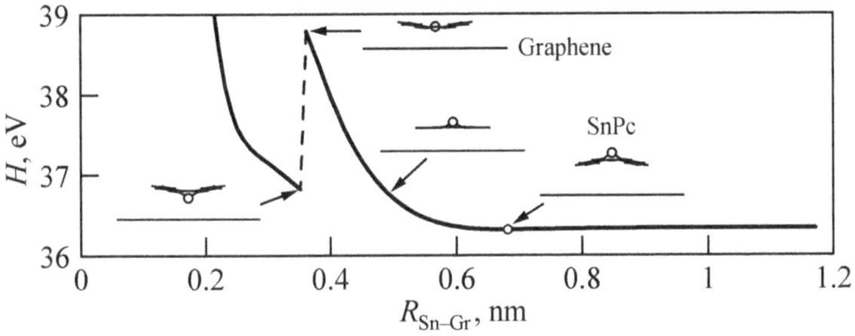

Figure 3. The heat of formation H as a function of the R_{Sn-Gr} distance from the tin atom in the SnPc molecule to the graphene for the configuration shown in Fig. 2. The side views of the molecule over the graphene plane are shown in points indicated by arrows.

Fig. 4 shows the sequence of the SnPc molecule configurations (the inversion process) when moving the tin atom to the graphene plane.

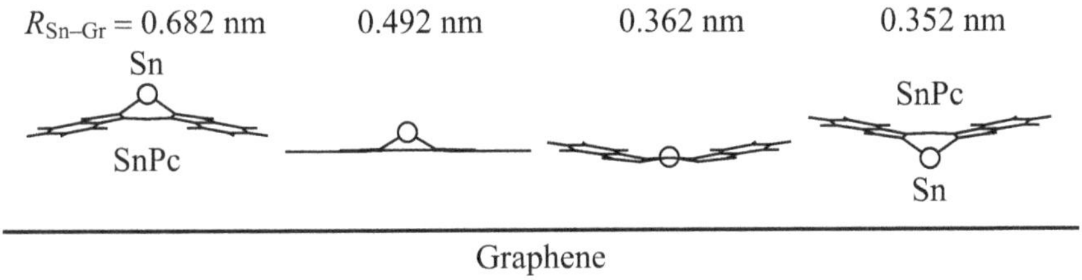

Figure 4. Inversion of SnPc molecule by moving (due to the external force) the tin atom to the graphene plane.

4. Conclusion

The equilibrium configuration of the funnel-shaped SnPc molecule near the fragment of the graphene plane is calculated by the PM3 method. We have studied the potential energy surface when moving the tin atom to the graphene plane and rotating the molecule around the vertical axis. For the configuration when the macrocycle SnPc molecule is located closer to the graphene than the tin atom, the equilibrium distance between the tin atom and graphene equals to 0.6818 nm. By moving the tin atom to the graphene plane, the abrupt inversion of the molecule occurs at a distance of 0.36–0.35 nm from the tin atom to the graphene plane. In order to invert the SnPc molecule it is necessary to overcome the energy barrier of 2.5 eV.

Acknowledgments

This work was partially supported by Belarusian Republican Foundation for Fundamental Research (grant F16R-107) and by the Belarusian Scientific Program "Convergence".

References

1. N. A. Poklonski, E. F. Kislyakov, S. A. Vyrko, S. V. Ratkevich, in: *Physics, Chemistry and Application of Nanostructures*, eds. V. E. Borisenko, S. V. Gaponenko, V. S. Gurin, C. H. Kam (Singapore, World Scientific, 2015), pp. 613–616.
2. N. A. Poklonski, in: *Proc. Int. Winter School on Semiconductor Physics 2015* (Ioffe Institute RAS, SPb., 2015), pp. 53–58 (in Russian).
3. N. A. Poklonski, E. F. Kislyakov, D. I. Sagaidak, A. I. Siaglo, G. G. Fedoruk, *Tech. Phys. Lett.* **27**, 180 (2001).
4. A. S. Nizovtsev, S. G. Kozlova, *J. Phys. Chem. A* **117**, 481 (2013).
5. K. Walzer, M. Hietschold, *Surf. Sci.* **471**, 1 (2001).
6. J. D. Baran, J. A. Larsson, *J. Phys. Chem. C* **116**, 9487 (2012).
7. N. A. Poklonski, S. A. Vyrko, S. V. Ratkevich, E. F. Kislyakov, in: *Proc. Int. Conf. NANO-2014, Minsk, 7–10 March 2014* (Belaruskaya navuka, Minsk, 2014), p. 309.
8. N. A. Poklonski, S. V. Ratkevich, S. A. Vyrko, E. F. Kislyakov, in: *Proc. Int. Conf. Materials and Structures of Modern Electronics, 8–9 Oct. 2014* (BSU, Minsk, 2014), pp. 225–228.
9. J. J. P. Stewart, *MOPAC2012* (Stewart Computational Chemistry, Colorado Springs, CO, USA, 2012), http://openmopac.net.
10. J. J. P. Stewart, *J. Comp. Chem.* **10**, 209 (1989).
11. C. Léonard, S. Carter, N. C. Handy, *Chem. Phys. Lett.* **370**, 360 (2003).

ROBUST ELECTRON-NUCLEAR SPIN SYSTEMS NV-^{13}C IN DIAMOND FOR QUANTUM TECHNOLOGIES

A. P. Nizovtsev[1], S. Ya. Kilin[1], A. L. Pushkarchuk[2,4],
V. A. Pushkarchuk[3], S. A. Kuten[4], F. Jelezko[5]
[1]Institute of Physics NASB, Nezavisimosti Ave. 68, 220072 Minsk, Belarus
apniz@ifanbel.bas-net.by
[2]Institute of Physical Organic Chemistry NASB, Surganova 13, 220072 Minsk, Belarus
[3]Belarusian State University of Informatics and Radioelectronics
P. Browka 6, 220013 Minsk, Belarus
[4]Institute for Nuclear Problems, Belarusian State University
Bobruiskaya 11, 220030 Minsk, Belarus
[5]Institute for Quantum Optics, Ulm University, Albert-Einstein Allee 11, 89069 Germany

Single NV centers in diamond coupled by hyperfine interaction (*hfi*) to neighbor/distant ^{13}C nuclear spins are now widely used in the emerging quantum technologies. Many of these applications benefit from dealing with the NV-^{13}C complexes characterized by the absence of *hfi*-induced nuclear spin flip-flops. Here we report on systematic search of such robust NV-^{13}C systems using density functional theory (DFT) to simulate *hfi* in the H-terminated cluster $C_{510}[NV]^{-}H_{252}$ hosting the NV center. Along with previously studied robust "NV- axial ^{13}C" systems wherein the ^{13}C nuclear spin is located on the NV axis, we found new positions for the ^{13}C nuclear spin exhibiting negligible *hfi*-induced flipping rates due to symmetric local spin density distribution resulting in zero non-diagonal elements of the respective *hfi* matrices. Spatially, these positions are located in the plane passing through the vacancy of the NV center and being perpendicular to the NV axis.

1. Introduction

Hybrid spin systems consisting of the electronic spin (e-spin) of single nitrogen-vacancy (NV) centers in diamond and neighbor nuclear spins (n-spins) of isotopic ^{13}C atoms are now widely used to implement numerous room-temperature applications in quantum information processing, quantum sensing and metrology (see *e.g.* recent review [1]). In the systems, the ^{13}C n-spins with their excellent coherence times serve as quantum memories accessed via the more easily controllable e-spin of the NV center that possess an inherent properties of optical alignment and spin-projection dependent fluorescence, allowing its initialization and readout even at room temperature. Currently, the techniques for creating of a given spin state of an e-n NV-^{13}C spin complexes as well as coherent manipulation of their states to implement one- and two-qubit operations are well established [1]. In particular, usage of dynamical decoupling methods to suppress unwanted background spin noise allows to observe experimentally the NV-^{13}C spin systems where the ^{13}C atom is disposed rather far from the NV center [2–4] and to probe stochastic dynamics of ^{13}C nuclear spins resulting from their *hfi* with the NV center [3].

At zero magnetic field the *hfi* induces the ^{13}C n-spin flip-flops with the rate proportional [2–4] to $\gamma_0 = A_{nd}^2 / \left(A_{nd}^2 + A_{ZZ}^2 \right)$, where $A_{nd} = \sqrt{A_{ZX}^2 + A_{ZY}^2}$ is off-diagonal part of the *hfi* matrix A_{KL} for some nuclear ^{13}C spin written in the principle axis system (PAS) of the NV center wherein the Z axis coincides with the NV center symmetry axis. Evidently, such stochastic dynamics that violates coherence of the ^{13}C n-spin in the NV-^{13}C system is absent in those specific cases when the part A_{nd} is zero or negligible. In particular, this is the case for the ^{13}C n-spin, located at the NV axis, because for such "axial" positions the quantization axes for e- and n-spins coincide and respective *hfi* matrices are intrinsically diagonal. Spin and spatial characteristics of such "NV-axial ^{13}C" systems were found recently [5] by DFT simulation of *hfi* in the H-terminated cluster $C_{510}[NV]^{-}H_{252}$ hosting the NV center. Moreover, the analysis of the calculated data revealed the presence of additional non-axial positions for the n-spin ^{13}C also exhibiting the absence of non-diagonal elements in respective *hfi* matrices A_{KL} and hence being robust.

Here we present calculated spatial and *hfi* characteristics for these non-axial robust NV-^{13}C spin systems and show that apparently some of them have already been observed experimentally [3,6-8].

2. Calculation details

Geometric structure of the $C_{510}[NV]^{-}H_{252}$ cluster was optimized and the spin density distribution was calculated by using the DFT/B3LYP/UKS/MINI/3-21G level of the theory. Calculations have been performed for negatively charged cluster in the triplet ground state (S=1). We have used the PC GAMESS (US) software package for geometry optimization of the cluster and the ORCA software package to calculate full *hfi* matrices A_{KL} for all possible positions of the ^{13}C atom in the cluster. Calculations have been done in the PAS of the NV center having the origin on the N atom of the center. Using the calculated matrices A_{KL} we found the parameter γ_0 values for all positions of the ^{13}C n-spin in the cluster.

3. Results and discussion

The results of calculations are shown in Fig. 1. One can see that in the cluster there are 22 positions in which a ^{13}C n-spin practically does not experience *hfi*-induced flip-flops ($\gamma_0 \leq 10^{-3}$). Spatial and *hfi* data for them are listed in Table 1.

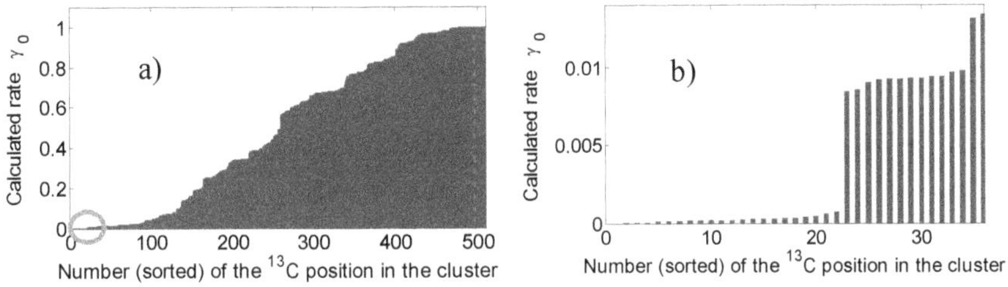

Figure 1. a) Calculated flipping rate γ_0 for all possible NV-^{13}C spin systems in the $C_{510}[NV]^{-}H_{252}$ cluster shown as sorted in ascending order of the parameter γ_0. b) The same for 36 positions exhibiting the lowest flipping rates γ_0.

Table 1. *Hfi* and spatial characteristics of stability positions of the ^{13}C n-spin in the studied cluster, exhibiting lowest flipping rates γ_0. Here Δ_0 is the zero-field splitting of the spin sublevels $m_S = \pm 1$ due to *hfi*, A_{ZZ} and A_{nd} are calculated *hfi* matrix elements, cos*Zz* are cosines between *Z* axis of the NV center PAS and *z* axis of the respective ^{13}C PAS wherein the *hfi* matrix is diagonal, R_{CN} is the distance from respective ^{13}C atom to the N atom of the NV center, R_{CZ} is *Z* coordinate of the ^{13}C atoms, R_{CXY} is the distance from the ^{13}C positions to the *Z* axis.

Position No	γ_0 ($\times 10^{-3}$)	Δ_0 (kHz)	A_{ZZ} (kHz)	A_{nd} (kHz)	cos*Zz*	R_{CN} (Å)	R_{CZ} (Å)	R_{CXY} (Å)
7(ax)	0.0030	183.1	183.1	0.3	1.000	6.47	6.47	0
8(ax)	0.0138	85.1	85.1	0.3	1.000	8.05	8.05	0.004
469(ax)	0.0020	99.9	99.9	0.1	1.000	4.58	-4.58	0.003
505(ax)	0	58.7	58.7	0	1.000	6.15	-6.16	0.004
222(K2)	0.1797	1005.7	-1005.5	13.5	-0.029	4.78	1.74	4.45
255(K2)	0.1916	1005.6	-1005.4	13.9	-0.030	4.78	1.73	4.46
260(K2)	0.2055	1011.7	-1011.5	14.5	0.031	4.77	1.73	4.45
223(Y)	0.1790	203.3	-203.3	2.7	0.004	5.81	2.26	5.36
225(Y)	0.2284	202.0	-202.0	3.1	0.004	5.82	2.26	5.36
256(Y)	0.1413	204.7	-204.7	2.4	-0.002	5.82	2.25	5.37
263(Y)	0.1647	203.9	-203.9	2.6	0.002	5.81	2.25	5.36
269(Y)	0.1376	206.0	-206.0	2.4	-0.002	5.82	2.25	5.37
275(Y)	0.1417	208.0	-208.0	2.5	-0.002	5.81	2.25	5.36
214(Ro3)	0.4465	52.9	-52.9	1.1	-0.008	7.78	2.22	7.45
267(Ro3)	0.5902	52.7	-52.7	1.3	-0.009	7.79	2.21	7.46
277(Ro3)	0.6902	53.0	-53.0	1.4	-0.010	7.77	2.20	7.46
212(Ro4)	0.3216	51.4	-51.4	0.9	-0.002	8.15	2.25	7.83
216(Ro4)	0.3713	51.1	-51.1	1.0	-0.002	8.15	2.25	7.83
254(Ro4)	0.2741	51.6	-51.6	0.9	0.000	8.15	2.24	7.84
264(Ro4)	0.3164	50.9	-50.9	0.9	-0.000	8.16	2.24	7.83
279(Ro4)	0.3179	51.7	-51.7	0.9	-0.001	8.15	2.24	7.84
286(Ro4)	0.3329	51.7	-51.7	0.9	-0.001	8.14	2.23	7.83

The analysis has shown that first four positions 7, 8, 469 and 505, exhibiting practically zero flipping rates γ_0, are the "axial" ones. The other 18 positions of ^{13}C stability are located near the plane passing through the vacancy of the NV center and being perpendicular to the NV axis, see Fig. 2. Among them, the positions 222, 255, 260 and 223, 225, 256, 263, 269, 275 belong to the families K2 and Y of equivalent positions for ^{13}C in the diamond lattice around the NV center, according to the classification done in [4] on the basis of DFT simulation of the smaller cluster $C_{291}[NV]^{-}H_{172}$. Respective NV-^{13}C spin systems exhibit rather large *hfi* splitting $\Delta_0 \sim 1$ MHz and ~205 kHz. Besides, in the present larger cluster $C_{510}[NV]^{-}H_{252}$ there are additional nine ^{13}C stability positions 212–286, which, according to their spatial characteristics, are also members of two new families which we will refer to the families Ro3 and Ro4. Zero-field *hfi* splittings Δ_0 for them are 51–53 kHz. Fig. 2a-c show spatial positions of ^{13}C spins belonging to the above non-axial "stability families". One can see, that in the cases of K2 and Y families they are located rather well inside of the cluster while those of the Ro3 and Ro4 families - at (or close to) the border of the cluster so that the simulated data for them can be influenced by the

Side view Top view

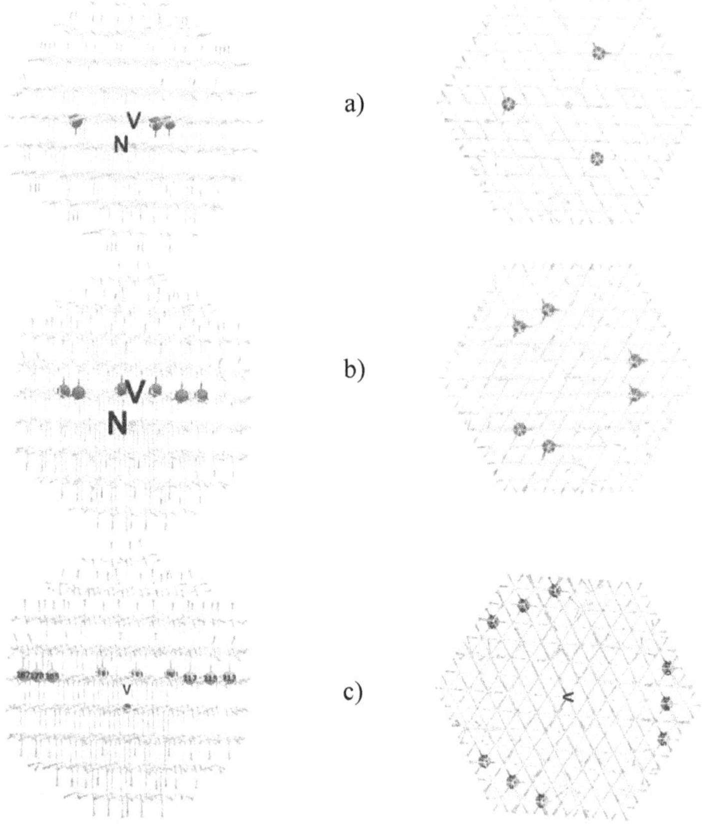

Figure 2. Spatial locations of lattice sites for robust ^{13}C spins belonging to the K2 (Fig. 2a), Y (Fig. 2b) and Ro3, Ro4 families (Fig. 2c), respectively.

cluster surface. Note that for all these non-axial robust ^{13}C positions the values of cosZz are close to zero indicating that quantization axes for such n-spins are perpendicular to the symmetry axis of the NV center. Analysis of the spin density spatial distribution over the cluster showed that the latter extends far enough from the Z axis just in the plane in which the above stability positions are located. Vanishing non-diagonal elements A_{nd} for ^{13}C atoms in these positions resulted from local symmetry of the spin density at respective locations of the ^{13}C n-spins.

Analysis of recent publications [3, 6–8] showed that, apparently, some of the above NV-^{13}C systems have already been observed experimentally: in these works experimentalists were able to find the robust ^{13}C spins exhibiting *hfi* splittings (A_{zz}) of 201 kHz [3], 89 kHz [6,7] and 50 kHz [8] which are close to the predicted values for the Y family, axial C8 position and for the Ro3, Ro4 families, respectively.

4. Conclusion

Using DFT we simulated the H-terminated cluster $C_{510}[NV]^-H_{252}$ hosting the NV center and found for a first time the *hfi* and spatial characteristics of new class of robust NV-^{13}C spin systems wherein the ^{13}C nuclear spin exhibits negligible *hfi*-induced flipping rates due to symmetric local spin density distribution. We hope, that the data of Table 1 will help to experimentalists to find, identify and use these robust spin systems NV-^{13}C in the emerging quantum technologies.

Support by National scientific program “Convergence-2020” and Belarusian Republican Foundation for Fundamental Research is gratefully acknowledged.

References

1. M. W. Doherty *et al.*, *Phys. Reports* **528**, 1 (2013).
2. A. Dréau *et al.*, *Phys. Rev. B* **85**, 134107 (2012).
3. A. Dréau *et al.*, *Phys. Rev. Lett.* **113**, 137601 (2014).
4. A. P. Nizovtsev *et al.*, *New J. Phys.* **16**, 083014 (2014).
5. A. P. Nizovtsev *et al.*, in: *Physics, Chemistry and Application of Nanostructures*, eds. V. E. Borisenko, S. V. Gaponenko, V. S. Gurin, C. H. Kam (Singapore, World Scientific, 2015), p. 24.
6. G. Waldherr *et al.*, *Nature* **506**, 204 (2014).
7. S. Zaiser *et al., Nature Commun.* 7, 12279 (2016).
8. Th. Unden *et al.*, *Phys. Rev. Lett.* **116**, 230502 (2016).

CORRELATED DEPHASING IN QUANTUM DOTS

D. S. Mogilevtsev, M. V. Korolkov
B. I. Stepanov Institute of Physics NASB, 220072 Minsk, Belarus

S. B. Cavalcanti
Instituto de Física, Universidade Federal de Alagoas, 57072-970 Maceió–AL, Brazil

Correlated dephasing between two electron-hole transitions of three-level quantum dots is theoretically investigated. Such dephasing occurs due to the interaction between the dots and the common reservoir of longitudinal acoustic phonons. For realistic structures the phononic dephasing is always correlated in the Markovian regime. Furthermore, the results indicate that the correlated dephasing leads to other important consequences *e.g.*, the suppression of non-Markovian effects in slow-light experiments.

1. Introduction

An electronic spectrum of a quantum dot (QD) is discrete with similar characteristics of the atomic system [1,2]. The structure of a discrete spectrum may be formed by tailoring artificially the geometry and the material of QD [3]. Experimental and theoretical investigations are revealing intriguing aspects of the electron state dynamics in QDs due to their interaction with acoustic phonons [4–6]. In particular, interactions with a common reservoir leads to the correlated dephasing and to appearance of decoherence-free subspaces. Recently, the correlated dephasing has been reported to suppress non-Markovian behavior causing, for example, the enhancement of the group velocity slow-down factor [6].

In the present paper, we consider the electron interaction with modes of acoustic phonons in three-level QDs, using a model of deformation potentials for electrons and holes. As a result, correlations between dephasing reservoirs of different transitions are shown to give rise to a non-Markovian behavior at very low temperatures (few K) and to a completely correlated Markovian relaxation for temperatures as low as 5–10 K.

2. Model

We will follow the approach presented in [5,6]. Let us describe the coupling matrix elements $g_{e/h}$ {q} for the electron and hole as a direct product of two factors

$$g_j^{e/h}(q) = G_j^{e/h}(q) \cdot F^{e/h}(q), \tag{1}$$

where the coefficients $G_j^{e/h}(q)$ depend on the specific coupling mechanism, and the form factor $F^{e/h}(q)$ are obtained from the wave functions $\Psi_{e/h}(x,y,z)$ of electrons 'e' and holes 'h' and phonon modes (with wave vector q) within the QD potential as

$$F^{e/h}(q) = \int d^3r \left|\Psi^{e/h}(r)\right|^2 \cdot \exp[iq \cdot r]. \tag{2}$$

For illustrative purposes, here we choose a simple QD model approximated by a cube, with edge length of a within infinite barriers. Then the wave functions and energies are well known [1,2]:

$$\Psi(x,y,z) = \sqrt{\left(\frac{8}{a^3}\right)} \sin\left(\frac{n\pi\, x}{a}\right) \cdot \sin\left(\frac{m\pi\, y}{a}\right) \cdot \sin\left(\frac{l\pi\, z}{a}\right). \tag{3}$$

$$E_{e,h} = E_0 \pm (\pi\hbar / a)^2 \cdot (n^2 + m^2 + l^2) / 2m^*_{e,h}, \tag{4}$$

where n, m, l = 1, 2 ,3 …, and the boundary condition is $\Psi(x, y, z \geq a) = 0$ for the infinite barriers. In the case of finite barriers, the wave function $\Psi(x, y, z \geq a)$ decreases faster or slower depending on the barrier height, and QD geometry [2].

Let us now consider the matrix elements that characterize the coupling between QD electrons and holes and longitudinal acoustic (LA) phonons due to the deformation potential. Using the coupling, $G_j^{e/h}(q)$, of the form given in [5], and typical parameters of GaAs QD with a deformational potential for electrons D_{LA}^e and holes D_{LA}^h and inserting the density of the semiconductor material $\rho = 5.37$ g/cm^3, we have obtained the following coupling matrix elements

$$g(q) = \frac{0.4745 \cdot 10^{-5} q \cdot \left[D_{LA}^e \cdot F^{e_J}(q) / a_e^{3/2} - D_{LA}^h \cdot F^h(q) / a_h^{3/2} \right]}{\left[\hbar\omega_j(q)\right]^{1/2}}, \quad j = 1, 2. \tag{5}$$

The correlation functions K_{ij} for the three level QD model may be written as functions of temperature T

$$K_{IJ}(\Delta\tau) = \sum_q g_J(q)^* \cdot g_I(q) \{ [n_T(\omega_{LA}) + 1] e^{-i\omega_{LA}\Delta\tau} + n_T(\omega_{LA}) e^{i\omega_{LA}\Delta\tau} \}, \tag{6}$$

where the average number of thermal phonons in the mode is $n_T(\omega_{LA})$. Three level schemes involve two types of the correlation function, the auto-correlation function K_{11} and the cross-correlation function K_{22}. Note that the imaginary part of the correlation function (6) does not depend on temperature. In the following, we study dephasing rates, γ_J, and frequency shifts, δ_I, ($J = 1, 2, 12$) in the long-time limit using the integrals of the correlation functions K_{IJ}

$$\gamma_{IJ} = \text{Re} \int K_{IJ}(\Delta\tau)\cdot d(\Delta\tau), \quad \delta_{IJ} = \text{Im} \int K_{IJ}(\Delta\tau)\cdot d(\Delta\tau). \tag{7}$$

3. Numerical results and discussion

A typical temperature dependence of the ratio γ_1/γ_2 between the dephasing rates of two different electronic transitions defined by the correlation functions K_{11} and K_{22} is shown in Fig. 1. Depicted in the same figure, the ratio between the cross-correlation function K_{12} and the correlation functions K_{11} and K_{22} (γ_{12}/γ_1 and γ_{12}/γ_2) are also shown.

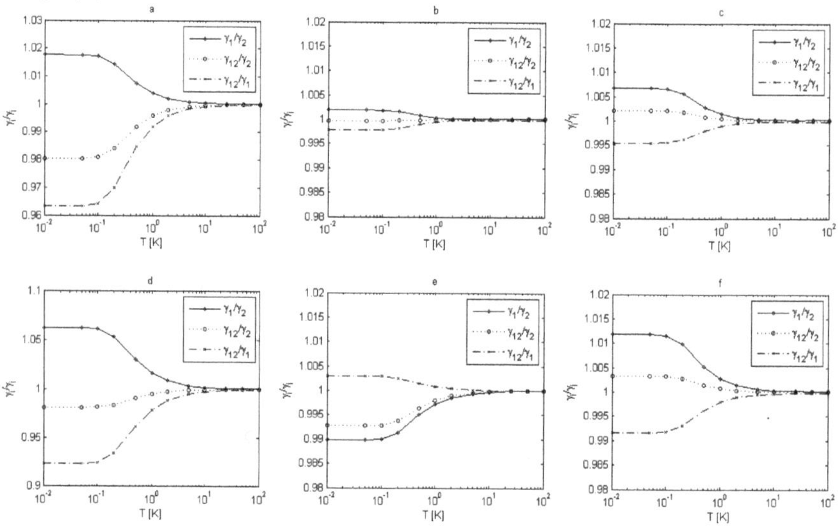

Figure 1. Temperature dependence of the ratio γ_J/γ_I of dephasing rates, for a QD with $a_e = a_h = 5$ nm (a–c) and $a_h = 0.7a_e = 3.5$ nm (d–e) for three sets of three level configurations $\Psi_h(nml)$, $\Psi_{e1}(nml)$, $\Psi_{e2}(nml)$: (a), (d) – $\Psi_h(111)$, $\Psi_{e1}(222)$, $\Psi_{e2}(333)$, (b), (e) – $\Psi_h(111)$, $\Psi_{e1}(123)$, $\Psi_{e2}(333)$, and (c), (f) – $\Psi_h(222)$, $\Psi_{e1}(123)$, $\Psi_{e2}(333)$.

The most important conclusion from Fig. 1 is that the ratio between the dephasing rates for different levels quickly tends to unity as the temperature increases. Ratios of the respective frequency shifts δ_J are also close to unity. An important characteristic of the process is that the ratio $|\gamma_J/\delta_J|$ is equal to one at $T \approx 2$ K , $T \approx 3$ K and $T \approx 6$ K, depending on the size of the QD $a = 20$ nm, $a = 10$ nm and $a = 5$ nm, respectively. However, the decay rates grow with increasing temperature, whereas the shifts remain constant. Also, the ratio of different decay rates tends to unity. One can conclude that in the Markovian limit (which arises beyond few Kelvins), for realistic QDs the nearly perfect correlated dephasing should be expected.

Thus, counterintuitively, the same effect that could lead to preservation of a specific quantum phenomenon, such as entanglement, occurs in a situation where the reservoir is commonly considered as the most harmful in keeping quantum correlations.

Acknowledgments

The authors gratefully acknowledge the support from EU Horizon 2020 project SUPERTWIN (id.686731). SBC acknowledges partial financial support from the Brazilian agency CNPq.

References

1. P. Harrison, *Quantum Wells, Wires and Dots* (Second Edition. John Wiley & Sons, Ltd. 2005).
2. S. L. Chuang, *Physics of Photonic Devices* (John Wiley & Sons, Inc., Hoboken, New Jersey, 2009).
3. S. Kiravittaya, A. Rastelli, O. G. Schmidt, *Rep. Prog. Phys*. **72**, 046502 (2009).
4. P. Kaer, T. R. Nielsen, P. Lodahl, A.-P. Jauho, J. Mørk, *Phys. Rev. B* **86**, 085302 (2012).
5. B. Krummheuer, V. M. Axt, T. Kuhn, *Phys. Rev. B* **65**, 195313 (2002).
6. D. Mogilevtsev, E. Reyes-Gómez, S. B. Cavalcanti, L. E. Oliveira, *Phys. Rev. B* **92**, 235446 (2015).

ENHANCEMENT OF ABSORPTION CROSS-SECTION AND LUMINESCENCE YIELD OF SEMICONDUCTOR QUANTUM DOTS AND UP-CONVERTING PARTICLES

J. Valenta, M. Greben, A. Fucikova

Department of Chemical Physics & Optics, Faculty of Mathematics & Physics
Charles University, CZ-12116 Prague, Czechia
jan.valenta@mff.cuni.cz

Advanced methods of optical spectroscopy were applied to study optical properties of semiconductor and lanthanide-doped up-converting nanoparticles and their enhancement by interaction with localized surface plasmons in metal nanoparticles. External quantum yield of luminescence was determined using a special set-up with an integrating sphere, while the absorption cross-section was obtained by intensity dependence of luminescence kinetics. Single-nano-object spectroscopy methods were applied to study scattering and luminescence of nanocomposites of gold nanorods with nanoparticles.

1. Introduction

Optical performance of many nanomaterials is limited by their low absorption cross section (ACS) and/or low emission quantum yield. As typical examples we can mention indirect band gap semiconductor nanoparticles (like silicon) or up-converting nanoparticles based on fluorides doped with lanthanide ions. Note that both these material systems are characterized by very slow decay kinetics. One of the most feasible ways to overcome these limitations is exploitation of the local field enhancement using metallic nanostructures. In this contribution we present application of gold nanorods to enhance optical properties of the above mentioned materials studied using the single-nano-object optical spectroscopy as well as the advanced quantitatively calibrated spectroscopy.

2. Experimental

Luminescence (external) quantum yield (QY) of both the down-converted and up-converted emission is obtained using a special set-up with an integrating sphere. The method is described in detail elsewhere [1]. The detection range covers an extraordinary broad region from UV to NIR by combining two detection branches which can work simultaneously [2]. As an example we present measurement of QY of up- and down-converted photoluminescence (PL) from particles of $NaLuF_4$:Yb,Er in Fig. 1.

ACS is determined via modulated PL kinetics described by Valenta *et al.* [3]. An acousto-optical modulator is employed to generate long square-shaped excitation pulses. The differential rate of onset and decay kinetics is measured as a function of excitation power. ACS is given by the slope of this intensity dependence. As the PL kinetics of the studied materials is quite slow (of the order of μs to ms) and often non-single-exponential a special care must be paid to the proper experimental settings as well as to the mathematical treatment [4].

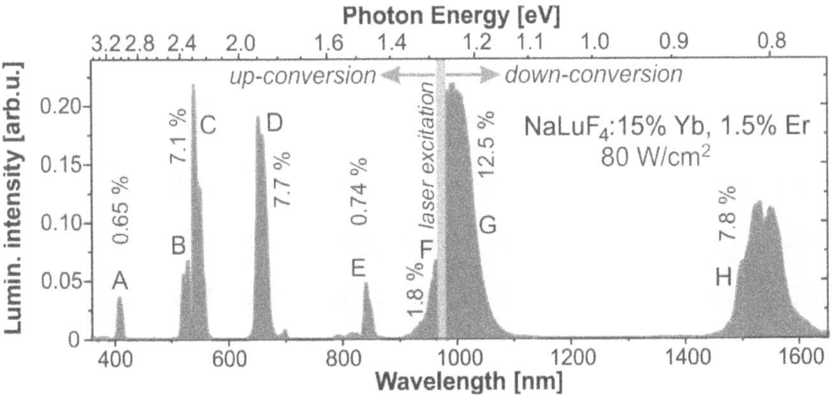

Figure 1. External QY spectrum of $NaLuF_4$: 15 % Yb, 1.5 % Er particles (this is the best performing material composition we found) under excitation with the 980-nm laser line. Spectrum includes both up- and down-conversion regions and the integrated QY for each band is given. The integrated power efficiency of up-converted and down-converted regions is about 18 % and 20 %, respectively.

Single-nano-object optical spectroscopy methods are based on the imaging micro-spectroscopy set-ups (for a recent review see [5]) with the epi-fluorescence wide-field excitation.

The studied materials include Si-based semiconductor nanoparticles [6] and fluorides doped with various lanthanides [7] (see Fig. 1) and their composite structures with noble-metal nanostructures (mainly Au nanorods). The effect of localized surface plasmon resonance (LSPR) is strongly dependent on the distance between metal nanostructures and a semiconductor or fluoride nanoparticle. The proper chemical linkage is still under development and the results will be presented during the meeting.

3. Conclusion

Advanced optical spectroscopy [8] techniques were developed for complex characterization of light-emitting nanostructures and their composites. The inherently limited absorption cross-section and internal quantum yield may be enhanced exploiting enhanced local field from localized surface plasmons in metal nanoparticles. The crucial point is a development of proper linkage between components of such nanocomposite materials.

Acknowledgments

This study was supported by the project 16-22092S from GACR. We acknowledge J. Lancok (Institute of Physics AVCR, Prague), D. Niznansky (Faculty of Natural Sciences, Charles University), F. Novotny (Czech Technical University, Prague), M. Fujii (Kobe University, Japan) and their co-workers for providing materials used in this work.

References

1. J. Valenta, *Nanoscience Methods* **3**, 11 (2014) (open access).
2. J. Valenta, M. Greben, *AIP Advances* **5**, 047131 (2015).
3. J. Valenta *et al.*, *Appl. Phys. Lett.* **108**, 023102 (2016).
4. M. Greben, J. Valenta, *Rev. Sci. Instr.* **87**, 126101 (2016).
5. I. Sychugov, J. Valenta, J. Linnros, *Nanotechnology* **28**, 072002 (2017).
6. T. Kanno *et al.*, *J. Appl. Phys.* **120**, 164307 (2016).
7. A. Repko, J. Valenta, D. Niznansky, The optimum composition and power-dependent up-conversion efficiency of hexagonal Na(Lu/Y/Gd)F_4: Yb, Er/Tm/Ho, to be published.
8. I. Pelant, J. Valenta, *Luminescence Spectroscopy of Semiconductors* (Oxford University Press, Oxford, 2012).

FEATURES OF PLASMON POLARITONS GENERATED IN ANISOTROPIC MAGNETODIELECTRIC METAMATERIALS

S. N. Kurilkina, V. N. Belyi, N. S. Kazak

B. I. Stepanov Institute of Physics NASB, Nezavisimosti Ave. 68, 220072 Minsk, Belarus
s.kurilkina@ifanbel.bas-net.by

The features of plasmon polaritons generated at the boundary of an anisotropic magnetodielectric metamaterial and an isotropic dielectric are considered. Formation of a specific surface-volume plasmonic field which is localized in the metamaterial and radiative in the dielectric is shown.

1. Introduction

The last decade has witnessed a huge activity in the development of metamaterials (MMs) [1] boosted by the possibility that such engineered media may give rise to novel optical properties in certain frequency ranges including negative refraction [2] and perfect lensing [3]. Such striking phenomena, inaccessible with natural materials, are all possible due to the magnetic activity built into metamaterials. It is generally desirable that MMs should be constructed from elementary cells which are smaller than the wavelength of the incident radiation. This fact often allows using an effective medium approximation to describe MMs in terms of an effective electric permittivity tensor ε and an effective magnetic permeability tensor μ, which incorporate the optical anisotropy of metamaterials.

A surface separating materials with dielectric permittivities of opposite signs can support the propagation of surface waves, also known as surface plasmon polaritons (SPPs). Note that SPPs propagate along the surface with periodicity lower than the wavelength of the electromagnetic radiation in vacuum of the same frequency, which is a very appropriate for the miniaturization of photonic devices. This feature and its resonant behavior make SPPs very attractive for numerous applications. The purpose of this paper is to analyze the features of SPP generation at the boundary of anisotropic magnetodielectric metamaterial.

2. Plasmon polaritons in the uniaxial magnetodielectric metamaterial

Consider the case of a planar interface between a dielectric with permittivity ε_1 and a uniaxial magnetodielectric medium whose electric and magnetic anisotropic directions coincide and are perpendicular to the interface. This is an example of a uniaxial medium with the optical axis coinciding with the anisotropic direction. For this case the electromagnetic tensors are diagonal in the coordinate system defined by the optical axis (Z axis) and any two orthogonal directions in the orthogonal plane (X and Y axes). Therefore, choosing the interface as the XY plane and the anisotropic medium being the half-space defined by $Z > 0$ (see Fig. 1), the permittivity and permeability tensors are given by

$$\boldsymbol{\varepsilon}_2 = \begin{pmatrix} \varepsilon_t & 0 & 0 \\ 0 & \varepsilon_t & 0 \\ 0 & 0 & \varepsilon_l \end{pmatrix}, \ \boldsymbol{\mu}_2 = \begin{pmatrix} \mu_t & 0 & 0 \\ 0 & \mu_t & 0 \\ 0 & 0 & \mu_l \end{pmatrix}. \tag{1}$$

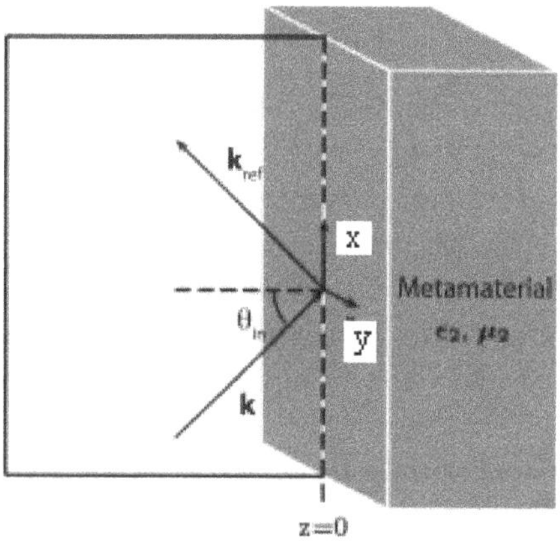

Figure 1. The light wave impinges upon the interface between the dielectric and the uniaxial magnetodielectric metamaterial.

Let us assume that a p-polarized plane wave with the wave vector $\mathbf{k}$ impinges upon the interface from $Z < 0$ towards the metamaterial. Given the rotational symmetry with respect to Z without loss of generality we can choose our coordinate system so that the plane of incidence (defined by $\mathbf{k}$ and Z) coincides with the XZ plane. Let us represent the field inside dielectric (d) and metamaterial (m) in the form:

$$\mathbf{F} = \begin{cases} \mathbf{F}^d e^{\kappa_d z + iqx - i\omega t}, & z < 0 \\ \mathbf{F}^m e^{-\kappa_m z + iqx - i\omega t}, & z > 0 \end{cases}, \quad \mathbf{F} = \begin{pmatrix} \mathbf{E} \\ \mathbf{H} \end{pmatrix}. \tag{2}$$

Here **E**, **H** are electric and magnetic vectors, respectively, $\kappa_{d,m}$ is the decay constant, q is the transversal wave number (transversal component of the wave vector) which is equal to the wave number of the surface plasmon polariton. It follows from the Maxwell's equations that for *p*-polarized incident wave the decay constants inside the dielectric and metamaterials are given by:

$$\kappa_d^2 = q^2 - k_0^2\varepsilon_1,\ \kappa_m^2 = (\varepsilon_t / \varepsilon_l)q^2 - k_0^2\varepsilon_t\mu_t. \tag{3}$$

Here $k_0 = \omega / c$ is the wave number in vacuum. Using the boundary conditions the equation for the reflection coefficient can be obtained from (2):

$$r^{TM} = \frac{\varepsilon_t\kappa_d - \kappa_m\varepsilon_1}{\varepsilon_t\kappa_d + \kappa_m\varepsilon_1}. \tag{4}$$

The dispersion equation, which determines the condition of plasmon polaritons excitation, is given by the relation $r^{TM} \to \infty$, and therefore

$$\varepsilon_t\kappa_d = -\kappa_m\varepsilon_1. \tag{5}$$

From (5) it follows

$$q = k_0\left[\varepsilon_l\mu_t\varepsilon_1(\varepsilon_1 - \varepsilon_t\mu_t)/\left(\varepsilon_1^2 - \mu_t^2\varepsilon_t\varepsilon_l\right)\right]^{1/2} = k_0\left[\varepsilon_{eff}\right]^{1/2}. \tag{6}$$

Consider localized plasmon polaritons at the interface between the dielectric and the structure experimentally realized in [4] as shown in Fig. 2.

Figure 2. Multilayered fishnet structure formed by 20 altering layers of MgF_2 (45 nm) and gold (30 nm).

The effective parameters of this material for $\lambda = 1.45$ μm are the following: $\varepsilon_t = -0.045 + i0.03$; $\varepsilon_l = 0.08 + i0.002$; $\mu_t = 0.06 + i0.13$; $\mu_t = 1$ [4]. Using (6) we obtain that plasmon polariton can exist at the interface between an air and the considered structure. The changes of the bordered dielectric permittivity influence weekly the phase velocity of this plasmon polariton. It should be noted that since the decay constant in the dielectric has both real and imaginary parts ($\kappa_d = \kappa_d' + i\kappa_d''$) (see Fig. 3), the excited plasmon polariton is characterized by the radiative flux into isotropic medium at the angle $\theta = \operatorname{atan}(\kappa_d'' / q')$ (here $q' = \operatorname{Re} q$) to the boundary. The radiative flux inside the metamaterial is negligible and practically independent on optical properties of dielectric.

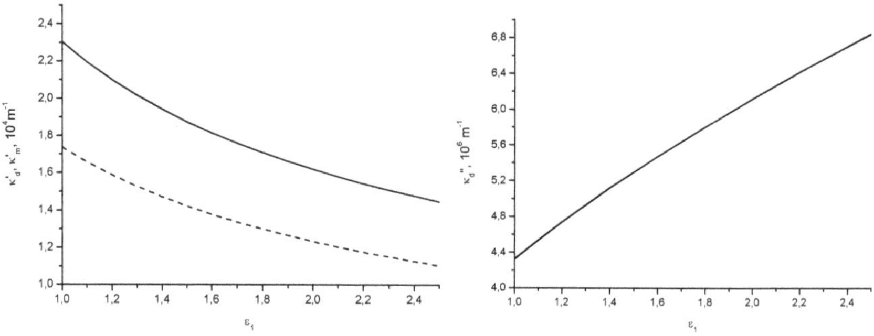

a b

Figure 3. Real (a) and imaginary (b) parts of the decay constant in dielectric (solid curves) and metamaterial (dashed curves) on the base of multilayered fishnet structure formed by 20 altering layers of MgF_2 (45 nm) and gold (30 nm). The wavelength is 1.45 μm.

3. Conclusion

The condition for SPP generation has been obtained for the boundary of metamaterial with dielectric and magnetic anisotropy. It is shown that the excitation of plasmon polariton field is possible if it is localized in the metamaterial and has radiative component in the dielectric.

References

1. V. M. Shalaev, A. Boardman, *J. Opt. Soc. Am. B* **23**, 386 (2006).
2. D. R. Smith, W. J. Padilla, D. C. Vier, S. C. Namat-Nasser, S. Schultz, *Phys. Rev. Lett.* **84**, 4184 (2000).
3. J. B. Pendry, *Phys. Rev. Lett.* **85**, 3966 (2000).
4. S. S. Kruk, Z. J. Wong, E. Pshenay-Severin, K. O'Brien, D. N. Neshev, Y. S. Kivshar, X. Zhang, *Nature Comm.* DOI: 10.1038/ncomms11329 (2016).

PURCELL EFFECT IN TAMM PLASMON STRUCTURES

A. R. Gubaydullin[1,2], C. Symonds[2], J. Bellessa[2], K. A. Ivanov[1,3], E. D. Kolykhalova[3,4], M. E. Sasin[4], A. Lemaitre[5], P. Senellart[5], G. Pozina[3], M. A. Kaliteevski[1,3,4]

[1]*St. Petersburg Academic University, Khlopina 8/3, 194021 St. Petersburg, Russia*
[2]*Univ Lyon, Université Claude Bernard Lyon 1, CNRS, Institut Lumière Matière F-69622 Lyon, France*
[3]*ITMO University, Kronverksky Ave. 49, 197101 St. Petersburg, Russia*
[4]*Ioffe Institute, Politekhnicheskaya 26, 194021 St. Petersburg, Russia*
[5]*Centre de Nanosciences et Nanotechnologies, CNRS Université Paris-Saclay Route de Nozay, F-91460 Marcoussis, France*

We have demonstrated experimentally that in metal/semiconductor Tamm plasmon based structures, where InAs quantum dots active area is sandwiched between GaAs/AlAs Bragg reflector and silver layer, the probability of spontaneous emission can be increased by the factor of more than one order of magnitude despite losses in metal. We have shown results of theoretically calculated modification of spontaneous emission rate by S-quantization formalism, showing good agreement with the experimentally observed photoluminescence pattern.

1. Introduction

The main obstacle which prevent the utilization of metallic elements in optical schemes of optoelectronic devices is the absorption of light in metals [1]. The optical absorption is accompanied by the metal heating, resulting in a catastrophic degradation of the mirrors and the surrounding material. Furthermore, this absorption is considered to be fatal for optical coherence and therefore prevent the development of even a small output laser. Modification of spontaneous emission rate in the structures can be considered as a consequence of interference in the structure [2], and since for Tamm plasmon structures interference is weakly influenced by absorption in metal the probability of spontaneous emission can be increased substantially.

Several major breakthrough in metallic optoelectronic devices have been achieved in recent years. For instance, macroscopic optical coherence, lasing [3], and single photon emission has been demonstrated in Tamm plasmon based structures [4]. A Tamm plasmon is a state of electromagnetic field, localized at the interface between metal and distributed Bragg reflector (DBR). The Tamm plasmon (TP) state can be tuned to provide positioning of the node of the electric field at the metallic layer therefore suppressing absorption of light [5].

2. Results and discussion

For experimental study we have created the following sample. The heterostructure was grown by molecular beam epitaxy on a GaAs [001] substrate. It is composed by an active distributed Bragg reflector formed by pairs of GaAs/AlAs quarter-wavelength stacked layers. In order to form the TP state, the metallic silver layer was thermally evaporated on top of the DBR, and is formed in the way that the silver is adjacent to the higher refractive index material GaAs. In the upper high refractive index quarter-wavelength layer an active medium is formed by embedding quantum dots (QDs) in GaAs matrix.

To perform angular resolved reflectometry and photoluminescence (PL) experiments, the sample was illuminated (pumped) from the top by means of a Ti:sapphire laser operating at 780 nm. We have measured good results at room temperature, which are illustrated on Fig. 1b, demonstrating that spontaneous emission is suppressed in all directions of propagation and frequencies except three modes characterized by parabolic dispersion: at low energy Tamm plasmon mode and two higher energy modes corresponding to distributed Bragg reflector.

For an analysis of modification of spontaneous emission rate we have compared a pattern of calculated modal Purcell [6] factor $F(\omega,\theta)$, which is defined as a ratio of probability of spontaneous emission for specific frequency and direction of emission to those in uniform media, and experimentally measured photoluminescence pattern.

Probability of spontaneous emission has been calculated using S-quantization [7] formalism, which is based on quantization of electromagnetic field by setting eigenvalues of scattering matrix of the system to unity. Such method provides a rigorous and self-consistent description of modes in non-uniform media, and allows to avoid the problems of divergence of integral describing effective mode volume. Therefore it provides a straightforward method for calculating probability of spontaneous emission for certain frequency of light, direction of propagation and polarizations of the emission.

Results are presented on Fig. 1c. Three modes characterized by parabolic dispersion could be observed: Tamm plasmon mode and two higher energy modes related to DBR. The dependence for different angles and frequencies is in a good agreement with the experimentally measured pattern.

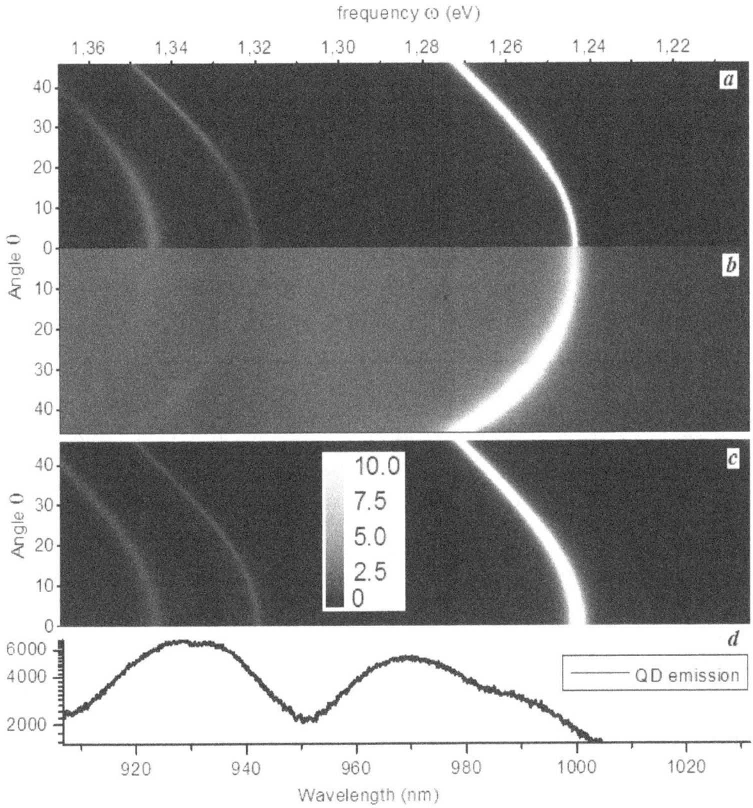

Figure 1. (a) Dependence of the product of calculated modal Purcell factor and experimentally measured emission spectra of quantum dots on frequency and angle of emission; (b) Experimentally measured photoemission spectra pattern of angle dependence; (c) Dependence of calculated modal Purcell factor on frequency and angle of emission; (d) Emission spectra of quantum dots.

In our structure the emitter is formed by layers of InAs quantum dots, which possess wide luminescence band and large ionization energy, illustrated on Fig. 1d. Charge carriers are captured by quantum dot (capture time is extremely fast-below 1 ps) and emitting at specific energy. They cannot escape QD, and they cannot be captured by another quantum dots, emitting light photon with energy, corresponding to another quantum dots. Therefore, the emission pattern of Tamm plasmon structure (see Fig. 1b) with quantum dots will be defined by modal Purcell factor and the emission spectrum of quantum dots $\rho(\omega)$ (Fig. 1d):

$$I(\omega,\theta) = F(\omega,\theta)\rho(\omega). \tag{1}$$

The product of the modal Purcell factor $I(\omega,\theta)$, shown in Fig. 1a, ideally coincides with the experimentally observed emission pattern shown in Fig. 1b, which confirms validity of the estimates provided by S-quantization model.

3. Conclusion

We have calculated spontaneous emission rate using S-quantization formalism and compared the result with the experimentally observed photoluminescence pattern, showing a good agreement, what provide the confirmation for S-quantization method. We have demonstrated that despite of the absorption in the metallic layer, the spontaneous emission rate for Tamm plasmon mode is substantially increased in respect to uniform media.

References

1. J. B. Khurgin, *Nature Nanotechnology* **10** (1), 2 (2015).
2. M. V. Rybin, S. F. Mingaleev, M. F. Limonov, Y. S. Kivshar, *Scientific Reports* **6**, 20599 (2016).
3. C. Symonds, G. Lheureux, J. P. Hugonin, J. J. Greffet, J. Laverdant, G. Brucoli, A. Lemaitre, P. Senellart, J. Bellessa, *Nano Lett.* **13**, 3179 (2013).
4. M. E. Sasin, R. P. Seisyan, M. A. Kalitteevski, S. Brand, R. A. Abram, J. M. Chamberlain, A. Yu. Egorov, A. P. Vasil'ev, V. S. Mikhrin, A. V. Kavokin, *Appl. Phys. Lett.* **92**, 251112 (2008).
5. M. A. Kaliteevski, A. A. Lazarenko, N. D. Il'inskaya, Yu. M. Zadiranov, M. E. Sasin, D. Zaitsev, V. A. Mazlin, P. N. Brunkov, S. I. Pavlov, A. Yu. Egorov, *Plasmonics* **10**, 281 (2015).
6. E. M. Purcell, H. C. Torrey, R. V. Pound, *Phys. Rev.* **69**, 37 (1946).
7. M. A. Kaliteevski, A. R. Gubaydullin, K. A. Ivanov, V. A. Mazlin, *Optics and Spectroscopy* **121**, 410 (2016).

THE FEATURES OF STRUCTURAL AND MAGNETIC CHARACTERISTICS OF Co/Cu/Co THIN-FILM SYSTEMS

E. E. Shalygina, A. V. Makarov, A. M. Kharlamova

Faculty of Physics, M. V. Lomonosov Moscow State University, 119991 Moscow, Russia
shal@magn.ru

G. V. Kurlyandskaya, A. V. Svalov

Ural Federal University, 620002 Ekaterinburg, Russia
kurlyandskaya@gmail.com

Structural and magnetic characteristics of Co/Cu/Co thin-film systems fabricated by magnetron sputtering are investigated. The Co thickness in all samples was 5 nm and the Cu thickness was varied from 0.5 to 4 nm. It was found that the saturation field, oscillates in magnitude with increasing Cu layer thickness with the period of the order of 1 nm. The obtained data are explained by the presence of exchange coupling between the ferromagnetic layers through the Cu spacer.

1. Introduction

Magnetic thin films have attracted the attention of researchers since the mid of the XX century due to the peculiarities of their physical properties. For example, the giant magnetoresistance and oscillating exchange interaction between ferromagnetic (FM) layers containing nonferromagnetic (NFM) spacers were discovered in multilayer systems consisting of alternating FM and NFM layers of submicron thicknesses [1–3]. Moreover, the multilayer systems (MS) are widely used in devices of micro- and nanoelectronics [4,5]. Despite the huge experimental data accumulated [6–8], the study of the influence of NFM layer on the magnetic properties of the thin-film magnetic systems is still relevant. This work is devoted the study of the influence of copper interlayer on the structural and magnetic properties of Co/Cu/Co thin-film systems.

2. Materials and methods

The Co/Cu/Co samples were grown by magnetron sputtering using Co and Cu targets and corning 2845 glass substrates. The Ta seed layers of 5 nm thick were deposited onto the glass substrates. The background pressure in the vacuum chamber was 4×10^{-7} mbar and the argon pressure was as high as 3.8×10^{-3} mbar. A constant magnetic field of 250 Oe was applied parallel to the substrate surface, H_{SUB}, in order to form an in-plane easy magnetization axis (EMA). The thickness of the Co layers, t_{Co}, was equal to 5 nm, and the Cu layer, t_{Cu}, was varied from 0.5 to 4 nm.

The structure of the Co/Cu/Co samples was studied by X-ray diffraction (XRD) method using CuKα radiation, and the surface morphology by atomic force microscope (AFM). The magnetic characteristics of the thin-film systems were measured employing a magneto-optical magnetometer by means of the transverse Kerr effect (TKE) and the vibration sample magnetometer (Lake Shore VSM 7400) with the sensitivity up to 10^{-7} G·cm^3. The hysteresis loops were recorded for two directions of the magnetic field, H. In particular, the orientation of H was parallel or perpendicular to the EMA (designated, respectively, as D1 and D2 directions).

3. Results and discussion

The results of XRD analysis of Co/Cu/Co samples and reference Co thin film deposited directly on the glass substrate are presented in Fig. 1.

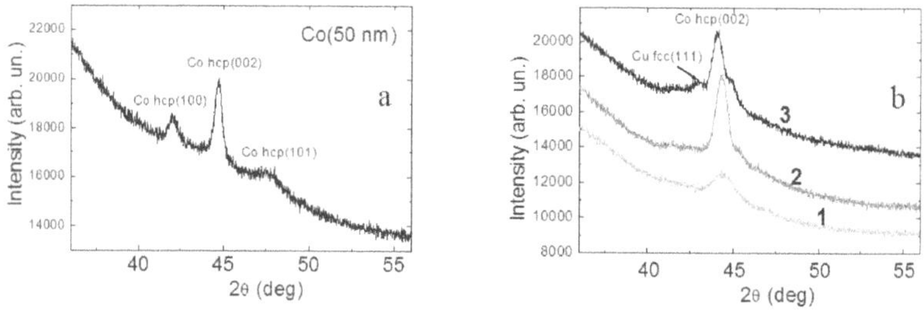

Figure 1. X-ray spectra observed for the reference 50 nm Co film (a), and the samples contained Ta, Co and Cu (b): curve 1 – Ta/Co(5 nm)/Ta, curve 2 – Ta/Co(5 nm)/Cu(1 nm)/Co(5 nm), curve 3 – Ta/Co(5 nm)/Cu(4 nm)/Co(5 nm).

The X-ray spectra obtained for the trilayer samples do not have peaks corresponding to the lines of Ta. The line fcc (111), corresponding to the lattice of Cu, becomes more noticeable on the spectrum of the sample with the thicker layer of copper. It was found that the size of the crystallites in the Co film is of the order of its thickness. According to AFM data, the average surface roughness, R_a, of the studied samples does not exceed 0.5 nm that confirms high quality of the samples surface.

The typical results of magnetic measurements are presented in Fig. 2. All measurements were carried out in open air at room temperature. The results of magnetic measurements of hysteresis loops allowed to receive the dependences of the saturation field H_S on t_{Cu} (Fig. 2e). The hysteresis loops measured in the magnetic fields, applied along the D1 and D2 directions, differ. It was observed for all samples, measured both by using magneto-optical and vibration techniques. This fact shows the presence of induced magnetic anisotropy with

EMA, parallel to the direction of the magnetic field, applied during the deposition process.

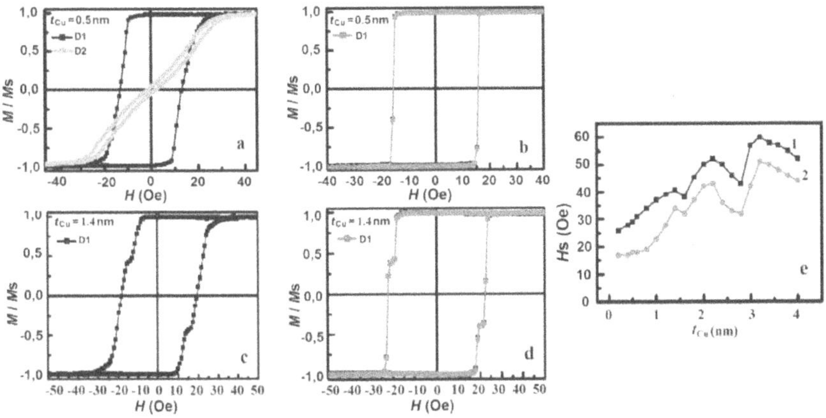

Figure 2. Hysteresis loops of Co/Cu/Co samples measured employing magneto-optical (a, c) and vibration sample (b, d) magnetometers; (e) dependences of $H_S(t_{Cu})$ obtained by using the data of magneto-optical (curve 1) and vibration sample magnetometers (curve 2).

According to the commonly accepted notions [9], the main mechanism of the induced magnetic anisotropy is pair ordering of atoms.

The shape of hysteresis loops observed for the Co/Cu/Co samples measured in the magnetic field applied along the D1 direction depends on t_{Cu}. In particular, it was found that some of the samples are characterized by almost rectangular hysteresis loops (Fig. 2a,2b) while others are characterized by more complex double-stage loops (Fig. 2c,2d). According to the calculations performed in [10], and the experimental results [6–8] the double-stage and almost rectangular hysteresis loops are observed at antiparallel and parallel orientations of magnetization components in Co layers. Thus, the obtained data can be explained by an exchange interaction between the FM layers through the Cu layer and its oscillating behavior with changing t_{Cu} (transition from FM to the antiferromagnetic (AFM) interaction). In the samples with the AFM-exchange interaction between the magnetic layers, the magnetization components in Co layers, parallel to the EMA, have the opposite directions, *i.e.*, these thin-film structures are similar to an uniaxial antiferromagnetic. The saturation field, H_S, of the samples with the AFM-exchange interaction between the magnetic layers is higher than H_S of the samples with the FM exchange. It is caused by an additional energy, which is necessary for overcoming the AF exchange between the magnetic layers.

One can see in Fig. 2e that the period of the oscillations of H_S, Λ, is of the order of 1 nm. At the same time according to the mechanism of the exchange interaction between the FM layers through the NFM interlayer via the RKKY interaction, the magnitude of Λ must be of the order of π/κ_F [11–13], where κ_F is the Fermi wave vector. For most metals π/κ_F is of the order of 0.3–0.4 nm. The found longer period of the oscillations of H_S can be explained by an influence of quantum size effects manifested in the change of the electronic structure of ultra-thin magnetic layer (the appearance of the so-called quantum well states) as compared with bulk samples [14,15].

4. Conclusion

The study of magnetic properties of the Co/Cu/Co thin film samples has shown that the dependence of the saturation magnetic field reveals an oscillating behavior as a function of t_{Cu}. The found period of the oscillations is consistent with the calculated value of Λ, obtained by taking into account quantum size effects, which manifests in the change of electronic structure of ultra-thin films as compared to the bulk sample.

This work was supported by the Russian Foundation of Basic Researches (grant 15-02-02077).

References

1. S. S. P. Parkin, Z. G. Li, D. J. Smith, *Appl. Phys. Lett.* **51**, 2710 (1991).
2. S. S. P. Parkin, N. More, K. P. Roche, *Phys. Rev. Lett.* **64**, 2304 (1990).
3. R. Coehoorn, in: *Handbook of Magnetic Materials*, ed. by K. H. J. Buschow (Elsevier, 2003), **15**, pp. 1–197.
4. A. V. Svalov *et al.*, *IEEE Trans. Magn.* **46**, 333 (2010).
5. S. I. Kasatkin, N. P. Vasilieva, A. M. Myraviev, *Spintronic Magnetoresistive Components and Devices Based on Them* (Institute of Control Sciences RAS, Moscow, 2005).
6. E. E. Shalygina *et al.*, *Solid State Phenom.* **190**, 914 (2012).
7. E. E. Shalygina *et al.*, *Solid State Phenom.* **233**, 653 (2015).
8. E. Shalygina *et al.*, *Jpn. J. Appl. Phys.* **55**, 07MF01-1-6 (2016).
9. S. Chikazumi, *Physics of Magnetism* (New York–London–Sydney: John Willey&Sons, Inc., 2001).
10. K. Zvezdin, *Phys. Solid State* **42**, 120 (2000).
11. M. A. Ruderman, C. Kittel, *Phys. Rev.* **96**, 99 (1954).
12. P. Bruno, C. Chappert, *Phys. Rev. Lett.* **67**, 1602 (1991).
13. Y. Yafet, *Phys. Rev. B* **36**, 3948 (1987).
14. M. D. Stiles, *Phys. Rev. B* **48**, 7238 (1993).
15. P. Bruno, *Phys. Rev. B* **52**, 411 (1995).

INVESTIGATION OF THE TUNNEL COMPONENT OF CONDUCTANCE IN HETEROSTRUCTURES WITH InGaAs/GaAs QUANTUM WELL BY ADMITTANCE SPECTROSCOPY

Y. V. Ivanova, V. I. Zubkov
St. Petersburg Electrotechnical University "LETI", Saint-Petersburg, Russia
yana@unix-server.su

For detection of the tunnel component in thermionic emission, high-quality heterostructures with a single quantum well (SQW) of $In_xGa_{1-x}As/GaAs$ were studied. Observation of two competing emission mechanisms can be fixed only for deep quantum wells with high content of In ($0.25 < x < 0.3$). The conductance value in direct ratio depends on the frequency of the test signal and strongly depends on the reverse bias. The dependence of thermionic peaks of conductance on the reverse bias has a resonance character.

1. Introduction

Quantum-dimensional structures are characterized by a discrete energy spectrum and may reveal other quantum-mechanical effects, *e.g.* the tunnel effect at specific conditions. The admittance method is the most prime and efficient research technique, both for bulk semiconductors and heterostructures with quantum wells (QWs) [1–4]. In doped heterostructures the bottom of conduction band (or top of valence band) is bent due to charge carriers redistribution forming additional barriers. They appear close to a triangular form. The detailed theoretical and experimental study of capture processes and charge carriers emission in heterostructures with QWs was presented [5]. It was shown that in the heterostructures containing QWs there are competing mechanisms of charge carriers emission: thermionic emission and tunneling which make different contributions to conductance at different temperatures.

In this paper, the isotype n-type heterojunctions with elastically strained $In_xGa_{1-x}As/GaAs$ SQW grown by MOCVD are investigated [1,2]. Thickness of the active layer was 6.0–9.5 nm. The parameters and quality of structures were controlled by high resolution X-ray diffraction (HRXRD) and local cathodoluminescence. The measurements were performed in the temperature range of 10–375 K, the voltage range of ±40 V, the frequency range of test signals of 20 Hz – 2 MHz.

2. Results

The experiment on the series of samples with $In_xGa_{1-x}As/GaAs$ SQW ($x = 0.065$–0.29) was carried out. At the small content of In in the solid solution, there is no features in the conductance spectra related to the tunnel emission, because too small band discontinuity giving the response which cannot be registered even at low temperatures.

The best qualitative characteristics were received for structure with $x = 0.29$ (limit of pseudomorphous growth). The results are presented in Fig. 1.

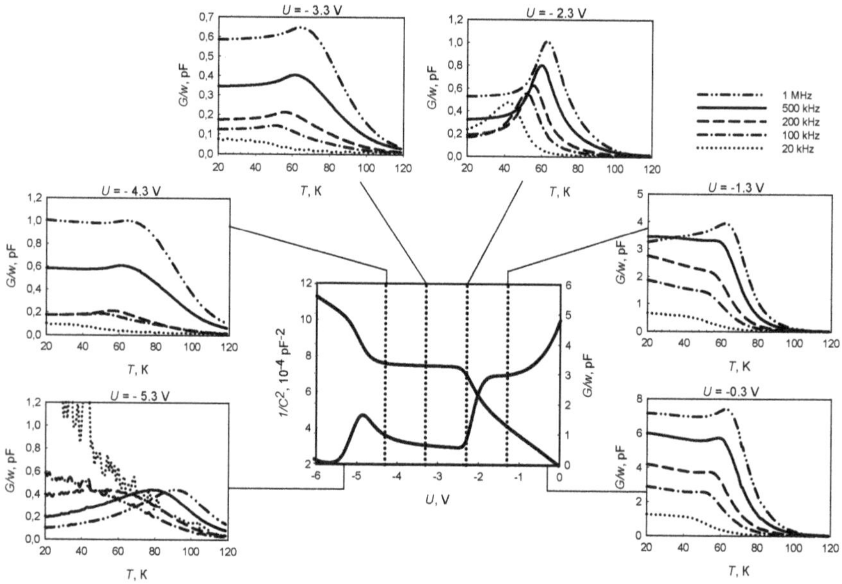

Figure 1. Generalization of the admittance measurements for $In_{0.29}Ga_{0.71}As/GaAs$ SQW. Central plot shows CV-GV dependences. Other subplots are the temperature spectra of conductance for the reference points of CV-curve.

It is noticeable that at low temperatures the delaying effects begin to play an important role, and at temperatures close to room temperature, the SQW manages to relax at test signal frequencies (the quasi-static mode of measurements is implemented). The conductance-voltage (GV) characteristics clearly demonstrates the complex resonance mechanism of formation of the active conductance for the structure containing SQW. The strong modification is experienced by temperature spectra of conductance depending of the reverse bias. Peaks in them are most brightly expressed at the reverse bias –2.3 V that corresponds to a part of capacitance-voltage (CV) characteristic adjoining the plateau from the small biases. This area is responsible for the beginning of intensive thermionic emission of charge carriers from quantization levels with the electric field entering QW.

The temperature shift of the conductance maximum confirms the thermally activation nature of the observed process. This phenomenon has a resonance character and exists in the narrow voltage range, about 0.5 V. The frequency dependence of peaks in the conduction spectra allows to process them as Arrhenius plots. The deviation from the linear dependence is observed that confirms the essential part of the tunnel component in emission [3,5]. As it was expected [5], in G-T experiments the plateau are observed at 20–50 K that evidences an existence of tunnel current. Thus, the emission net activation energy of charge carriers from QW should be determined by the high-temperature part of the Arrhenius curve, minus the tunnel component. The activation energy deviates from 36 to 70 meV depending on the reverse bias.

For adequate definition of the activation energy we will address to the data in Fig. 1. In the temperature spectra of G/w peaks of the thermionic nature (at –2.3 V), the activation energy determined by the Arrhenius curve reaches the maximal value. The experimental points of the Arrhenius curve fit exactly a straight line. At other reverse biases where the thermionic component is poorly expressed against the tunnel background, the experimental points of the Arrhenius curve significantly deviate from a straight line. That shows considerable decrease of the activation energy.

3. Discussion

The study of high-quality heterostructures with SQW conducted by admittance spectroscopy confirms the existence of two competing mechanisms for charge carriers emission: thermionic and tunnel mechanisms. Thermionic peaks in the conduction spectra have a resonance character with maxima in the narrow region of the biases. Plateaus in the spectra corresponding to the tunnel component at reverse biases 0–3.0 V attain the maximum at –0.3 V followed by the sharp fall. Such behavior of the tunnel component can be explained on the basis of resonance tunneling features.

It is necessary to bear in mind that the top part of a doped heterostructure with a QW on the band diagram represents a two-barrier structure in energy region of quasi-resonances ($E > 0$). In [6] it was shown that the structure with identical barriers has the greatest probability for transmission of carriers. In this case, the probability of resonance tunneling gets maximum. A minimum relative change of the parameters of one of the barriers leads to a sharp recession of the transfer coefficient. This qualitative conclusion, made on the basis of analytical calculations, allows explaining the above experimental results. At small reverse biases the external electric field does not reach a space charge region, and the SQW remains with symmetric barriers, identical in form. At the same time the plateaues on the temperature spectra of conductance get maximum. Increasing the reverse bias results in the rise of the left-hand edge of the space charge region. Hence, the symmetry of barriers is broken and the tunnel transfer probability sharply falls. In the experimental results, this effect is expressed by sharp falling of the plateau at –1.3 V.

References

1. V. I. Zubkov, *Diagnostics of Semiconductor Nanoheterostructures by Admittance Spectroscopy Method* (Saint-Petersburg, Publishing house St. Petersburg Electrotechnical University "LETI", 2011), p.162.
2. V. I. Zubkov, *Semiconductors* **41**, 320 (2007).
3. C. M. A. Kapteyn, *Carrier Emission and electronic Properties of Self-organized Semiconductor Quantum Dots*: Dissertation (Berlin: Mensch&Buch Verlag, 2001), p.156.
4. V. I. Zubkov, O. V. Kucherova, I. N. Yakovlev, A. V. Solomonov, *Russian Microelectronics* **44**, 203 (2015).
5. K. Schmalz, I. N. Yassievich, H. Rücker, H. G. Grimmeiss, H. Frankenfeld, W. Mehr, H. J. Osten, P. Schley, H. P. Zeindl, *Phys. Rev. B* **50**, 14287 (1994).
6. V. I. Zubkov, I. N. Yakovlev, V. G. Litvinov, A. V. Ermachihin, O. V. Kucherova, V. N. Cherkasova, *Semiconductors* **48**, 923 (2014).
7. A. Yu. Aladyshkin, *Tunnel Phenomena in Nanophysics* (Nizhny Novgorod, University of Nizhny Novgorod, 2011), p. 32.

OPTICAL PROPERTIES OF THE $CaSi_2$/Si(111) AND Si(111)/$CaSi_2$/Si(111) HETEROSTRUCTURES

N. G. Galkin, K. N. Galkin, A. M. Maslov

Laboratory of Optics and Electrophysics, Institute of Automation and Control Processes, Far Eastern Branch RAS, Vladivostok, Russia
galkin@iacp.dvo.ru

D. B. Migas, V. O. Bogorodz

Belarusian State University of Informatics and Radioelectronics, Minsk, Belarus
migas@bsuir.by

Optical properties of a $CaSi_2$ film and a $CaSi_2$ layer embedded in silicon matrix have been studied by experimental and theoretical methods. According to the experimental data $CaSi_2$ layer displays degenerate semiconductor properties with strong absorbance at photon energies higher 2.3 eV. First principal calculations predict $CaSi_2$ to be a semimetal independent of stressed $CaSi_2$ lattice parameters.

1. Introduction

$CaSi_2$ in the Ca-Si system is a compound with a maximum silicon content [1]. According to the theoretical calculations [2] $CaSi_2$ is a metal, while experimental works suggest semimetal properties to characterize the bulk polycrystalline $CaSi_2$ samples [3]. At atmospheric pressure $CaSi_2$ has a rhombohedral structure with the $R\bar{3}m$ space group and two modifications (namely hR3 and hR6) [4]. The polycrystalline silicide film formed by reactive deposition on Si(111) substrate at $T = 500$ °C with a thickness of about 120 nm [5] has displayed the photoluminescence peak at about 0.95 eV at $T = 5$ K [6], which is not typical of semimetal $CaSi_2$ [3]. Recently, by analyzing data of HR TEM on cross-sections of a Si/$CaSi_2$/Si system [7] it is revealed that the formed silicide film has the hR6 modification with the compressed a and stretched c lattice parameters. Comparison of experimental and theoretical optical data for $CaSi_2$ with stressed lattice parameters has not been done yet.

2. Experimental methods and details of calculations

Two samples containing Si/$CaSi_2$/Si double heterostructures (DHS) and one sample having a $CaSi_2$/Si(111) heterostructure (HS) with silicide thickness of about 114 nm were grown in the ultra-high vacuum "VARIAN" chamber with a base pressure 2×10^{-10} Torr. A $CaSi_2$ layer was grown by reactive deposition epitaxy (RDE) at 500 °C for all heterostructures. Si capping was carried out at 500 °C using solid phase epitaxy (SPE) with fixed thickness of about 50 nm.

Optical reflectance and transmittance spectra of the grown heterostructures were studied at room temperature on the Hitachi U-3010 spectrophotometer and Fourier spectrometer Bruker Vertex 80v in the photon energy range of 0.05–6.2 eV (with the integrated sphere at 1.5–6.2 eV).

The electronic band structures and optical properties of rhombohedral $CaSi_2$ under variable compression and stretching were determined by means of the first principles calculations utilizing total energy projector-augmented wave method within the generalized gradient approximation.

3. Results and discussion

The calculated band structure of $CaSi_2$ indicates the presence of some segments along the high symmetry directions of the hexagonal Brillouin zone, where the Fermi level does not cross any band pointing out semimetal properties. Since the corresponding band structures of $CaSi_2$ with fully relaxed and strained lattice parameters are very similar, we present only the case with the fully relaxed parameters (Fig. 1).

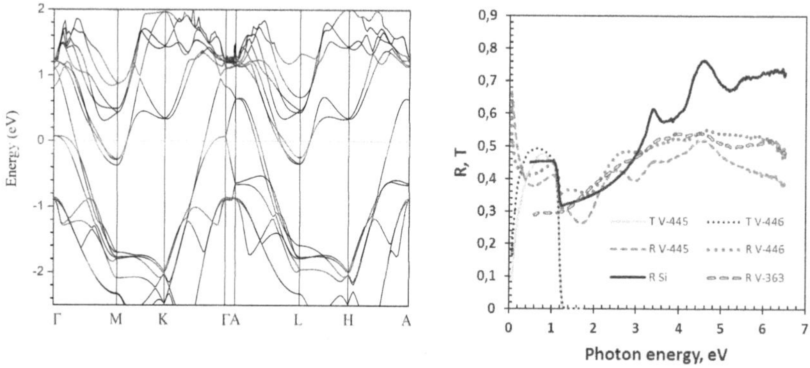

Figure 1. The left panel: the band structure of hR-6 $CaSi_2$. Zero at the energy scale corresponds to the Fermi level. The right panel: transmittance and reflectance spectra of $CaSi_2$/Si HS (V-363), Si/$CaSi_2$/Si DHS (V-445 and V-446) heterostructures and Si.

Transmittance and reflectance spectra of the two DHS with $CaSi_2$ thicknesses of 24 nm (V-446) and 50 nm (V-445) and the HS are presented in Fig. 1. In the transmittance spectra the strong increase of absorbance appears at photon energies smaller than 0.8 eV for both samples. In addition to this strong decrease of the transmittance, the reflectance minimum with the following increase in reflectance when decreasing the photon energies are observed as well. Such a behavior usually corresponds to the plasma reflectance on free carriers in a sample. Free carries can only appear in the embedded $CaSi_2$ layer in the grown DHS. The high energy reflectance peaks (2.5–2.6, 3.7–3.9 and

6.0–6.1 eV) with near equal energy positions for the HS and the DHS (Fig. 1) were observed.

However, some unusual features (local minimum (1.8–2.0 eV) and maximum (1.4–1.5 eV)) were determined in the reflectance spectra of the DHS. The observed decrease of the transmittance and the value of the low energy reflectance maximum (Fig. 1) are proportional to the thickness of the grown $CaSi_2$ film. So, we can say that the increase of the $CaSi_2$ film embedded in the silicon matrix leads to the increase of a free carrier concentration. However, if we assume only one type of free carriers it is impossible to fit experimental data. Thus, two types of free carriers with strongly different effective masses can be included in the model of plasma resonance in the grown DHS. Since the initially formed $CaSi_2$/Si bi-layer with total thickness (74–100 nm) in the DHS has sufficient transmittance in the energy range of 1.2–2.0 eV, the interference fringes were realized in the reflectance spectra (Fig. 1).

Calculations of the refraction functions for $CaSi_2$ layers in the HS and the DHS were performed from reflectance spectra using the Kramers-Kronig integral relations. In the DHS the system $CaSi_2$ layer/Si top layer was analyzed as a single layer with some effective parameters, whereas the $CaSi_2$ layer in the HS was treated as a single phase without Si contributions. Results of calculations are presented in Fig. 2a.

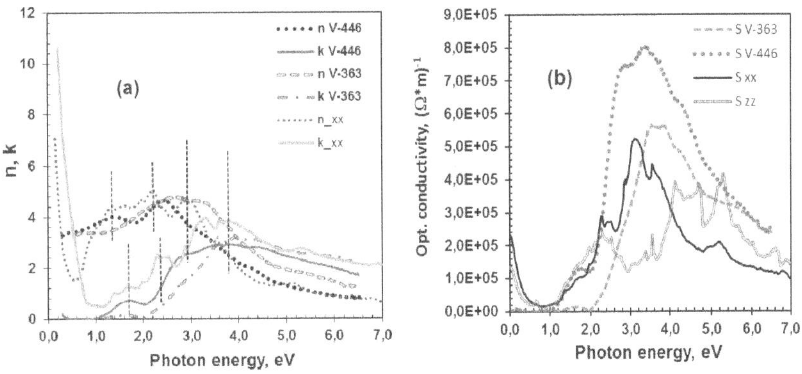

Figure 2. (a) – the real (n) and imaginary (k) parts of the refraction function for $CaSi_2$/Si HS (V-363), Si/$CaSi_2$/Si DHS (V-446) calculated from the Kramers-Kronig integral relations using reflectance spectra and theoretical n_xx and k_xx values (for the E||*a* light polarization) for hR-6 $CaSi_2$ bulk; (b) the optical conductivities $\sigma_{opt}(\omega) \sim 2\, n{\cdot}k{\cdot}\varepsilon_o{\cdot}\omega_o$ (where ε_o is the permittivity of vacuum, ω_o is the circle frequency) for both samples and theoretical values of $\sigma_{opt}(\omega)$ multiplied by 40 (for better comparison) for *E*||*a* and *E*||*c* light polarizations (xx and zz, respectively) for hR-6 $CaSi_2$ bulk.

Two energy bands are seen at 1.0–2.0 eV and 2.5–5.5 eV for the both samples. The maximum in refraction for the V-446 sample at 1.35 eV and the corresponding minimum at 2.0 eV apparently associated with an interference in the bi-layer. Therefore, the first band for the DHS is distorted by contributions from interference features. However for the $CaSi_2$ film with 114 nm thickness the interference fringes are absent (Fig. 1), while the refraction coefficient is slowly increased in the energy range of 1.0–2.0 eV (Fig. 2a) due to the presence of first interband transitions in the $CaSi_2$ film along with an existence of the small density of states at these energies.

Calculations of the absorption coefficient allow determination of the main direct transitions with higher density of states in the $CaSi_2$ film and in the DHS. They are 2.75 eV and 2.3 eV, accordingly. Transitions with smaller energies cannot be calculated due to small $CaSi_2$ layer thickness. In the conditions of preservation of the square momentum matrix element and parabolicity of band dispersion, the optical conductivity ($\sigma_{opt} = 2nk\varepsilon_0\omega$) is proportional to the state density function ($G(\omega)$) divided by the circle frequency (ω) in the energy bands. The optical conductivity (OC) for the HS and the DHS are presented in Fig. 2b indicating very small contributions in the first band at 1.4–2.0 eV (as we have suggested above) and major contributions at 2.1–6.5 eV. In the DHS an additional band is observed at 2.7–2.8 eV, but other peaks in OC have near the same energy positions. Thus, we could estimate contributions of Si capping layer in the DHS to be marginal. Calculations and analysis of the optical functions showed that the $CaSi_2$ layer embedded in the DHS with the distorted lattice parameters [7] is characterized as a degenerate semiconductor with two types of free carriers (at least) and small density of states up to 1.4–2.0 eV, while strong interband transitions start from 2.1–2.3 eV.

The results of first principles calculations in the case of the refraction functions have shown a good quantitative match with experimental optical spectra in the energy range of 1.0–6.5 eV and a complete mismatch at lower energies (Fig. 2a). The latter is attributed to the strong absorption due to very high concentration of free carriers in the theoretical model. Similar issues can be traced for the optical conductivity (Fig. 2b).

4. Conclusion

Experimentally determined optical functions for the $CaSi_2$ film and $CaSi_2$ layer embedded in the silicon matrix have shown the presence of degenerate semiconducting properties with strong absorbance at photon energies higher 2.3 eV and small contribution from the free carrier absorption in the low energy range. First principal calculations of hR-6 $CaSi_2$ bulk with and without lattice distortion have demonstrated the semimetal energy band structure with very high absorption due to free carriers at the low energy range (0–0.8 eV). However, a good match with experimental data has been observed at higher photon energies. This discrepancy at low photon energies can stem from the generalized gradient approximation used in our calculations and a hybrid functional is desirable to correctly account for the exchange-correlation interaction.

Acknowledgments

This work is supported by RFBR (grant 16-52-00074_Bel_a) and Belarusian Science Foundation for Fundamental Research (grant F16R-048).

References

1. P. Manfrinetti, M. L. Fornasini, A. Palenzona, *Intermetallics* **8**, 223 (2000).
2. O. Bisi, L. Braikovich, C. Carbone, I. Lindau, A. Iandelli, G. L. Olcese, A. Palenzona, *Phys. Rev. B* **40**, 10194 (1989).
3. M. Affronte, O. Laborde, G. L. Olsece, A. Palenzona, *J. Alloys Comp.* **274**, 68 (1998).
4. S. Lebegue, B. Arnaud, M. Alouani, *Phys. Rev. B* **72**, 085103 (2005).
5. S. A. Dotsenko, K. N. Galkin, D. A. Bezbabnyy, D. L. Goroshko, N. G. Galkin, *Physics Procedia* **23**, 41 (2012).
6. N. G. Galkin, D. A. Bezbabnyi, K. N. Galkin, S. A. Dotsenko, E. Zielony, R. Kudrawiec, J. Misiewicz, *Phys. Stat. Sol. (c)* **10**, 1819 (2013).
7. N. G. Galkin, S. A. Dotsenko, K. N. Galkin, L. Dózsa, I. Cora, B. Pécz, in: *Proceedings of SPIE* (*APAC-Silicide 2016*) (Fukuoka, Japan, 2016) (accepted in print).

NOVEL UP-CONVERSION IN OXYFLUORIDE GLASS-CERAMICS WITH (Eu,Tb,Yb):PbF_2 NANOCRYSTALS

G. E. Rachkovskaya, G. B. Zakharevich

Glass and Ceramics Technology Department, Belarusian State Technological University, Sverdlova 13a, 220006 Minsk, Belarus
rach_halina@mail.ru

E. V. Vilejshikova, K. V. Yumashev

Center for Optical Materials and Technologies, Belarusian National Technical University, Nezavisimosti Ave. 65/17, 220013 Minsk, Belarus
vilejshikova@gmail.com

Oxyfluoride glass-ceramics containing nanosized (5–8 nm) (Yb^{3+},Eu^{3+},Tb^{3+}):PbF_2 crystals was synthesized in the SiO_2–PbO–PbF_2–CdF_2 glass system. Under excitation at $\lambda = 960$ nm, the visible yellow-green cooperative up-conversion luminescence was observed. The prepared glass-ceramics is a promising material for yellow-green phosphors.

1. Introduction

Transparent oxyfluoride glass-ceramics (GCs) doped with trivalent rare-earth ions (RE^{3+}) are perspective low-phonon luminescent materials. Such GCs usually consist of two phases, one of which (nanocrystalline) is fluoride that preferentially incorporates RE^{3+} ions. The small size of these nanocrystals (<10 nm) is responsible for low scattering losses and good optical quality of such materials. Crystalline hosts containing Eu^{3+} and Tb^{3+} ions provide intense red and green luminescence due to electronic transitions from their metastable $^5D_0$ (Eu^{3+}) and $^5D_4$ (Tb^{3+}) states into lower-lying ^7F_J states (J = 0, …,6). The long-lived $^5D_0$ (Eu^{3+}) and $^5D_4$ (Tb^{3+}) states of these ions are isolated from ^7F_J states by the relatively large energy gap (>12000 cm^{-1}). For this reason, luminescence of Eu- and Tb-containing materials is characterized by a high quantum efficiency, which is responsible for their wide use as red and green phosphors and luminescent labels [1,2] in chemistry and biology [3]. In the present paper, we report on the synthesis and spectroscopic study of the novel oxyfluoride GCs triply doped with Yb^{3+}, Eu^{3+} and Tb^{3+} ions.

2. Synthesis and structure

Initial glass was synthesized by the conventional glass technology using the SiO_2–PbO–PbF_2–CdF_2 glass system and doped with YbF_3 (1.0 mol%), Eu_2O_3 (1.0 mol%), and Tb_2O_3 (1.0 mol%). The raw components were homogenized, mixed, and melted in corundum crucibles (25 ml) in an electric furnace at 900 ± 50 °C for 0.5 h. The fused glass was cast on a steal plate and annealed at 300 °C for 2 h. Then the glass samples were heat treated at temperatures close to the glass-transition temperature T_g in order to produce transparent nanosized GCs.

Phase transitions in this system were studied by differential scanning calorimetry (DSC) with a Netzsch DSC 404 F3 Pegasus calorimeter. The determined glass transition temperature T_g was 385 °C. The first crystallization peak was observed at T_p = 508 °C, with the thermal stability factor $\Delta = T_p - T_g$ of 123 °C. Thus, the GCs was produced at 400–420 °C for 5–45 h in air. X-ray powder diffraction analysis confirmed amorphous character of the glass. The reflections in X-ray diffraction patterns occurred after the secondary heat-treatment. The positions and relative intensities of them indicated the cubic phase of β-PbF_2 (JCPDS 06-0251) to be formed. According to the Scherrer formula, the average crystal size was 5–8 nm. Experimental samples for optical studies were transparent yellowish plates of ~2 mm thick.

3. Optical absorption

Fig. 1 shows absorption spectra of the initial oxyfluoride glass and the sample heat-treated at 400°C.

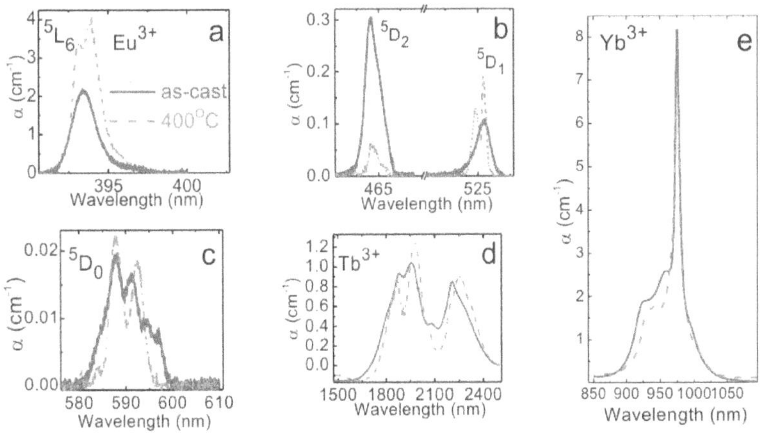

Figure 1. Absorption spectra of Eu^{3+} (a–c), Tb^{3+} (d) and Yb^{3+} (e) ions in glass and GCs: solid and dashed curves correspond to the as-cast glass and the GCs treated at 400 °C for 10 h, respectively.

The Eu^{3+} ion has the normal electronic configuration with unfilled $4f^6$ shell. Its electronic states are characterized by the low-energy ^7F_J manifold and the high-energy states ^5D_J, ^5H_J, ^5G_J, and ^5L_J. The Tb^{3+} ion in the normal electronic configuration [Xe]4f ^{8} conjugated with Eu^{3+} has the same system of electronic states, but with reversed fine structure (direction of change of *J*). In the short-wavelength region of the absorption spectra of the initial glass and GCs with Eu^{3+}, narrow bands belonging to transitions from the ground $^7F_0$ and thermally populated $^7F_1$ states to $^5D_0$ (Fig. 1c), $^5D_1$ and $^5D_2$ (Fig. 1b), and $^5L_6$ (Fig. 1a) states are observed. Relatively strong absorption of Eu^{3+} and Tb^{3+} due to transitions between ^7F_J states is observed in the near-IR region as a wide structured band spanning from 1700 to 2400 nm (Fig. 1d). The intense absorption band of Yb^{3+} ions spans in the 0.8–1.0 μm spectral range (Fig. 1e). At the wavelength of 960 nm, corresponding to the emission of InGaAs laser diodes, the absorption coefficient is ~2.6 cm^{-1}. A comparison of the spectra of the initial glass and GCs evidences that the heat treatment induces splitting of the absorption bands accompanied by intensity redistribution that confirms incorporation of RE^{3+} ions into the PbF_2 phase.

4. Up-conversion luminescence

Eu^{3+} and Tb^{3+} ions have no electronic states with energies close to the only excited state of Yb^{3+} ion (~10000 cm^{-1}). This prevents up-conversion processes via the sequential sensitization or direct energy transfer, $Yb^{3+} \rightarrow Eu^{3+}(Tb^{3+})$. A possible up-conversion mechanism in such materials is the cooperative energy transfer from excited pair Yb^{3+}-Yb^{3+} to single Eu^{3+} or Tb^{3+} ions (Fig. 2a). High local concentration of Yb^{3+} is necessary for the electrostatic coupling of Yb^{3+}–Yb^{3+} pairs and, consequently, for the cooperative up-conversion. The volume crystallization of the RE^{3+}:PbF_2 results in the local concentration of Yb^{3+} ions to be high enough for the cooperative up-conversion process.

In the experiment, up-conversion luminescence (UCL) was excited by a commercial InGaAs diode at 960 nm ($^2F_{7/2} \rightarrow {}^2F_{5/2}$ transition of Yb^{3+} ions). The Eu^{3+} UCL is observed as the bands at 590 nm ($^5D_0 \rightarrow {}^7F_1$), 612 nm ($^5D_0 \rightarrow {}^7F_2$), and 700 nm ($^5D_0 \rightarrow {}^7F_4$) (Fig. 2b). The intense bands at ~488 and 543 nm are related to transitions $^5D_4 \rightarrow {}^7F_6$ and $^5D_4 \rightarrow {}^7F_5$ of the Tb^{3+} ions, respectively. Structuring and narrowing of emission bands of all RE ions, along with intensity redistribution for the bands at ~612 nm and ~590 nm, are related to the electric-dipole (ED) transition of Eu^{3+}, $^5D_0 \rightarrow {}^7F_2$. It is hyper-sensitive to local field symmetry, while the magnetic dipole (MD) transition, $^5D_0 \rightarrow {}^7F_1$, is insensitive to the symmetry. These facts indicate the entering of rare-earth ions RE^{3+} in the GCs under study into the cubic PbF_2 phase.

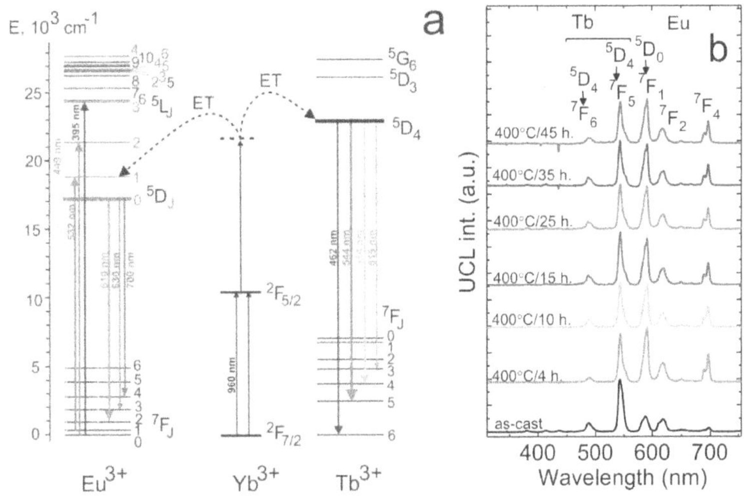

Figure 2. (a) cooperative up-conversion mechanisms; (b) UCL spectra of the glass and GCs treated at 400 °C.

The intensity redistribution changes the color of UCL. For the initial glass CIE coordinates are $x = 0.383$ and $y = 0.558$ (yellow green), and for the sample heat-treated at 400 °C they are $x = 0.448$ and $y = 0.501$ (greenish yellow). The high value of color purity parameter (p~80–100 %) makes the synthesized transparent GCs to be good candidates for efficient near-IR-to-visible converters.

5. Conclusion

The oxyfluoride GCs containing PbF_2 nanocrystals triply doped with Yb^{3+}, Eu^{3+} and Tb^{3+} ions was synthesized by the conventional melt-quenching technique. Under the excitation at 960 nm, visible yellowish green UCL was observed. The studied transparent GCs are promising materials as up-conversion phosphors for the wide applications.

References

1. J. Dhanaraj, R. Jagannathan, T. R. N. Kutty, C. H. Lu, *J. Phys. Chem. B* **105**, 11098 (2001).
2. R. P. Rao, *J. Electrochem. Soc.* **150**, H165 (2003).
3. K. Binnemans, C. Gorller-Walrand, *J. Rare Earths* **14**, 173 (1996).

BORON CARBIDE PARTICLE SIZE EFFECTS ON THERMAL NEUTRON ATTENUATION BEHAVIOUR OF BORON CARBIDE-TITANIUM DIBORIDE COMPOSITES

B. Buyuk, A. B. Tugrul, E. Demir
Nuclear Researches Department, Istanbul Technical University, Energy Institute
34469 Istanbul, Turkey
buyukbu@itu.edu.tr

S. Aktop, A. O. Addemir
Metallurgical and Materials Engineering Faculty, Istanbul Technical University
34469 Istanbul, Turkey

Boron carbide-titanium diboride composites were fabricated by hot pressing and reaction sintering methods. Micro- and nano-sized boron carbide particles were used to produce these composites. Thermal neutron attenuation behaviors of the composites were investigated. A decrease of the particle size was found to increase neutron attenuation.

1. Introduction

Boron carbide (B_4C) is one of the essential materials for nuclear technology. It has high hardness and corrosion resistance, chemical stability and high neutron cross section [1–3]. On the other hand, it is hard to reach high dense pure boron carbide materials without additives. Therefore some additives such as TiB_2, AlF_3, W_2B_5 and carbon were used to get high dense composites [4,5]. In this study, boron carbide-titanium diboride composites which were produced from different boron carbide particle sizes were investigated on thermal neutron attenuation.

2. Experimental procedure and materials

Neutron transmission technique was used in the experiments. A Pu-Be Neutron Howitzer which had 10^5 n/cm^{2}s neutron flux was used as a neutron source. There were two channels of Howitzer: thermal and epithermal ones. Beer Lambert's equation was used to determine the total macroscopic cross-section of the composites for thermal neutrons in the narrow beam geometry [4]:

$$I = I_0 e^{-\Sigma_{tot} x}, \tag{1}$$

where I and I_0 represent transmitted and incident thermal neutron counts, respectively. Σ_{tot} is the total macroscopic cross section of the material and x is the thickness of the sample. In the experiments PM1401K model He-3 neutron detector was used to measure neutron counts. Finally the relative count ((I/I_0)-material thickness graphs were drawn and fitted exponentially to carry out total macroscopic cross-sections.

Boron carbide-titanium diboride (B_4C-TiB_2) composites were produced by hot pressing and reaction sintering methods. Average particle sizes of boron carbide and titanium diboride particles were 3.997 μm and 4.365 μm, respectively. The boron carbide particles were milled in Spex 8000 mill for 30 min with WC balls. The average particle size of milled boron carbide was 400 nm. Particle size distributions of unmilled and milled boron carbide are given in Fig. 1. All composite materials were sintered at 2250 °C for 1 h under 50 MPa pressure.

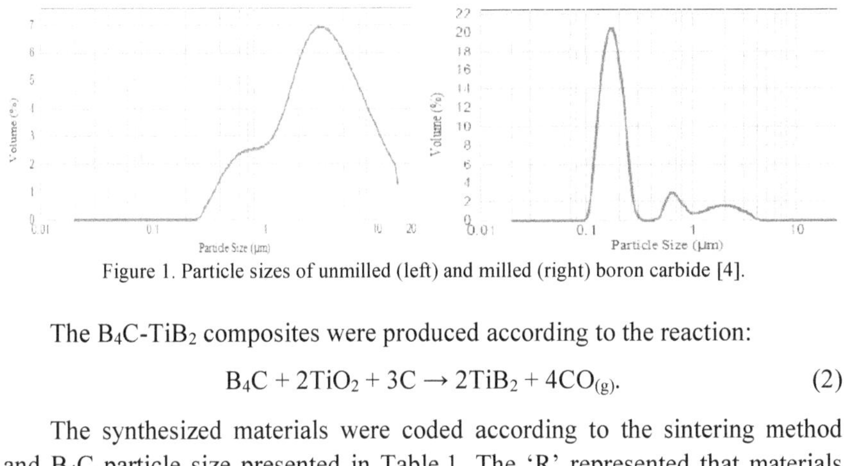

Figure 1. Particle sizes of unmilled (left) and milled (right) boron carbide [4].

The B_4C-TiB_2 composites were produced according to the reaction:

$$B_4C + 2TiO_2 + 3C \rightarrow 2TiB_2 + 4CO_{(g)}. \quad (2)$$

The synthesized materials were coded according to the sintering method and B_4C particle size presented in Table 1. The 'R' represented that materials were produced by reaction sintering method.

Table 1. The volumetric percentages of components in B_4C-TiB_2 composites.

Material (code)	B_4C (3.997 μm)	B_4C (3.997 nm)	TiB_2	TiO_2	C
8515_m	85	-	15	-	-
8515_n_m	42.5	42.5	15	-	-
8515_n	-	85	15	-	-
R8515_m	77	-	-	15	8
R8515_n_m	38.5	38.5	-	15	8
R8515_n	-	77	-	15	8

3. Experimental results and discussion

The relative count- material thickness values of the boron carbide-titanium diboride composites for thermal neutron were given in Table 2.

Table 2. The relative count-thickness values of B_4C-TiB_2 composites for thermal neutrons.

Material (code)	Material thickness (cm)	Relative count (I/I_0)	Material (code)	Material thickness (cm)	Relative count (I/I_0)
	0	1		0	1
8515_m	0.4755	0.8571	R8515_m	0.5071	0.8447
	0.9311	0.7687		1.0118	0.7437
	1.384	0.6799		1.517	0.6255
	0	1		0	1
8515_n_m	0.4583	0.8113	R8515_n_m	0.4666	0.8571
	0.9094	0.7478		0.9158	0.7560
	1.3877	0.648		1.3670	0.6484
	0	1		0	1
8515_n	0.4486	0.8036	R8515_n	0.4817	0.8578
	0.8973	0.7292		0.9691	0.7366
	1.7972	0.5779		1.4547	0.6252

The relative count-material thickness graphs are given in Fig. 2.

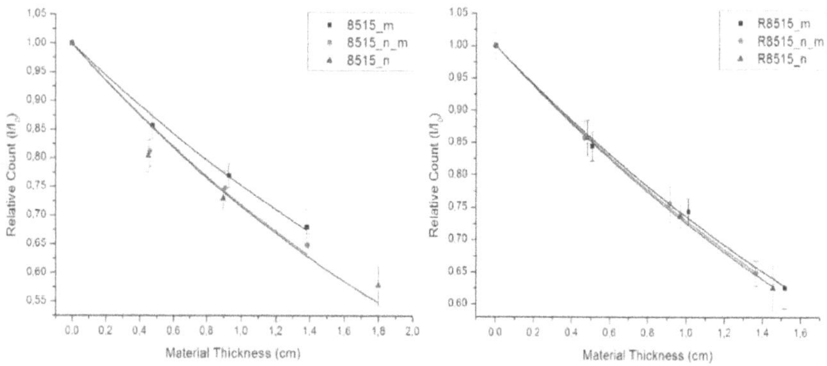

Figure 2. Thermal neutron transmission graphs for hot pressed (left) and reaction sintered (right) B_4C-TiB_2 composites.

The 8515_n boron carbide-titanium diboride composite has better thermal neutron attenuation coefficient than 8515_m and 8515_n_m ones. In addition the R8515_n sample has bigger thermal neutron attenuation property than R8515_m and R8515_n_m.

The total macroscopic cross-sections of the boron carbide-titanium diboride composites are summarized in Table 3.

Table 3. The total macroscopic cross-sections of the B_4C-TiB_2 composites.

Material (code)	Total macroscopic cross-section, cm^{-1}	Standard deviation	Material (code)	Total macroscopic cross-section, cm^{-1}	Standard deviation
8515_m	0.285	0.008	R8515_m	0.306	0.007
8515_n_m	0.331	0.025	R8515_n_m	0.315	0.004
8515_n	0.335	0.030	R8515_n	0.320	0.002

Both, hot pressed and reaction sintered B_4C-TiB_2 composites have thermal attenuation getting higher by using smaller B_4C particles. Decreasing B_4C particle size results in a higher material density and subsequently in a higher thermal neutron absorption unit volume. There was no significant difference between hot pressed and reaction sintered boron carbides by the view of thermal neutron attenuation.

4. Conclusion

Boron carbide-titanium diboride composites were produced by using boron carbide and titanium diboride where the second was produced by reaction sintering technique. In addition micro- and nano-sized boron carbide particles were used to produce composites. Decreasing particle size from micro- to nanoscale increases the thermal neutron attenuation coefficients due to reaching higher material density. In addition the thermal neutron attenuation coefficients of hot pressed and reaction sintered boron carbide-titanium boride composites have similar values. The use of nanoscale boron carbide particles increases the total macroscopic cross-sections providing better thermal neutron attenuation properties.

References

1. F. Thevenot, *J. Europ. Ceram. Soc.* **6**, 205 (1990).
2. B. Buyuk, A. B. Tugrul, *Rad. Physics Chem.* **97**, 354 (2014).
3. B. Buyuk, A. B. Tugrul, *Ann. Nucl. Energy* **71**, 46 (2014).
4. B. Buyuk *et al.*, *Acta Physica Polonica A* **123**, 177 (2013).
5. H. O. Pierson, *Handbook of Refractory Carbides and Nitrides* (Noyes, Publications, NJ, 1996).

THE ADVANCED ANALYSIS OF SLOW NON-STRETCHED EXPONENTIAL DECAY KINETICS FROM Si NANOCRYSTALS

M. Greben, J. Valenta

Department of Chemical Physics & Optics, Faculty of Mathematics & Physics
Charles University, Prague, CZ-12116, Czechia
leibnits@gmail.com

P. Khoroshyy

Institute of Organic Chemistry and Biochemistry, Czech Academy of Sciences
Flemingovo namesti 2, 16000 Prague 6, Czechia
khoroshyy@gmail.com

The procedure of reliable extraction of average decay lifetimes was demonstrated with an example of slow non-stretched exponential photoluminescence kinetics from multilayers of Si nanocrystals and three different fitting models, namely: 1) multi-exponential, 2) combination of mono- and stretched exponential as well as 3) lognormal distribution of decay rates. All three decay models gave the same average decay lifetime 68 ± 1 μs.

1. Introduction

The stretched exponential (SE) line shape became classical footprint of silicon nanocrystals (Si NCs) that demonstrate slow non-exponential decay:

$$I(t) = I(0)e^{-(\frac{t}{\tau_0})^{\beta}} + background \tag{1}$$

where τ_0 is the lifetime parameter and β is the dispersion factor. This decay law is relatively simple and describes quite well the photoluminescence (PL) decays of colloidal Si NCs [1] or porous silicon. Unfortunately, the decay kinetics of Si NCs in a solid matrix (for instance, SiO_2) do not always reveal SE behavior [2]. An exact procedure to analyze such complex decays has not been developed yet.

2. Results and discussion

Recently we have discussed [3] special requirements that restrict the experimental measurements of slow PL kinetics in μs to ms scale namely: low excitation power, low pulse repetition rate, and long excitation pulses. Those conditions ensure that NCs are not overexcited, PL decays are not truncated and a signal to noise (S/N) ratio is high enough. However, the above given points make any experiment of great challenge and in some cases a curve truncation becomes unavoidable.

As an example in Fig. 1 we present PL decay of Si NC multilayers in high quality PE-CVD sample [2]. The SE fit is not ideal (Fig. 1a) and it might be noticed a characteristic S-shape of fit residuals (Fig. 1b). Keeping into account slight truncation of the curve one could expect even more pronounced disagreement of SE fit in case of longer curve. Therefore SE fit cannot be used for the reliable analysis of such decay. Instead, we applied a set of other fitting models and compared the results (Fig. 1c,d).

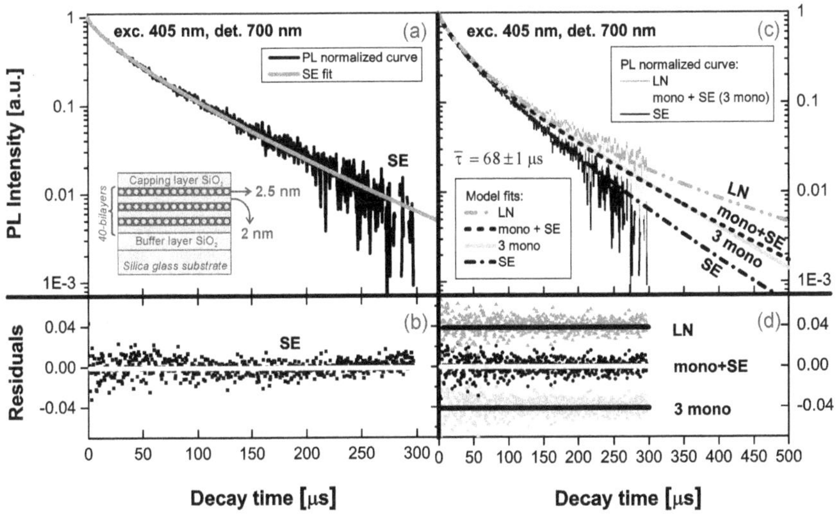

Figure 1. Normalized PL decay curves (a, c) and stretched exponential (SE) fits (a) or fits by different other models (c): combination of three mono-exponentials (3 mono), combination of mono- and stretched-exponentials (mono+SE), log-normal (LN) distribution of rates. (b, d) residuals to corresponding fits.

The most straight-forward approach is to try the multi-exponential (ME) fit ($N = 3$ in our case):

$$I(t) = \sum_{i=1}^{N} A_i e^{-(\frac{t}{\tau_i})} + background \tag{2}$$

where A_i and τ_i are the amplitude and lifetime parameters, accordingly.

Another approach is based on fitting by combination of mono- and SE (MSE) [4]:

$$I(t) = A_i e^{-(\frac{t}{\tau_1})} + A_2 e^{-(\frac{t}{\tau_2})^{\beta}} + background \tag{3}$$

Finally the model proposed by van Driel *et al.* [5] was applied, which imply the lognormal (LN) distribution of decay rates:

$$I(t) = I(0)\int_0^\infty f(k)e^{-(kt)}dk + background \tag{4}$$

where $f(k)$ is LN distribution of the decay rate k. It is given by:

$$f(k) = Ae^{-\left(\frac{\ln(k/k_m)}{\sinh^{-1}(k_d/2k_m)}\right)^2} \tag{5}$$

where k_m and k_d are the fitting parameters.

The results of fitting as well as fit residuals of experimental PL decay curves by the above mentioned models are presented in Fig. 1c,d. The normalized PL curve is varying in dependence on a chosen model (Fig. 1c) due to the model-dependent fit amplitude $I(0)$ and background:

$$I_{norm}(t) = (I(t) - background)/(I(0) - background) \tag{6}$$

All models fit the curve pretty well though the resulting fits are different. This difference provides an evidence to the fact that a good fit does not necessarily mean that a chosen model is correct and unique. Obviously, if one would apply an inverse Laplace transform to different fitting models quite different lifetime distributions will be obtained. Hence, the detailed analysis of a lifetime distribution is meaningless. Therefore we should focus on extraction of average PL decay lifetimes.

The average lifetimes of SE (Eq. (7)), ME (Eq. (8)), MSE (Eq. (9)) and LN (Eq. (10)) were calculated by the following equations:

$$\bar{\tau}_{SE} = \Gamma(\frac{2}{\beta})/\Gamma(\frac{1}{\beta})\tau_0 \tag{7}$$

$$\bar{\tau}_{ME} = \sum_{i=0}^{3} A_i\tau_i^2 / \sum_{i=0}^{3} A_i\tau_i \tag{8}$$

$$\bar{\tau}_{MSE} = (A_1\tau_1^2 + A_2\frac{\tau_2^2}{\beta}\Gamma(\frac{2}{\beta}))/(A_1\tau_1 + A_2\frac{\tau_2}{\beta}\Gamma(\frac{1}{\beta})) \tag{9}$$

$$\frac{1}{\bar{\tau}_{LN}} = k_m \exp(\frac{3}{4}(\sinh^{-1}(k_d/2k_m))^2) \tag{10}$$

The obtained average PL decay lifetime by ME, MSE and LN methods is $\bar{\tau} = 68 \pm 1$ μs. In contrast, the average decay lifetime obtained by SE model is expectedly different and equals to $\bar{\tau}_{SE} = 58$ μs. The average lifetime of our

Si NCs in SiO_2 is roughly 1.5 times shorter than the average lifetime of Si NCs in colloidal suspensions. This indicates that non-radiative processes are more pronounced in solid samples.

3. Conclusion

The procedure of reliable extraction of average decay lifetimes was demonstrated with an example of slow non-stretched exponential PL kinetics with three different fitting models. Though the lifetime distributions are different for the applied models, the average lifetime is the same for all of them. It demonstrates that the average decay lifetime is model-free and could be reliably determined even in the case of slightly truncated curves. On the other hand, we should note that the results are highly disputable in the case of good fits by a set of models and different average lifetimes until the true decay model is known.

In addition, the shorter is the curve (truncation), the higher is the probability of an successful fit by a model though the subsequently extracted average decay lifetime is not necessarily correct. By applying different fit models one could confirm the correctness of extracted average values.

Acknowledgments

The authors acknowledge S. Gutsch, D. Hiller, and M. Zacharias (IMTEK, Freiburg University, Germany) for providing high quality PE-CVD samples. We would like to thank also to A. Meldrum (University of Alberta, Canada) for the topic discussion.

References

1. F. Sangghaleh *et al.*, *ACS Nano* **9**, 7097 (2015).
2. J. Valenta *et al.*, *Appl. Phys. Lett.* **108**, 023102 (2016).
3. M. Greben, J. Valenta, *Rev. Sci. Instr.* **87**, 126101 (2016).
4. J. Linnros *et al.*, *J. Appl. Phys.* **86**, 6128 (1999).
5. A. F. van Driel *et al.*, *Phys. Rev. B* **75**, 035329 (2007).

TEMPERATURE EFFECTS AND EXCITON-PHONON COUPLING IN SEMICONDUCTOR CdSe/ZnS QUANTUM DOTS

E. Zenkevich, T. Kononova
Department of Information Technologies and Robotics, Belarusian National Technical University, 220013 Minsk, Belarus
zenkev@tut.by

A. Stupak
B. I. Stepanov Institute of Physics NASB, 220072 Minsk, Belarus
stupak@imaph.bas-net.by

C. von Borczyskowski
Institute of Physics, Chemnitz University of Technology, 09107 Chemnitz, Germany
borczyskowski@physik.tu-chemnitz.de

Experimental results on temperature dependent absorption and photoluminescence (PL) properties for TOPO- and AM-capped CdSe/ZnS quantum dots (QDs) in methylcyclohexane/toluene mixture (6:1) in the temperature range of 77–293 K were analyzed using existing models for exciton phonon coupling in semiconductor nanostructures. Based on a combination of PL and absorption data the temperature dependence of the Stokes shift has been evaluated, which in turn reveals temperature dependent Huang-Rhys factors. The analysis of our findings shows that absorption is more sensitive to core CdSe LO optical phonons while PL also contains coupling to ZnS phonons of the shell. It means also that QD PL emission is to a large extent due to QD surface states.

1. Introduction

One long standing question in the physics of semiconductor CdSe/ZnS QDs is the effect of quantum confinement on the strength of exciton-phonon coupling and phonon frequency [1,2]. The strength of exciton-phonon coupling bears relevance to several fundamental issues of nanoscale physics: exciton relaxation dynamics, homogeneous line widths, and electronic structure. The direct optical access to exciton-phonon coupling is based on two types of experimental observations: i) the intensity ratios of the spectrally resolved pure (phonon-less) electronic transition intensities and phonon side bands, and ii) overall optical transition intensities, contributions to the homogeneous spectral line widths, PL decay times and energies of absorption and/or PL emission together with experimentally determined Stokes shifts. Here, we discuss temperature dependent exciton-phonon coupling in CdSe/ZnS QDs capped by trioctylphosphine oxide (TOPO) or amine (AM) ligands and embedded into a glass forming liquid (methylcyclo-hexane/toluene mixture 6:1 in the temperature range of 77–293 K). The analysis is based on experimental absorption and PL data for CdSe/ZnS QDs.

2. Results and discussion

Notably, while in most of reported experiments PL data (energy; band width, or Stokes shifts) have been used to determine electron-phonon coupling parameters we used in parallel temperature dependent absorption data which provided in combination with PL emission the identification of the temperature dependence of the Stokes shift, which in turn reveals temperature dependent Huang-Rhys factors. Temperature dependent data from absorption are slightly but definitely different from those obtained by PL (Fig. 1). We explained this finding by the analysis that absorption is more sensitive to core CdSe LO optical phonons while PL also contains coupling to ZnS phonons of the shell. This supports the earlier assumption [3] that PL emission is to a large extent due to QD surface states. From the temperature dependent Stokes shift we found that exciton-phonon coupling below the phase transition temperature T_{crit} for TOPO capping layer is governed more efficiently by coupling to ZnS phonons, while above coupling to CdSe increases. This finding is supported by the observation that the phase transition (size dependent) appears at –10 K lower temperatures in absorption (predominant excitation of small QDs) as compared to the one detected in PL from dominantly surface states.

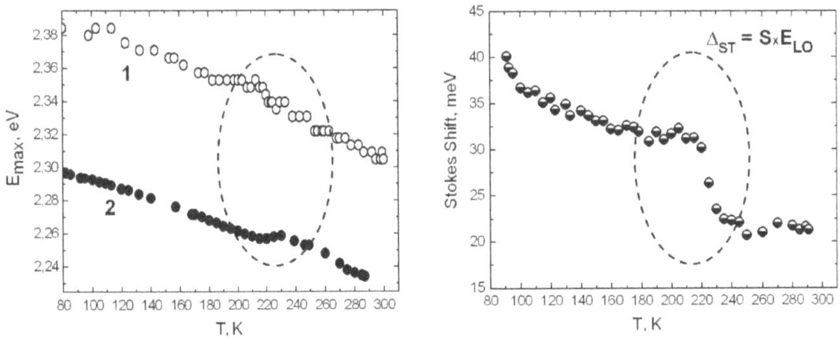

Figure 1. Stokes shift analysis for TOPO-capped CdSe/ZnS QDs in the solvent mixture (methylcyclohexane: toluene = 6:1). Left: Comparison of temperature dependence of (1) the first absorption peak (mean weighted data for two Gaussians lines) and (2) PL maximum at λ_{exc}=450 nm. Right: Temperature dependent Stokes shift being obtained between the PL band fitted by 1 Gaussian component and lowest absorption band. Ellipses show the range of "phase transition" of the capping TOPO layer.

With respect to methodological aspects we state, that determination of exciton-phonon coupling via Stokes shifts needs a temperature dependent investigation to reduce (i) contributions from the thermalisation of the excitonic fine structure levels, (ii) influence of differently contributing near-band-edge states and shallow traps and (iii) identification of the "true" first absorption band. Neglecting these influences will result in too large Stokes shifts and exciton-phonon coupling constants.

Experimental temperature dependences of the absorption and PL data for TOPO-capped CdSe/ZnS QDs have been modeled by the Debye (1) or Einstein (2) models for electron-phonon coupling according to

$$E_g(T) = E_g(0) - \alpha T^2 / (T + \theta_D), \quad (1)$$

where $E_g(0)$ is the band gap at 0 K, α is a constant and θ_D is the Debye temperature, and

$$E_g(T) = E_g(0) - k / [\exp(\theta_E / T) - 1], \quad (2)$$

where k is a constant, and θ_E is the Einstein temperature. Some results are presented in Table 1.

Table 1. Temperature dependent absorption and PL data for CdSe/ZnS (TOPO) according to Einstein model.

QD/diameter	CdSe/ZnS 3 nm	CdSe/ZnS 3 nm	CdSe/ZnS 3 nm	CdSe/ZnS 3 nm
Parameter	PL total	PL LT	Abs LT	Abs HT
E_0, eV	2.386	2.305 (2)	2.378 (7)	2.378 (4)
κ, meV	236	63 (10)	61 (39)	
Θ_E, K	207	180 (23)	158 (86)	165 (7)

Notes: LT and HT correspond to fits across the low and high temperature range, respectively, while "total" refers to the total temperature range neglecting effects at the phase transition temperature. Errors of the last digit are given in brackets. Excitation at 450 nm.

The obtained results are consistent with the corresponding data known for bulk CdSe, where $\theta_D \sim 181$–315 K and $\alpha \sim (2.8$–$4.1)\times10^{-4}$ eV/K. This means that the energy shift upon temperature lowering is due to the temperature-dependent phonon induced band-gap shrinkage of the CdSe core material.

Typically, exciton-phonon coupling adds to dephasing processes resulting in homogeneous line width broadening. In case that the intrinsic (temperature dependent) inhomogeneous broadening of the full width at half-maximum

(FWHM) is comparable to these homogeneous contributions; exciton-phonon coupling has been deduced from experimental temperature dependent PL as

$$\mathrm{FWHM}(T) = \Gamma(T) = \Gamma_{inh} + \sigma T + \Gamma_{LO}/\,[\exp(E_{LO}/k_B T) - 1]. \tag{3}$$

Γ_{inh} is the inhomogeneous temperature-independent broadening due to fluctuations in size, shape, composition, Γ_{LO} is the exciton-LO-phonon coupling strength and E_{LO} is LO-phonon energy. A linear coupling to temperature follows a coupling constant σ. Finally, assuming a temperature-independent radiative lifetime due to the strong confinement regime and the temperature dependent PL decrease as a superposition of the temperature activated increase of non-radiative channels, we have analyzed the temperature dependence of the QD PL intensity using the following approach [3]

$$I_{PL}(T) = \frac{I_0}{1 + a\left(e^{-E_a/k_B T}\right) + b\left(e^{E_{LO}/k_B T} - 1\right)^{-m}}, \tag{4}$$

where I_0 is the PL intensity at T=0, E_a is the activation energy for excitation of non-radiative traps, E_{LO} is the energy of either CdSe or ZnS phonons, and m is the number of phonons needed to populate the next higher excitonic QD state which promotes exciton-phonon induced radiationless transitions; a and b correspond to the relative contribution of these two processes.

In conclusion, the investigation of temperature dependent QD PL intensities and FWHM values reveals that exciton-phonon coupling is not the only mechanism to cause changes in optical properties. As discussed with respect to Stokes shifted data these additional relaxation processes might also influence PL energies and line widths. The activation energies are in a range of trap state distributions as determined very recently [3]. Most interesting, however, is the finding that contrary to results reported in literature, temperature activated coupling to optical phonons proceeds not via excitation to the second excitonic state but to a recently identified hole state which has often been overlooked in several reported investigations.

References

1. T. J. Liptay, L. F. Marshall, P. S. Rao, R. J. Ram, M. G. Bawendi, *Phys. Rev.* B **76**, 155314 (2007).
2. C. Lin, K. Gong, D. F. Kelley, A. M. Kelley, *ACS Nano* **9**, 8131 (2015).
3. E. Zenkevich, A. Stupak, C. von Borczyskowski, in: *Tuning Semiconducting and Metallic Quantum Dots: Spectroscopy and Dynamics,* eds. C. von Borczyskowski and E. Zenkevich (Pan Stanford Publishers, Singapore, 2016), pp. 103–135.

CONTROL OF THE POLARIZATION STATE OF LIGHT BEAMS BY ANODIC ALUMINUM OXIDE NANOPOROUS FILMS

V. A. Dlugunovich, A. Yu. Zhumar
B. I. Stepanov Institute of Physics NASB, Nezavisimosti Ave. 68, 220072 Minsk, Belarus
a.zhumar@dragon.bas-net.by

N. I. Mukhurov, I. V. Gasenkova
State Research and Production Association
"Optic, Optoelectronic and Laser Technique"
Nezavisimosti Ave. 68, 220072 Minsk, Belarus
n.mukhurov@ifanbel.bas-net.by

Stokes polarimetry and coherence matrix methods were used to study the form-anisotropy birefringence phenomenon arising during transmission of the light beam through the anodic aluminum oxide nanoporous films with a view to creating on their basis achromatic retardation plates with a variable phase difference from 0 to $\lambda/2$.

1. Introduction

Anodic aluminum oxide (AAO) nanoporous films fabricated by electrochemical oxidation of aluminum in acidic electrolytes have great potential to be used as various optical devices [1]. The pores in the films are oriented perpendicular to the sample surface, therefore during irradiation at angles differing from normal incidence a birefringence caused by the form-anisotropy is observed [2]. It is determined by the difference in the refractive indices of the material for the light polarized along and perpendicular to the pores [3–5]. Radiation scattering on the structural inhomogeneities is also occurred [4]. However, there is a depolarization of the transmitted radiation, as well as the amplitude anisotropy of the material previously not considered at the estimation of the phase anisotropy of nanoporous materials.

The purpose of this work is to study with Stokes polarimetry and coherence matrix methods the form-anisotropy birefringence phenomenon arising during transmission of the light beam through the AAO nanoporous films with a view to creating on their basis achromatic retardation plates with a variable phase difference from 0 to $\lambda/2$.

2. Experimental procedure and data processing

AAO nanoporous films were fabricated in the potentiostatic mode at 40 V and 10 C in 3 % oxalic acid electrolyte by using a two-stage procedure at different etching intervals [6].

Stokes polarimeter based on liquid-crystal plates with a variable phase retardation was used to study phase and amplitude anisotropy introduced by the investigated films and depolarization of the transmitted radiation [7].

The radiation reflected (scattered) or transmitted through a material can be represented as an aggregate incoherent beam described by four parameters forming a four-dimensional Stokes vector parameter **S** [8,9], which is written in the row $\{I, P_1, P_2, P_3\}$. All Stokes parameters I, P_1, P_2, P_3 have the dimension of radiation intensity and fully characterize the polarization state of the radiant flux. The parameter I characterizes here the complete intensity of the radiation. The remaining vector components represent: P_1 is the parameter of the predominantly horizontal polarization, P_2 is the parameter of the predominant polarization at the angle of 45 deg, P_3 is the parameter of the predominantly right-circular polarization.

Along with the Stokes parameters the polarized radiation is described by using other polarization characteristics such as a degree of polarization P and an angle of ellipticity ε of radiation which are related with the Stokes parameters through the expressions [8,9]:

$$P = \sqrt{P_1^2 + P_2^2 + P_3^2} / I, \ \sin 2\varepsilon = P_3/IP. \tag{1}$$

The most universal method to describe polarization properties of both partially coherent and partially polarized light involves the use of a coherence matrix **J**. Using the coherence matrix and Stokes parameters for the description of the partially polarized radiation transmitted by the investigated amplitude- and phase-anisotropic AAO nanoporous films one can obtain an expression that relates the phase difference Δ between the orthogonally polarized components of the transmitted radiation and its angle of ellipticity ε [9]:

$$\sin \Delta = \frac{\sqrt{(1+qP\cos 2\varphi)^2 - (1-q^2)(1-P^2)}}{\sqrt{1-q^2}\, P \sin 2\varphi} \sin 2\varepsilon, \tag{2}$$

where $q = (\tau_x - \tau_y)/(\tau_x + \tau_y)$ is the polarizance (diattenuation [8]) of the object; τ_x and τ_y are the transmittance of the investigated films for orthogonal polarized components of the radiation; φ is the azimuth of the polarization of the incident radiation.

The refractive indexes n_e and n_o of the investigated films for radiation with different wavelengths polarized along and perpendicular to the pores, respectively, were estimated with the least square method for sampling of Δ values determined experimentally.

For determining the porosity of the films the effective medium theory, in particular the Maxwell-Garnett model for material phase anisotropy, was used [10,11]. Calculations of the porosity of the films was performed for each of the four wavelengths of the incident radiation (400, 633, 808 and 980 nm) and then the average porosity was calculated.

3. Experimental results

It was established that the porosity of the investigated films was between 0.100±0.010 and 0.225±0.017. The pore size was varied from 20.0±0.5 to 30.7±1.2 nm. Other parameters are summarized in Table 1.

Table 1. Porosity, transmittance T of the AAO nanoporous films, degree of polarization of the transmitted radiation P for incidence angles α, at which the phase difference Δ between the orthogonally polarized components of the transmitted radiation, correspond to $\lambda/4$ or $\lambda/2$.

Porosity	$\alpha(\Delta = \lambda/4)$, deg.	$T(\Delta = \lambda/4)$	$P(\Delta = \lambda/4)$	$\alpha(\Delta = \lambda/2)$, deg.	$T(\Delta = \lambda/2)$	$P(\Delta = \lambda/2)$
			400 nm			
0	60	0.70	0.94			
0.1±0.01	35	0.62	0.98	56	0.52	0.99
0.143±0.012	22.5	0.60	0.96	33	0.55	0.97
0.195±0.012	20	0.55	0.95	29	0.50	0.96
0.225±0.017	17.5	0.28	0.94	26	0.30	0.92
			633 nm			
0	65	0.07	0.99			
0.1±0.01	43	0.80	0.98			
0.143±0.012	26	0.84	0.97	40.5	0.81	0.97
0.195±0.012	24	0.84	0.98	35	0.83	0.96
0.225±0.017	23	0.68	0.97	34	0.65	0.98
			980 nm			
0.225±0.017	30	0.95	0.99	42	0.9	0.98
0.195±0.012	34	0.96	0.97	49	0.88	0.96
0.143±0.012	35	0.92	0.97	54	0.75	0.96

4. Conclusion

It is shown that AAO nanoporous films can create miniature devices for control a polarization state of light beams, including quarter-wave or half-wave plates with a high transmittance and degree of polarization of the transmitted radiation in the spectral range from 0.4 to 1.0 μm. The transition from one polarization state to another is realized by varying the incidence angle of radiation.

References

1. D. Losic, A. Santos (eds.), *Nanoporous Alumina: Fabrication, Structure, Properties and Applications* (Springer Series in Materials Science, **219**, 2015).
2. M. Born, E. Wolf, *Principles of Optics* (Pergamon Press Oxford, London, 1970).
3. V. A. Dlugunovich, A. Yu. Zhumar, S. N. Kurilkina, N. I. Mukhurov, *J. Appl. Spectroscopy* **82**, 824 (2015).
4. A. A. Lutich, I. S. Molchan, N. V. Gaponenko, S. V. Gaponenko, *Proc. SPIE* **6258**, 62580J (2006).
5. L. A. Golovan, V. Yu. Timoshenko, P. K. Kashkarov, *Phys. Usp.* **50**, 595 (2007).
6. N. I. Mukhurov, I. V. Gasenkova, I. M. Andrukhovich, *J. Mater Sci. Nanotechnol* **1**, 110 (2014).
7. V. N. Snopko, *Measurement Techniques* **51**, 1284 (2008).
8. D. H. Goldstein, *Polarized Light* (CRC Press, New York, 2011).
9. V. N. Snopko, *Polarizing Characteristics of Optical Radiation* (Navuka i Tekhnika, Minsk, 1992).
10. C. F. Bohren, D. R. Huffman. *Absorption and Scattering of Light by Small Particles* (WILEY-VCH Verlag GmbH & Co, Weinheim, 2004).
11. R. Atkinson W. R. Hendren, G. A. Wurtz, W. Dickson, A. V. Zayats, P. Evans, R. J. Pollard, *Phys. Rev. B* **73**, 235402 (2006).

LIGHT EMISSION AND STRUCTURAL PROPERTIES OF Al_2O_3:SiOC NANOCOMPOSITES

†D. V. Kysil[1], A. V. Vasin[1,2], S. V. Sevastianov[3], Yu. P. Piryatinsky[4], S. P. Starik[5], V. A. Tertykh[3], A. N. Nazarov[1], V. S. Lysenko[1]

[1]Lashkaryov Institute of Semiconductor Physics of the NAS of Ukraine
Nauki Ave. 41, 03680 Kyiv, Ukraine
†kdmitr93@gmail.com
[2]National Technical University of Ukraine "KPI"
Peremohy Ave. 37, 03056 Kyiv, Ukraine
[3]Institute of Surface Chemistry of the NAS of Ukraine
General Naumov Str. 17, 03056 Kyiv, Ukraine
sevostyanov.s@gmail.com
[4]Institute of Physics, NAS of Ukraine, Nauki Ave. 46, 02000 Kyiv, Ukraine
[5]Bakul Institute of Superhard Materials of the NAS of Ukraine
Avtozavodska Str. 2, 04074 Kyiv, Ukraine

Structural and luminescent properties of synthesized Al_2O_3:SiOC nanocomposites were studied. It is demonstrated that an increase of annealing temperature results in formation of silica precipitates on the surface of alumina particles. No pyrolitic carbon precipitation was found in the powder under thermal pyrolysis of phenyl groups up to 600 °C.

1. Introduction

Previous research demonstrated that silica nanoparticles with phenyltrimetoxysilane (PhTMS) groups grafted to the surface exhibited strong visible photoluminescence (PL) after thermal treatment at about 600 °C [1]. Other carbonized nanocomposite materials based on silica (SiO_2:C) have been also reported to demonstrate the broadband PL in all visible spectral range [2–4]. The authors of [4] suggested that emission centers in SiO_2:C nanostructures are associated with electron states in carbon related structural sites on the surface of silica. In this model the SiO_2 is used as host matrix for carbon related emission centers. Due to its properties [5–7] we assumed, that alumina nanopowder is a good candidate as an alternative host material. This report is a first attempt to study the effect of thermal treatment on structure and PL properties of alumina nanoparticles modified by PhTMS.

2. Methods

Fumed alumina (specific surface area of about 89 m^2/g) and phenyltrimethoxysilane (PhTMS) were used as a morphological template and carbon precursor, respectively. Alumina was treated with a PhTMS toluene solution (1.73 ml of PhTMS per 10 ml of toluene) at 70 °C for 4 h in the presence of a catalyst. The reaction product (hereafter "phenyl-alumina") was dried and subjected to annealing at 400 °C, 500 °C, 600 °C for 30 min in a flow of pure nitrogen at atmospheric pressure.

Transmission infrared spectra were recorded at room temperature using Nicolet 6700. Photoluminescence was studied under 290 nm excitation.

3. Results and discussion

Aluminum oxide structural matrix in pristine alumina is represented by the main absorption band at 830 cm^{-1} as illustrated in Fig. 1a. Strong broad absorption band in the range of 980–1200 cm^{-1} is due Si–O–Si bonds. Broadening of the band can be caused by crosslinking of siloxane bonds in chains and rings [8]. Formation of this broad band is possibly explained by the formation of polymer-like siloxane compounds on the surface of Al_2O_3 particles during the chemical treatment procedure.

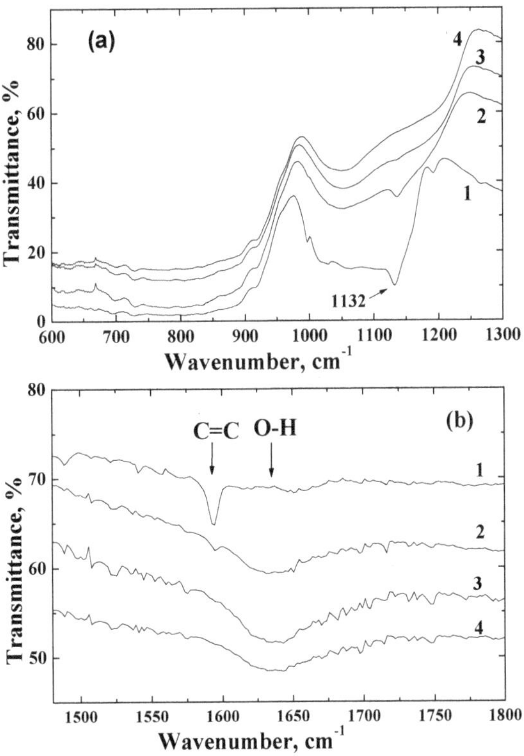

Figure 1. FTIR spectra of "phenyl-alumina" before (1) and after annealing at 400 °C (2), 500 °C, (3) 600 °C (4) in the spectral range: (a) 600–1300 cm^{-1}, (b) 1400–1800 cm^{-1}. Spectra are offset along the ordinate axis for clarity.

Increasing of the annealing temperature causes the decrease of the intensity of the narrow absorption band at 1594 cm^{-1} (Fig. 1b) which is assigned to C=C stretching vibrations in the phenyl rings. It is worth noting that destruction of the benzene rings does not cause formation of amorphous pyrolitic carbon typically characterized by a broad absorption band about 1600 cm^{-1}. It can be explained by a thermally activated carbon diffusion from surface inside Al_2O_3 particles during annealing in an inert ambient [9].

In the PL spectrum of the original sample there are three emission bands clearly seen (Fig. 2, spectrum 1): with maxima at about 335 nm, 390–400 nm and 470 nm which are possibly related to different types of centers: F^+-centers [7], P^--centers or F_S^+-centers [10], respectively. Unfortunately we could not interpret the origin of the third band.

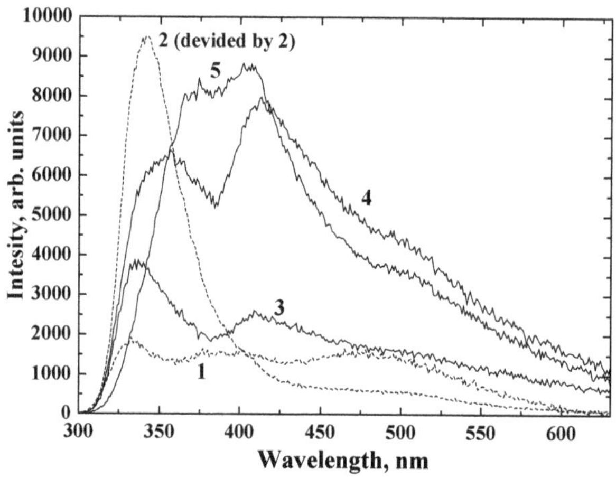

Figure 2. PL spectra under 290 nm excitation: (1) pristine fumed alumina and "phenyl-alumina" before annealing (2) (denoted by dashed lines); "phenyl-alumina" after annealing at 400 °C (3), 500 °C (4) and 600 °C (5) (denoted by solid lines).

The intense PL band with the maximum at 340 nm appears in Al_2O_3 after chemical treatment (Fig. 2, spectrum 2). This band is presumably associated with excimer states in closely located phenyl groups grafted to alumina surface [11,12]. Decomposition of the phenyl groups during annealing leads to disappearance of this band. As it was demonstrated by IR study, the structure of these samples can be represented as Al_2O_3/SiO_2 composite. Hence, it is reasonable to expect a contribution of silica related emission centers in the PL band. Unfortunately, at present time we are not able to identify correctly the evolution of the PL band of these samples.

4. Conclusion

It is demonstrated that an increase of the annealing temperature in the range of 400–600 °C results in pyrolysis of phenyl groups and formation of silica precipitates. No carbon precipitation was detected after pyrolisis of organosilicon groups. It is suggested that pyrolitical carbon is dissolved in alumina particles.

Acknowledgments

The authors acknowledge the Ministry of Education and Science of Ukraine for the support (project F2904).

References

1. D. Savchenko, E. Kalabukhova, A. Sitnikov, A. Vasin, S. Starik, O. Gontar, G. Rudko, A. Nazarov, V. Lysenko, V. Tertykh, *Advanced Materials Research* **854**, 99 (2014).
2. S. Hayashi, M. Kataoka, K. Yamamoto, *Jpn. J. Appl. Phys* **32**, L274 (1993).
3. W. H. Green, K. P. Le, J. Grey, T. T. Au, M. J. Sailor, *Science* **276**, 1826 (1997).
4. A. Vasin, I. Verovsky, V. Tyortykh, Yu. Bolbukh, D. Kisel, G. Rudko, Ye. Gule, Yu. Piryatinsky, S. Starik, A. Nazarov, V. Lysenko, *J. Nano Research* **39**, 80 (2016).
5. R. H. French, H. Müllejans, D. J. Jones, *J. Am. Ceram. Soc.* **81(10)**, 2549 (1998).
6. M. Choi, A. Janotti, Ch. G. Van de Walle, *Appl. Phys.* **113**, 044501 (2013).
7. T. V. Perevalov, O. E. Tereshenko, V. A. Gritsenko, V. A. Pustovarov, A. P. Yelisseyev, Ch. Park, J. Hee Han, Ch. Lee, *J. Appl. Phys.* **108**, 013501 (2010).
8. Y.-S. Li, Y. Wang, S. Ceesay, *Spectrochimica Acta Part A* **71**, 1819 (2009).
9. M. C. Guenette, M. D. Tucker, M. Ionescu, M. M. M. Bilek, D. R. McKenzie, *J. Appl. Phys.* **109**, 083503 (2011).
10. S. V. Gorbunov, A. F. Zatsepin, V. A. Pustovarov, S. O. Cholakh, V. Yu. Yakovlev, *Phys. Solid State* **47**, 733 (2005).
11. G. A. George, *Pure & Appl. Chem.* **57(7)**, 945 (1985).
12. A. F. Fehervary, L. C. Kagumba, S. Hadjikyriacou, F. Chan, R. A. Gaudiana, *J. Appl. Polymer Science* **87**, 1634 (2003).

LUMINESCENT AND OPTICAL PROPERTIES OF DOPED AND UNDOPED ULTRATHIN TITANIUM OXIDE FILMS

M. V. Rudenko, T. I. Orekhovskaya
Belarusian State University of Informatics and Radioelectronics
P. Browka 6, 220013 Minsk, Belarus
nik@nano.bsuir.edu.by

A. V. Mudryi
SSPA Scientific and Practical Materials Research Center NASB
P. Browka 19, 220072 Minsk, Belarus

We describe fabrication of ultrathin (3–4 nm thick) sol-gel derived titania films undoped and doped with erbium on silicon. The erbium-doped films reveal room temperature photoluminescence (PL) with maximum at 1546 nm after annealing at 1000 °C.

1. Introduction

Titanium oxide reveals an interest as a material for different applications. Undoped and rare earth doped [1–3] titania presents an extensive area of investigations in optics, photocatalyses, microelectronics and nanophotonics. In this paper, we describe synthesis and optical properties of erbium-doped ultrathin titania sol-gel derived films.

2. Experimental method

Titanium oxide (titania, TiO_2) films were formed by sol-gel method on silicon and quartz substrates and structures of porous anodic alumina formed on monocrystalline silicon. Titanium ethoxide ($Ti(CH_3CHO)_4$) and erbium nitrate ($Er(NO_3)_3 \cdot 5H_2O$) were used for the sol preparation. Acetic acid was used as the solvent. Acetylacetone and acetone were added as stabilizers. Finally, the obtained solution represented homogeneous liquid transparent orange colored sol. The total concentration of the dissolved species in the synthesized solution was 10 mg/ml. The film was deposited by spinning of the sol. After spinning and drying the film was subjected to final heat treatment at 1000 °C for 30 min.

The morphology of the samples was examined by scanning electron microscopy (SEM).

A 0.6 m focal length single grating monochromator and the 532 nm solid state of a 300 mW laser were used for the photoluminescence measurements. After the monochromator the photoluminescence signal was detected by an InGaAs p-i-n photodiode sensitive from 0.9 to 1.7 μm. Conventional lock-in-amplifier with mechanical chopping at a frequency of about 20 Hz was used for signal recovery.

3. Results and discussion

According to SEM data (Fig. 1) the film thickness of the single layer of titania xerogel is 3.3 nm.

Figure 1. SEM image of undoped titania xerogel film formed on silicon after annealing at 450 °C.

The same film fabricated on the fused silica substrate reveals high transmission of about 97 % in the wide optical range as it is shown in Fig. 2.

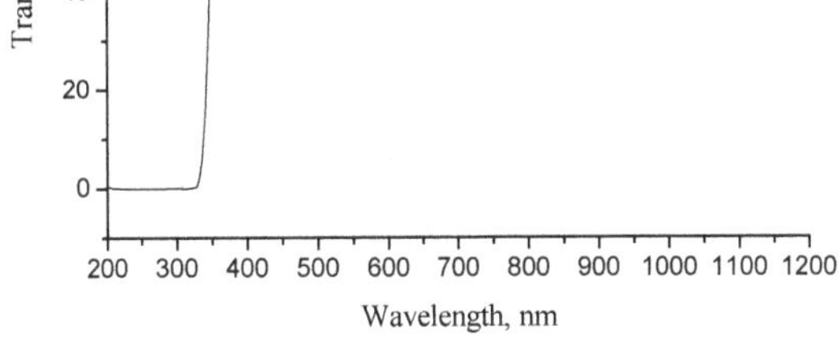

Figure 2. Optical transmission spectrum of undoped titania xerogel film formed on the fused silica substrate.

Emission spectra of xerogel (Fig. 3) reveal PL band at 1.54 μm, corresponding to the transition between the first excited state $^4I_{13/2}$ and the ground state $^4I_{15/2}$ of Er^{3+} ions in TiO_2 matrix [2]; such PL was detected at room temperature after fabrication of one layer onto porous anodic alumina and monocrystalline silicon. The PL intensity at 1.54 μm increases and additional band with the maximum at 987 nm is resolved corresponding to the transition $^4I_{11/2} \rightarrow {}^4I_{15/2}$ when using porous anodic alumina formed on silicon. The relatively broad PL band at 1.14 nm (1.08 eV) is related to interband recombination in silicon substrates.

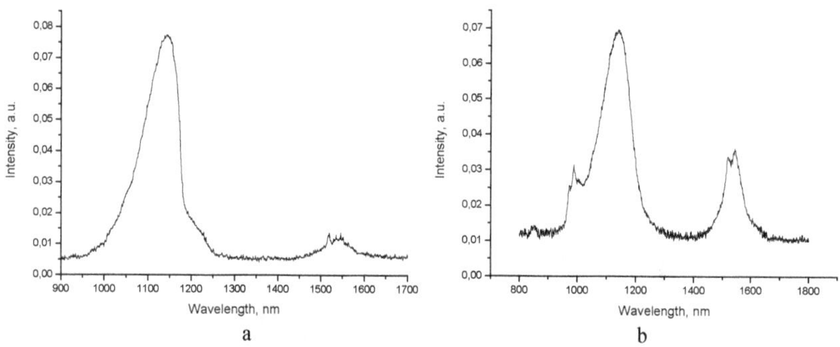

Figure 3. The luminescence spectra of erbium doped titania xerogel formed on silicon (a) and in porous anodic alumina formed on monocrystalline silicon (b).

4. Conclusion

The method of fabrication of ultrathin TiO_2 xerogel films of about 4 nm thick is proposed. Erbium-doped titania xerogel processed on porous anodic alumina exhibits room-temperature erbium-related PL at 1.54 μm.

References

1. Y. Badr *et al.*, *Indian Journal of Pure & Applied Physics* **46**, 706 (2008).
2. N. V. Gaponenko *et al.*, *Optical Materials* **28**, 688 (2006).
3. S. R. Johannsen *et al.*, *Optical Materials Express* **6**, 1678 (2016).
4. V. B. Sreedhar *et al.*, *J. Molecular Structure* **1130**, 1001 (2017).

EUROPIUM LUMINESCENCE OF AMORPHOUS YTTRIUM ALUMINA FILMS

J. C. Villegas Brito, N. V. Gaponenko, K. S. Sukalin
Belarusian State University of Informatics and Radioelectronics
P. Browka 6, 220013 Minsk, Belarus
nik@nano.bsuir.edu.by

T. F. Raichenok
B. I. Stepanov Institute of Physics, NASB, Nezavisimosti Ave. 68, 220072 Minsk, Belarus

We investigate photoluminescence (PL) of sol-gel derived films corresponding to the composition of yttrium alumina garnet ($Y_3Al_5O_{12}$) doped with trivalent europium. The film of about 220 nm in thickness was deposited on a fused silica substrate by spinning from the corresponding sol followed by drying and high temperature annealing. The fabricated film was amorphous and revealed the strong PL band at 710 nm along with the weaker bands ranging from 550 to 720 nm corresponding to $^5D_0 \rightarrow {}^7F_j$ (j=0,..,4) transitions of trivalent europium ions (Eu^{3+}). Photoluminescence excitation (PLE) spectrum was obtained with the set of assigned transitions and possible application of the films is discussed.

1. Introduction

Optical materials transparent in visible and near infrared ranges doped with light emitting impurities receive an interest as light convertors of UV radiation improving efficiency of silicon solar cells. Particularly, trivalent europium ions are characterized by strong visible luminescence [1] that may be used for the light convertors.

Luminescence properties of Eu-doped materials, particularly PL and PLE spectra, depend on the host material, synthesis conditions, annealing temperature and europium concentration. For major host materials, the band at 610 nm is most intensive as compared to the band at 710 nm. However, for silicon-based photodetectors, the conversion of UV radiation into the emission at 710 nm is of great importance considering their spectral sensitivity. Recently, strong PL of Eu^{3+} with efficient 710 nm emission was observed from aluminates codoped with Eu and Bi [2]. In this paper, we study the emission of Eu^{3+} ions from amorphous yttrium alumina film deposited on the fused silica substrate.

2. Experimental

The components for the sol were ethanol, citric acid, $Al(NO_3)_3 \cdot 9H_2O$ (98.5 %), $(Eu(NO_3)_3 \cdot 5H_2O$ (99.9 %) $Y(NO_3)_3 \cdot 5H_2O$ (99.999 %). The concentration of europium in the sols was 0.03 mol/l. The sol was deposited onto a quartz substrate by spinning at the rate of 2700 rpm for 30 s. After drying of the first layer at 200 °C the second layer was deposited and dried at the same conditions. Finally, the sample was heat treated at 1000 °C for 30 min.

PL and PLE spectra were recorded using a spectrofluorometer CM2203. A high pressure xenon arc lamp (150 W) was used as an excitation source. X-ray diffraction (XRD) analysis was performed on a diffractometer D8 Advance, Bruker AXS (Germany), using CuKα-radiation. Morphological analysis of the films was done using a HITACHI S-4800 scanning electron microscope (SEM).

3. Results and discussion

According to SEM and XRD data the fabricated xerogel films were of about 210 nm thickness and amorphous. Vague images of grains can be slightly resolved in SEM images.

Fig. 1 shows the PL spectrum of the film for the excitation wavelength of 260 nm. There are bands corresponding to optical transitions between the terms of Eu^{3+} ions: $^5D_0 \rightarrow {}^5F_0$ (545–585 nm), $^5D_0 \rightarrow {}^7F_1$ (585–600 nm); $^5D_0 \rightarrow {}^7F_2$ (610–630 nm); $D_0 \rightarrow {}^7F_3$ (640–660 nm); $^5D_0 \rightarrow {}^7F_4$ (680–710 nm) [3].

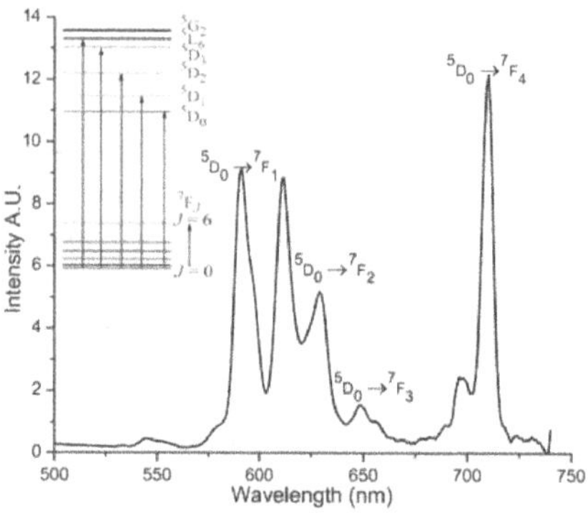

Figure 1. PL spectrum of the Eu-doped yttrium alumina film after annealing at 1000 °C for 30 min. Excitation wavelength $\lambda_{ex} = 260$ nm.

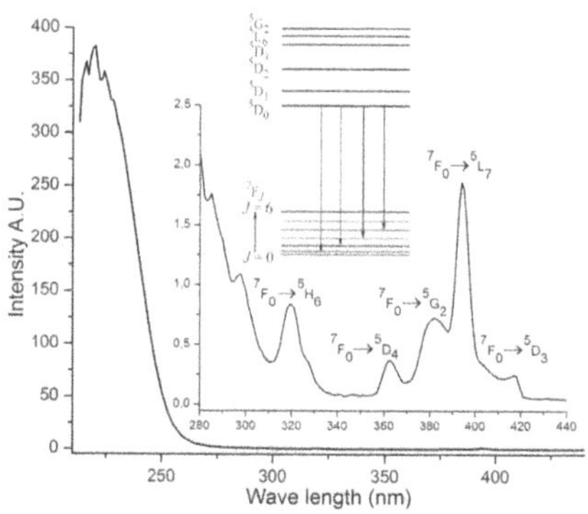

Figure 2. PLE spectrum of Eu-doped yttrium alumina film for the emission wavelength 710 nm. Inset depicts PLE spectrum in the range 280–440 nm.

PLE spectra shown in Fig. 2 were recorded for the most intensive $^5D_0 \rightarrow {}^7F_4$ band (emission wavelength 710 nm). The observed bands in the PLE spectra correspond to a series of Eu^{3+} transitions between the terms: $^7F_0 \rightarrow {}^5H_6$ (318 nm), $^7F_0 \rightarrow {}^5D_4$ (365 nm), $^7F_0 \rightarrow {}^5G_2$ (380 nm), $^7F_0 \rightarrow {}^5L_6$ (393 nm), $^7F_0 \rightarrow {}^5D_3$ (413 nm) [4]. The intensive wide band at 214–280 nm is associated with the absorption due to the charge transfer Eu^{3+}-O^{2-} [5].

It should be noticed that the fabricated Eu-doped films can find application as transparent convertors of UV radiation into visible range [6], raising efficiency of silicon solar cells, converting UV radiation into visible due to reducing electron thermalization and degradation of solar cells. Eu-doped amorphous films are of interest also in research of spontaneous emission properties when embedded in microcavities [7], photonic crystals [8] and other matrices with anisotropy of photonic density of states [9,10].

4. Conclusion

Sol-gel synthesis of Eu-doped yttrium alumina films was described. The films comprising two layers deposited by spinning on fused silica substrates were amorphous after conventional furnace annealing at 1000 °C. The films reveal the visible Eu-related PL with the bands $^5D_0 \rightarrow {}^7F_1$ (588 nm); $^5D_0 \rightarrow {}^7F_2$ (608 and 629 nm); $^5D_0 \rightarrow {}^7F_3$ (654 nm); $^5D_0 \rightarrow {}^7F_4$ (710 nm) excited with UV radiation in the range of 200–394 nm. They allow efficient conversion of UV radiation into visible with the most intensive PL band of 710 nm.

References

1. K. Binnemans, *Coordination Chemistry Review* **295**, 1 (2015).
2. G. E. Malashkevich, G. P. Shevchenko, Yu. B. Bokshitz, A. A. Kornienko, P. P. Pershukevich, *Opt. Spectrosc.* **98**, 190 (2005).
3. M. L. Sandro, H. da C. A. Luis, *Opt. Express*. **20**, 12658 (2012).
4. A. Ling, Zh. Li, J. Liu, in: *2nd Int. Conf. on Modeling, Identification and Control* (2015).
5. A. Esparza, M. Garcia, C. Falcony, *Thin Solid Films* **325**, 14 (1998).
6. A. Potdevin, S. Lechevallier, G. Chadeyron, *Opt. Mat.* **517**, 4610 (2009).
7. J. Bellessa, S. Rabaste, J. C. Plenet, J. Dumas, J. Mugnier, O. Marty, *Appl. Phys. Lett.* **79**, 2142 (2001).
8. E. P. Petrov, V. N. Bogomolov, I. I. Kalosha, S. V. Gaponenko, *Phys. Rev. Lett.* **81**, 77 (1998).
9. A. A. Lutich, S. V. Gaponenko, N. V. Gaponenko, I. S. Molchan, V. A. Sokol, V. Parkhutik, *Nano Lett.* **4**, 1755 (2004).
10. N. V. Gaponenko, *Acta Phys. Polon.* **112**, 737 (2007).

THERMAL BEHAVIOR OF BIOACTIVE MAGNETIC NANOCOMPOSITES

G. P. Aleksandrova, G. F. Prozorova, B. G. Sukhov, B. A. Trofimov
A.E. Favorsky Irkutsk Institute of Chemistry SB RAS
Favorsky Str. 1, 664033 Irkutsk, Russia
alexa@irioch.irk.ru

Thermal properties of bionanocomposites containing magnetite nanoparticles in the arabinogalactan matrix were investigated by means of thermic measurements. The thermogravimetric parameters of the thermal degradation of the nanocomposites are estimated. The base concentration of iron oxide within the polymer matrix greatly influences the properties of the nanocomposites.

1. Introduction

Iron oxide nanoparticles (NPs) and their based nanocomposites are of great interest due to their unique physicochemical and biological properties [1,2]. Nanodispersed magnetic materials, which are polymer coated magnetic nanoparticles, have attracted considerable attention [2]. The biomedical application of nanomaterials was demonstrated by an example of nanosized iron oxides [1,3]. To increase the bioaccessibility and transport of iron compounds to hemopoietic organs, some methods of their immobilization onto a membranotropic natural polysaccharide arabinogalactan have been developed [2]. Based on this matrix, water soluble nanocomposites consisting of ferrite structures encapsulated in an arabinogalactan macromolecule have been synthesized. Owing to the synergism of the properties of the polysaccharide matrix and the specific biological properties of iron, they exhibit a pronounced antianemic activity [3]. Other important characteristic for their practical applications is thermal stability. The aim of the present work was to study the thermal stability of multipurpose nanocomposites based on arabinogalactan and containing magnetite nanoparticles.

2. Experimental

Nanobiocomposites of ferroarabinogalactans (FAG) were synthesized according to the procedure described previously [2,4]. The iron content in the nanocomposites was determined by atomic absorption spectroscopy. Thermogravimetric analysis of the nanocomposites was performed using a Q1500 derivatograph (MOM) at maximum temperature of 600 °C, heating rate of 10 °C/min, and DTA sensitivity of 1/10 (in air).

3. Results and discussion

The polysaccharide arabinogalactan (AG) from Siberian larch was used as the matrix providing self-organization of the system, resulting in self-assembly of nanoparticles to form ordered nanosized composites. Owing to natural polysaccharide with hydrotropic properties as a stabilizing agent, the products are water-soluble. A series of ferroarabinogalactans with total iron content varying from 3.5 to 10.5 % and the controlled dispersion of iron, oxide NPs were prepared. Occurring processes can be described by the scheme which involves hydrolysis of the salts, the formation of a mixed iron oxide due to the increased pH by addition of aqueous NaOH in the presence of arabinogalactan. The structures of the nanocomposites containing highly dispersed magnetite Fe_3O_4 were determined by X-ray diffraction and TEM [4]. The size of the resulting magnetite NPs varies in the range of 4–7 nm depending on the total iron content in the nanocomposites [4].

To establish thermal stability of the metal-containing nanocomposites, their thermal behavior was investigated with a linear rise in temperature over a wide range (up to 600 °C) via differential thermal (DTA) and thermogravimetric (TGA) analysis. The thermal stability of the nanocomposites was found in the range of 20–200 °C. Typical TGA and DTA thermograms for the initial AG and FAG nanocomposites are shown in Fig. 1. Three thermal effects, different in both polarities and values, were detected in the temperature range of thermal activity for FAG (20–410 °C) and AG (20–480 °C).

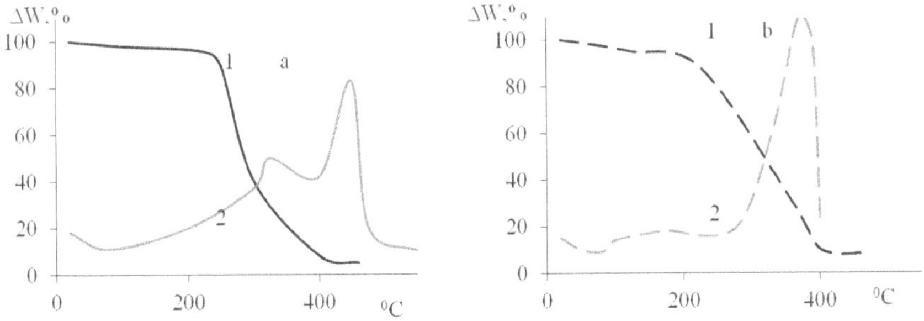

Figure 1. TGA (1) and DTA (2) thermograms for arabinogalactan (a) and ferroarabinogalactan (b) with the iron content of 3.5 %. (ΔW is the weight loss).

This extended range (20–200 °C) of the nanocomposites thermal stability is of great importance, since it allows the use of these materials at high temperatures in particular for thermal sterilization of specimens under standard conditions.

Each thermal effect was accompanied by a corresponding change in the sample weight during the linear rise of the temperature (Fig. 1). The first thermal effect of the FAG at 20–200 °C was endothermic and corresponded to the evaporation of adsorbed water (4–8 %) from the bulk of the samples without their destruction. All subsequent exothermic effects in the DTA curves were above 200 °C (Table 1). They were associated with the gradual oxidation and destruction of AG down to low molecular compounds.

Table 1. Thermal characteristics of ferroarabinogalactan nanocomposites (TSR - thermal stability range, TAR - thermal activity range).

Thermal effect	AG	FAG (3.5 % Fe)	FAG (10.5 % Fe)
TAR, °C	20–480	20–410	20–380
TSR, °C	20–230	20–200	20–200
E_a TSR, kJ/mol	4.4	3.8	3.8
E_a TAR, kJ/mol	29.5	26.1	23.3
Size NP, nm	-	4.1	7.4

The final stage of AG thermal destruction is characterized by an abrupt decrease in weight and the release of a great amount of heat, which is visible in the differential curve as a high intensity exothermic effect (220–290 °C). The high thermal stability of wood polysaccharides is due to their high degree of structural ordering as a result of highly developed hydrogen bond networks. The pyrolysis of polysaccharides, particularly cellulose, has been thoroughly studied as well. The hemicelluloses of coniferous wood, of which AG is one example, thus exhibit higher stability during thermal destruction (pyrolysis) than xylan or cellulose, which coincides with the ranges of AG thermal activity up to 480 °C.

The fabricated nanocomposites are characterized by a very intense termal effect in the temperature range of 290–390 °C (Fig. 1b). The complete destruction of FAG stops at 380, 410 °C (Table 1). At a constant rate of heating, the transformation of polysaccharides usually proceeds as a first order reaction. Activation energy E_a at each stage of thermal destruction was calculated using the Arrhenius equation in combination with a double logarithm. The average activation energy of the first stage (dehydration) displayed almost no dependence on the type of the sample and varied between 4.4 and 3.8 kJ/mol. The average activation energy E_a of thermal destruction for AG and FAGs were 29.5 and 26.1–23.3 kJ/mol, respectively.

Analysis of the DTA and TGA parameters revealed an important effect that Fe_3O_4 NPs have on the thermal stability of AG based nanocomposites. The calculation results showed that introducing metal NPs into the AG matrix notably lowers the average activation energy of thermal destruction (Table 1), resulting in a pronounced drop of 70–100 °C in the temperature range of nanocomposites destruction, relative to the initial AG matrix (480 °C). The degree of structural modifications during the effect of NPs on polymer, particularly the presence and length of the metal-polymer interfaces, depends on the concentration and sizes of NPs [1]. Magnetite NPs are distributed uniformly in the AG matrix, as can be seen in the TEM images. The presence of Fe_3O_4 NPs (3.5–10.5 %) favors an energy barrier of 23.3 kJ/mol and induces a great reduction in the range of thermal destruction (down to 410–380°C).

Along with the exothermic effect in the range of 220–290 °C due to thermal degradation of arabinogalactan (Fig. 1) very intense effect exists in the temperature range of 290–390 °C. This effect is due to an intense oxidation of iron oxide (II), which is part of magnetite due to the diffusion of oxygen from the air.

4. Conclusion

Thermal stability ranges for nanocomposites of magnetite nanoparticles dispersed in an polysaccharide matrix were determined. The effect of magnetite nanoparticles in nanocomposites with arabinogalactan was determined as the reduced activation energy of thermal destruction and the shrinking range of their thermal stability as compared to the initial polysaccharide.

References

1. S. P. Gubin, Yu. A. Koksharov *et al.*, *Russ. Chem. Rev.* **74**, 539 (2005).
2. G. P. Aleksandrova, L. A. Grishchenko *et al.*, *Nanotechnics* **3**, 11 (2008).
3. G. P. Aleksandrova *et al.*, *Russ. J. Bioorg. Chem.* **37**, 829 (2011).
4. G. P. Aleksandrova *et al.*, *Russ. Chem. Bull.* **59**, 2318 (2010).

SANS AND WXRD STUDY OF PbSe-DOPED SILICATE GLASSES

V. S. Gurin[1], G. E. Rachkovskaya[2], G. B. Zakharevich[2], S. E. Kichanov[3]

[1]*Research Institute for Physical Chemical Problems, Belarusian State University Minsk, Belarus*

gurin@bsu.by

[2]*Belarusian State Technological University, Minsk, Belarus*

[3]*Joint Institute of Nuclear Research, Dubna, Russia*

PbSe-doped silicate glasses were studied by the two scattering methods: wide-angle X-ray diffraction (WXRD) and small-angle neutron scattering (SANS) those can monitor structural features of a solid at different scales. PbSe nanoparticles with the bulk-like cubic crystal structure (but $a > a_{bulk}$) were detected by WXRD. SANS measurements evidenced formation of complicated structures like fractal aggregates in line with single nanoparticles.

1. Introduction

Glasses with quantum sized semiconductor nanoparticles are a novel class of optical materials with advanced features for non-linear optical elements, passive laser locks, selective spectral filters, *etc.* [1]. Lead chalcogenides (PbS and PbSe) are featured by explicit manifestation of quantum size effects even for ten-nm-scale because they possess the large value of the Bohr exciton radius (18 and 46 nm, respectively). The nanoparticles can be successfully localized within a glass matrix (various types: silicate, phosphate, borate). For the size range of 2–10 nm their excitonic absorption peaks enter the near IR range that is of great interest for applications in lasers for telecommunication and medicine [1]. Meanwhile, the nanoparticles synthesized through soft chemistry methods may have typically the less size with the optical absorption in the visible range up to UV. In contrast to colloidal chalcogenides, the particles in glass matrices are stable with respect to environment as well to intense laser irradiation that provide pathways to use such glasses as a non-linear optical media with a proper control of spectral properties via the variation of lead chalcogenide particle state (size, concentration, aggregation).

Detailed structural studies of nanoparticles in a glass are troubled because of low particle concentration. Also, any extraction of particles from a glass matrix is impossible without their essential modification. Thus, indemolition techniques are of great interest to get new information on the state of such systems. In the present paper, we combine SANS with WXRD for exploration of silicate glasses with PbSe nanoparticles in order to shed a light onto their structural features.

2. Experimental

The samples of PbSe glasses have been fabricated according to the technique described earlier [2]. It is based on the conventional two-step method used for semiconductor-doped glasses: (1) a melting-cooling cycle of glass-forming oxides (SiO_2, Na_2O, K_2O, ZnO, Al_2O_3, and PbO) with additional fluoride component and selenium; (2) the secondary heat treatment near T_g of the glass (480–530 °C). Nanoparticles are nucleated and growing during the step (2). Temperature and duration of this step are used to control a final state of the nanoparticles-glass system.

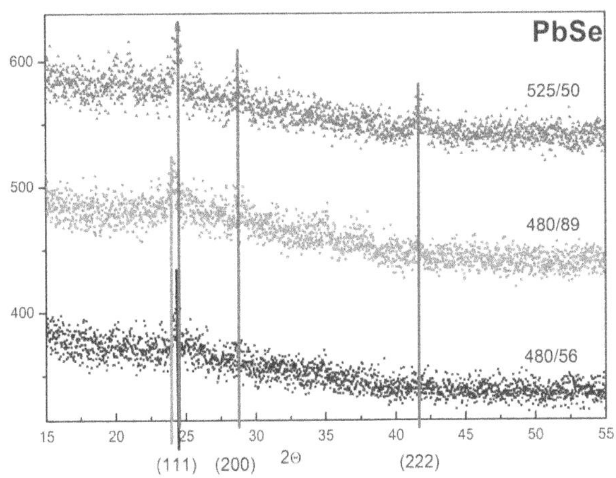

Figure 1. Diffraction patterns for glasses with PbSe nanoparticles.

3. Results and discussion

WXRD was performed here to discover particles as nanocrystalline phases while the glass appeared as a typical amorphous constituent generating wide haloes and noises in WXRD patterns. Fig. 1 displays the diffraction patterns for the glasses with PbSe, that can be decoded as cubic PbSe (JCPDS 06-0354), for all samples under study. Small size of nanoparticles and overall low concentration do not allow good diffraction picture and pronounced peaks, but the main positions may be used for the identification and estimation of unit cell parameters (Table 1). Approximate particle sizes were evaluated through the conventional Scherrer method to be ≤ 10 nm.

Table 1. Unit cell parameters of PbSe nanoparticles determined from WXRD measurements for a series of glasses after different secondary heat treatment and the Scherrer-derived particle diameter.

Temp/time, °C/h	a, Å	$(a-a_{ref})/a_{ref}$, %; a_{ref} = 6.1213 Å	Diameter, nm
480/56	6.3	+2.9	
480/89	6.131	+0.16	5–10
525/50	6.156	+0.57	

Measurements of neutron scattering were made using a time-of-flight spectrometer UMO at the reactor IBR-2 (Dubna, Russia). The scattering parameter range covered 0.05–2.5 nm^{-1} allowing to get information on structures at the scale of 2–100 nm. The raw data (Fig. 2) display the SANS curves for the couple of glasses treated at 480 and 525 °C, however, they appear rather different that may require a use of individual scattering models for each glass.

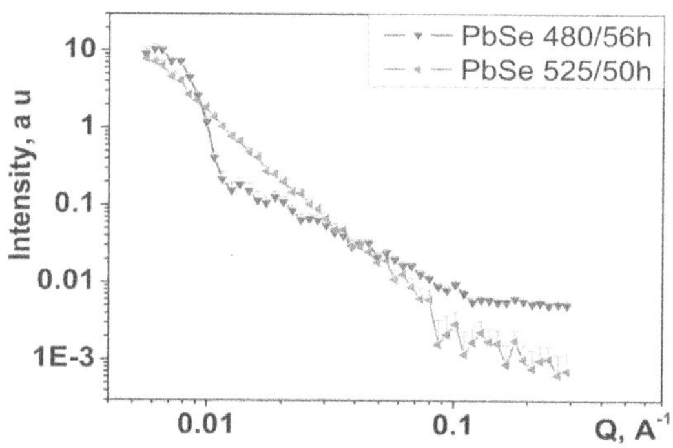

Figure 2. Raw data of SANS measurements for two samples of PbSe-doped glasses.

In order to simulate the experimental data and elaborate adequate structure models we divide the measurement interval into $Q < 0.02$ and $0.02 \leq Q \leq 0.3$ as far as fitting of single model to the full range was problematic. We use SASfit software [3] and incorporate into the simulation both fractal aggregates of overlapping spheres and isolated spheres. For the simulation of fractal aggregates the scattering function is:

$$I(Q) = 4\pi \int_0^\infty g(R) R^2 \frac{\sin(QR)}{QR} dR,$$

where $g(R) = R^{D-d}((1+R/4\xi)(1-R/2\xi)^2)$ for $R < 2\xi$ and $g(R) = 0$ for $R \geq 2\xi$. Within the framework of this model the value of 2ξ characterizes an aggregate diameter, d is the dimension of space (=3), and D is the fractal dimension. The fitting is shown in Fig. 3a,c for two PbSe-doped glasses.

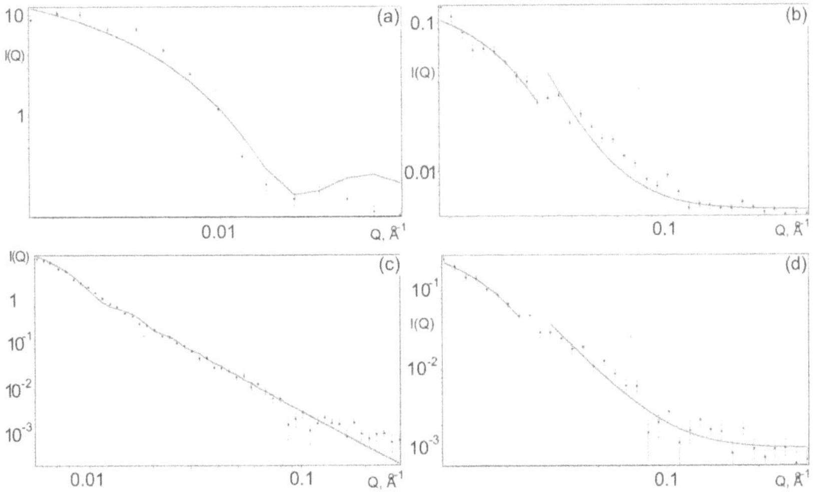

Figure 3. Fitting of the SANS data for two samples of PbSe-doped glasses: (a–b) 480/56 and (c–d) 525/50 (temp, °C/time of the treatment, h).

In the next range of scattering parameters, $0.02 \leq Q \leq 0.1$, we use the Guinier approximation:

$$I(Q) = (4\pi^2 N^2/3)\exp(-Q^2 R_g^2/3),$$

where N is the particle concentration and R_g is the gyration radius which is related to particle radius for the spherical case as $R=(5/3)^{1/2}R_g$. The result of this fitting is given in Fig. 3b,d (the part of smaller Q) and decodes the presence of spherical particles of a certain radius (R). Fitting at a larger Q in these plots corresponds to the Porod approximation and can describe objects of large-scale ordering. Detailed analysis of this part is out of the scope of this publication.

Table 2 summarize the data obtained as a result of the above analysis of SANS data for two glass samples. They indicate that structure of the glasses may be presented as a combination of single particles (R = 6–8 nm) and fractal aggregates of different dimensions $D < 3$. The latter are of the order of tens of nm and comprised of several nanoparticles, R = 8–10 nm. D is close to 3 (dense aggregate) for 480/56 sample while 525/50 sample reveals less dense aggregate, *i.e.* there is the growth of PbSe nanoparticles (480/56 → 525/50) accompanied by disintegration of the aggregates. Separate growing PbSe particles with size-controlling by the heat treatment provides the quantum-sized excitonic absorption familiar for lead chalcogenides of this size range [2].

Table 2. Parameters of the models obtained through fitting procedures of SANS data for two samples of PbSe-doped glasses.

Sample	ξ, nm	D	R_g, nm	R, nm
480/56	27.6	2.92	6.13	7.9
525/50	24.5	2.53	8.11	10.5

4. Conclusion

Silicate glasses with PbSe nanoparticles have been studied with WXRD and SANS. The results provide structural features of the particles and glass described above. Particles parameters of both techniques enter the size range corresponding to formation of quantum dots in a strong quantization mode.

Acknowledgements

This work was supported by the Belarusian Republican Foundation for Fundamental Research and the State Program "Nanotech".

References

1. A. M. Malyarevich *et al.*, *J. Appl. Phys.* **103**, 081301 (2008).
2. P. A. Loiko *et al.*, *J. Non-Cryst. Solids* **358**, 1840 (2012).
3. http://kur.web.psi.ch/sans1/SANSSoft/sasfit.html

EFFECT OF MANGANESE IONS ON PHOTOLUMINESCENCE OF NANOPOROUS ANODIC ALUMINA

I. V. Gasenkova, N. I. Mukhurov, S. P. Zhvavyi, E. E. Kolesnik

State Research and Production Association "Optics, Optoelectronics and Laser Technology", Minsk, Belarus

n.mukhurov@ifanbel.bas-net.by

A. P. Stupak

B. I. Stepanov Institute of Physics NASB, Nezavisimosti Ave. 68, 220072 Minsk, Belarus

We have demonstrated a possibility of porous anodic alumina (PAA) doping with Mn atoms and studied the effect of this impurity on optical properties of PAA. The difference of photoluminescence spectra of doped and undoped specimens is explained by the change in the valence of manganese ions in the complex defects in alumina.

1. Introduction

An interest in optical properties of PAA is continued due to the prospects of its use in optoelectronics and nanooptics [1]. When PAA is exposed to UV radiation, photoluminescence (PL) is observed in the visible range. PL features depend on electrolyte composition, regime of anodizing and thermal treatment modes [2]. In spite of a large number of researches on PL of PAA, no unambiguous view has been attained so far on mechanisms of PL origin. To reveal the role of oxygen vacancies and impurity atoms in the features of PAA PL spectra, we have investigated undoped and Mn-doped PAA specimens.

2. Specimen preparation and experimental methods

PAA specimens were prepared by electrochemical oxidation of aluminum foil (99.99 %) in an electrolyte based on oxalic acid (0.3 M) and in the same electrolyte with addition of $KMnO_4$ (0.8 g/l) and annealed at 200–1300 °C. The Al foil used contained impurities among which Mn and Cr can contribute into luminescence spectra.

PL spectra were recorded with SDL-2 automated spectrofluorometer consisted of a high-aperture excitation monochromator MDR-12 and a recording monochromator MDR-23. Correction of the MDR-23 – PMT recording system was performed by using a standard tungsten tape lamp TRSH 2850. The excitation wavelengths were 275 and 325 nm.

3. Results and discussion

The PL spectra (Fig. 1) of the undoped and doped (PAA:Mn) original specimens and those annealed up to 600 °C are rather similar and represent the broad bands in the range of 350–650 nm with the maximum at 450–500 nm. These bands are attributed to intrinsic defects in alumina such as oxygen vacancies (F- and F_2-type centers) in different charge states [3].

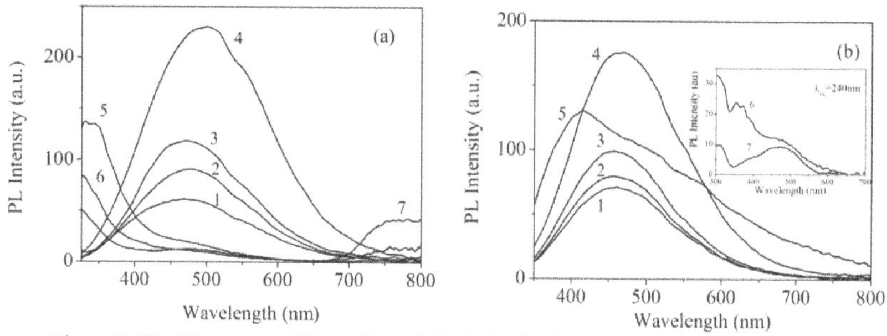

Figure 1. The PL spectra of PAA (a) and PAA:Mn (b) from the original specimens (1) and specimens annealed at 200 °C (2), 400 °C (3), 600 °C (4), 800 °C (5), 900 °C (6) and 1000 °C (7) for excitation at 275 nm.

As the annealing temperature increases, the PL intensity rises reaching the maximum at 600 °C. The ratio of areas under PL curves for the specimens annealed at 600 °C and the original one is $S_{600}/S_{\mathrm{am}} \sim 3$ for PAA:Mn, $S_{600}/S_{\mathrm{am}} \sim 4$ for PAA. For $T_a \geq 800$ °C, the PL spectra observed from PAA and PAA:Mn specimens show a notable difference: for PAA there is a shift to the shorter wavelengths with an appearance of the second band of about 760 nm. For PAA:Mn specimens annealed at 800 °C, the noticeable decrease in the PL intensity is also observed and the bands become significantly wider with the maximum at 415 nm for $\lambda_{ex} = 275$ nm and at 500 nm for $\lambda_{ex} = 325$ nm. For $T_{\mathrm{a}} = 900$ and 1000 °C, PAA:Mn specimens did not show any PL under excitation at the same wavelengths. For excitation with the shorter wavelength, $\lambda_{ex} = 240$ nm, the low PL signal was observed (inset in Fig. 1b).

Fig. 2 shows the PL spectra of the PAA and PAA:Mn specimens annealed at 1300 °C providing significant changes in the structure of alumina [3]. In the blue-green region, the weak bands appear with the maxima near 500 nm and 400 nm for excitations at 275 and 325 nm. In the red region, there are two intense R-lines due to Mn^{4+} and Cr^{3+} ions. For PAA:Mn specimens, the peak intensity at 677 nm is higher than intensity at 693 nm regardless of the excitation wavelength. In contrast to PAA specimens, PAA:Mn ones have a pinkish tint, indicating the presence of Mn^{3+} ions.

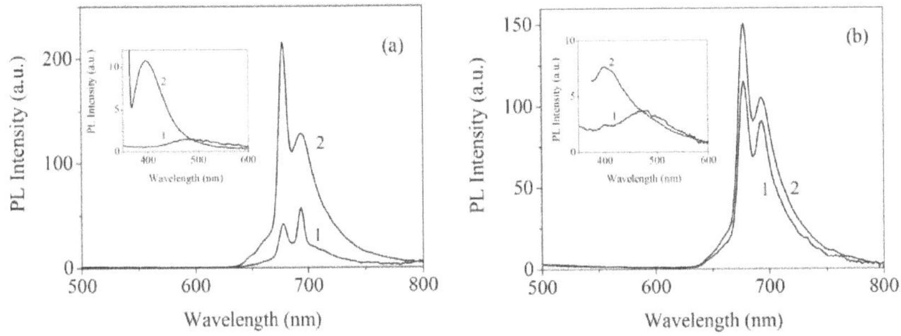

Figure 2. The PL spectra of PAA (a) and PAA:Mn (b) specimens annealed at 1300 °C for excitation at 275 (1) and 325 (2) nm wavelengths.

In the course of experiment on UV-stimulated changes of PL, it has been found that irradiation of the PAA:Mn specimens annealed at 600 °C resulted in a decrease of PL with the irradiation time (Fig. 3). Long UV exposure of PAA:Mn also lead to the reduced intensity of R-lines of Mn^{4+} and Cr^{3+} (Fig. 3) and disappearance of the pink tint.

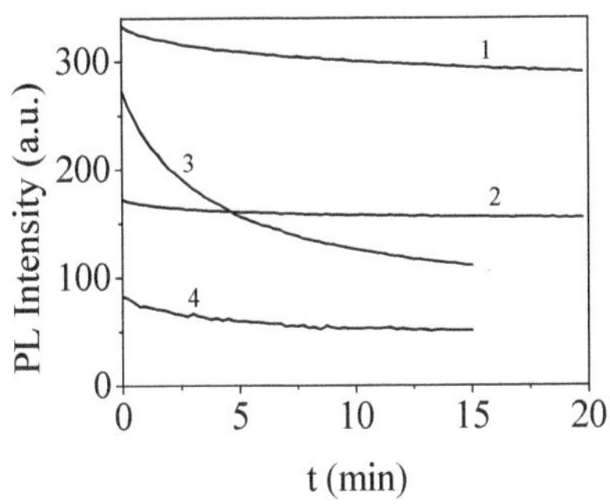

Figure 3. The PL intensity at 450 nm from the PAA:Mn specimens annealed at 600 °C (1, 2) and the intensities of R-lines of Mn^{4+} (3) and Cr^{3+} (4) ions versus the irradiation time at the excitation wavelengths of 325 (1, 3) and 275 nm (2, 4).

In order to discuss the behavior of PAA:Mn specimens the multivalent nature of manganese should be considered [4]. Charge compensation of Mn ions, as they are embedded into the PAA structure, can occur due to the formation of both charge-compensating pairs Mn^{2+}-Mn^{4+} and Mn^{2+}-V_o-Mn^{4+} and more intricate defect complexes also comprising oxygen vacancies such as Mn^{3+}-F^{+}-Mn^{4+}, Mn^{4+}-F-Mn^{4+}, Mn^{4+}-F_2^{2+}-Mn^{4+}. When UV radiation is absorbed, part of oxygen vacancies in the complex defects may participate in the change of manganese ions valence state owing to photoionization of F- and F_2-centers ($F + h\nu \rightarrow F^{+} + e^{-}$), *e.g.*, according to the scheme: Mn^{4+}-F-Mn^{4+} $+ h\nu_{ex} \rightarrow$ Mn^{4+}-F^{+} $+e^{-}$-Mn^{4+} $\rightarrow$ Mn^{3+}-F^{+}-Mn^{4+}.

The PL excitation wavelengths of the PAA and PAA:Mn annealed at 1300 °C located in the region of absorption bands of F^+-, F_2-, F_2^+- and F_2^{2+}-centers. Thus, Mn^{4+} ions in the complex defects can be excited as a result of energy transfer from oxygen vacancies and reabsorption of radiation due to F^+-centers ($\lambda_{em} = 480$ nm) by Mn^{4+} ions in the absorption band at 470 nm. In addition, when UV radiation is absorbed by F- and F_2-centers in the complex defects, changes of the manganese ions valence state and the charge state of oxygen vacancies may occur. In this case, disappearance of a pink color of the specimens due to lower concentration of Mn^{3+} ions ($Mn^{3+} \rightarrow Mn^{2+}$) should be observed together with a fall of the R-line intensity of manganese ions as a function of UV exposure time. Excitation of Cr^{3+} ions in the absorption bands at 410 and 560 nm can occurs via reabsorption of radiation by F- and F_2^{2+}-centers. With a prolonged UV exposure, the reduction of concentration of individual F- and F_2^{2+}-centers and these centers in complex defects can proceed in accordance with the following reactions $F + h\nu \rightarrow F^+ + e^-$, $F_2^{2+} + e^- \rightarrow F_2^+$. This process leads to the reduced R-line intensity of the chromium ions.

4. Conclusion

PL features of Mn-doped PAA have been studied in comparison with undoped PAA. Variation of PL for the specimens annealed ay different temperatures as well as the effect of UV irradiation upon PL have been monitored and explained by interaction of multivalent manganese ions with oxygen vacancies of different types.

References

1. N. Liu *et al.*, *Nature Mater.* 7, 31 (2008).
2. S. Stojadinovic *et al.*, *App. Surf. Sci.* **256**, 763 (2009).
3. N. I. Mukhurov *et al.*, *J. Appl. Spectrosc.* 77, 549 (2010).
4. D. T. Sviridov, R. K. Sviridova, Yu. F. Smirnov, *Optical Spectra of Transition Metal Ions in Crystals* (Nauka, Moscow, 1976).

MICRO-RAMAN STUDY OF $TlInS_2$ NANOCRYSTAL FORMATION IN Tl–In–As–S GLASS UNDER LASER ANNEALING

Yu. M. Azhniuk, A. V. Gomonnai, V. V. Lopushansky

Materials for Functional Electronics, Institute of Electron Physics Ukr. Nat. Acad. Sci.
Universytetska Str. 21, 88000 Uzhhorod, Ukraine
yu.azhniuk@gmail.com

O. O. Gomonnai

Department of Optics, Uzhhorod National University
Pidhirna Str. 46, 88000 Uzhhorod, Ukraine
center.uzh@gmail.com

V. M. Rubish

Uzhhorod Scientific and Technology Center, Institute for Information Recording
Ukr. Nat. Acad. Sci., Zamkovi Skhody 4, 88000 Uzhhorod, Ukraine
center.uzh@gmail.com

D. R. T. Zahn

Semiconductor Physics, Chemnitz University of Technology
D-09107 Chemnitz, Germany
zahn@tu-chemnitz.de

Raman spectra of Tl–In–As–S glasses measured at low laser power densities (P_{exc}≤0.4 MW/cm^2) confirm their amorphous structure. At P_{exc}≥ 0.6 W/cm^2, $TlInS_2$ nanocrystals are formed in the glass during the micro-Raman measurement. That is revealed by the appearance of the $TlInS_2$ phonon bands which finally dominate in the Raman spectra. This crystallization is irreversible and caused by local heating under the laser beam.

1. Introduction

$TlInS_2$ is a layered indirect-gap crystalline semiconductor which undergoes phase transitions to incommensurate and ferroelectric phases at low temperatures. It is interesting in view of applications in optoelectronics, in particular, due to its high photosensitivity in the visible spectral range enabling memory effects [1]. To our knowledge, no studies of $TlInS_2$-type nanocrystals (NCs) have been reported so far. Meanwhile, nanoscale ferroelectric materials are important in the view of both fundamental and practical aspects [1]. Recently, SbSI [2] and $TlInSe_2$ [3] ferroelectric NCs were formed in chalcogenide glass matrices by laser irradiation. In the present paper, we analyze

Raman scattering data directly confirming the formation of $TlInS_2$ NCs in Tl–In–As–S glass under laser illumination.

2. Experimental

Tl–In–As–S glasses were prepared from pre-synthesized As_2S_3 and $TlInS_2$. The details can be found in [3,4]. Ampoules with the mixture were heated to 760 K at the rate of 50 K/h, kept at 760 K for 5 h, heated again to 890–940 K at the rate of 70 K/h and kept at this temperature for 48 h with occasional stirring. Then, the melts were cooled down in air.

Micro-Raman spectra were recorded at 295 K using a Dilor XY800 spectrometer and a Kr^+ laser (λ_{exc} = 647.1 nm) at the laser power density P_{exc} up to 1 MW/cm^2. P_{exc} could not be higher due to a strong photoplastic effect (mass transfer of the sample material from the laser spot and formation of a pit on the surface [5]) resulting in the laser beam defocusing during the measurements.

3. Results and discussion

The Raman spectra of a series of $(As_2S_3)_{1-x}(TlInS_2)_x$ samples recorded at low P_{exc} are shown in Fig. 1a along with the spectra of As_2S_3 glass and $TlInS_2$ crystal. The dominating broad maximum at 335–345 cm^{-1} in the spectra of As_2S_3 glass consists of several components, the major contribution resulting from stretching vibrations of $AsS_{3/2}$ pyramids (AsS_3 pyramids linked together by As–S–As bonds) [6] while some of the constituents are related to As_4S_4 structural units [7]. The lower-frequency maxima at 187 and 233 cm^{-1} result from As–As vibrations in As_4S_4 units [7].

The Raman features observed for $(As_2S_3)_{1-x}(TlInS_2)_x$ glasses are broad confirming their amorphous structure. For $x \leq 0.4$ the spectra do not differ from that of As_2S_3 glass (Fig. 1a) while for higher x two intense broad features at 288 and 371 cm^{-1} emerge at the background of the most intense maximum. Comparison with the Raman spectra of $TlInS_2$ crystal (Fig. 1a) shows that the peak at 288 cm^{-1} is most likely related to In–S bonds (in $TlInS_2$ the corresponding intense peaks are ascribed to internal vibrations of In_4S_{10} tetrahedra [8]). The maximum at 371 cm^{-1} corresponding to stretching vibrations of $AsS_{3/2}$ pyramids [7] is strongly pronounced because of a decreasing intensity of other constituents which in As_2S_3 glass contribute to the broad maximum near 340 cm^{-1} (peaks at 346, 352, and 361 cm^{-1} are related to As–S bonds in As_4S_4 units [7]). These changes, as well as the observed gradual vanishing of the bands at 187 and 233 cm^{-1}, indicate a decrease of the amount of As_4S_4 structural units in the glass with $x > 0.4$.

With increasing P_{exc}, the Raman spectra of the $(As_2S_3)_{1-x}(TlInS_2)_x$ samples with $x \leq 0.4$ remain unchanged. However, for $(As_2S_3)_{0.5}(TlInS_2)_{0.5}$ rather sharp peaks in the low-frequency range appear at $P_{exc} \geq 0.6$ MW/cm^2 (Fig. 1b). Their spectral positions being close to those of crystalline $TlInS_2$ (Fig. 1a) related to translational vibrations [8]. Besides, the features at 288 and 371 cm^{-1} corresponding to In–S bonds, are clearly resolved at the background of the much less intensive Raman signal from As–S vibrations at 340 cm^{-1} (the topmost curve in Fig. 1b). A similar behavior is observed for $(As_2S_3)_{0.4}(TlInS_2)_{0.6}$. One may conclude that the laser illumination at sufficiently high $P_{exc} \geq 0.6$ MW/cm^2 induces the formation of $TlInS_2$ nanocrystals in the Tl–In–As–S glass.

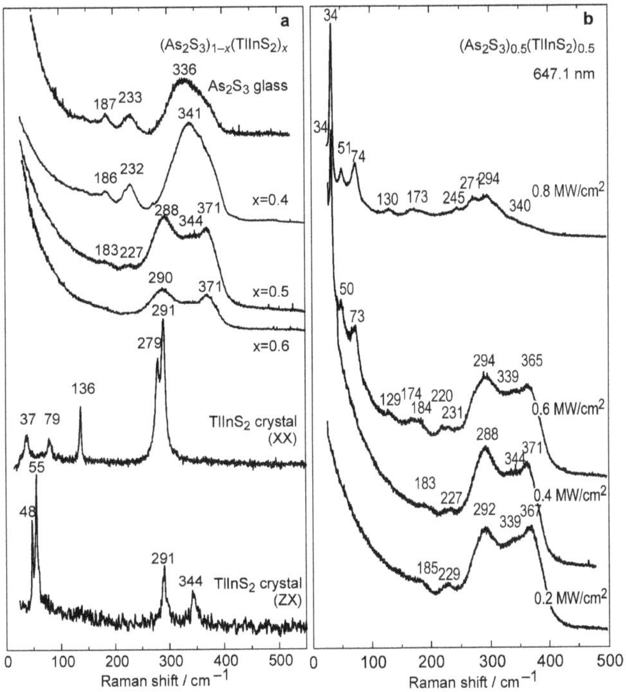

Figure 1. (a) Raman spectra of $(As_2S_3)_{1-x}(TlInS_2)_x$ samples measured at $P_{exc} < 0.4$ MW/cm^2 and the spectra of As_2S_3 glass and $TlInS_2$ crystal; (b) Raman spectra of $(As_2S_3)_{0.5}(TlInS_2)_{0.5}$ measured at different P_{exc}. For the spectra of $TlInS_2$ crystal $\lambda_{exc} = 514.5$ nm and for all other spectra $\lambda_{exc} = 647.1$ nm were used.

After the laser beam illumination was stopped and the second Raman measurement made from the same sample spot, the spectrum of the crystallized part remains the same with $TlInSe_2$ nanocrystal-related bands clearly resolved. Hence, the crystallization process of $TlInSe_2$ is irreversible. Evidently, the formation of $TlInS_2$ nanocrystals from the Tl–In–As–Se glass is caused by the local temperature increase in the spot subjected to the laser beam irradiation.

4. Conclusion

During the micro-Raman measurement of Tl–In–As–S glasses with the laser power density $P_{exc} \geq 0.6$ W/cm^2, $TlInS_2$ nanocrystals were formed in the amorphous glass matrix. This is clearly revealed by the appearance and increase of the $TlInS_2$-related phonon bands which finally dominate the Raman spectra both in the low-frequency range (translational vibrations of $TlInS_2$ lattice) and the high-frequency range (internal vibrations of In_4S_{10} tetrahedra in $TlInS_2$ structure). The observed crystallization effect is irreversible and caused by the local heating under the laser beam.

Acknowledgments

Y.M.A. thanks to DFG Research Unit SMINT (FOR 1713) for the financial support.

References

1. A. M. Panich, *J. Phys.: Condens. Matter* **20**, 293202 (2008).
2. Yu. M. Azhniuk *et al.*, *Mater. Res. Bull.* **47**, 1520 (2012).
3. Yu. M. Azhniuk *et al.*, *J. Phys. Chem. Solids.* **74**, 1452 (2013).
4. A.V. Gomonnai *et al., J. Crystal Growth* **367**, 35 (2013).
5. S. N. Yannopoulos, M. L. Trunov, *Phys. Status Solidi (b)* **246**, 1773 (2009).
6. S. N. Yannopoulos, K. S. Andrikopoulos, D. Th. Kastrissios, G. N. Papatheodorou, *Phys. Status Solidi (b)* **249**, 2005 (2012).
7. D. G. Georgiev, P. Boolchand, K. A. Jackson, *Phil. Mag.* **83**, 2941 (2003).
8. N. S. Yuksek, N. M. Gasanly, A. Aydinli, *J. Raman Spectrosc.* **35**, 55 (2004).

MEASUREMENTS OF ELECTRICAL PROPERTIES OF NANOSTRUCTURES

W. Nawrocki, M. Maćkowski

Faculty of Electronics and Telecommunications, Poznan University of Technology
60-965 Poznan, Poland
nawrocki@et.put.poznan.pl

We studied and measured electrical conductance quantization in metals and semiconductors. We propose to use the measurement of electrical conductance or resistance for characterization of geometrical dimensions of nanostructures.

1. Introduction

Very widely used nanostructures are elements of digital integrated circuits (ICs) of large scale of integration, such as processors in computers and smartphones. Transistors, resistors and conductive paths inside the chip are fabricated in a form of thin films of smaller and smaller thickness. At present (2016) microprocessors of fabrication technology nodes of 14 nm are manufactured. The technology of 10 nm nodes will be implemented next year and the technology of 5 nm ones is expected in nearest 5 years. The dimension of 10 nm or 5 nm means the length of a gate of MOSFET transistor inside the chip. Studies of nanostructures and testing them in a fabrication process can be helpful in the development of semiconductor industry.

2. Conductance quantization in nanostructures

Electric proprieties of electronic devices with sizes of nanometers are described by quantum theories of conductance. The quantum unit of electrical conductance $G_0 = 2e^2/h$ was predicted by R. Landauer in his theory of electrical conductance [1]. Parameters characterizing a nanosystem are the Fermi wavelength λ_F and the mean free path Λ ($\lambda_F = 2\pi/k_F$, where k_F is the Fermi wave-vector). For metals like gold $\lambda_F \approx 0.52$ nm is shorter than free electron path Λ ($\Lambda_{Au} = 40$ nm). For semiconductors the Fermi wavelength depends on the intensity of doping (see Table 1 [2]). If the length of the system is shorter than the free electron path, the impurity scattering is negligible, so the electrons transport is ballistic. If a nanostructure has a cross-section dimension comparable with the Fermi wavelength λ_F, and its length L is less than Λ, the system can be regarded as one-dimensional (1D), the electron regarded as a wave, and one can expect quantum effects illustrated in Fig. 1. A long Fermi wavelength in Si determined the limit for the 1st step of conductance quantization in the conducting paths inside ICs.

Table 1. Fermi wavelength λ_F and Fermi energy E_F in silicon doped from 10^{16} cm^{-3} to 10^{20} cm^{-3} (at 300 K) [2].

Material	Silicon	Silicon	Silicon	Silicon	Silicon	Gold
Doping n_o, cm^{-3}	10^{16}	10^{17}	10^{18}	10^{19}	10^{20}	no
Fermi energy E_F, eV	173×10^{-6}	0.80×10^{-3}	3.72×10^{-3}	17.3×10^{-3}	80×10^{-3}	5.53
Fermi wavelength λ_F (n_o), nm	93	43	20	9.3	4.3	0.52

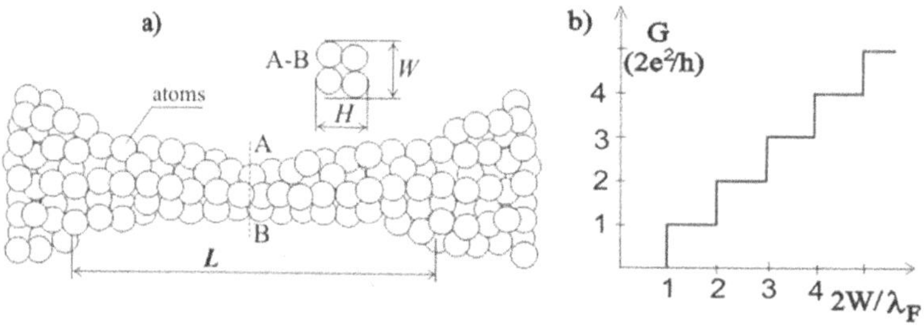

Figure 1. Conductance quantization in a nanowire (conductor with length $L<\Lambda$ and width W and thickness H comparable with the length of Fermi wave λ_F: a) a nanostructure formed by atoms; b) conductance quantization G versus width W.

The total electrical conductance of a nanowire is given by

$$G = \frac{2e^2}{h} N, \tag{1}$$

where N is the number of transmission channels. For 1D system, with thickness $H \leq \lambda_F$, the number of channels N depends on the width of the structure,

$$N = \mathrm{Int}(2W/\lambda_F). \tag{2}$$

For 2D system, with H, $W \geq \lambda_F$,

$$N = \mathrm{Int}(W \times H/\lambda_F^2), \tag{3}$$

where Int(A) means the integer of A. However, defects, impurities and irregularities of the shape of the conductor can induce scattering, then conductivity is given by the Landauer equation:

$$G = \frac{2e^2}{h} \sum_{i,j=1}^{N} t_{ij}, \tag{4}$$

where t_{ij} denotes probability of the transition from a jth to i^{th} state. In the absence of scattering thus (4) is reduced to (1). Fig. 1 presents a picture of a path (nanowire) in the form of a constriction in an electrical conductor with the dimension W (width), H (thickness) and L (length).

Let's consider a silicon path with the length $L = 20$ nm ($L<\Lambda_{Si}$), width $W = 12$ nm, thickness $H = 4$ nm and $\lambda_F = 4.3$ nm. Such path creates a nanowire with the conductance: $G_P = (2e^2/h)\times N$; $N = \text{Int}\,(W\times H/\lambda_F^2) = 2$; $G_P = 7.75\times10^{-5}\times2 = 15.5\times10^{-5}$ [A/V], accordingly to (3). The resistance of the path $R_P = 1/G_P = 6.5$ kΩ, thus it is surprisingly high. An electrical capacity of the path is 2 pF/cm, so for $L = 20$ nm $C_P = 4\times10^{-18}$ F.

Accordingly to (2), the conductance would be zero (number of channels $N<1$) if the width of the sample is $W\leq\lambda_F\,(n_o)/2$. The last inequality creates the important limit for miniaturization of devices (MOSFETs) and conducting paths in integrated circuits. Only silicon (or other semiconductor) with a heavy doping, $n_o \geq 10^{19}$ cm^{-3}, enables to apply a technology of 5 nm or smaller.

In our experiments the conductance was measured between two metalic electrodes, moved to contact by the piezoelectric tube actuator [2]. All experiments were performed at room temperature and ambient pressure. For conductors and semiconductors the conductance quantization in units of $G_0 = 2e^2/h$ was measured in many experiments. The quantization of conductance was evident. All characteristics showed the same steps equal to $2e^2/h$.

3. Measurements of geometrical dimensions of nanostructure

The effect of quantization of electrical conductance (resistance) with the quantum $G_0 = 2e^2/h$ occurs for nanostructures. This effect can be used to measure a width of nanostructures, or rather to estimate the width. It is an inverse problem in relation to the measurement of electrical conductance. Electrical measurements of geometric sizes of nanostructures are potentially very important for nanotechnology. They are relatively simple in general and the results are more accurate when nanostructures are small. The conditions necessary to perform the measurements of the geometric dimensions of nanostructures using electrical measurements are:

- the material of the nanostructure is an electrical conductor
- the nanostructure (sample) has a constriction between two wide terminals
- the thickness and the width of the nanostructure are comparable with the Fermi wavelength λ_F and the length of the constriction is shorter than the mean free path Λ in the bulk material.

A nanowire of the smallest size forms a chain of atoms connected one to another by path, $L < \Lambda$. The chain creates one transmission channel for free electrons. Thus, the conductance $G_0 = 7.75\times10^{-5}$ [A/V] of the nanowire indicates that the nanowire contains only a single atom at its narrowest point. The cross-sectional area of such a nanowire is the area of a circle defined by $2R$, where R is the radius of the atom. The conductance of $2G_0$ (or NG_0) of a nanowire means that in the nanowire there are 2 transmission channels (or N channels), respectively. The width of the nanowire at its narrowest point is W. On the basis of conductance G (or resistance R) and Eqs. (2) and (3) the width W of the nanostructures can be determined by the formula [2]:

$$W = 0.5\lambda_F\,[\mathrm{Int}\,(G/G_0) + 1]. \qquad (5)$$

For example, a gold nanowire shows resistance $R = 4.4$ kΩ, it means its conductance $G = 1/R = 22.7\times10^{-5}$ [A/V]. For gold, the Fermi wavelength $\lambda_F = 0.52$ nm. We calculate the width of the nanowire from (5): $W = 0.5\lambda_F\,[\mathrm{Int}\,(22.7/7.75) + 1] = 3\times0.52 = 1.56$ nm. We can estimate the length L of the nanowire as well. The length L is shorter than mean free path, $L < 40$ nm.

4. Conclusion

Estimations of electrical conductance in metallic and semiconductor nanostructures have been presented. Electrical conductance in metallic nanostructures was measured. The new method of measurement of geometrical dimensions of nanostructures was proposed in this paper.

Acknowledgments

The paper was prepared with the financial support of the project No 08/83/DSPB/4719 at Poznan University of Technology, Poland.

References

1. R. Landauer, *J. Phys.: Condensed Matter* **1**, 8099 (1989).
2. W. Nawrocki, Yu. Shukrinov, *Int. J. Electron. Electric. Eng.* **4**, 301 (2016).

ENERGY SPECTRUM OF ELECTRONS TUNNELING TO THE SURFACE STATES IN NANOCOMPOSITE BASED ON TiO_2

T. N. Sidorova, A. L. Danilyuk

Belarusian State University of Informatics and Radioelectronics, Minsk, Belarus
sharsu_antea@bk.ru

A simulation of tunneling of electrons excited by the sunlight in TiO_2 to surface states is performed. The surface states appear due to absorption of impurities and organic compounds on the TiO_2 nanocomposite in a gas environment. Dependences of the transmission coefficient of tunneling electrons are explained by the interference of transmitted and reflected waves under conditions of the complex potential relief.

1. Introduction

Composites with titanium dioxide (TiO_2) nanoparticles (nanocomposite) have the widest prevalence in heterogeneous photocatalytic processes of organic compounds oxidation [1]. These processes are realized by the electron-hole pairs generation excited by the sunlight in TiO_2 and their transfer to the surface states created by the organic compounds.

A spectrum of non-equilibrium electrons photogenerated in the conduction band of TiO_2 is resulted from the dependence of sunlight intensity on the photon wavelength [2]. A transfer of generated electrons to the surface states, which is determined by the tunneling transmission coefficient, depends on the surface potential reliefs created by the organic compounds on the TiO_2 surface. A real potential relief has a very complex form and includes alternated potential barriers and wells. As a result, the energy spectrum of tunneling electrons to the surface states of absorbed organic compounds depends on the sunlight intensity. Consequently, the photocatalytic activity and effectiveness of organic compounds oxidation on the TiO_2 surface is also determined by the sunlight intensity. Therefore, in order to evaluate and predict the photocatalytic activity of the TiO_2 surface it is necessary to estimate the spectrum of electrons tunneling to the surface states in dependence on the sunlight intensity. For this aim it is necessary to determine transmission coefficient of the tunneling electrons.

To determine transmission coefficients of the tunneling electrons, we developed the model based on the phase functions method. The model takes into account bulk and interface barrier parameters and the image force potential and allows to include the potential relief of the wide-gap semiconductor. The principal feature of the phase functions method is the ability to find out the transmission coefficients. According to this method, only wave function variation, as a result of potential actions, is calculated rather than the proper wave function. This process is time-consuming for complex potentials. Besides, it is difficult to estimate the faults of results derived.

The aim of this paper is modeling of the tunneling transmission of electrons excited by the sunlight through barriers generated by the surface states potential relief produced due to absorption of impurities and organic compounds.

2. Model

The tunneling transmission coefficient through a barrier of the width d which is described by the potential $U(x)$ is equal to: [4]

$$D(E) = \exp\left[\frac{1}{k}\int_0^d U(x)\left[b(x)\cos(2kx) - a(x)\sin(2kx)\right]dx\right], \tag{1}$$

where $k(E) = (8\pi^2 m^* E / h^2)^{1/2}$ is the tunneling electron wave vector, m^* and E are the effective mass and energy of the tunneling electron, h is the Planck constant. Functions $a(x)$ and $b(x)$ are defined from

$$\frac{da(x)}{dx} = \frac{U(x)}{2k}\left[-\sin(2kx) - 2b + (a^2 - b^2)\sin(2kx) - 2ab\cos(2kx)\right], \tag{2}$$

$$\frac{db(x)}{dx} = \frac{U(x)}{2k}\left[\cos(2kx) + 2a + (a^2 - b^2)\cos(2k\,x) - 2ab\sin(2kx)\right]. \tag{3}$$

To describe the potential barrier $U(x)$ the following equation is used:

$$U(x) = \left(8\pi^2 m^* / h^2\right)\left(U_0 - q\phi(x) \pm V_s(x)\right), \tag{4}$$

where U_0 is the maximal height of the TiO_2 surface potential barrier, $\varphi(x)$ is the image force potential, V_s is the potential which takes into account the surface irregularity. The above model was implemented for tunneling electron transmission from the bulk to the absorbed layer on TiO_2 surface illuminated by the light in neutral gas environment with organic compounds. For the generated electron-hole pairs division, TiO_2 nanocomposite is positioned on a Si substrate.

3. Results and discussion

Two forms of the potential relief are presented in Fig. 1. Curves 1–3 display the potential relief which consists of two potential barriers with one potential well between them. Curves 4, 5 display the potential relief consisting of one potential barrier.

In the first case (Fig. 2, curve 1), the dependence of the transmission coefficient on sunlight intensity demonstrates two peaks. The first one is located at about 200 W/m^2. The second peak appears in the range of 600–800 W/m^2; at the higher intensity $D(I)$ does not change (Fig. 2, curve 1). Within the framework of our model, by simultaneous varying several parameters of the relief one may obtain the dependences with two peaks or without peaks.

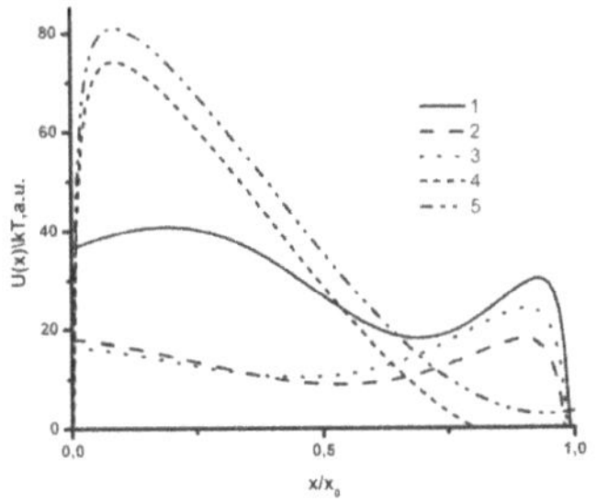

Figure 1. The forms of the potential barriers.

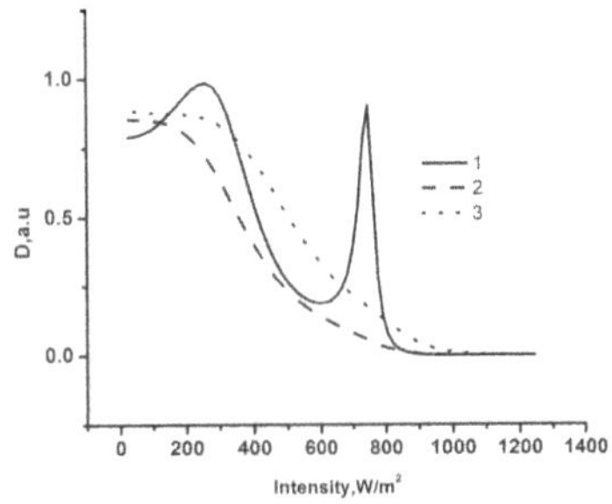

Figure 2. The dependence of the transmission coefficient on intensity of illumination.

When the potential relief consists of one potential barrier only (Fig. 1, curves 2, 3), the transmission coefficient changes monotonously. $D(I)$ decreases gradually and at the high sunlight intensity it drops sharply (Fig. 2, curves 2, 3).

When the potential relief consists of potential barrier displayed in Fig. 1, curves 4, 5, the transmission coefficient dependence contains only one peak at low intensities.

4. Conclusion

The modeling of tunneling transmission at the surface states of electrons excited by the sunlight was performed for TiO_2/Si nanostructure. The tunneling transmission coefficient of the surface states created by organic compounds was calculated by phase functions method.

The tunneling transmission energy spectrum of the TiO_2 surface states with organic compounds can be both monotonous and non-monotonous. It depends on the parameters and form of the potential relief. By modifying the potential relief, it is possible to obtain the dependence $D(I)$ both with one and two peaks. The non-monotonous behavior is characterized by the appearance of one peak (in the area of high intensities) or two peaks (at low and high intensities). The behavior of $D(I)$ can be explained by the interference of transmitted and reflected waves, a wave of over barrier transmission and a reflected wave in the conditions of complex potential barrier.

The results obtained by the simulation allow formulation of new recommendations for efficient control of photocatalytic processes of organic compounds oxidation on the surface of TiO_2 nanostructures under sunlight irradiation.

References

1. M. Barrera, J. Pla, C. Bocchi, A. Migliori, *Sol. Energy Mater.Sol. Cells* **92**, 1115 (2008).
2. T. Soga, *Nanostructured Materials for Solar Energy Conversion* (Amsterdam, Elsevier, 2006).
3. V. V. Babikov, *Phase Function Method in Quantum Mechanics* (Nauka, Moscow, 1976) – in Russian.
4. T. N. Sidorova, A. L. Danilyuk, V. E. Borisenko, F. Arnaud d'Avitaya, J.-L. Lazzari, *Proceedings of SPIE* 7377, 737708 (2009).
5. T. Sidorova, A. Danilyuk, *Mater. Phys. Mechan.* **20**, 106 (2014).

NANOELECTROMAGNETICS

INVITED

TERAHERTZ AND INFRARED PHOTODETECTORS BASED ON VERTICAL GRAPHENE VAN DER WAALS HETEROSTRUCTURES: CONCEPTS, FEATURES OF OPERATION AND CHARACTERISTICS

V. Ryzhii*[1], T. Otsuji[1], M. Ryzhii[2], D. Svintsov[3], V. Leiman[3], V. Mitin[4], M. S. Shur[5]

[1]*Research Institute of Electrical Communication, Tohoku University 980-8577 Sendai, Japan*

v-ryzhii@riec.tohoku.ac.jp

[2]*Department of Computer Science and Engineering, University of Aizu 965-8580 Aizu-Wakamatsu, Japan*

[3]*Laboratory of 2D Materials Optoelectronics, Moscow Institute of Physics and Technology, 141700 Dolgoprudny, Russia*

[4]*Department of Electrical Engineering, University at Buffalo, Buffalo 1460-1920 New York, USA*

[5]*Department of Electrical, Computer, and Systems Engineering and Department of Physics, Applied Physics, and Astronomy, Rensselaer Polytechnic Institute, Troy 12180 New York, USA*

We discuss the concept and evaluate performance of the terahertz (THz) and infrared (IR) photodetectors based on the recently proposed vertical graphene layer van der Waals heterostructures. The proposed graphene layer infrared photodetectors (GLIPs) are compared with other types of THz/IR phototodetectors.

1. Introduction

The gapless energy spectrum in graphene layers (GLs) enables their use in photodetectors and radiation sources operating in a rather wide spectral range, including the terahertz (THz) and infrared (IR) bands. Several types of the photodetectors using interband transitions in the GLs have been proposed and evaluated (see, for example [1–5]) in particular, THz/IR p-i-n photodiodes based on the lateral single- or multiple (non Bernal stacked) GL heterostructures [1]. Modification of the energy spectrum (opening of the energy gap) in the patterned GLs, such as the arrays of the graphene nanoribbons (GNRs) allows to diversify the spectral characteristics and vary other properties of the devices using such arrays.

The van der Waals (vdW) heterostructures [6–14] based on stacking of two-dimensional (2D) crystals and GLs open up additional opportunities in THz/IR optoelectronics. Figs. 1a and 1b show the sketches of the GL-vdW

* On leave from Institute of Ultra-High-Frequency Semiconductor Electronics of RAS, Moscow 117105, Russia

based device structures operating as the THz/IR photodetectors using the interband inter-GL tunneling transitions assisted by photons and using the inter-GL transitions stimulated by the electron heating due to the intra-GL photon (Drude-like) absorption, respectively [15,16]. Fig. 1c demonstrates the GL-infrared photodetector (GLIP) based on the GL-vdW heterostructure and a fragment of its band diagram. This device employs the interband generation of the electron-hole pairs in the GLs with the sufficiently energetic electrons followed by their tunneling to the continuum states above the inter-GL barriers.

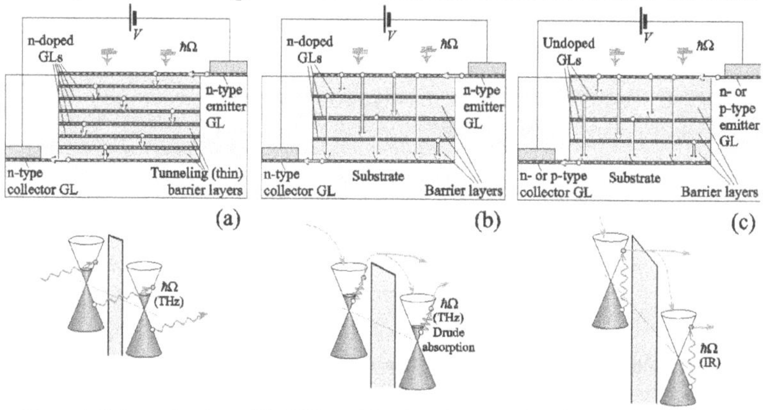

Figure 1. Schematic view of the THz and IR photodetectors based on vertical GL-vdW heterostructures (upper panels) with (a) interband tunneling inter-GL bound-to-bound tunneling transitions [15], (b) the thermionic emission from GLs of the electrons heated by THz radiation due to intraband (Drude) absorption [16], and (c) interband bound-to-continuum tunneling-assisted transitions of the photoexcited electrons. Lower panels show the fragments of the pertinent band diagrams. Wavy arrows correspond to the photon absorption, solid and dashed arrows indicate the escape and capture processes, respectively.

In this paper, we consider the concept of the recently proposed vertical GLIPs [17] (see also [18]), evaluate these photodetectors, and compare their characteristics with those of the lateral p-i-n GL photodiodes (GLPDs) and quantum-well photodetectors (QWIPs) based on the standard QW heterostructures.

2. The GLIP structure and its operation principle

As seen from Fig. 1c, the GLIP in question comprises a number of the GLs separated by the barrier layers. The extreme doped GLs serve as the emitter and collector. The inner (undoped) GLs play a role of the elements absorbing the incident radiation and producing the photoelectrons. Figs. 2a and 2b show the GLIP band diagrams without and with the THz/IR irradiation, respectively. The

GLIP operation is associated with the interband photogeneration of the electrons and holes in the GLs, tunneling escape of the photoelectrons from the GLs into the continuum states above the inter-GL barriers, propagation of the equilibrium electrons and photoelectrons injected from the emitter GL and the photoelectrons escaped from the inner GLs, capture of the propagating electrons into the inner and collector GLs.

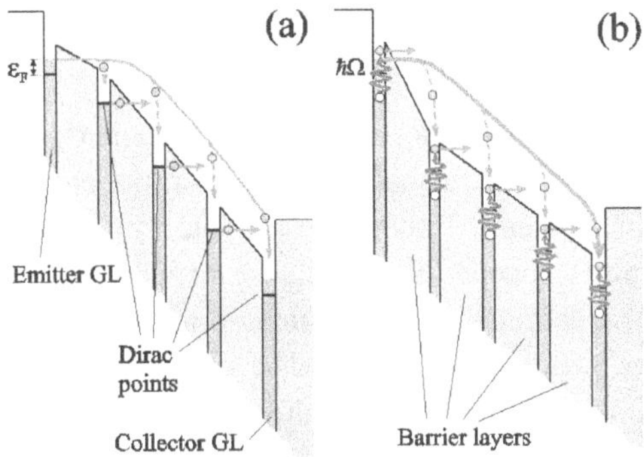

Figure 2. Band diagrams of a GLIP at a bias voltage (a) in the absence of irradiation and (b) under a strong irradiation.

In the dark conditions, the terminal current is associated with the electrons injected from the emitter GL (primarily from the states near the Fermi energy in the emitter GL) and the electrons injected from the inner GLs (primarily from the states near the Dirac points in these GLs). A fraction of the injected electrons is captured back into the GLs. In the steady-state situation, the electron injection from the inner GLs is compensated by the capture into the GLs. This electron balance corresponds to a nonuniform potential distribution across the structure with generally different electric fields, E_E and E_B, in the emitter barrier and the barriers in the structure bulk. Since the injection rate from the GLs is determined by the electric field in the respective barrier layer, the values of E_E and E_B appear to be such as to provide the balance mentioned above.

The THz/IR irradiation results in an increase of the electron injection from the inner GLs. This leads to the potential redistribution with an increasing electric field. Such a situation is shown in Fig. 2, where $E_E < E_B$ without irradiation and $E_E > E_B$ under a sufficiently strong irradiation. The variation of the current under the irradiation (via the potential redistribution) includes the variation of the current injected from the emitter substantially exceeding the contribution of the electrons photoexcited from the inner GLs (corresponding to the photoelectric gain or amplification).

One important feature of the GLIPs is that their operation does not require the mutual angular alignment of the GLs. To some extent, the operation mechanism of the GLIPs resembles that of the quantum-well infrared photodetectors (QWIPs) [20,21], although the specifics of the GL-vdW heterostructures and their characteristics provide marked distinctions in the operation and the enhanced performance.

3. The GLIP device model and characteristics

The GLIP device model includes the Poisson equation governing the electric potential distribution and the equations governing the electron balance in each GL under the steady-state irradiation. The electron balance in the GLs is described using the concept of the capture efficiency as a parameter. This corresponds to the electron capture rate being proportional to the current density j and the capture efficiency [20–24]. The capture efficiency $p \leq 1$ or even $p << 1$. In the framework of our model, the photocurrent density j_{photo} in a GLIP with N (N = 1, 2, 3, ...) inner GLs at not too strong radiation intensity I is presented as [16]

$$j_{photo} = \frac{e\beta I}{(\gamma^{3/2} + N)} \frac{\left[\gamma^{3/2} + \frac{(1-\beta)[1-(1-\beta)^N]}{p\beta}\right]}{1 + \frac{\tau_{esc}}{\tau_{relax}} \exp\left[\frac{\eta^{3/2}(\gamma^{3/2} + N)dE_{tunn}}{V}\right]}. \quad (1)$$

Here $\beta = \pi\alpha/\sqrt{\epsilon}$ is the probability of the interband absorption in GLs embedded in the medium with the refraction index ε, $\alpha = e^2/\hbar c \simeq 1/137$ is the fine structure constant, e and $\hbar$ are the electron charge and the reduced Planck constant, τ_{relax} and τ_{esc} are the time of the energy relaxation of a photoexcited electron in a GL and the time of its escape from a GL, respectively, $E_{tunn}\sqrt{\Delta}$ is the characteristic tunneling field for a triangular barrier of the height Δ, where Δ is the conduction band offset at the GL/barrier layer interface, d is the thickness of the inter-GL barrier layers, $\gamma = (\Delta - \varepsilon_F)/\Delta$, $\eta = (\Delta - \hbar\Omega/2)/\Delta$, ε_F is the Fermi energy in the emitter GL, $\hbar\Omega$ is the energy of the incident photons, and V is the bias voltage. Since normally $\gamma^{3/2}p << 1$, for the GLIPs with not too large number of the GLs ($N < \beta^{-1} \sim 40$), Eq. (1) can be rewritten as

$$j_{photo} \simeq \frac{\frac{e\beta I}{p}\frac{N}{(\gamma^{3/2} + N)}}{1 + \frac{\tau_{esc}}{\tau_{relax}} \exp\left[\frac{\eta^{3/2}(\gamma^{3/2} + N)dE_{tunn}}{V}\right]}. \quad (2)$$

It is instructive that the photocurrent density given by (1) and (2) is larger than the current created by the electrons photoexcited from the emitter GL by a factor p^{-1}, which can be very large. This is because the photoemission from the inner GLs stimulates a strong injection of the thermalized electrons from the emitter GLs resulting in a substantial amplification of the photocurrent (the effect of the photoelectric gain).

Using (1) or (2), one can find the GLIP detector responsivity $\Re$, which is defined as $\Re = Aj^{photo}/\hbar\Omega\Im$. Here A is the device area and $\Im$ is the incident IR radiation flux, so that, $\Im = I(1-r)$, where r is the coefficient of the IR radiation reflection from the GLIP top. As a result, we obtain

$$\Re \simeq \overline{\Re}\frac{\dfrac{N}{(\gamma^{3/2}+N)}}{1+\dfrac{\tau_{esc}}{\tau_{relax}}\exp\left[\dfrac{\eta^{3/2}(\gamma^{3/2}+N)dE_{tunn}}{V}\right]}. \tag{3}$$

Here $\overline{\Re} = \overline{R}\xi$, where $\overline{\Re} = e\pi\alpha/\hbar\Omega\, p$ and $\xi = (1-r)/\sqrt{\epsilon} = 4/\sqrt{\epsilon}(\sqrt{\epsilon}+1)^2$.

As seen from (3), the GLIP responsivity is weakly dependent on the number of the GLs N that due to the relative smallness of the emitter factor $\gamma^{3/2}$ if the average electric field $V/N\,d$ is fixed.

In GLIPs, the coefficient $\pi\alpha$ is relatively large and the capture efficiency p is relatively small ($p \simeq 0.5\,\%$ [20]) compared to the standard III-V based QW heterostructures. Therefore, the characteristic responsivity $\overline{R}$ can be fairly large. In particular, for $\hbar\Omega = 0.1-1.0$ eV and $p = 0.01$, we obtain $\overline{R} \simeq$ (2.3–23.0) A/W and somewhat lower values of $\Re$ (due to the radiation reflection and transmission effects).

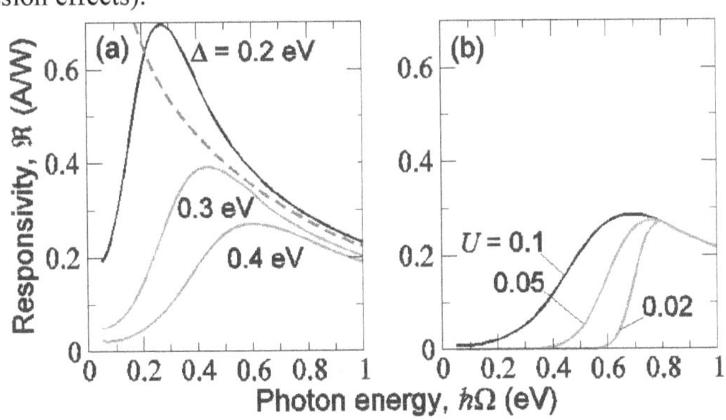

Figure 3. The responsivity $\Re$ versus photon energy $\hbar\Omega$ calculated for the GLIPs with (a) $N = 1$, $\tau_{esc}/\tau_{relax} = 0.1$, $p = 0.01$, $U = 0.1$, and $\Delta = 0.2-0.4$ eV (dashed line corresponds to $\Delta = 0.4$ eV and $U = 0.25$) and (b) $N = 2$, $\Delta = 0.4$ eV, and normalized voltages $U = 0.02-0.10$ ($\tau_{esc}/\tau_{relax} = 0.2$, and $p = 0.02$).

Fig. 3 shows the examples of the spectral dependences of the GLIP responsivity $\Re$ calculated using (3) for different structure parameters and the normalized bias voltages $U = V/dE_{tunn}$. As seen from Fig 3, the responsivity can be relatively large not only at $\hbar\Omega \simeq 2\Delta$ but in the subthreshold range $\hbar\Omega < 2\Delta$ as well. An increase in U leads to a marked increase in the responsivity in this spectral range. This provides an opportunity to control the GLIP spectral characteristics by the bias voltage and to use GLIPs in the THz range even at the relatively large Δ.

The dark current limited detector detectivity D^* is expressed via the responsivity $\Re$, the dark current density j_{photo}, and the gain g as [19,20]: $D^* = \Re/\sqrt{4egj_{dark}}$. Using this definition, for the GLIPs with $\gamma^{3/2} < N < \beta^{-1}$ one obtains [16]

$$D^* \propto \sqrt{N}\frac{\exp\left(\frac{NdE_{tunn}}{2V}\right)}{1+\frac{\tau_{esc}}{\tau_{relax}}\exp\left(\frac{N\eta^{3/2}dE_{tunn}}{V}\right)} = \sqrt{N}\frac{\exp\left(\frac{N}{2U}\right)}{1+\frac{\tau_{esc}}{\tau_{relax}}\exp\left(\frac{N\eta^{3/2}}{U}\right)}. \quad (4)$$

Eq. (4) immediately shows that the GLIP detectivity increases with increasing the number of the inner GLs N. The spectral dependence of D^* repeats that of $\Re$, shown in Fig. 3. The detectivity D^* markedly increases with the decreasing voltage V. This is due to the pertinent decrease in the dark current. However, since at small V, the tunneling current from the emitter GL can become smaller than the thermionic current, there is an optimum value of U, at which the detectivity reaches its maximum.

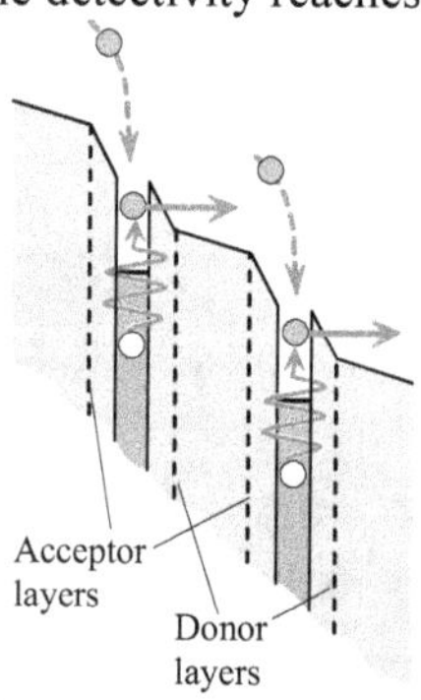

Figure 4. Fragment of the band structure of a GLIP with the selective doping of the barrier layers.

The GLIP responsivity and detectivity are sensitive to the shape of the tunneling barriers. One can expect that both the responsivity and detectivity can be increased in the GLIPs in which the undoped inter-GL barriers are replaced by the properly doped ones, in particular, by the selectively doped barriers as shown in Fig. 4. An interesting feature of the GLIPs with the selectively doped barriers is their ability to operate in a relatively low range of photon energies, in particular, in the range of several THz.

4. Comparison with other IR photodetectors

One can single the lateral p-i-n GLPDs out among a variety of GL and GNR photodetectors. Fig. 5 schematically shows the device structures of these photodiodes. Their operation is associated with an interband photogeneration of electron-hole pairs in the i-region depleted due to the reverse bias between the p- and n- regions. The built-in electric field forces the electrons and holes to move along the GLs creating the terminal photocurrent. The features of the p-i-n GLPDs (apart from their sensitivity to the radiation with the THz frequencies and much higher frequency radiation) are the relatively high interband absorption coefficient, the possibility of a fairly high speed of the electron transport, slow recombination and interband thermal generation due to high optical phonon energies in the GLs. Using the multiple GL structures (non-Bernal stacked) as the photosensitive elements allows reaching absorption efficiency close to unity and increasing the photodetector responsivity by a factor close to the number of the GLs N.

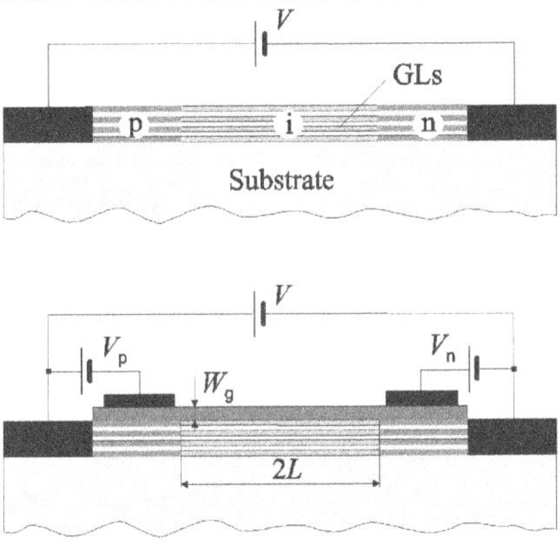

Figure 5. Schematic views of THz and IR p-i-n photodiodes based on lateral multiple-GL and -GNR heterostructures with chemically doped (upper panel) and electrically induced (lower panel) p- and n-regions.

Eqs. (3) and (4) from [1] show that the responsivity, $\Re$, of a GLIP and the responsivity, $\Re_{RGLPD}$, of a GLPD with a lateral p-i-n junction are approximately given by the following formulas (at $\tau_{esc} << \tau_{relax}$):

$$\Re \simeq \frac{e\beta}{\hbar\Omega p}, \qquad \Re_{GLPD} \simeq \frac{2e\beta N}{\hbar\Omega}, \tag{5}$$

respectively. Hence

$$\frac{\Re}{\Re_{GLPD}} \simeq \frac{1}{2pN}. \tag{6}$$

Since the capture efficiency p can be very small, the above ratio can be quite large.

As the electron transit time in the GLIPs (which are vertical transport devices) can be shorter than in the GLPDs (using the lateral transport), the GLIPs can surpass the GLPDs in speed.

The comparison of GLIPs and QWIPs (based, in particular, on InGaAs/AlGaAs and other heterostructures with the n-type QWs) operating in the mid- and near IR ranges, shows that GLIPs can exhibit the following advantages: (a) sensitivity to the normally incident radiation (no radiation coupling structures are needed); (b) no GL doping required, (c) a higher probability of the electron interband photoexitation in GLs and, hence, a higher photo-assisted escape to the continuum states above the barriers due to a relatively large universal constant $\pi\alpha \simeq 2.3$ % (compare with the pertinent value for a GaAs QW [20] ~ 0.5 % with the typical electron density $\sim 5\times10^{11}$ cm^{-2}); (d) a smaller capture efficiency of the electrons propagating over the barriers into the GLs (up to the orders of magnitude as compared the data from [20] and [23]). These advantages imply that the GLIP responsivity can exceed that of QWIPs. Since the modulation bandwidths of both GLIP and QWIPs is determined by the vertical electron transport across the barrier layers, one might assume that these bandwidths could be closed.

The GLIPs can also surpass or compete with some other photodetectors, for example, the uni-travelling-carriers PDs (UTC-PDs) [25]. The photodetectors with the structures similar to the GLIPs can also be based on the HgCdTe heterostructures with multiple zero-gap QWs. However, the features of these heterostructures require a separate consideration.

5. Conclusion

We have considered the GLIPs (infrared photodetectors based on vdW heterostructures) with the GLs serving as an emitter, collector and absorbing layers. We showed that due to small capture probability of the excited electrons into the GLs and a relatively strong interband photon absorption, the GLIPs can exhibit elevated photoelectric gain and responsivity. An increase in the number of GLs in the GLIP heterostructure can provide relatively high values of the GLIP detectivity. Among other GLIP advantages there are highly conducting transparent contact GL layer, sensitivity to the normally incident radiation and a high speed operation. The GLIP performance can exceed that of other already realized and newly proposed photodetectors for the imaging and optical communication systems and THz photomixers.

Acknowledgments

The work was supported by the Russian Scientific Foundation (grant 14-29-00277) and the Japan Society for Promotion of Science, KAKENHI (grant 16H06361). The work by D.S. and V.L. was supported by the Russian Foundation for Basic Research (grant 16-37-60110). V.M. and M.S. acknowledge the support by the RIEC Nation-Wide Cooperative Research Project and by the US ARL Cooperative Research Agreement, respectively.

References

1. V. Ryzhii *et al.*, *J. Appl. Phys.* **107**, 054512 (2010).
2. T. Mueller, F. N. A. Xia, P. Avouris, *Nat. Photon.* **4**, 297 (2010).
3. F. Bonaccorso, Z. Sun *et al.*, *Nat. Photon.* **4**, 611 (2010).
4. V. Ryzhii, N. Ryabova *et al.*, *Opto-Electron. Rev.* **20**, 15 (2012).
5. A. Tredicucci, M. S. Vitielo, *IEEE J. Sel. Top. Quantum Electron.* **20**, 130 (2014).
6. C. H. Liu *et al.*, *Nat. Nanotechnol.* **9**, 9273 (2014).
7. A. K. Geim, I. V. Grigorieva, *Nature* **499**, 419 (2013).
8. L. Britnell, R. V. Gorbachev *et al.*, *Nano Lett.* **12**, 1707 (2012).
9. T. Georgiou, R. Jalil *et al.*, *Nat. Nanotechnol.* 7, 100 (2013).
10. L. Britnell, R. V. Gorbachev *et al.*, *Nat. Commun.* **4**, 1794 (2013).
11. G. W. Mudd, S. A. Svatek *et al.*, *Advanced Mat.* **27**, 3760 (2015).
12. F. Xia, H. Wang, Di Xiao *et al.*, *Nat. Photon.* **8**, 899 (2014).
13. V. Ryzhii, T. Otsuji *et al.*, *Appl. Phys. Lett.* **104**, 163505 (2014).
14. Q. Ma, T. I. Andersen, N. L. Nair *et al.*, *Nat. Physics* **12**, 455 (2016).
15. V. Ryzhii, T. Otsuji, M. Ryzhii *et al.*, *2D Materials* **2**, 025002 (2015).
16. V. Ryzhii, A. Satou, T. Otsuji *et al.*, *J. Appl. Phys.* **116**, 114504 (2014).
17. V. Ryzhii, M. Ryzhii, D. Svintsov *et al.*, *arXiv:* 1609.01381 (2016).
18. V. Ryzhii, M. Ryzhii, V. Mitin *et al.*, *J. Appl. Phys.* **113**, 174505 (2013).
19. K. K. Choi, *Physics of Quantum Well Infrared Photodetectors* (World Scientific, Singapore, 1997).
20. H. Schneider, H. C. Lui, *Quantum Well Infrared Photodetectors* (Springer, Berlin, 2007).
21. V. Ya. Aleshkin *et al.*, *J. Phys. Soc. Japan* **84**, 094703 (2015).
22. V. Ryzhii *et al.*, *J. Appl. Phys.* **117**, 154504 (2015).
23. E. Rosencher *et al.*, *IEEE J. Quantum Electron.* **30**, 2975 (1994).
24. H. C. Liu, *Appl. Phys. Lett.* **60**, 1507 (1992).
25. T. Ishibashi *et al.*, *IEICE Trans. Electron.* **E83**, 938 (2003).

AVOIDING NON-RADIATIVE DECAY PATHWAYS IN EMITTER-NANOANTENNA SYSTEMS FOR LARGE-SCALE TECHNOLOGICAL IMPLEMENTATIONS

A. Rakovich, R. Berté, J. Cambiasso, S. A. Maier
Department of Physics, Imperial College London, London, SW7 2AZ, United Kingdom
a.rakovich@imperial.ac.uk

Nanoantennas have long been at the centre of attention for enhancement of radiative properties of fluorophores. However, in many cases, the efficiency of such a system is limited by the concurrent enhancement of non-radiative decay pathways, restricting the operational window of the device to a narrow spectral region. Furthermore, recent years saw a drive towards large-area implementation of nanoantenna-based devices, adding further limitations on the design and performance of these systems, owing mostly due to the requirement of achieving a reproducible as-designed performance for each device. Here, the implications of these restrictions are discussed, and approaches to overcome them are proposed, in the context of developing efficient and reproducible nanostructure-based devices.

1. Introduction

Within the domain of nanosized photonic devices for implementation into current and future technologies, plasmonic nanoantennas have shown considerable promise due to their extraordinary properties of light confinement and enhanced light-matter interactions in the near field, *e.g.* surface enhanced fluorescence (SEF) of fluorophores when placed in antenna hotspot. The implementation of plasmonic devices, however, has been hindered by the characteristic and sometimes large losses, and lack of consistent behaviour, among others.

With respect to latter, the issue of reproducibility of emitter-plasmonic systems has been previously discussed [1]. In summary, the main cause of the irreproducibility in these systems is the inconsistency of the fabrication method when it comes to small-gap systems (which tend to be the most efficient antennas). This is aggravated by the size-dispersion when colloidal semiconductor quantum dots (QDs) are utilized as an emissive entity, or the lack of localization techniques for the case of molecular fluorophores leading to inconsistent numbers of molecules interacting with the antenna. One possible solution to both these issues is an employment of large-gap plasmonic antenna, which allows exploitation of the collective behaviour of semiconductor QDs. The consistency of this approach has been shown in our previous work, where the radiative behaviour of QDs selectively placed inside plasmonic ring cavities (PRCs) was investigated [1,2]. Of note, however, is the fact that very little fluorescence enhancement (585 nm) was observed in this system, both due to small electric field enhancements by such a plasmonic system and, most importantly, due to the significant enhancement of non-radiative rates by the coupling of QDs to the PRC.

2. The challenge of the non-radiative pathways

The modification of the radiative behaviour of a fluorophore, when in the presence of a plasmonic antenna, is typically described as a two-part process [3]. Firstly, the excitation of the fluorophore is increased proportionally to the enhancement of the electric field intensity in the nanoantenna hotspot. Secondly, the emission process is modified through coupling to the plasmonic structure, due to an increased local density of states. Overall, the enhancement of the fluorescence can be estimated using:

$$SEF = \left(\frac{\mu E}{\mu_0 E_0} \right)^2 \frac{\eta}{\eta_0}, \tag{1}$$

where the $E(E_0)$ is the local electric field, and $\mu(\mu_0)$ and $\eta(\eta_0)$ denote the dipole orientation factor and the quantum yield of the fluorophore in the presence (absence) of antenna. It is noteworthy that, in the case of a permanently fixed dipole, the dipole orientation factor is unlikely to change.

However, there is a great importance of the fact that the modification of the quantum yield of the fluorophore by the antenna is a superposition of the changes to both its radiative (γ_r) and non-radiative rates (γ_{nr}), that is:

$$\frac{\eta}{\eta_0} = \frac{\gamma_r}{\gamma_{r0}} \frac{\gamma_{r0} + \gamma_{nr0}}{\gamma_r + \gamma_{nr}}. \tag{2}$$

In practical terms, this limits the spectral range where overall radiative enhancement can be obtained. In fact, it has been shown that for a fluorophore with a high initial quantum yield, the corresponding factor in (2) is always less than 1. Thus fluorescence can only be enhanced by an increased excitation of the fluorophore [4]. Irrespective of the initial quantum yield, however, minimization of the non-radiative decay pathways is a pre-requisite for an efficient photonic system.

In emitter-PRC systems, the non-radiative rates are enhanced much more than the radiative counterparts up to ~650 nm. However, in this spectral region the electric field confinements are moderate, *i.e.* not large enough to overcome the reduction in the quantum yield (Figs. 1C and 2A). On the other hand, above 650 nm, the radiative and non-radiative enhancements become comparable, but no "bright" modes exist for the PRC in this spectral region due to a lack of a dipolar character of these modes (Figs. 1B-D). Both issues can be addressed by insertion of a disk off-centre and inside the PRC.

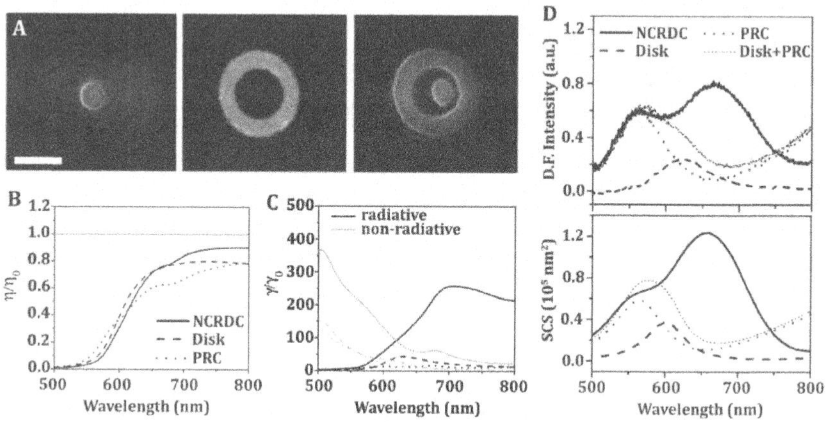

Figure 1. Comparison between disks, PRCs and NCRDCs. (A) SEM images of the 40 nm high disk, PRC and NCRDC on a quartz substrate, from left to right respectively. The white bar indicates 200 nm length. (B) and (C): Fluorophore's quantum yield and decay rates' modification by antennas in (A), as calculated by FDTD method. (D) shows experimental scattering spectra (top panel) and theoretical scattering cross-sections (bottom panel) of the fabricated structures.

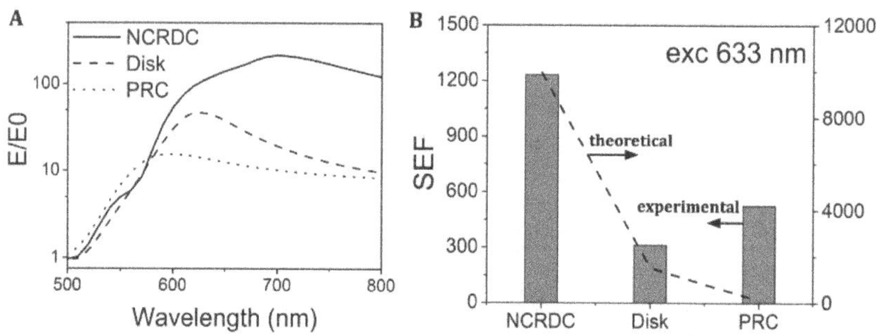

Figure 2. Electric field (A) and fluorescence enhancement (B) for PRCs, disks and NCRDCs. In (B), the theoretical fluorescence enhancement (dashed line) was calculated for a single position of the dipole (10 nm away from PRC and disk, and in the middle of 20 nm gap of NCRDC). Experimental SEF is an average of several QDs placed within a small area close to the plasmonic structure.

The non-concentric geometry increases the local electric field strength and improves radiative rate enhancements (Figs. 1B-C and 2A). Additionally, it imposes bright-character to the dark modes of the PRC making it possible for these modes to propagate to the far field (Fig. 1D), and thus obtain reasonable fluorescence enhancements for these structures (Fig. 2B) at wavelengths above 600 nm.

3. Conclusion

When developing plasmonic systems for technological implementation, the greatest challenge is presented by the same properties of the nanoantenna as those that are the most advantageous, namely strong field confinements and enhanced interactions. A reproducibility of emitter-nanoantenna system can be achieved in large-hotspot geometries, however, an enhancement of non-radiative decay rates limits the fluorescence enhancement. Inclusion of an additional plasmonic element with a strong dipolar character can improve the enhancement of fluorescence in a large-hotspot antenna, while maintaining behaviour deviation to a minimum level.

Acknowledgments

The authors would like to acknowledge EPSRC Active Plasmonics and Reactive Plasmonics programmes, the Leverhume trust and the Royal Society (UF150542) for funding the research presented here. Dr. Pablo Albella, Dr. Emiliano Cortes, Dr. Vincenzo Giannini and Dr. Tyler Roschuk are also acknowledged for their support.

References

1. A. Rakovich, P. Albella, S. A. Maier, in: *Physics, Chemistry and Applications of Nanostructures*, eds. V. E. Borisenko, S. V. Gaponenko, V. S. Gurin, C. H. Kam (World Scientific, Singapore, 2015), pp. 65–69.
2. A. Rakovich, P. Albella, S. A. Maier, *ACS Nano* **9**, 2648 (2015).
3. V. Giannini, A. I. Fernandez-Domínguez, S. C. Heck, S. A. Maier, *Chem. Rev.* **111**, 3888 (2011).
4. V. Giannini, J. A. Sánchez-Gil, O. L. Muskens, J. Gómez Rivas, *J. Opt. Soc. Am. B* **26**, 1569 (2009).

TUNING THz TRANSITIONS IN A QUANTUM RING WITH TWO GATES

T. P. Collier[1], V. A. Saroka[1,2], M. E. Portnoi[1]
[1]*School of Physics, University of Exeter, Exeter EX4 4QL, United Kingdom*
tpc207@exeter.ac.uk
[2]*Institute for Nuclear Problems, Belarusian State University*
Bobruiskava 11, 220030 Minsk, Belarus

We investigate theoretically the system of a quantum ring between two parallel charged wires acting as lateral gates. This system is shown to be a tuneable double quantum well. The energy splitting between the ground and first excited state for an electron in the quantum ring corresponds to the THz frequency range and is shown to be tuneable by the voltage applied to the wires. We study the corresponding optical transitions and show that their polarization properties are controlled by the sign of the voltage on the lateral gates.

1. Introduction

Bridging the THz gap of the EM-spectrum for which practical technologies lack the ability to produce or detect coherent radiation, is one of the most formidable problems of modern applied physics. The literature provides a variety of proposals for practical THz devices. The most of proposed solutions are based on semiconductor nanostructures hosting multiple quantum wells, such as optically induced double quantum wells [1] and quantum cascade lasers [2]. Promising candidates are also non-simply connected nanostructures, such as carbon nanotubes [3] or quantum rings [4] (QRs). The appeal in using these latter structures, in lieu of those former, lies in their tuneability with external fields. Subjecting a QR to external electric and magnetic fields results in a range of fascinating phenomena [4–6], predominantly resulting from the Aharonov-Bohm (AB) effect. However exploiting the AB properties of a QR requires several Tesla, limiting somewhat our ability to propose compact and portable QR-based THz devices.

In this paper, we theoretically investigate the system of an infinitely-narrow, single-electron QR under the influence of two charged wires, and show that this system is a tuneable double quantum well. The control of energy separations is therefore achieved via tuning the double well parameters with experimentally attainable applied voltage. We consider the corresponding transition dipole matrix elements between the lowest energy states and discuss the related THz transitions.

2. Quantum ring between two charged wires

Consider a system of a QR sandwiched in-between two charged wires, modelled as infinitely long, uniformly charged and perfectly parallel (Fig. 1). Under an appropriate expansion whereby the QR is much smaller than the system dimensions, $R/d \ll 1$, the dimensionless potential along the QR may be expressed as

$$V(\varphi) = 2\beta\frac{d}{R}(\gamma-1)\cos(\varphi) + \beta\gamma\cos(2\varphi), \tag{1}$$

where $\beta = k\lambda R^2/2d^2\varepsilon_1(0)$ and $\varepsilon_1(0) = \hbar^2/2m_e R^2$ lies in the THz range, with m_e representing the electron effective mass. The angle φ is taken from the horizontal axis as shown in Fig. 1. We have defined the asymmetry term γ as the ratio of the charge densities divided by the ratio of the distances $\gamma = \lambda_1 d_2/\lambda_2 d_1$ and dropped the subscripts, $\lambda_1 \equiv \lambda$ and $d_1 \equiv d$.

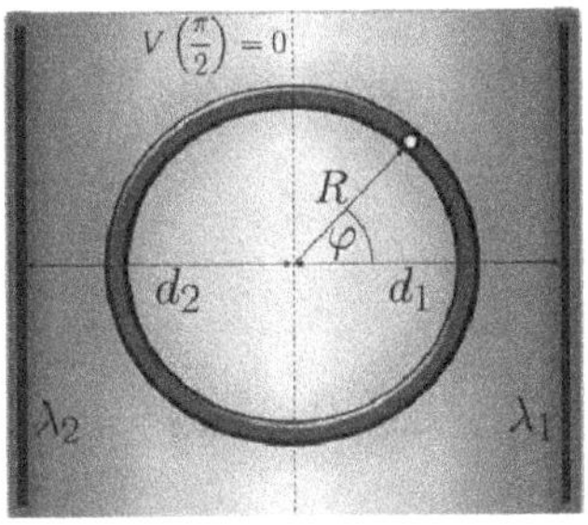

Figure 1. Geometry of the considered QR system.

Substituting (1) into the 1D stationary Schrödinger equation for a particle on a ring, we seek solutions as a linear combination of the 2π-periodic basis functions

$$\Psi_n(\varphi) = \sum_{m=-\infty}^{m=+\infty} c_m^{(n)} e^{im\varphi}, \tag{2}$$

where m is the angular momentum quantum number and n labels the nth eigenstate. The corresponding infinite set of simultaneous equations for the coefficients $c_m^{(n)}$ reads as

$$\left(m^2 - a_n\right)c_m^{(n)} + \beta\frac{d}{R}(\gamma-1)\left(c_{m-1}^{(n)} + c_{m+1}^{(n)}\right) + \frac{\beta\gamma}{2}\left(c_{m-2}^{(n)} + c_{m+2}^{(n)}\right) = 0. \tag{3}$$

Here $a_n = \varepsilon_n/\varepsilon_1(0)$ is the dimensionless n^{th} energy eigenvalue. The lowest two levels, with the corresponding wave functions superposed on top, are shown in the left panel of Fig. 2 ($\gamma = 1$ and $\beta = 0.9$). The right panel shows the energy splitting between the levels, calculated by truncating Eq. (3). at $|m| = \pm 13$ and diagonalising the resulting matrix.

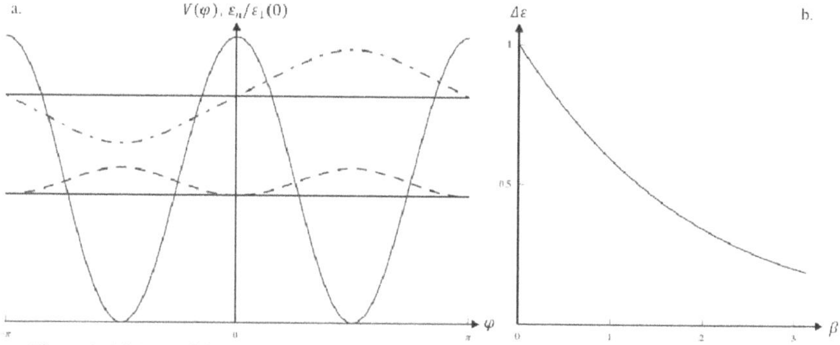

Figure 2. (a) Potential and lowest two energy levels for $\beta = 0.9$ and $\gamma = 1$, with wave functions sketched on top. (b) Energy splitting between lowest two energy levels as a function of β with $\gamma = 1$.

For small potential strengths, β<<1, the dimensionless energy difference between the ground state and the first excited state up to 2nd order in perturbation theory is

$$\Delta\varepsilon = 1 - \frac{\beta\gamma}{2} + \frac{3\beta^2\gamma^2}{32} + \frac{5d^2}{3R^2}\beta^2(\gamma-1)^2. \tag{4}$$

In the opposite limit of deep wells and high potential barriers, control can be shown to arise from barrier-dependent tunneling between the wells. Should the carrier in the QR be created via optical excitation, the potential well for an electron would correspond to a potential barrier for a hole. The exciton for such a system then becomes long-lived, and a QR between two charged wires may be used for exciton storage.

3. Transition dipole matrix elements

The transition rate between the electron ground state and first excited state is governed by the transition dipole matrix element

$$P_{01}(\theta) = eR\int \Psi_1^* \Psi_0 \cos(\theta - \varphi)\,d\varphi, \tag{5}$$

where e is the elementary charge and θ is the angle between the projection of linearly-polarized radiation onto the plane of the QR and the normal to the wires. Considering that Ψ_0 is a symmetric function whereas Ψ_1 is antisymmetric, it can readily be shown that $P_{01}(\theta)$ has a sinusoidal dependence. The transition rate induced by linearly polarized radiation reaches its maximum at $\theta = \pi/2$, conversely at $\theta = 0$ optical transitions are forbidden. Changing the voltage signs on the wires exchanges the positions of barriers and wells, resulting in a maximum transition rate induced by light polarized perpendicular to the wires at $\theta = 0$, and forbidden transitions for $\theta = \pi/2$.

4. Conclusion

We show that the system of a QR between two charged wires is a tuneable double quantum well. The energy splitting between the ground state and first excited state for an electron in a QR can therefore be tuned via the charge densities of the wires. In the limit of small potential strength this can be expressed using perturbation theory and in the opposing limit of large potential this tuneability can be expressed using WKB theory. As this energy splitting lies in the THz regime we have shown that THz transitions can be realized with this QR system. Furthermore, changing the energy and polarization of such transitions does not require the presence of large magnetic fields which would otherwise limit the portability of any proposed QR-based THz device. Finally, such a system holds implications for exciton storage.

Acknowledgments

This work was supported by the UK EPSRC CDT in Metamaterials XM^2 (TPC), EU FP7 ITN NOTEDEV (VAS), FP7 IRSES projects CANTOR and INTERNOM, and H2020 RISE project CoExAN.

References

1. H. G. Roskos, M. C. Nuss, J. Shah, K. Leo, D. A. B. Miller, A. M. Fox, S. Schmitt-Rink, K. Köhler, *Phys. Rev. Lett.* **68**, 2216 (1992).
2. J. Faist, F. Capasso, D. D. L. Sivco, C. Sirtori, A. L. Hutchinson, A. A. Y. Cho, *Science* **264**, 553 (1994).
3. M. E. Portnoi, M. Rosenau da Costa, O. V. Kibis, I. A. Shelykh, *Int. J. Mod. Phys. B* **23**, 2846 (2009).
4. A. M. Alexeev, M. E. Portnoi, *Phys. Rev. B* **85**, 245419 (2012).
5. A. M. Fischer, V. L. Campo, M. E. Portnoi, R. A. Roemer, *Phys. Rev. Lett.* **102**, 2 (2009).
6. K. L. Koshelev, V. Y. Kachorovskii, M. Titov, *Phys. Rev. B* **92**, 235426 (2015).

TERAHERTZ TRANSITIONS IN NARROW-GAP CARBON NANOTUBES AND GRAPHENE NANORIBBONS

V. A. Saroka
School of Physics, University of Exeter, Exeter, EX4 4QL, UK
v.saroka@exeter.ac.uk

R. R. Hartmann
Physics Department, De La Salle University, Manila, Philippines

M. E. Portnoi
School of Physics, University of Exeter, Exeter, EX4 4QL, UK

Interband dipole transitions are calculated in metallic single-walled carbon nanotubes and armchair graphene nanoribbons. It is shown that the curvature effects for the tubes and the edge effects for the ribbons result not only in a small band gap opening, corresponding to THz frequencies, but also in a significant enhancement of the transition probability rate across the band gap. Such enhancement makes these nanostructures promising for sources and detectors of THz radiation.

1. Introduction

Creating reliable, portable, tunable sources and detectors of terahertz radiation is one of the most challenging tasks of contemporary applied physics. One of the recent trends in bridging the so-called THz gap is a use of carbon-based nanostructures [1]. Several original schemes utilizing unique electronic properties of carbon nanotubes (CNTs) and graphene for THz applications were brought forward by our group [2–4]. These schemes include THz generation by hot electrons in quasi-metallic CNTs, frequency multiplication in chiral-nanotube-based superlattices controlled by a transverse electric field, tunable THz radiation detection and optically-pumped emission in metallic CNTs in a strong magnetic field, and using graphene *p-n* junctions for sub-wavelength polarization-sensitive THz detection. In this paper, we focus on direct interband dipole transitions in narrow gap CNTs and graphene nanoribbons.

2. Carbon nanotubes

Carbon nanotubes are cylinders made from a single sheet of graphene rolled along a particular direction specified by the chiral vector (n, m) [5]. Within a simple tight-binding model a CNT can either be metallic or semiconducting. If

$n - m = 3p$, where p is an integer, the tube is metallic and has a linear energy dispersion, with the conduction and valence bands touching at a point called the Dirac point. For zigzag CNTs $n = 3p$, $m = 0$ and the Dirac point is positioned in the center of the Brillouin zone (BZ). This simple model, however, does not take into account the effect of curvature, which plays an important role for small diameter tubes (1–2 nm).

It is now well-established that curvature effects encompass three main contributions: the C-C bond contraction, π-orbital tilting and π- and σ-orbitals mixing. All three contributions can be treated within the tight-binding model by introducing corrections to the hopping integrals. With the exception of armchair CNT (n, n), these corrections result in a small band gap opening at the Dirac point of metallic CNTs. This means that these CNTs are in fact quasimetallic and have band gaps up to 30 meV. These tiny band gaps do not have much of an effect on the CNTs' transport properties at room temperature because real samples of CNTs are always spuriously doped by chemicals used in the sample's preparation. However, the situation changes drastically for optical properties of quasimetallic tubes.

Neglecting the effect of curvature, the probability rate of interband transitions between the closest valence and conduction subbands has a linear dependence on the electron wave vector measured from the Dirac point. This means that in the vicinity of the Dirac point interband transitions are suppressed. In contrast, if the curvature effects are taken into account, the interband transitions become strongly allowed within a narrow region around this point. As one can see from Fig. 1, there is a profound peak at the Dirac point for the interband matrix element of the velocity operator along the nanotube axis, v_{cv}, as a function of the electron wave vector k. The peak height is independent of the tube's chirality and equal to the Fermi velocity of electrons in graphene ($v_F \approx 10^6$ m/s). The shape of the peak is described by the second (Lorentzian-type) term in the expression:

$$v_{cv}(k) = iv_F\left(\frac{a_0 \cos 3\theta}{4}k - \frac{\Delta k}{\sqrt{\Delta k^2 + k^2}}\right), \tag{1}$$

where $\Delta k = E_g/(2\hbar v_F)$, with E_g being the band gap, $a_0 = 0.142$ nm is the distance between the nearest carbon atoms and $\theta = \arctan[\sqrt{3}m/(2n+m)]$ is the chiral angle.

3. Graphene nanoribbons

Graphene nanoribbons (GNRs) are narrow graphene strips. The most commonly considered GNRs are classified by their edge shape as zigzag or armchair. Within each of these two classes a ribbon is uniquely specified by the number of carbon atoms pairs N in its unit cell. The basic tight-binding model results in all zigzag ribbons (ZGNR) being metallic, whereas only armchair ribbons (AGNRs) with $N = 3p - 2$, where p is an integer, are metallic. The dispersion in the lowest conduction subband of metallic AGNRs is linear and very similar to that of metallic CNTs, while the dispersion in ZGNRs is almost flat due to edge states. However, the *e-e* interaction in ZGNR and the C-C bond length change at the edge of AGNR lead to a band gap up to 50 meV.

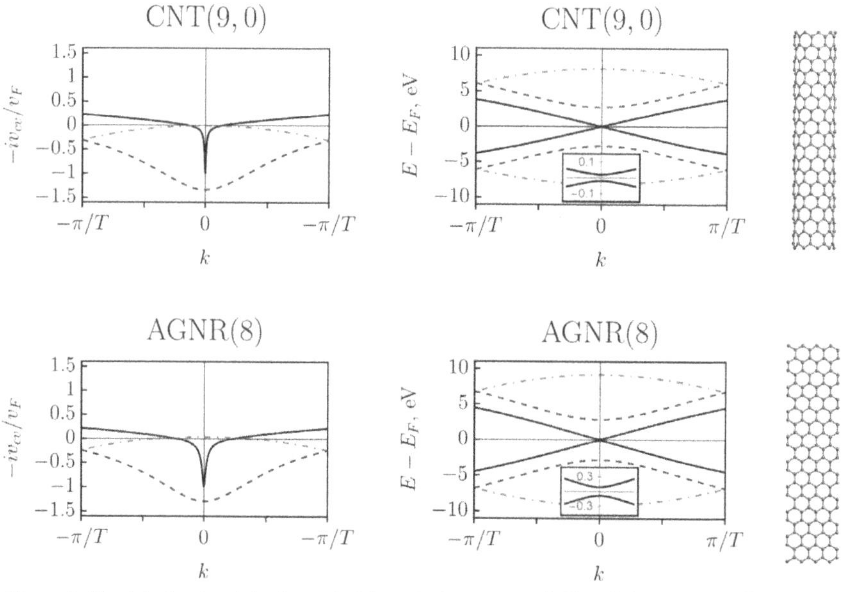

Figure 1. The interband matrix element of the velocity operator (left) and the corresponding bands (right) plotted in the first Brillouin zone for CNT(9,0) and AGNR(8). The matrix elements are plotted using the same line style as the bands between which transitions occur. The insets show the band gap openings.

Here we consider quasimetallic AGNRs, for which the edge effect results in a drastic enhancement of interband transition rate across the band gap. The band structure of AGNRs can be obtained from that of graphene by a zone folding technique used for dealing with CNTs. In fact, it has been shown that there is an equivalence between the electronic properties of AGNR(N) to those of the zigzag CNT (N+1,0) [6]. We show that this equivalence extends to optical transition selection rules. The aforementioned edge effect in the armchair ribbons can be incorporated into the tight-binding model as small corrections to the hopping integrals [7].

Our calculations, presented in Fig. 1, show that the edge effect for quasimetallic armchair GNRs results in a peak similar to that for CNTs. The peak has the same height and its shape is well described by the second term of Eq. (1).

4. Conclusion

Curvature effects in quasi-metallic single-walled CNTs and edge effects in gapless armchair GNRs not only open band gaps, which are typically in the THz range, but also result in a drastic enhancement of the probability of optical transitions across these gaps. The matrix element of the velocity operator for these transitions has a steep dependence on the electron wave vector, which cannot be neglected as it is conventionally done for semiconductors. The dependence looks like a sharp peak, which has a universal height equal to the graphene Fermi velocity for all quasimetallic tubes and ribbons. This peak in conjunction with the van Hove singularity at the band gap edge make these systems promising candidates for active elements in coherent THz radiation emitters, with a possibility of the frequency control by structural modifications [8] or external fields [4,9].

Acknowledgments

This work was supported by the EU FP7 ITN NOTEDEV; FP7 IRSES projects QOCaN, INTERNOM, CANTOR and H2020 RISE project CoExAN.

References

1. R. R. Hartmann *et al.*, *Nanotechnology* **25**, 322001 (2014).
2. O. V. Kibis, M. Rosenau da Costa, M. E. Portnoi, *Nanolett.* 7, 3414 (2007).
3. M. E. Portnoi et al., Superlattices Microstruct. **43**, 399 (2008).
4. M. E. Portnoi *et al.*, *Int. J. Mod. Phys. B* **23**, 2846 (2009).
5. R. Saito, G. Dresselhaus, M. Dresselhaus, *Physical Properties of Carbon Nanotubes* (ICP, London, 1998).
6. C. T. White *et al.*, *Nanolett.* **7**, 825 (2007).
7. H. Zheng *et al.*, *Phys. Rev. B* **75**, 165414 (2007).
8. V. A. Saroka *et al.*, *Phys. Solid State* **56**, 2135 (2014).
9. V. A. Saroka *et al.*, *J. Phys. Condens. Matter* **27**, 145305 (2015).

APPLICATION OF METAL GRATING TO ENHANCE ABSORPTION OF ELECTROMAGNETIC RADIATION IN GRAPHENE

M. A. Yakovleva, K. G. Batrakov

Institute for Nuclear Problems of Belarus State University, 220050 Minsk, Belarus
yakovlevmarin@gmail.com

An ability to enhance significantly absorption of electromagnetic radiation in graphene combined with metallic grating compared to a freestanding graphene film is demonstrated. The choice of metal grating parameters significantly influences the level of the absorption. We show a possibility of almost complete absorption of electromagnetic radiation by the system for two different polarizations.

1. Introduction

Since the discovery of graphene, its remarkable properties make this 2D material and graphene based structures very promising for material-based applications and nanoelectronics. The unusual electronic properties of graphene result in high electron mobility and, consequently, high conductivity which are very important for various applications. As a result, graphene has a high absorption in microwave and terahertz ranges that may be used for shielding, which is important for graphene practical applications such as sensing and detecting of electromagnetic radiation. One graphene monolayer absorbs up to 30 % of the incident electromagnetic radiation the microwave and terahertz range. Using dielectric layers with graphene gives a possibility to increase wave absorption in graphene.

This paper discusses an additional way to increase absorption of electromagnetic radiation by using metal periodic grating. When an electromagnetic wave diffracts on the grating, it generates modes propagating along the surface. The interaction time of graphene with such modes increases, that leads to an additional absorption in graphene. Moreover, metal gratings enhance reflection of electromagnetic waves and prevent radiation passing through the system.

2. Calculation method

The problem was solved for both the transverse magnetic (TM) and transverse electric (TE) waves that allow one to evaluate the case of incident wave with an arbitrary polarization, which can be represented as a mixture of TM and TE polarizations. It is illustrated in Fig. 1.

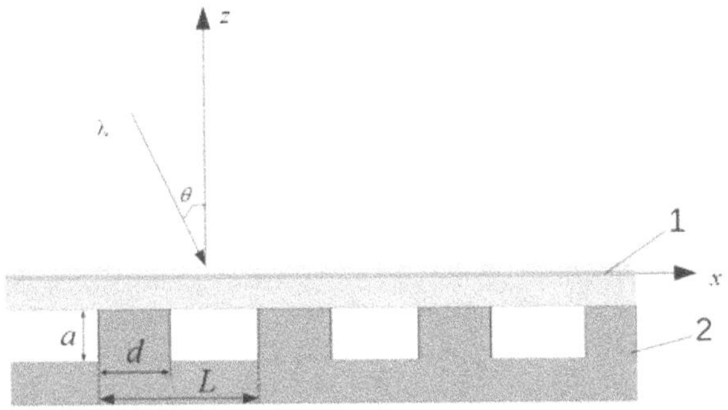

Figure 1. Schematic representation of the studying system: 1 – graphene monolayer, 2 – periodic metallic grating, where L is the grating periodicity, a is the groove height, and d is its length.

The Floquet EM functions for this structure were calculated using boundary conditions for tangential components of the electric and magnetic fields at each boundaries. The obtained system of linear equations was solved numerically for both polarizations. The reflection matrix was found for each case and the absorption coefficient was calculated from it. The result is the reflectance R out of which the absorption coefficient can be found as A=1–R (as the system is opaque to the radiation).

3. Numerical results

The absorption coefficient for graphene with grating can exceed that for a virgin graphene film. In what follows next parameters, for which the maximum of the absorption is achieved, are found for frequency 1 THz. Such parameters can be obtained for any frequency and 1 THz was chosen only for demonstration.

For TM waves the maximum absorbance was found to be 0.93 at 1 THz for L = 0.02 cm, d = 0.01 cm, a = 0.02 cm, with the top dielectric layer thickness of 0.025 cm and $\varepsilon_1 = 3.5$, the bottom dielectric layer thickness of 0.03 cm and $\varepsilon_2 = 3.0$. The dielectric constant of the material filling the grating groove $\varepsilon_3 = 4.4$. For TE waves the maximum of absorbance was found to be 0.96 at 1 THz for L = 0.035 cm, d = 0.0175 cm, a = 0.01 cm. The over graphene dielectric has the thickness of 0.025 cm and ε_1=3.5, under graphene dielectric has the thickness of 0.025 cm and ε_2=3.0. The dielectric constant of the material filling the grating groove $\varepsilon_3 = 5.2$.

Fig. 2 shows angular and frequency dependence for optimal system parameters, which are different for TM and TE polarizations.

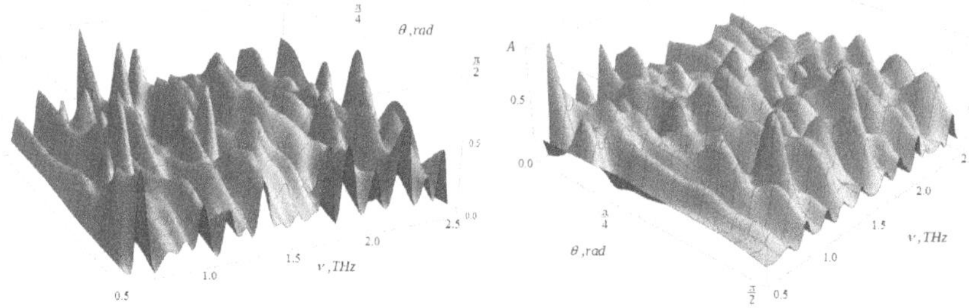

a b

Figure 2. The frequency and angular dependence of absorbance for (a) TM and (b) TE polarizations in the case of optimal system parameters.

4. Conclusion

The diffraction of the incident electromagnetic waves on the metal grating leads to the emergence of the modes that propagate along the surface of the graphene and, therefore, to a prolonged interaction between the electromagnetic wave and graphene. This is the first contribution to an increased absorption. Another possible contribution is due to an increase of the electric field at the graphene location by correct selection of the distance between the graphene monolayer and the grating, as well as the period, parameters of the grooves and dielectrics, which result in an enhanced absorption.

The large absorption in the terahertz frequency range can be utilized in various applications requiring energy concentration in a small volume such as radiation sensing and other scientific, technical and medical applications

References

1. J. Baringhaus *et al.*, *Nature* **506**, 349 (2014).
2. Bo Zhao, Z. M. Zhang, *ACS Photonics* **11**, 1611 (2015).
3. K. G. Batrakov *et al., Sci. Rep.* **4**, 7191 (2014).
4. K. G. Batrakov *et al., Appl. Phys. Lett.* **108**, 123101 (2016).

ELECTROMAGNETIC RESPONSE OF A ONE-DIMENSIONAL CHAIN OF INTERCONNECTED CARBON NANOTUBES IN THE SUB-TERAHERTZ RANGE

A. V. Melnikov, M. V. Shuba

Institute for Nuclear Problems of Belarus State University
Bobruiskaya 11, 220050 Minsk, Belarus

Electromagnetic response from a finite-length one-dimensional chain of interconnected identical single-walled carbon nanotubes has been theoretically studied in the sub-terahertz range. It has been shown that the polarizability spectrum of this chain has two peaks. The high frequency peak originates from a finite length effect in each tube of the chain, whereas the low frequency peak is a result of a collective interaction of the tubes with the electromagnetic radiation.

1. Introduction

Single walled carbon nanotubes (CNTs) are actively applied as inclusions for composite materials having a high conductivity and low percolation threshold [1]. CNT-based composites demonstrate strong shielding effect in the gigahertz range [1].

According to quantum mechanical consideration, the axial conductivity of the single-walled CNT satisfies Drude law in the terahertz and sub-terahertz ranges [2]. The polarizability of a finite-length CNT can be found by solving Hallen's integral equation for the current excited by the incident electromagnetic radiation [3]. The electromagnetic properties of the composite materials have been analyzed previously using the effective medium approach [4] for uncoupled CNTs, *i.e.* at concentrations below the percolation threshold.

Above the percolation threshold, both electronic and electromagnetic coupling of CNTs should be taken into account. As the first step to investigate the composite materials comprising coupled CNTs, we consider one-dimensional chain of metallic single-walled CNTs electrically connected by tunnel junctions, referred to as a nanotube chain (NC). We will consider straight NC, where CNTs are coupled by their tips as depicted in Fig. 1a. The polarizablity of the NC will be compared with the polarizability of (i) the same chain comprising electrically uncoupled CNTs (Fig. 1b), and (ii) hypothetical individual tube with diminished conductivity (Fig. 1c).

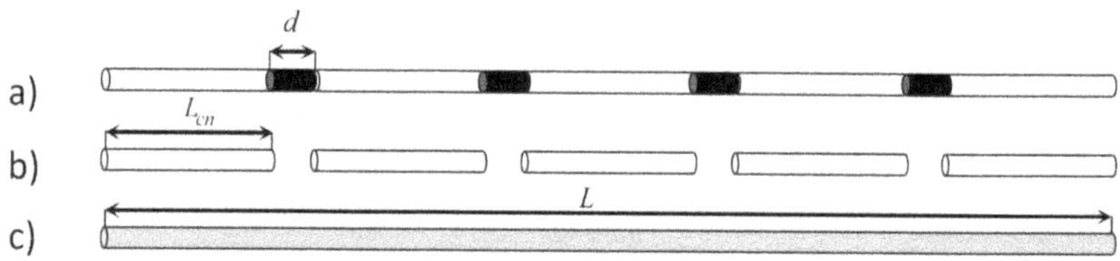

Figure 1. Nanotube chains comprising of electrically coupled (a) and uncoupled (b) CNTs. Dark sections indicate tunnel junctions between adjacent CNTs; (c) hypothetical individual CNT with diminished conductivity.

2. Calculation method

Let us consider an NC comprising of N identical metallic CNTs with a length L_{cn} and a radius R. CNTs are connected by the identical tunnel junctions, and aligned parallel to the z axis of Cartesian coordinate system. Each nanotube can be modeled as a conductive cylinder and the only azimuthally symmetric electric current densities can be excited in the CNTs [3]. We assume that the tunnel current through each junction between the adjacent CNTs has the same azimuthal distribution as the current on the nanotube surface. Under these assumptions we can model the NC as a single nanotube of total length $L = NL_{cn}$ and having a z-dependent axial conductivity. The polarizability of this model nanotube can be found by solving the Hallen's integral equation for an axial current excited by the incident electromagnetic radiation [3]. The axial surface conductivity of the CNT σ_0 has a Drude-like behavior with the electron relaxation time τ as follows from the quantum transport theory [2]. For the case of identical CNTs and the same velocity of the electrons near the Fermi energy in all the conductive channels, the conductance of the tunnel junction $G(\omega)$ can be derived from the Landauer formula [5]:

$$G(\omega) \approx G(0) = \frac{e^2}{\pi\hbar} \cdot N_c \cdot \sum_{i=1}^{N_c} T_i \cdot \frac{1}{\sum_{i=1}^{N_c} R_i}, \qquad (1)$$

where e is the electron charge, $\hbar$ is the reduced Planck constant, N_c is a number of the conductive channels in the CNT. T_i and $R_i = 1 - T_i$ are, respectively, the total transmission and refection coefficients for the electron with the Fermi energy in the channel i. We assume that the conductance of the junction is pure real and determined by (1) in the frequency range of 10^8–10^{13} Hz. Then the effective conductivity of the junction with the length d is $\sigma = G(0)\cdot d/2\pi R$. We found that the polarizability of the NC weakly depends on the value of d and is determined mainly by $G(0)$. To simplify numerical calculations with the

Hallen's integral equation we will assume d to be much larger than the realistic junction size.

3. Numerical results and discussion

Fig. 2 demonstrates the frequency dependence of the polarizability a_{zz} for the NC comprising of electrically coupled and uncoupled CNTs shown in Figs. 1a and 1b, respectively. The polarizability of the hypothetical individual CNT of the length L and axial conductivity σ_d are shown as well. For the calculations, we used the following parameters: $L = 10$ μm, $L_{cn} = 1$ μm, $N = 10$, $R = 0.5$ nm, $\tau = 50$ fs, $d = 20$ nm, $T_1 = T_2 = 0.005$, $N_c = 2$, $\sigma_d = \sigma_0/18.46$. The conductance of the hypothetical individual CNT equals to the conductance of NC.

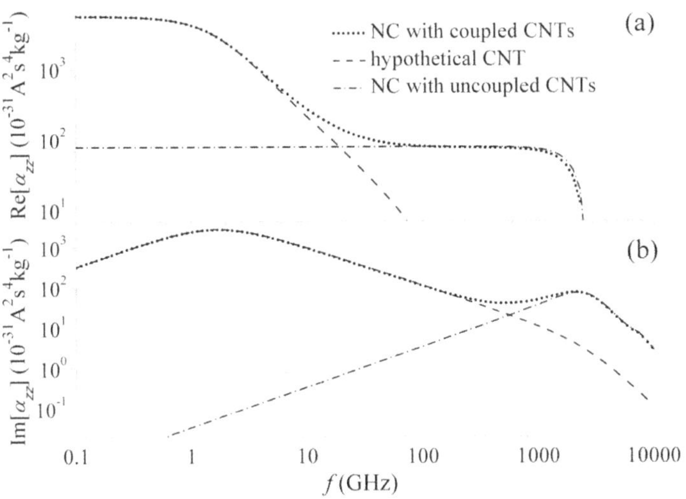

Figure 2. The frequency dependence of the real (a) and imaginary (b) parts of the NC polarizability comprising of coupled (dotted line) and uncoupled (dot-dashed line) tubes. The polarisability of hypothetical CNT is shown by the dashed line.

In Fig. 2, we can indicate three regimes of NC interaction with the electromagnetic wave:

- In the quasi-static regime (below 1 GHz), the tunnel junctions do not prevent charge propagation along the tubes. In this regime, the real part of the NC polarizability is the same as for the infinitely conductive nanotube of the same length L as the NC. The imaginary part of the NC polarizability is much higher for the case of electrically coupled than uncoupled CNTs. This means that the tunnel junctions change the phase between the current and total field on the tube surface providing stronger energy dissipation. Thus, coupling leads to a weakening of the finite-length effect in CNTs.

- In the range of 0.5–10 THz, the junctions between adjacent CNTs in the chain suppress charge propagation along the NC. Therefore the polarizability of the NC of the coupled CNTs coincides with that for uncoupled tubes. The length effect in the CNTs is strong resulting in the terahertz peak in the polarizability spectra of the CN (see Fig. 2b).
- In the range of 1–500 GHz, the tunnel junctions partly suppress propagation of the charges along the chain. Collective interaction of the coupled tubes with the electromagnetic field leads to a low frequency peak at 2 GHz in the polarisability spectra of the NC.

4. Conclusion

The polarizability of finite-length NC has been calculated and analyzed in the sub-terahertz frequency range taking into account both electromagnetic and electronic coupling of the CNTs. Two peaks emerge in the polarizability spectra. Low frequency peak is a result of the collective interaction of the coupled CNTs with the electromagnetic field. The high frequency peak is due to the finite length effect in each tubes of the chain.

Acknowledgments

This research was funded by Ministry of Education and Science of Russian Federation (project ID RFMEFI57714X0006).

References

1. P. P. Kuzhir *et. al*, *Thin Solid Films* **519**, 4114 (2011).
2. G. Ya. Slepyan, S. A. Maksimenko, A. Lakhtakia, O. Yevtushenko, A. V. Gusakov, *Phys. Rev. B* **60**, 17136 (1999).
3. G. Ya. Slepyan, M. V. Shuba, S. A. Maksimenko, A. Lakhtakia, *Phys. Rev. B* **73**, 195416 (2006).
4. G. Ya. Slepyan, M. V. Shuba, S. A. Maksimenko, C. Thomsen, A. Lakhtakia, *Phys. Rev. B* **81**, 205423 (2010).
5. M. Büttiker, Y. Imry, R. Landauer, S. Pinhas, *Phys. Rev. B* **31**, 6207 (1985).

FULLERENE-BASED NANOANTENNA AND ITS PROPERTIES

M. A. Britch
Heat and Mass Transfer Institute NASB, P. Browka 15, 220072 Minsk, Belarus
mabritch@hmti.ac.by

An interaction of electromagnetic fields with a nanoantenna containing quantum dots in a form of fullerenes has been simulated. Due to resonant tunneling, the fullerenes have highly nonlinear current-voltage characteristics. As a result, the radiation scattered by such antennas has a wide frequency spectrum. Characteristics of nonlinear antennas can be tuned by means of the bias voltage variations. Due to their small size nanoantennas can be used in a variety of micro- and nanodevices.

1. Introduction

Nanoantennas are essential components of electronic and photonic devices [1,2]. Due to their small size, nanoantennas can be used in a variety of micro- and nanodevices (particularly for monitoring the state of an organism). When an antenna works in a passive regime (that is, when scattering the signal, produced by an external source of radiation), it does not require an autonomous source of energy, and so, the total size of the device can be reduced significantly.

On the other hand, because of its small size, the nanoantennas have low efficiency that leads to a problem of weak scattered signal registration. To overcome this problem, the approach can be proposed to introduce the nonlinear element in the nanoantenna that results in the generation of higher harmonics, and so it will facilitate the registration of the scattered signals. Also, the usage of nonlinear elements is considered as a most suitable approach to make nanoantennas with tunable working characteristics [1].

The nonlinearity can be achieved by means of the nonlinear optical response of metals and semiconductors [1], or by means of the tunneling of electrons [3]. Nevertheless, for metals and semiconductors, and in a case of tunneling, the nonlinear effects become pronounced only at a very high intensity of the incident field [3], which restricts the significance of this mechanism in real systems.

The degree of nonlinearity can be made much more extensive (and, accordingly, an antenna can become much more sensitive with respect to nonlinear effects) when the tunneling is assisted with quantum dots featuring resonant electrons tunneling. Particularly, such a property is intrinsic to fullerene C_{60} molecules [4]. These fullerenes have well-defined regular structure that is essential for fabrication of devices with reproducible electrical parameters. Also, they have advantages of relatively low cost and suitability for mass production and manipulation. Fullerenes possess the ability to the self-assembly resulting in the formation of the fullerite film or crystal.

In our previous works [4,5], calculations of the tunneling current through the fullerene C_{60} molecules have been performed and the highly nonlinear current-voltage characteristics have been revealed. In [6], the interaction of electromagnetic field with several types of nanoantennas containing fullerenes has been simulated. Here, we consider one more design of a nanoantenna which is more suitable for fabrication by means of the vapor deposition technology.

2. Nanoantenna based on a fullerite film

Fullerenes possess an ability to self-assembling resulting in the formation of the fullerite molecular crystal, stabilized by weak interactions. A thin film composed of the fullerenes inherits the property of the resonant tunneling intrinsic to a single fullerene molecule, and as a consequence it demonstrates nonlinear current-voltage characteristics.

The modeling of interaction of electromagnetic radiation with the construction shown in Fig. 1 under different bias voltages U have been performed.

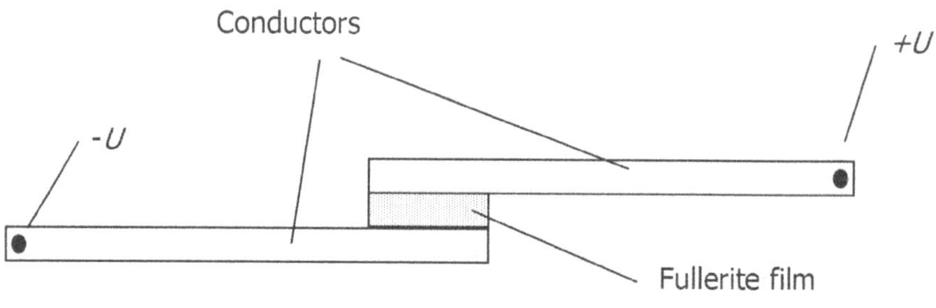

Figure 1. A sketch of the nanoantenna with a fullerite film placed between two metal electrodes.

The results have demonstrated the efficiency of such an antenna with respect to higher harmonics generation. It is illustrated by the data presented in Fig. 2. For example, the amplitude ratio for the second harmonic can be approximated with the Gaussian function. It is evident that the spectral characteristics of the nanoantenna can be tuned by variations in the bias voltage.

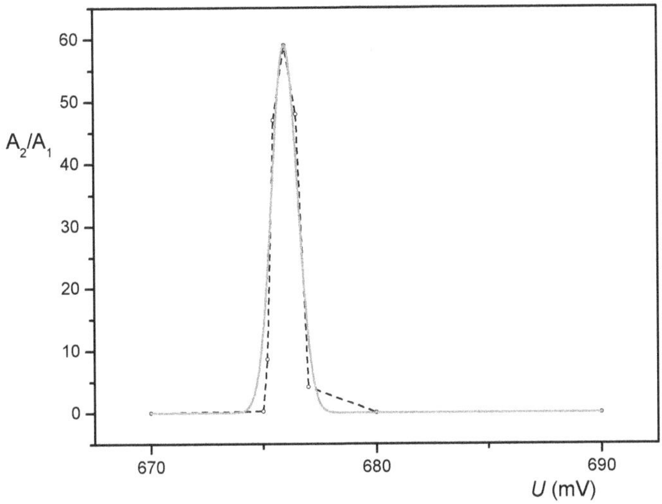

Figure 2. The ratio of the second harmonic amplitude (A_2) to that of the first one (A_1) for the radiation scattered by a nanoantenna with a fullerite film (Fig. 1) as a function of the bias voltage. The circles, connected with the dashed lines are results of calculation. Solid line is Gaussian approximation of the results.

The small size of nanoantennas based on a fullerite film allows them to be used, for example, in devices for monitoring the state of an organism. It can be either made as a separate construction (similar to the capsule endoscope), or integrated with a living cell.

References

1. A. E. Krasnok *et al.*, *Physics – Uspekhi* **56**, 539 (2013).
2. P. Bharadwaj *et al.*, *Advances in Optics and Photonics* **1**, 438 (2009).
3. D. C. Marinica *et al.*, *NanoLett.* **12**, 1333 (2012).
4. M. A. Britch, K. V. Dobrego, L. I. Krasovskaya, *J. Nanophotonics* **7**, 073098 (2013).
5. M. A. Britch, K. V. Dobrego, L. I. Krasovskaya, in: *15th Int. Workshop on New Approaches to High-Tech: Nano-Design, Technology, Computer Simulations* (Minsk, 2013), pp. 55–56.
6. M. Britch, in: *Physics, Chemistry and Application of Nanostructures*, eds. V. E. Borisenko, S. V. Gaponenko, V. S. Gurin, C. H. Kam, (Singapore, World Scientific, 2015), pp. 252–254.

BROADBAND ELECTRICAL PROPERTIES OF CARBON NANOTUBES-EPOXY RESIN COMPOSITES

I. Kranauskaite, J. Macutkevic, J. Banys
Radiophysics Department, Physics Faculty, Vilnius University
Sauletekio Ave. 9, LT-10222, Vilnius, Lithuania
ieva.kranauskaite@ff.vu.lt

D. Bychanok, D. Meisak
Research Institute for Nuclear Problems of Belarusian State University
Bobruiskaya 11, 220030 Minsk, Belarus
dzmitrybychanok@ya.ru

Epoxy resin composites with multiwalled carbon nanotubes (MWCNT) were produced with concentrations up to 2 wt.%. Electrical measurements were performed in the wide frequency range (20 Hz – 2 THz) at different temperatures. The results showed very high dielectric permittivity and conductivity of the MWCNT composites. The percolation threshold was estimated as 0.1 wt.%. Analysis of dielectric measurements for the composites reveals a significant effect of polymer matrix and MWCNT distributions on the composite broadband electric properties.

1. Introduction

Carbon nanotubes (CNT) attract attention of many researchers due to their excellent thermal and mechanical properties and high electrical conductivity [1–3]. Combining conductive CNTs and insulating polymers as a matrix, new materials with properties different from each component separately can be designed [4,5]. Electrically percolative polymer composites are of interest due to potential applications as antistatic and sensitive materials and also electromagnetic coatings [6].

Despite of outstanding properties, CNTs are fairly expensive. Production of the composites requires as small amount of CNTs as possible. That means that the percolation threshold should be as low as possible.

According to the excluded volume theory, the percolation threshold is inversely proportional to CNTs aspect ratio [7]. However, the value of experimental percolation threshold could be much higher or lower than theoretical estimations. Conductivity and percolation properties depend on filler parameters, dispersion and even composite preparation technology [8]. An impact of these factors on broadband electrical properties of composites is not clearly understood and research is still needed in this field.

In this paper, epoxy resin composites filled with MWCNTs were produced and electrical properties in line with the percolation phenomenon were investigated in the frequency range of 20 Hz – 2 THz.

2. Experimental

The epoxy resin composites were produced with MWCNTs concentration of 0–2 wt.%. The average diameter of the nanotubes was 30–40 nm, the length was up to 100 μm. First, the resin was degassed under vacuum (13 mbar) for 12/14 h and heated in an oven at 65 °C. Next, MWCNTs were dispersed in ethanol using an ultrasonic tip for 30 min. Then, the solution was mixed with the resin and sonicated for 60 min with ultrasonic tip at the temperature near 80 °C. After this step, the dispersion was completed and alcohol evaporated. The curing agent (TETA) was added to the mixture of resin and MWCNTs by shear mixing during several min. The mixture was then poured into a mold for 20 h curing followed by heating in an oven at 80 °C for 2 h.

The dielectric properties of composites were measured in the frequency range from 20 Hz to 1 MHz at the room temperature using a LCR meter (HP4284) measuring capacitance and loss tangent. In the frequency range of 1 MHz – 3 GHz, the measurements were done using a vector network analyzer Agilent 8714ET. In the microwave frequency range of 8–53 GHz, a home-made waveguide spectrometer was used. The thin rod method in the waveguide was used. Silver paste created electrical contacts. In the terahertz frequency range (100 GHz – 2 THz), a terahertz time domain spectrometer (Ekspla Ltd) based on a femtosecond laser was used. The real part of electrical conductivity was calculated using the formula: $\sigma' = \varepsilon_0\varepsilon''\omega$, where ε_0 is the vacuum permittivity, ε'' is the imaginary part of the dielectric permittivity and ω is the frequency.

3. Results and discussion

Frequency dependences of the dielectric permittivity and the electrical conductivity of the composites and pure epoxy resin are presented in Fig. 1. The pure epoxy resin is an insulator with low dielectric permittivity no *DC* conductivity at the room temperature. Addition of small amount of MWCNT (0.25 wt.%) into epoxy resin results in the higher dielectric permittivity (below 10 kHz) on several orders of magnitude and *DC* conductivity is 0.1 mS/m. The dielectric permittivity and electrical conductivity rapidly increase with MWCNT concentration, particularly at lower frequencies (below 10 kHz).

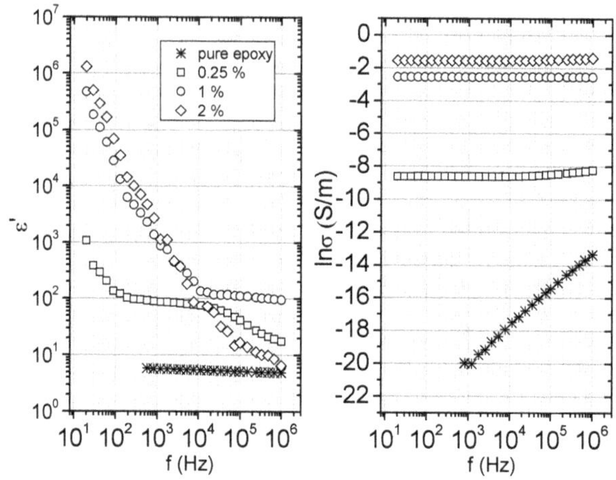

Figure 1. Frequency dependences of the dielectric permittivity and the electrical conductivity of the MWCNT composites and pure epoxy resin at room temperature.

The electrical conductivity as a function of the MWCNT concentration is presented in Fig. 2. Above and near the percolation threshold the conductivity is to follow the power law [9]:

$$\sigma(p) \propto (p - p_c)^t, \tag{1}$$

where p_c is the percolation threshold, t is a critical exponent.

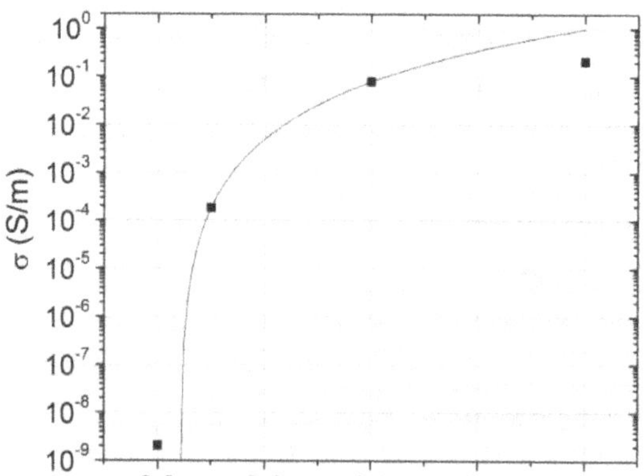

Figure 2. The concentration dependence of electrical conductivity for composites with different concentrations of f MWCNT. Solid line is the power law fitting.

From the above measurement, *DC* conductivity and the critical frequency were estimated. The percolation threshold value was estimated as 0.1 wt.%. The electrical transport in the composites was established to be determined by the mechanisms of electron hopping inside MWCNT and tunneling between MWCNT clusters. At temperatures above 400 K the contribution of electrical conductivity of polymer matrix becomes also important.

4. Conclusion

The dielectric properties of MWCNT composites were investigated in the frequency range of 20 Hz – 2 THz. Dielectric permittivity and electrical conductivity of the epoxy resin without MWCNT are very low while addition of small amount of conductive MWCNTs increases the electrical conductivity by several orders of magnitude. The percolation threshold was estimated as the rather low value for the composite, 0.1 wt.%.

Acknowledgments

This research was partially supported by Lithuanian Science Council according to the Lithuanian-Belarusian collaboration program.

References

1. R. H. Baughman, A. A. Zakhidov, W. A. De Heer, *Science* **297**, 787 (2002).
2. S. Iijima, T. Ichihashi, *Nature* **363**, 603 (1993).
3. M. M. J. Treacy, T. W. Ebbesen, J. M. Gibson, *Nature* **381**, 678 (1996).
4. T. Kuilla, S. Bhadra, D. Yao, N. H. Kim, S. Bose, J. H. Lee, *Progr. Polymer Sci.* **35**, 1350 (2010).
5. R. J. Young, I. A. Kinloch, L. Gong, K. S. Novoselov, *Comp. Sci. Technol.*72, 1459 (2012).
6. F. Qin, C. Brosseau, *J. Appl. Phys.* **111**, 061301 (2012).
7. I. Balberg, C. H. Anderson, S. Alexander, N. Wagner, *Phys. Rev. B* **30**, 3933 (1984).
8. W. Baunhofer, J. Z. Kovacs, *Comp. Sci. Technol.* **69**, 1486 (2009).
9. D. Stauffer, A. Aharony, *Introduction to Percolation Theory* (London, 1994).

SYNERGETIC IMPACT OF GRAPHITE NANOPLATELETS AND TRIGLYCINE SULPHATE PARTICLES ON THE FERROELECTRIC PROPERTIES OF EPOXY COMPOSITES

A. Plyushch, J. Macutkevic, V. Samulionis, J. Banys
Vilnius University, Universiteto St. 3, 01513 Vilnius, Lithuania
artyom.plyushch@ff.vu.lt

P. Kuzhir
Institute for Nuclear Problems, Belarus State University
Bobruiskaya 11, 220030 Minsk, Belarus

V. Fierro, A. Celzard
IJL–UMR Université de Lorraine–CNRS, 7198, ENSTIB, 27 Rue Philippe Seguin, CS 60036, 88026 Epinal Cédex, France

Dielectric, acoustic, and piezoelectric properties of composite materials made of epoxy resin filled with 30 wt.% of triglycine sulfate (TGS) and small additions (up to 1 wt.%) of graphite nanoplatelets (GNP) are presented in the 300–380 K temperature range. All composites demonstrated a ferroelectric behavior with a phase transition temperature close to 322 K. The TGS-based composites enriched with GNP presented better dielectric and piezoelectric properties due to strong synergy effects between the different kinds of particles.

1. Introduction

Triglycine sulfate (TGS) is one of the most well-studied ferroelectric crystal with a second-order ferroelectric phase transition close to 322 K [1]. The ferroelectric phase transition is of an order-disorder type with a strong dielectric dispersion in the microwave frequency range and a typical critical slowing-down effect [2]. TGS crystals have also huge values of dielectric permittivity, piezoelectric and pyroelectric coefficients, enabling a wide range of applications, including very sensitive infrared sensors. It is also known that polycrystallinity [3], moisture [4,5], and organic or inorganic doping [6] may significantly affect TGS properties. It was therefore assumed to be interesting to investigate a possibility of designing composite materials based on epoxy resin filled with both TGS particles and conductive inclusions. As for the latter, graphite nanoplatelets (GNP) were utilized. The main ideas are the following: (i) being covered with epoxy resin, TGS can be protected from the ambient moisture; (ii) by introducing various amounts of conductive filler, the dielectric and ultrasonic properties of the composites can be controlled.

2. Materials

Monocrystals of triglycine sulfate (TGS) were first dissolved in distilled water and, after evaporation, TGS powder was obtained which was further ground in an agate mortar. Graphite nanoplatelets (GNP) were obtained by a succession of grinding and sonication steps applied to a suspension of exfoliated graphite in cyclohexane. Isopropyl alcohol was used as a dispersion medium for sonication of TGS and GNP suspensions. During the sonication process, the alcohol was exchanged by epoxy resin (Buehler EpoThin Resin) and, after complete evaporation of alcohol, a curing agent was added. Composites with 30 wt.% of TGS and 0, 0.5, 1 wt.% of GNP were prepared.

3. Characterization

In the 20 Hz – 1 MHz frequency range, dielectric properties were measured with a LCR HP4284A meter. For temperature-dependent measurements, the samples were placed in a home-made oven. In the 10 MHz – 1 GHz frequency range, the measurements were performed with a coaxial dielectric spectrometer using an Agilent E8363B vector network analyzer. Silver paste was applied for contacts. Ultrasonic investigations were performed by an automatic computer-controlled pulse-echo method [7].

4. Results and discussion

The temperature dependence of the complex dielectric permittivity of the TGS/epoxy composites at different frequencies is presented in Fig. 1a. The maximum of dielectric permittivity close to 322 K indicates a ferroelectric phase transition. At lower frequencies (below 1 MHz), the dielectric properties are slightly influenced by electrical conductivity, which appears due to the random distribution of TGS crystallites inside the polymer matrix. At higher frequencies (above 1 MHz), the dielectric dispersion is caused by the relaxation soft mode, similarly as in pure TGS crystals [1]. However, no critical slowing-down effect is observed, obviously due to the broader distribution of relaxation times in comparison with pure TGS crystals.

The addition of GNP increases both real and imaginary parts of the dielectric permittivity of the TGS/epoxy composites in a wide temperature range, also close to the phase transition temperature (see Fig. 1b). The ferroelectric behavior of the composites was confirmed by acoustic and piezoelectric measurements (Fig. 2). The appearance of the piezoelectric signal close to the phase transition temperature clearly indicates the ferroelectric origin of the phase transition. Just like in pure TGS, the phase transition is a second-order one.

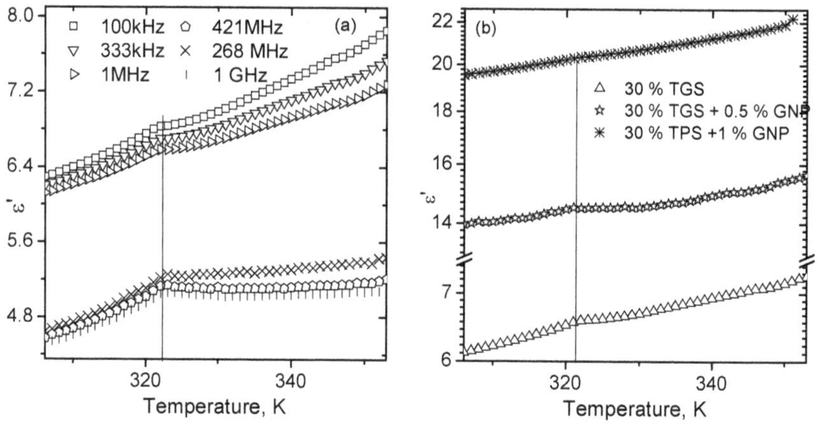

Figure 1. Temperature dependence of the real part of dielectric permittivity of TGS/epoxy composites at different frequencies (a); TGS+GNP/epoxy composites with different GNP concentrations at 1 MHz (b).

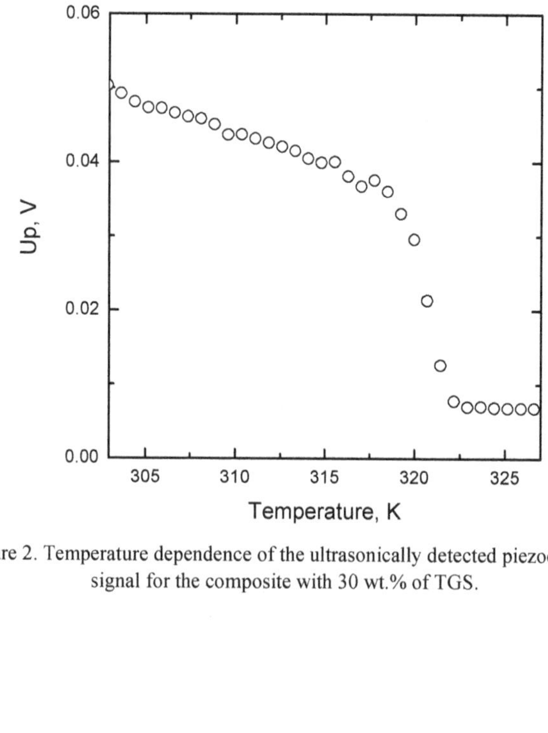

Figure 2. Temperature dependence of the ultrasonically detected piezoelectric signal for the composite with 30 wt.% of TGS.

5. Conclusion

The epoxy resin filled with TGS particles (with the concentration not less than 30 wt.%) and with (up to 1 wt.%) or without addition of GNP present a second-order ferroelectric phase transition close to 322 K. At frequencies above 1 MHz, the dielectric dispersion is similar to that of pure TGS crystals, except the critical slowing-down effect. The addition of GNP particles to TGS composites substantially improved dielectric and piezoelectric properties of the composites due to a strong synergy effect.

Acknowledgments

This research was supported by NAMICEMC and partially supported by Lithuanian-Belarussian collaboration program project.

References

1. R. M. Hill, S. K. Ichiki, *Phys. Rev.* **128,** 1140 (1962).
2. R. Blinc, S. Detoni, M. Pintar, *Phys. Rev.* **124**, 1036 (1961).
3. M. Amin, K. A. Darwish, S. S. Ibrahim, *Ferroelectrics* **76**, 33 (1987).
4. V. E. Khutorsky, S. B. Lang, *J. Appl. Phys.* **82**, 1288 (1997).
5. V. E. Khutorsky, S. B. Lang, in: *Applications of Ferroelectrics*, 1994. (ISAF'94., Proceedings of the Ninth IEEE International Symposium on. IEEE, 1991).
6. N. Theresita Shanthi, P. Selvarajan, C. K. Mahadevan, *Current Applied Physics* **9**, 1155 (2009).
7. V. Samulionis, V. Valevicius, J. Banys, A. Brilingas, *Journal de Physique IV* **6**, 405 (1996).

ELECTROPHYSICAL PROPERTIES OF COMPOSITE FILMS BASED ON GRAPHENE NANOPLATELETS COVERED BY METAL NANOPARTICLES

A. V. Kukhta
Institute for Nuclear Problems, Belarusian State University, 220030 Minsk, Belarus
al.kukhta@gmail.com

A. V. Misevich, A. E. Pochtenny
Belarusian State Technological University, 220006 Minsk, Belarus
pae@tut.by

DC conductivity study of thin films based on graphene and graphene containing copper or cobalt nanoparticles showed that they have a metallic type of conductivity, but thin films consisting of graphene with magnetite nanoparticles, have a semiconductor type of conductivity. Formation of graphene-Fe_3O_4 films in the magnetic field results in the dielectric-metal transition that can be caused by spin effects in graphene. Adsorbed oxygen in the films based on graphene acts as scattering centers decreasing mobility of charge carriers and, as a consequence, conductivity of the films.

1. Introduction

Pure graphene possesses unique electrical, thermal and mechanical properties, however it is rather expensive [1]. Graphene nanoplatelets (GNP) can be fabricated by diversity of cheap chemical and physical methods, though their properties appear to be worse. GNP covered by metal or semiconductor nanoparticles are attractive due to improved properties [2,3]. In this paper, the results of electrophysical studies of thin films made of pure GNP and GNP covered by Cu, Co and Fe_3O_4 are presented in order to determine the conductivity features depending on external conditions and oxygen adsorbed from surrounding air.

2. Experimental

GNP were fabricated by the physical tool without any admixtures and chemical defects. Nanoparticles were synthesized and deposited on GNP surface by corresponding reactions made in Research Institute for Physical Chemical Problems of Belarusian State University. Cu and Fe_3O_4 nanoparticles were spherical with diameters of 13.5 and 20 nm, respectively. Co nanoparticles were both spherical (5 nm) and needle-like (up to 50–100 nm long). Thin films of 100–500 nm thickness were obtained by deposition of water-propanol emulsion on a ceramic substrate with nickel electrodes followed by drying in a box at

100 °C. Fe_3O_4 films were also fabricated under 0.15 T magnetic field (perpendicularly to the film surface). Morphology and composition of the films has been studied with electron microscope LEO-906 (0.1 nm resolution) and LEO-1455VP. *DC* conductivity of the films was determined with an electrometer B7-57/1 (Belvar).

3. Results and discussion

The temperature dependences of *DC* conductivity were monitored by the cyclic thermo-desorption [4] in 10^{-2} Pa vacuum. A sample was heated up to a certain temperature resulting in the decrease of the adsorbed oxygen amount in the film down to some fixed level. Then, the sample was cooling with the same concentration of the adsorbed oxygen, and the temperature dependence of conductivity was measured during this process. Then, this sample was heated again up to the higher temperature, and measurements were repeated at lower concentration of the adsorbed oxygen. As a result, we obtain a series of the temperature dependences with decreasing adsorbed oxygen in the film for each subsequent heating-cooling cycle.

GNP had typical width of 10–20 μm and thickness up to 3 nm. The high resolution electron microscopy shows that the nanoparticles are physically adsorbed on the graphene surface randomly and located enough homogeneously.

Fig. 1 illustrates the temperature dependences of conductivity for thin films based on graphene and graphene with Cu and Co nanoparticles. The resistance of the graphene films is growing (correspondingly, conductivity, *G*, is decreasing, Fig. 1) with the temperature increase with the slight super-linearity. Such behavior is typical for semimetals containing admixtures acting as centers of charge carriers scattering.

During the oxygen desorption the resistance of the graphene film decreases that can be caused by the lower concentration of the scattering centers and increase of the charge carriers mobility. The curves obtained for thin films made of graphene and graphene with copper or cobalt nanoparticles, are linear in the conductivity-temperature coordinates. The presence of copper or cobalt nanoparticles on the graphene surface does not affect essentially the character of the temperature dependences of conductivity, but results in the weaker temperature dependence. This effect can be due to injection of the charge carriers (*e.g.* electrons) from these nanoparticles.

Fig. 2 illustrates the temperature dependences of the conductivity of the graphene thin films with Fe_3O_4 nanoparticles formed in the magnetic field and without the field. These results indicate that magnetic field applied to the samples during the formation process, strongly affects the electrical properties of these films.

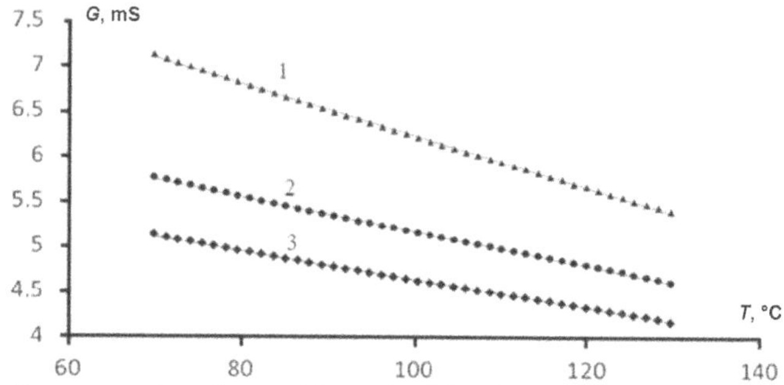

Figure 1. Temperature dependences of the conductivity of the thin films based on graphene (1), graphene with copper nanoparticles (2), and graphene with iron oxide nanoparticles (3) measured in vacuum under cooling from 130 °C.

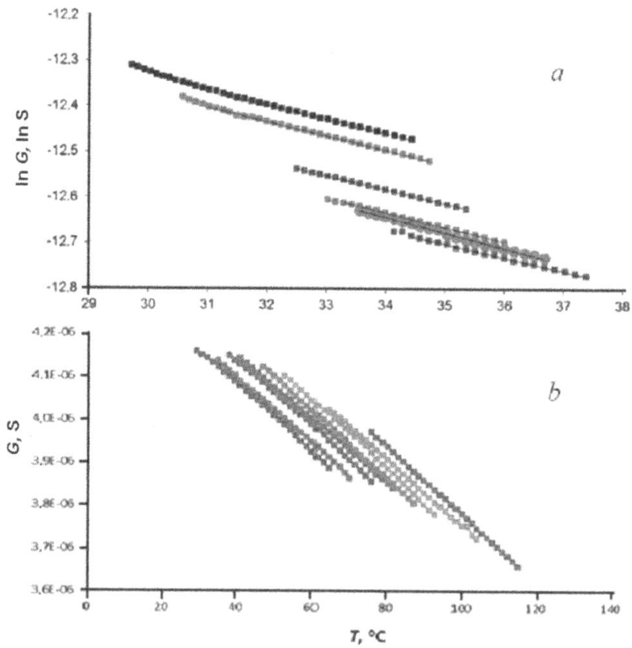

Figure 2. Temperature dependences of the conductivity and resistance of the thin films of graphene with iron oxide nanoparticles formed without (a) and in the magnetic field of 0.15 T (b).

Graphene-Fe_3O_4 films formed without magnetic field have an activation character of the temperature-conductivity dependence with no essential changes of the activation energy in the range of 0.029–0.034 eV. The graphene-Fe_3O_4

films formed in the magnetic field show the temperature dependences of conductivity similar to that of graphene and graphene containing Cu or Co nanoparticles. Thus, the magnetic field acting during the graphene-Fe_3O_4 film formation results in the dielectric-metal transition.

The physical nature of this transition can be explained assuming that the conductivity activation energy in graphene-Fe_3O_4 film formed in the magnetic field is the same as in bare magnetite. It can be simply assumed that the formation of graphene-Fe_3O_4 film without the magnetic field results in an appearance of a continuous conducting cluster chain consisting of graphene nanoplatelets, while in the magnetic field, a structure consisting of graphene bundles separated by magnetite nanoparticles occurs. However, electron microscopy data for both samples show the same structure. Only the structure formed in the magnetic field is more porous. Thus, this assumption is not enough justified. The more likely explanation of the observed phenomena can be an assumption about partial reorientation of the electron spins in graphene under the magnetic field.

In the film obtained without magnetic field, electrons have to overcome the potential barrier at the graphene-Fe_3O_4 interface, which equals to a half of the width of the magnetite band gap, that results in an appearance of the activated character of the temperature dependence of conductivity. The magnetic field acting during the film formation can results in redistribution of the electrons energy in graphene. The part of electrons with spins coinciding with the magnetic field vector increases and these electrons have energies near the Fermi level. The width of this band is determined not only by thermal energy, but also by magnetic field energy for electron spin reorientations. After the magnetic field switching off, the electron spins can be retained owing to the ferromagnetic nature of Fe_3O_4. As a result, in the graphene-Fe_3O_4 film formed in the magnetic field, redistributed electrons with energies higher than the magnetite conductivity band are responsible for the non-activation charge transfer.

References

1. A. K. Geim, *Science* **324**, 1530 (2009).
2. K. S. Subrahmanyam, A. K. Manna, S. K. Pati, C. N. R. Rao, *Chem. Phys. Lett.* **497**, 70 (2010).
3. S. Das, R. Fullerton, D. Cole, K. D. Behler, F. Irin, D. Parviz, M. N. F. Hoque, Z. Fan, M. J. Green, *Carbon* **72**, 192 (2014).
4. A. E. Pochtenny, A.V. Misevich, *Tech. Phys. Lett.* **29**, 26 (2003).

MICROWAVE-ABSORBING PROPERTIES OF PHOSPHATE CERAMICS FILLED WITH CARBON NANOTUBES, $BaTiO_3$ AND Fe_3O_4

K. Piasotski, D. Bychanok, G. Gorokhov, D. Meisak, A. Plyushch, P. Kuzhir

Research Institute for Nuclear Problems Belarusian State University
Bobruiskaya Str. 11, 220030 Minsk, Belarus
kirillpias@gmail.com

A. Sokol, K. Lapko

Research Institute for Physical Chemical Problems of the Belarusian State University
Leningradskaya Str. 14, 220030 Minsk, Belarus

A. Sánchez-Sánchez, V. Fierro, A. Celzard

Institut Jean Lamour - UMR Université de Lorraine - CNRS 7198, ENSTIB
27 rue Philippe Seguin, CS 60036, 88026 Epinal Cedex, France

C. P. Gallagher, A. P. Hibbins, F. Y. Ogrin

University of Exeter, Exeter EX4 4QL, United Kingdom

C. Brosseau

Universite Europeenne de Bretagne, Université de Brest, Lab-STICC,CS 93837, 6 avenue Le Gorgeu, 29238 Brest Cedex 3, France

We report on the modelling and characterization of microwave-absorbing materials specially designed for the 26–37 GHz frequency range (Ka-band). Composite materials based on carbon nanotubes/$BaTiO_3$/Fe_3O_4 in a phosphate ceramic matrix were fabricated and their electromagnetic response was investigated. Both theoretical and experimental results demonstrate that this material can absorb up to 100 % of the power of an incident plane wave at the normal incidence angle. The physics underlying such absorption level is discussed in terms of refractive index of materials.

1. Introduction

In the present work, the parameters of microwave radiation absorption in composites based on a phosphate ceramic matrix were analyzed. To reduce the total thickness of the absorbing material, we considered the composite located on a metal plate (composite with a back reflector) [1]. We develop the description of the reflection/absorption coefficients of such composite and investigate their dependencies on several physical parameters. Next, we experimentally measured the frequency dependence of the absorption coefficient of composites based on multi-walled carbon nanotubes (MWCNTs)/$BaTiO_3$/Fe_3O_4 particles embedded in a phosphate ceramic matrix.

2. Effective absorbing composite modelling

All the results below are given in SI units and we assume an $\exp[i(k_z - \omega t)]$ dependence of the incident electric field. Let us consider normal scattering of an incident plane wave by a plane-parallel layer of a composite material located on a metal plate. The first-principles equations describing the electromagnetic response of such system in free space are well-known [2,3]. The amplitude of the reflected signal in the waveguide is defined as:

$$S_{11}(\lambda, \tau, n) = -\frac{k_z(\exp[2i\tau k_{2z}] - 1) + k_{2z}(1 + \exp[2i\tau k_{2z}])}{k_z(1 - \exp[2i\tau k_{2z}]) + k_{2z}(1 + \exp[2i\tau k_{2z}])}, \quad (1)$$

with

$$k_z = \frac{\pi}{\lambda a}\sqrt{4a^2 - \lambda^2}, k_{2z} = \frac{\pi}{\lambda a}\sqrt{4n^2a^2 - \lambda^2}, n = \sqrt{\varepsilon\mu}, \quad (2)$$

where τ is the composite thickness, a is the waveguide width (set to 7.2 mm), c is the vacuum light velocity, v is the frequency, $n = n' + in''$ is the refractive index, ε is the complex (relative) permittivity and μ is the (relative) magnetic permeability of the composite. Eq. (1) can be also used in free space. In this case, wave vectors $k_z = 2\pi/\lambda$ and $k_{2z} = 2\pi n/\lambda$ should be used. From (1) we can calculate the absorption coefficient as $A = 1 - |S_{11}|^2$. To achieve the absorption maximum, one can vary the composite thickness. Additionally, by changing the type of the filler and its composition, it is also possible to change the refractive index in order to obtain the maximal absorption at a fixed thickness and frequency.

The dependence of the absorption maximum on the thickness at constant frequency in the waveguide and in free space is presented in Figs. 1a and 1b, respectively.

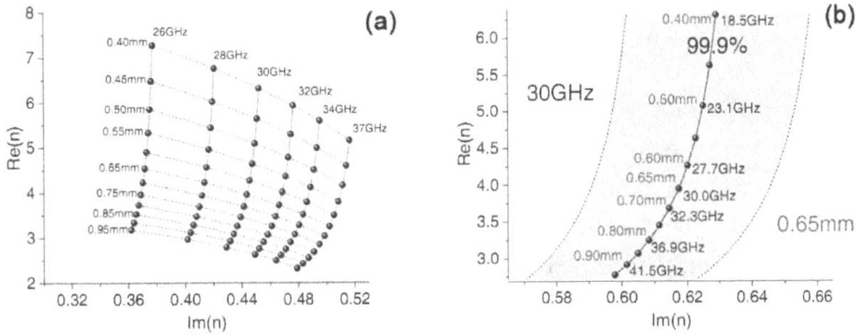

Figure 1. Position of absorption maximum ($S_{11} = 0$) as a function of two variables Re(n), Im(n) and two parameters (frequency and thickness): (a) inside the waveguide; and (b) in free space.

3. Results and discussion

Composite materials with a refractive index close to the absorption maximum displayed in Fig. 1 can be fabricated by mixing magnetic particles (magnetite Fe_3O_4), piezoelectric particles (barium titanate $BaTiO_3$), and carbon nanotubes in the phosphate ceramic matrix [2,4].

The components were mixed together and pressed into pellets under the pressure of 20 MPa. After 12 h of curing in air at room temperature, the samples were heat-treated in an oven at 200 °C. All measurements were performed in a 7.2×3.4 mm waveguide system using a scalar network analyzer. The spectral variation of S-parameter of the 0.65 mm-thick composite without MWCNT is presented in Fig. 2a. It can be seen that these data are well matched by (1) with $n = 3.95 + i0.45$ (solid line in Fig. 2a). As it can be seen in Fig. 2a, the absorption varied from 65 to 100 %, reaching its maxima around 30 GHz. Additionally, the absorption maximum could be easily varied by changing the sample thickness.

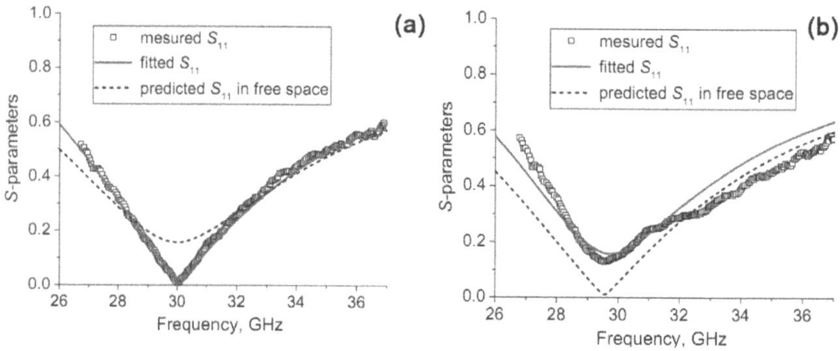

Figure 2. (a) Frequency dependence of measured S_{11}-parameters of the 0.65 mm-thick composite in the waveguide (squares). The solid line is the fit of experimental data by (1) with n =3.95+i0.45. The dashed line represents the predicted S_{11}-parameters of the 0.65 mm-thick composite in free space. (b) Frequency dependence of the measured S_{11}-parameters of the 0.55 mm-thick composite in waveguide (squares). The solid line is the fit to experimental data by (1) with n = 4.70+i0.61. The dashed line represents the predicted S_{11}-parameters of the 0.55 mm-thick composite in free space.

Using (1) with $k_z = 2\pi/\lambda$ and $k_{2z} = 2\pi n/\lambda$, the free space electromagnetic response of composites located on a metal plate can be estimated. The data for the 0.65 mm-thick composite with $n = 3.95 + i0.45$ are presented in Fig. 2a (dashed line). The free space absorption coefficient remains quite large but differs significantly from 100 %. Due to the difference of wave vectors in free space and in the waveguide, the higher values of the imaginary part of the refractive index need for getting the higher absorption coefficients. For this purpose, addition of MWCNTs in the composite sample can significantly change its electromagnetic properties [5].

The measured frequency dependence of S_{11} for the 0.55 mm-thick composite containing 0.1 wt.% of MWCNTs located on the metal plate is presented in Fig. 2b. The data were fitted by (1) with $n = 4.70 + i0.61$ (solid line in Fig. 2b). The absorption coefficient in free space is very close to 100 % near the frequency of 30 GHz (dashed line in Fig. 2b).

4. Conclusion

The results of this study demonstrate good potential for constructing surfaces absorbing 100 % of the incident microwave radiation in a waveguide and free space. The materials fabricated are thermally stable, thin, cost-effective and easy to produce. They have an importance for aerospace technologies. The highly absorbing region was found around the interference minimum. Within the 26–37 GHz frequency range, the absorption level varied from 65 to 100 %. The absorption maximum can be shifted by varying the sample thickness. Addition of small amount (below percolation limit) of MWCNTs to the ceramic matrix allowed tuning the refractive index of the composite for wide civilian and military applications.

References

1. K. Gaylor, *Materials Research Labs Ascot Vale* (Australia), No. MRL-TR-89-1 (1989).
2. D. Bychanok *et al.*, *Progress in Electromagnetics Research C* **66**, 77 (2016).
3. J. Baker-Jarvis *et al.*, *NIST Technical Note*, 1355 (1993).
4. A. Plyushch *et al.*, *Phys. Status Solidi (a)* **211**, 2580 (2014).
5. http://nano.bsu.by/products/mwcnt

FRONTIERS OF MOLECULAR DIAGNOSTICS WITH NANOSTRUCTURES

INVITED

SURFACE ENHANCED RAMAN SPECTROSCOPY OF ORGANIC MOLECULES ON METALIZED POROUS SILICON

H. V. Bandarenka

Belarusian State University of Informatics and Radioelectronics
P. Browka 6, 220013 Minsk, Belarus
h.bandarenka@bsuir.by

Open structure of porous silicon has driven scientists to introduce into pores different "foreign" substances, such as metals, semiconductors, dielectrics, organic and biological species, pursuing a goal to create composites and hybrid materials with outstanding properties. Figuratively speaking, under this approach porous silicon plays a role of the "host" material, which takes the other one as a "guest". Among various prospective "guests" in porous silicon, metals are some of very attractive due to their specific electrical, magnetic and optical properties. For more than 15 years, porous silicon has been considered as a template for formation of metallic nanostructures with precise control of their shape, dimensions and spatial location aiming the goal to fabricate highly sensitive and cost-effective substrates for surface enhanced Raman scattering. The present review is focused on the critical consideration of the porous silicon role in the analysis of organic molecules by surface-enhanced Raman scattering in order to evaluate the results achieved and define the prospects of this approach.

1. Introduction

It is known that activity in surface-enhanced Raman scattering (SERS) is demonstrated by metallic nanostructures which are generally accepted to be called SERS substrates. An electromagnetic field of the metallic nanostructures increases under laser excitation during Raman scattering due to an effect of surface plasmon resonance (SPR). Fabrication of the metallic nanostructured films allows to improve their SERS activity because of multiplication of the electromagmetic field in the nanogaps between metallic nanostructures (so-called "hot spots"). Despite all metals show SPR, noble metals are mostly used in SERS substrates as they are characterized by SPR in the visible range and demonstrate high stability. Thus, the most critical aspect of SERS is the development of new noble metal-based substrates.

SERS substrates need not only to have a rough nanoscale features, but also should demonstrate high sensitivity, reproducibility, stability, ease of preparation and compact size. Traditional SERS substrates include electrodes roughened by an oxidation-reduction cycle and aggregated colloidal nanoparticles (NPs). Rough metallic electrodes are stable and reproducible, but they have low sensitivity and unhandy for many applications. Contrariwise aggregated colloids can be easily prepared and often provide the strong Raman enhancement. However, such substrates are typically unstable, irreproducible and can be hardly used in practice due to the liquid basis. Therefore greater attention has been paid to the ordered arrays of metallic nanostructures on a solid support such us glass plate, silicon wafer and so on.

A lot of solid SERS substrates were proposed during the last decade due to the progress in nanotechnologies. The growing interest to the nanofabrication is obvious since SERS effect depends on the size, shape, and interspacing of noble-metal nanostructures [1]. There are several methods for the formation of such surface-confined nanostructures, including electron-beam lithography [2], nanosphere lithography [3], and films over nanospheres [4]. However, practically all these materials are costly, require special technique and trained personals for their fabrication.

In the beginning of this century porous silicon (PS) has attracted an attention of scientists as a template for formation of effective SERS substrates [5]. The use of PS allows to combine solid silicon support and precise control of parameters of the metallic nanostructures on the surface of porous matrix by the managing regimes of PS formation and metal deposition.

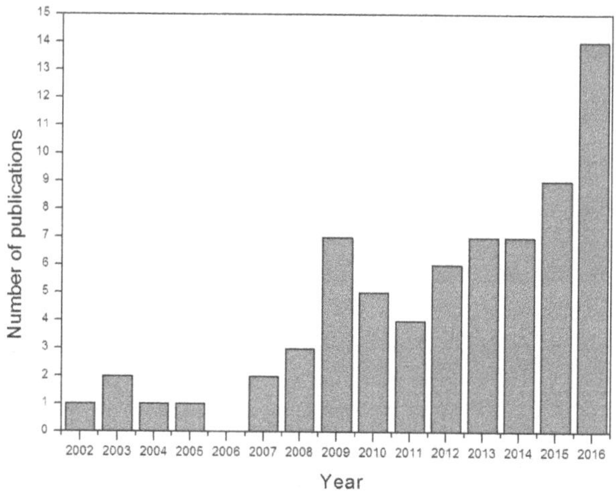

Figure 1. Publications on SERS substrates based on PS by years.

Growing interest to this approach in fabrication of the SERS substrates is observed from Fig. 1 showing graduate increase of the number of publications on PS in SERS. This review is devoted to the role of PS in SERS and includes parts on fabrication of the SERS substrates based on PS, description of the resulting metallic nanostructures in dependence of the PS type (micro-, meso- and macroporous) and their SERS activity.

2. Fabrication of SERS substrates based on porous silicon

The fabrication process generally includes formation of PS and deposition of metal on the surface of porous matrix. Proper choice of the PS type and method of metal deposition provide creation of appropriate SERS substrates.

2.1. *Formation of porous silicon*

Since the very beginning of the introduction of PS into SERS [5] an electrochemical anodization of monocrystalline silicon has been mostly applied for the PS fabrication aiming to create SERS substrates. This method is favorable for forming porous matrix with geometry that allows to control the morphology of further depositing metallic structures. PS parameters are defined by variation of anodization regimes, electrolyte composition and type of the silicon wafer. Such approach has been reported for the fabrication of the supports for SERS substrates based on the all types of PS: micro- [6], meso- [5,7–11] and macroporous silicon [12–15]. Great attention has been paid to mesoporous material formed in highly doped silicon wafers [5,7,9,11] as it leads to the growth of ordered arrays of the pore channels and silicon crystallites of nanoscaled dimensions. Following metal deposition on such mesoporous silicon allows to control accurately the size, shape and gaps between metallic structures to match the conditions for the best SPR.

Despite widely used anodization, some other techniques can be also applied for fabrication of PS templates for SERS substrates such as reactive ion etching [16], hydrothermal etching [17], metal-assisted etching [18], plasma chemical vapor deposition [19].

2.2. *Deposition of metallic nanostructures*

The most popular technique for coating of PS with metallic nanostructures is their *electroless* deposition from aqueous solutions of metal salts. The method is very simple because there is no need to apply voltage or activators/reducers unlike in the other "wet" techniques. In this case the deposition of metallic atoms on PS occurs due to redox reactions between silicon atoms and metallic cations. A preference has been given to this method because namely the strongest plasmonic metals (noble metals and copper) have a positive redox potential and thus can be deposited on PS from aqueous solutions of their salts.

The family of the metallic structures that have been formed by electroless deposition of metals on PS is very rich including nanoparticles [6,11,20], dendrites [7] and nanothorns [21]. An intensive study of the electroless deposition has already led to the selective fabrication of SERS-active areas by inkjet printing which is prospective for the mass production [22,23].

An advantage of *electrodeposition* is in a possibility to cover external and internal surfaces of the template with conformal metallic films as it was realized for macroporous silicon in [14,18]. On the other hand it is difficult enough for micro- and mesoporous silicon because of weak conductivity of the silicon nanocrystals [24].

Thermal decomposition has been rarely used for the fabrication of the SERS substrates based on PS. Nevertheless, it should be noted because this technique has been proposed at the dawn of development of the metallized PS for SERS [5]. Prior to the metallic nanostructures formation PS has to be oxidized and then impregnated with the solution of metal salt. Following heating of the soaked PS sample (for instance, in case of silver nitrate temperature is increased up to 500 °C) leads to the decomposition of the salt and formation of the metallic atoms. This approach provides uniform distribution of the metallic nanostructures over the entire PS volume.

Deposition from colloids on PS for SERS analysis was reported just for Au NPs [10,25]. To immobilize gold NPs on the substrate PS was oxidized and functionalized with 3-aminopropyltriethoxysilane molecules.

"Dry" techniques have been prevalently used for the formation of SERS substrates based on silicon nanopillars. Evaporation typically leads to the location of the depositing metal on the bottom of pores and top of silicon matrix avoiding coating pore walls [16]. For example, evaporation of gold on silicon pillars in [26] resulted in fabrication of the gold film of 10–20 nm thickness on their tops. At the same time sputtering yields conformal coatings of the silicon matrix [16,19]. By nowadays the SERS substrates formed by evaporation/sputtering of the noble metals on silicon nanopillars have already been commercialized [27].

3. Parameters of the SERS substrates based on porous silicon

Enhancement factor (EF) characterizes increase of Raman signal that SERS substrate provides and presents a ratio of the intensities of the SERS and Raman signals for a single molecule. Frankly speaking, known ways to estimate EF can be hardly applied for solid nontransparent SERS substrates including PS-based materials. There are two problems: (i) incorrect counting of molecules in the laser spot and (ii) difficulty in detection of the non-enhanced Raman spectrum for the monolayer of molecules. To overcome these problems for the porous substrates the numbers of molecules in the Raman scattering volume and adsorbed molecules in the volume for the SERS measurements have to be accurately counted [28,29]. Recently it has been also proposed to evaluate EF using fluorescence and Raman cross-sections of the analyte molecules that are known parameters [30]. In [31] so-called external amplified Raman efficiency was used instead of EF to facilitate evaluation of the effectiveness of the SERS substrate based on PS. This parameter is a ratio of the analyte concentration threshold detectable on the metallized PS to the minimum concentration of the detected analyte on the metal-free PS.

Limit of detection (LOD) is the lowest concentration of a substance that can be found with SERS substrate.

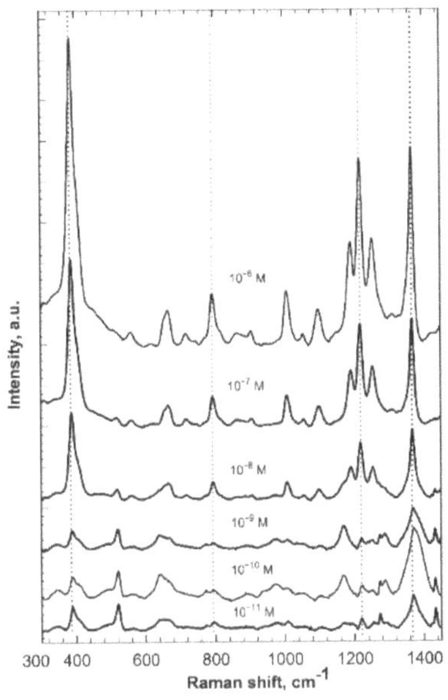

Figure 2. SERS spectra and detection limit of the CuTMpyP4 molecules adsorbed on the Ag/PS substrates.

For example, Fig. 2 shows SERS spectra of cationic Cu(II)-tetrakis(4-N-methylpyridyl) porphyrin (CuTMpyP4) in different concentrations deposited on the SERS substrate based on silvered PS [26]. Following this figure, LOD of CuTMpyP4 is 10^{-11} M because it is the lowest concentration when the less intensive analyte related peak is still resolved at the noise level.

LOD is very useful to evaluate effectiveness of SERS substrates for objective comparing the results from different research groups and characterization of the commercial SERS substrates. This parameter can be taken into account independently on the morphology of the active structure and thus is widely used to describe PS-based SERS substrates.

4. Application of SERS substrates based on porous silicon

SERS substrates based on PS open prospects of study of a number of substances including organic dyes, porphyrins, proteins, DNA, explosives, nerve gases, *etc.* Table 1 shows a brief observation of some analytes which have been studied with the SERS substrates based on different types of PS. The best reported LOD is 10^{-15} M of R6G [6,17]. At the same time EF reaches 10^{11} coinciding with that of known commercially available SERS substrates.

One more benefit of PS-based substrates is in long storage stability because electrons from the metallic atoms are strongly attracted by free states on the developed PS surface. Moreover, metallic nanostructures on PS inherit their ordered morphology providing repeatability of the SERS substrate properties. Therefore SERS signal deviation is very small in the range of one sample and in the batch of the samples and can vary in the range of 7–14 % [32,33].

Table 1. Examples of SERS substrates based on PS and their characteristics.

SERS substrate	Analyte	LOD	EF	Reference
Ag nanoparticles/ microporous silicon	rhodamine 6G	10^{-9} M	–	[6]
Ag dendrites/ mesoporous silicon	adenine	10^{-9} M	–	[7]
Ag nanoparticles/ mesoporous silicon	horseradish peroxidize	10^{-8} M	–	[9]
	crystal violet	10^{-10} M	$1.3\cdot10^{6}$	[34]
	cyanine dye	10^{-14} M		[35]
	p-thiocresol	$5.2\cdot10^{-9}$ M	$2.8\cdot10^{8}$	[36]
	trinitrotoluene	$1.1\cdot10^{-7}$ M		
Au nanoparticles/ mesoporous silicon	bovine serum albumine	10^{-8} M	–	[26]
	benzenethiol	10^{-6} M	–	[10]
Au nanothorns/ macroporous silicon	crystal violet	10^{-12} – 10^{-15} M	10^{8}	[21]
Ag nanoparticles/ macroporous silicon	rhodamine 6G	10^{-15} M	–	[6]
Ag nanoparticles/ silicon nanopillars	thiophenol (gas)	–	$2.1\cdot10^{11}$	[16]
	rhodamine 6G	10^{-15} M	–	[37]
	acetone	0.0037 ng	10^{11}	[38]
	BPE	10^{-9} M	10^{8}	[32]

In conclusion, SERS substrates based on PS coated with nanostructures of plasmonic metals are fabricated by simple and cost-effective methods. The substrates are characterized by repeatability and stability of structural and optical properties as well as strong enhancement of the Raman signal. Therefore PS-based substrates meet requirements for SERS-active materials for practical applications in medicine, biology, forensics, pharmaceutics, ecology, *etc.* Moreover, the fabrication process is compatible with the silicon technology opening an opportunity to develop lab-on-chip systems with such substrates.

References

1. C. L. Haynes, R. P. V. Duyne, *J. Phys. Chem. B* **105** (2001).

2. N. Felidj *et al.*, *Phys. Rev. B* **65**, 075419 (2002).
3. J. C. Hulteen, R. P. V. Duyne, *J. Vac. Sci. Technol. A* **13** (1995).
4. L. Baia *et al.*, *J. Phys. Chem. B* **110** (2006).
5. S. Chan *et al.*, *Adv. Mater.* **15**, 19 (2003).
6. M. Kosovic *et al.*, *Appl. Spectrosc.* **69**:12 (2015).
7. H. Lin *et al.*, *J. Phys. Chem. B* **108** (2004).
8. A. Yu. Panarin *et al.*, *Proc. of SPIE* **6728**, 672828 (2007).
9. F. Giorgis *et al.*, *Appl. Sur. Sci.* **254** (2008).
10. Y. Jiao *et al.*, *Appl. Phys. Lett.* **97**, 153125 (2010).
11. K. Artsemyeva *et al.*, *222nd ECS Fall Meeting* (Honolulu, 2012).
12. T. Ignat *et al.*, *Rom. J. Inf. Sci. Tech.* **12**, 4 (2009).
13. S. N. Terekhov *et al.*, *Nanomeeting*, Minsk, Belarus (2009).
14. K. Artsemyeva *et al., ECS Trans* **53,** 11 (2013).
15. H. V. Bandarenka *et al.*, *Nanoscale. Res. Lett.* **11**, 262 (2016).
16. M. S. Schmidt *et al.*, *Adv. Mater.* **24** (2012).
17. W. F. Jiang *et al.*, *J. Phys: Condens. Matter.* **22**, 415105 (2010).
18. B. Kiraly *et al.*, *Nanotechnology* **24** (2013).
19. S. Chattopadhyay *et al.*, *Chem. Mater.* **17** (2005).
20. S. Niauzorau *et al.*, *Phys. Stat. Solidi C* **13**, 4 (2016).
21. K. J. Khajehpour *et al.*, *Chem. Commun.* **48** (2012).
22. C. Novara *et al.*, *Nanoscale Research Letters* **9**, 527 (2014).
23. A. Chiolerio *et al.*, *Nanoscale Research Letters* **7**, 502 (2010).
24. K. Fukami *et al.*, *Materials* **4** (2011).
25. H. Dridi *et al.*, *J. Porous Mater.* **22**, 1 (2014).
26. M. S. Pio *et al.*, *Mater. Res. Soc. Symp. Proc.* **729** (2002)
27. Silmeco ApS http://www.silmeco.com/
28. E.C. Le Ru *et al.*, *J. Phys. Chem. C* **111** (2007).
29. E. Galopin *et al.*, *Appl. Mater. Inter.* **1**, 7 (2009).
30. A. Yu. Panarin *et al.*, *J. Appl. Spectrosc.* **76**, 2 (2009).
31. A. Virga *et al.*, *J. Raman Spectrosc.* **43** (2011).
32. K. Wu *et al.*, *J. Phys. Chem. C* **119** (2015).
33. H. Bandarenka *et al. Phys. Status Solidi C* **10**, 4 (2013).
34. F. A. Harraz *et al.*, *Appl. Surf. Sci.* **331** (2015).
35. A. Virga *et al.*, *J. Phys. Chem. C* **117**, 39 (2013).
36. Y. Lai *et al.*, *Analyt. Lett.* **47** (2014).
37. Y. Q. Wang *et al.*, *Appl. Surf. Sci.* **258** (2012).
38. C. L. Wong *et al.*, *Analyt. Chim. Acta* **844** (2014).

SERS ACTIVITY OF SILVER NANOSTRUCTURES WITH DIFFERENT SHAPE IN PORES OF SiO_2 TEMPLATE ON *n*-Si

D. V. Yakimchuk, E. Yu. Kaniukov, S. E. Demyanov
Division of Cryogenic Research, Scientific-Practical Materials Research Center NASB
220072 Minsk, Belarus
dim2yakim@gmail.com

G. M. Arzumanyan
Department of Raman Spectroscopy (Center «Nanobiophotonics»), Frank Laboratory of Neutron Physics, Joint Institute for Nuclear Research
Joliot-Curie Str. 6, 141980 Dubna, Russia
arzuman@jinr.ru

V. Sivakov
Functional Interfaces Department, Leibniz Institute of Photonic Technology
Albert Einstein Str. 9, 07745 Jena, Germany
vladimir.sivakov@leibniz-ipht.de

The study of the SERS effect on the silver nanostructures with different shape grown in the pores of SiO_2 template with diameters of 350–800 nm on *n*-silicon substrate was carried out. Possibility of the SERS-detection of 10^{-6} M R6G dye at 2.5 μW laser power has been shown.

1. Introduction

Optical properties of nanostructures (NSs) depend on their size, shape, type of metal and interparticle interaction [1]. Interparticle interaction due to the formation of "hot spots" leads to the significant increase of local electric fields. This effect is widely used for detection of ultralow concentrations of adsorbed impurities by surface-enhanced Raman spectroscopy (SERS) on nanostructured plasmonic substrates [2,3]. The NSs on the base of silver are mainly used in plasmonics due to their unique enhancing properties and the possibility of relatively long term exploitation in air conditions. The selection of optimum shape and interposition of silver NSs allows to obtaining the maximum enhancement factor. The easiest way to control the morphology of plasmonic surfaces is the use of porous templates at the stage of NSs creation [4]. In connection with this, in our work investigations of the silver NSs with different shapes in the pores of the SiO_2 templates on *n*-silicon substrate were carried out. The dependence of SERS-effect on morphology at different laser powers was analyzed.

2. Experimental

Ag NSs were deposited onto ion-track porous matrices [5] on the base of single-crystalline silicon (KEF-4.5, *n*-type, (100) orientation) wafers with the SiO_2 amorphous dielectric layer with 350–800 nm pore diameters. The deposition was carried out by electroless galvanic displacement method using the solution of 0.02 M $AgNO_3$ and 5 M HF at ambient conditions. The surface morphology of the *n*-Si/SiO_2(Ag) samples was characterized by scanning electron microscopy (SEM) using a JEOL JSM-7000F microscope.

Raman signal enhancement was studied using a CARS (Coherent Anti-Stokes Raman Scattering) microscope (SOLAR TII). Red laser radiation with the wavelength of 633 nm and power of 2.5–50 μW within the exposure time of 0.5 s was used in the experiment. A 100× lens with the numeric aperture NA = 0.95 provided the magnification of 2000 times was applied. R6G dye with the concentration of 10^{-6} M was used as an analyte.

3. Results and discussion

The silver NSs formed by galvanic displacement deposition into pores of SiO_2 template with various diameters are shown in Fig. 1.

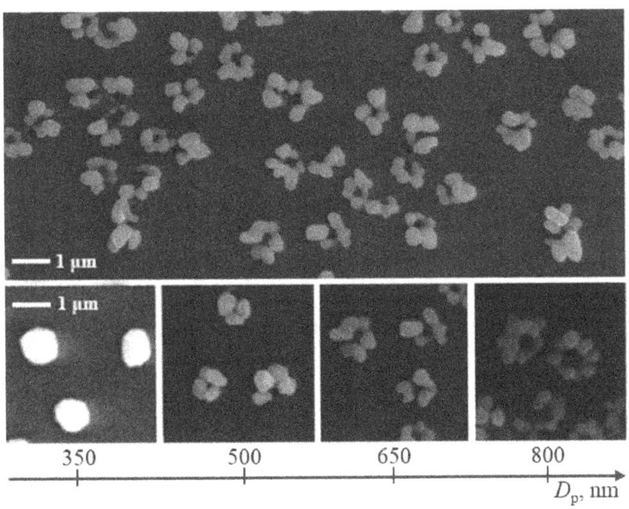

Figure 1. SEM image of the surface of SiO_2 template with silver NSs grown in the pores with 650 nm diameters. The images below show the evolution of the silver NSs shape by changing the pore diameter in the range of 350–800 nm. The scale is the same for each images in line.

The NSs with different forms have been synthesized from the solution of 0.02 M $AgNO_3$ and 5 M HF at ambient conditions. At the pore diameters

D_p = 350 nm individual hemispherical metal crystallites with the characteristic dimensions of 800 nm were formed. The increase of D_p to 500 nm leads to formation a few (3–4) elongated crystallites with the thickness of 250 nm and the length of about 1 μm. The D_p = 650 nm allows the creation of 5–8 crystallites into pores with metal precipitate *evolution* to the sunflower-like structures at D_p = 800 nm. It should be noted that the dependence of the shape of silver precipitates on the SiO_2 template parameters was mentioned at the first time in our previous publication [6].

The SERS spectra of 10^{-6} M R6G achieved at the 50 μW laser power on the substrates with different shapes of silver NSs (Fig. 2a) have similar quality of the spectra (noise level) and signal intensity. This indicates an absence of the SERS signal dependence on the NSs shape.

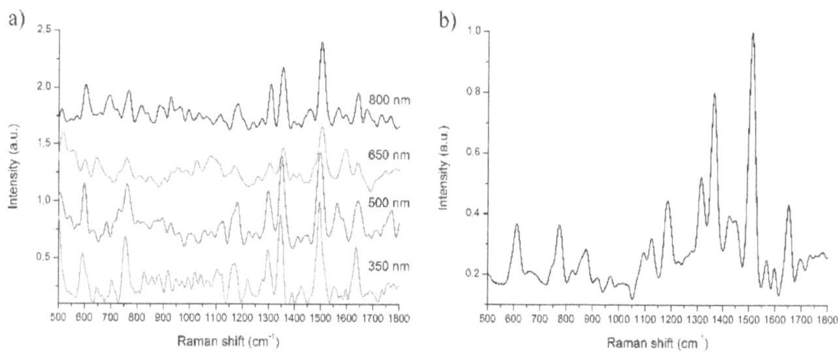

Figure 2. The characteristic SERS spectra of 10^{-6} M R6G dye recorded at the 50 μW power of laser radiation with 633 nm wavelength on silver nanostructures of different forms grown in the pores of SiO_2 template with the diameters of 350–800 nm (a); SERS-spectrum of sunflower-like structure at the 2.5 μW laser power (b).

In the experiments with varying of the laser power, it was found that the low branching NSs (grown in pores with smaller diameters) with decreasing of the light intensity the useful signal cannot be defined. At the 2.5 μW laser power a signal enhancement was observed only from the sunflower-like structures. At the same time the spectrum becomes even more visible than at 50 μW (Fig. 2b). It may be due to analyte burning when using the laser with a higher power. As a result, the signal from a lower analyte concentration is recorded. Thus, it can be concluded that the structures with the sunflower shape are more attractive for SERS studies with a use of low-power lasers.

4. Conclusion

Silver NSs with different shapes were formed in the pores of the SiO_2 templates with 350–800 nm pore diameters. The silver precipitate evolution from individual crystallites to sunflower-like structures at the pore diameter changing is shown. The study of the SERS effect of 10^{-6} M R6G at the 50 μW laser power did not allow identification the dependence of spectral characteristics on the shape of silver NSs. The differences in the SERS spectra for different surface morphologies occur at a light intensity change: at 2.5 μW laser power the signal enhancement was observed only at the most branched NSs having sunflower-like structures. These results demonstrate perspectives of applying such surfaces with silver NSs into pores of the SiO_2 template for detection of small doses of sensitive to the laser power analytes, such as biological objects.

Acknowledgments

This work was supported by the Scientific-technical program “Technology-SG” (№3.1.5.1), Grant of National Academy of Science of Belarus (№216-26-129).

References

1. C. Noguez, *J. Phys. Chem. C* **111**, 3806 (2007).
2. H. Wei, H. Xu, *Nanoscale* **5**, 10794 (2013).
3. G. Mettela, S. Siddhanta, C. Narayana, G. U. Kulkarni, *Nanoscale* **6**, 7480 (2014).
4. J.-H. Han, E. Khoo, P. Bai, M. Z. Bazant, *Sci. Rep.* **4**, 1 (2014).
5. E. Yu. Kaniukov *et al.*, *Nanotechnology* **27**, 115305 (2016).
6. V. Sivakov *et al.*, *J. Cryst. Growth* **400**, 21 (2014).

MODELLING OF REAL-TIME PHYSICAL AND BIO-NANOSENSORS FOR MEDICAL APPLICATIONS AND ECOLOGICAL MONITORING

Yu. Shunin[1,6], D. Fink[2], A. Kiv[3], A. Mansharipova[4], R. Muhamediyev[4], Yu. Zhukovskii[1], T. Lobanova-Shunina[5], N. Burlutskaya[6], V. Gopeyenko[6], S. Bellucci[7]

[1]Institute of Solid State Physics, University of Latvia, LV-1063 Riga, Latvia
yu_shunin@inbox.lv, yuri.zhukovskii@gmail.com
[2]Departamento de Fisica, Universidad Autónoma Metropolitana-Iztapalapa
PO Box 55-534, 09340 México, D.F., México; fink@xanum.uam.mx
[3]Ben-Gurion University, PO Box 653, Beer-Sheva 84105, Israel; kiv@bgu.ac.il
[4]Kazakh-British Technical University, Faculty of Information Technologies
050000 Almaty, Kazakhstan; dralma@mail.ru, ravil.muhamedyev@gmail.com
[5]Riga Technical University, Faculty of Mechanical Engineering, Transport and Aeronautics, LV-1019, Riga, Latvia; busus@inbox.lv
[6]ISMA University, LV-1019, Riga, Latvia
viktors.gopejenko@isma.lv, natalja.burlucka@isma.lv
[7]INFN-Laboratori Nazionali di Frascati, Frascati-Rome, Italy
Stefano.Bellucci@lnf.infn.it

Two convergent areas of real-time control nanosystems - for ecological monitoring and medical applications are discussed. Functionalized CNTs and GNRs nanostructures serve as a basis for the creation of physical pressure and temperature nanosensors. The development of bio-nanosensors (*e.g.*, glucose biosensors) based on polymer nanotracks with various enzymes is also considered. Both nanosensor models are analyzed in comparison with the experiment.

1. Physical nanosensors based on CNTs- and GNRs-epoxy resin nanocomposites

We consider physical nanosensors (pressure and temperature) based on functionalized CNTs and GNRs nanostructures. The model of nanocomposite materials based on carbon nanocluster suspension (CNTs and GNRs) in dielectric polymer environments (*e.g.*, epoxy resins) is regarded as a disordered system of fragments of nanocarbon inclusions with different morphologies (chirality and geometry) in relation to a high electrical conductivity in a continuous dielectric environment (Fig. 1). The electrical conductivity of a nanocomposite material depends on the concentration of nanocarbon inclusions (in fact, carbon macromolecules). The basic conductivity mechanism for the considered nanocomposites is the hopping conductivity (Eq. 1, Fig. 2). This mechanism can usually be observed, for example, in amorphous semiconductors or some disordered solids. Various nanocomposite morphologies are considered and computer simulation results for pressure and temperature nanosensor models are discussed in comparison with the experimental device prototype [1–3].

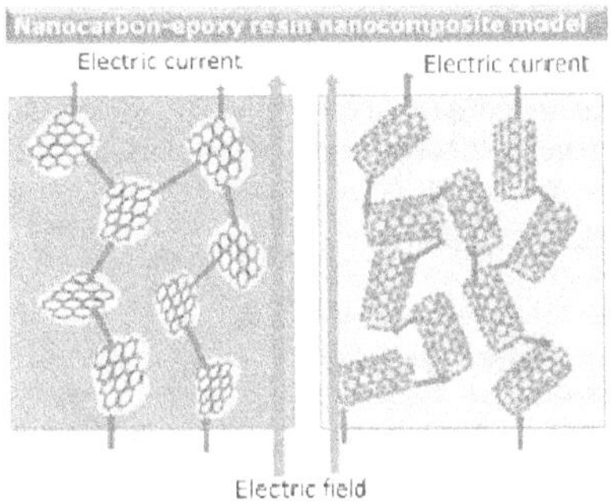

Figure 1. Model of composite polymer material with carbon nanocluster inclusions of GMRs- and CNTs- types.

$$\sigma = A\sigma_0 \exp(-\frac{4}{3}\left(\frac{4\alpha r_{tun}}{a}\right)^{3/4}\left(\frac{W_0}{\kappa T}\right)^{1/4}), \qquad (1)$$

where σ_0 is the normalization constant, which means the conductivity of monolithic dielectric medium.

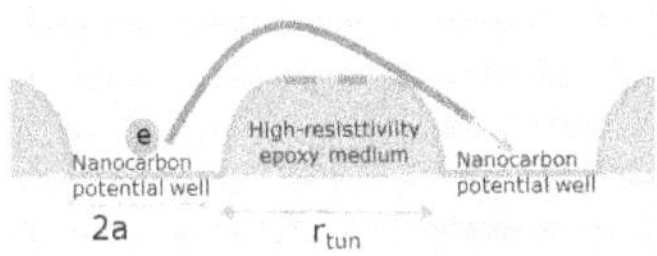

Figure 2. Potential wells model for hopping in polymer nanocomposites, where $2a$ is the characteristic size of nanocarbon inclusion, r_{tun} is the length of the tunnel 'jump' of the electron.

When testing the pressure sensor, the load ranged from 0 to 500 N, which corresponds to the change in pressure from 0 to 30 Bars. The typical dependence of the sensor resistance on the pressure changes, as compared with the simulation results, is shown in Fig. 3a. Small deviations are connected with technological problems in the reproduction of perfect morphology, which reduces the percolation limit of the nanocomposite. The typical dependence of the temperature sensor in the range of 27 to 90 °C compared with the simulation results is shown in Fig. 3b. The discrepancy in the behavior between the experimental and theoretical dependencies is associated with morphological imperfections of the real sensor induced the orientation dispersion of CNTs. This effect can diminish the hopping mechanism efficiency, especially, for the higher temperatures.

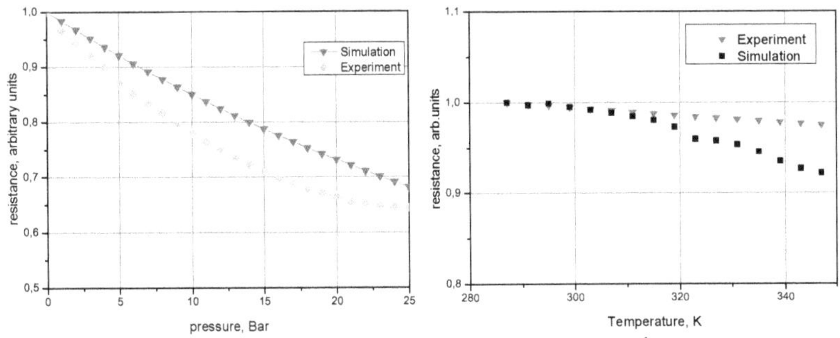

a b

Figure 3. Comparison of real pressure and temperature nanosensor indications [4] with more adequate models morphologies simulation: a) pressure nanosensor; b) temperature nanosensor.

Modelling of resistivity dependence via weight fraction of CNTs is presented in Fig. 4, where the sharp changes of resistivity within the weight interval (0.08-0.12) is observed.

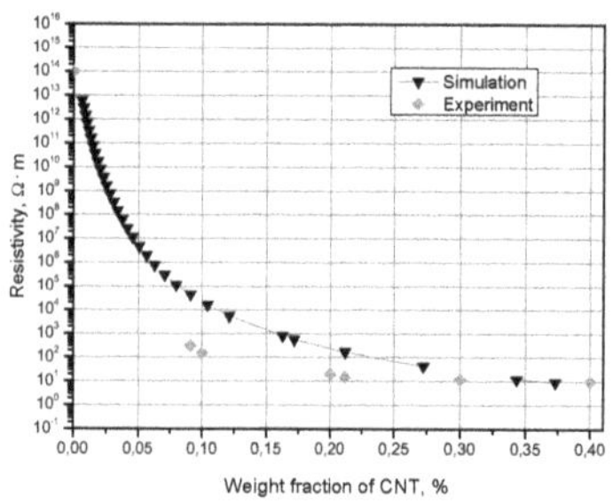

Figure 4. Comparison of resistivity of developed model CNT-based nanocomposite via weight fraction of CNT with the experimental data [5].

2. Polymer nanoporous structures based bionanosensors

We developed bio-nanosensors based on polymer nanoporous structures (nanotracks) with various enzymes, which provide the corresponding biocatalytic reactions and give reliably controlled ion currents [6,7] (Fig. 5).

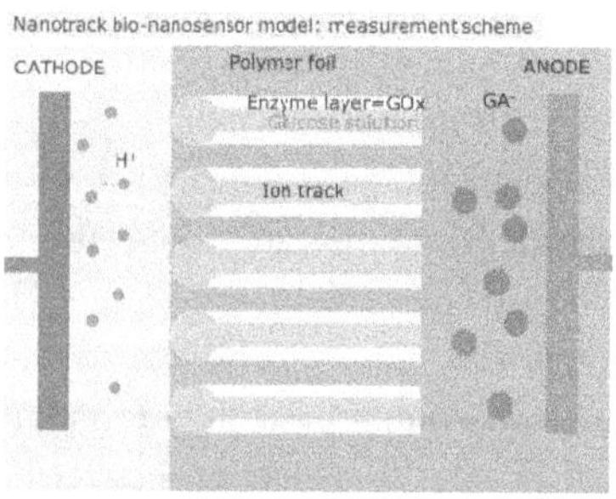

$$C_6H_{12}O_6 + O_2 \xrightarrow{GOx} C_6H_{12}O_7 + O$$

$$O + H_2O \rightarrow H_2O_2$$

$$H_2O_2 \xrightarrow{catalaza} \frac{1}{2}O_2 + H_2O$$

$$C_6H_{12}O_7 \rightarrow C_6H_{12}O_7^- + H^+$$

Anode: $H_2O \rightarrow \frac{1}{2}O_2 + 2H^+ + 2e^-$

Cathode: $2H^+ + 2e^- \rightarrow H_2$

Ultimate reaction:

$$C_6H_{12}O_6 + H_2O \rightarrow C_6H_{12}O_7 + H_2$$

a

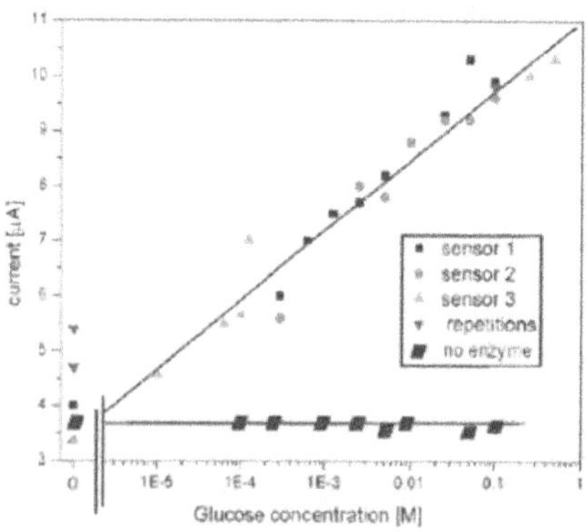

b

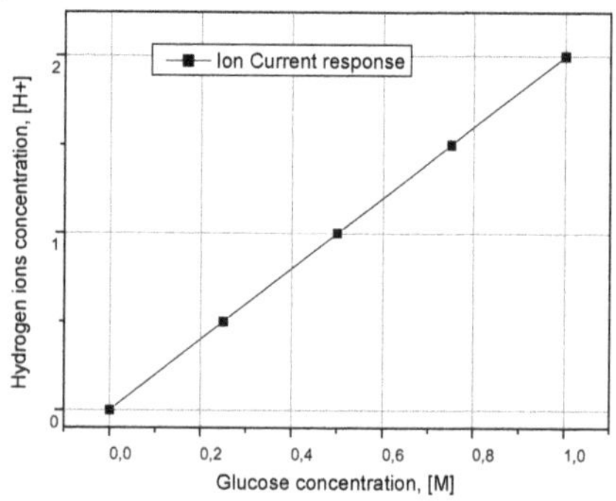

c

Figure 5. a) General model of glucose detection process on the ion track-containing foils embedded in electrolyte and basic set of biochemical reaction; b) Experimental calibration dependence; c) Theoretical model of the calibration dependence based on chemical kinetics.

In conclusion, we describe a concept for a glucose biosensor based on the enzyme glucose oxidase (GOx) covalently linked to nanopores of etched nuclear track membranes. Using simulation of chemical kinetics glucose oxidation with glucose oxidase we have obtained theoretical calibration dependences, when the concentration of H^+ is proportional to the concentration of the detected glucose. Experimental and theoretical calibration dependences demonstrate similar trends. The proposed device can serve to detect physiologically relevant glucose concentrations. The catalytic sensor can be made re-usable due to the production of diffusible products from the oxidative biomolecular recognition event. Moreover, we can develop a multi-agent packet nanosensor, which can be used for human breathing analyzer in relation to cancer detection, hepatitis, and other applications.

Acknowledgments

This research has been partially supported by Belarus-Latvia Bilateral Project 'Correlation of electromagnetic, mechanical and heat properties of aerogels and polymer composites with nanocarbon inclusions' (2014-2015), grant 'Nanostructures for bacteria detection and study' (NANOBAC) (01.10.2015 – 31.12.2017) Ministry of Education and Science of the Republic of Kazakhstan (2015-2017) and Ventspils Higher School Research Division.

References

1. Yu. N. Shunin *et al.*, in: *Nanodevices and Nanomaterials for Ecological Security*, Series: NATO Science for Peace Series B - Physics and Biophysics, eds. Yu. Shunin, A. Kiv (Springer Verlag, Heidelberg, 2012), pp. 237–262.
2. Yu. N. Shunin *et al.*, in: *Physics, Chemistry and Application of Nanostructures*, eds. V. E. Borisenko, S. V. Gaponenko, V. S. Gurin, C. H. Kam, (Singapore, World Scientific, 2013), pp. 250–253.
3. Y. Shunin *et al.*, *Computer Modelling and New Technologies* **19**, 14 (2015).
4. R. R. Abdrakhimov, S. B. Sapozhnikov, V. V. Sinitsin, *Bulletin of the South Ural State University Series "Computer Technologies, Automatic Control, Radio Electronics* **13**, 16 (2013).
5. Y. X. Zhou *et al.*, *EXPRESS Polymer Letters* **2**, 40 (2008).
6. D. Fink *et al.*, *Biosensors and Bioelectronics* **24**, 2702 (2009).
7. D. Fink *et al.*, *Computer Modelling and New Technologies* **19**, 7 (2015).

SERS-ACTIVE SUBSTRATES BASED ON SILVERED POROUS SILICON COVERED WITH GRAPHENE

K. Girel, N. Kovalchuk, I. Komissarov, H. Bandarenka
Belarusian State University of Informatics and Radioelectronics
P. Browka 6, 220013 Minsk, Belarus
k.girel@bsuir.by

Silver nanostructures have been formed on the surface of porous silicon by immersion deposition. Silver has been found to form a quasi-continuous film composed of nanoparticles with sizes varying in the range of 100–500 nm which are located at the distance less than 10 nm from each other. The CVD method has been applied for the synthesis of graphene. The Ag/porous silicon samples have been covered with graphene by wet chemical transfer. Resulting substrates demonstrate an improved activity in the surface enhanced Raman spectroscopy for detection of low concentrations of organic molecules in comparison with the graphene-free Ag/porous silicon.

1. Introduction

Surface enhanced Raman spectroscopy (SERS) is a powerful tool to characterize structure of molecules at extremely low concentrations down to the single molecule level [1]. Various metallic nanostructures with tunable plasmonic properties have been explored as SERS-active substrates. Traditionally, nanorough surfaces of the noble metals such as Ag, Au have been used for the SERS applications because of their intensive surface plasmon resonance. At present SERS technique is actively studied for structural analysis of many bioorganic substances including DNA, microRNA, proteins, peptides and others. However, the noble metals have weak biological compatibility and SERS measurements of bioorganic molecules are often irreproducible because of this feature. Such a problem is especially acute at extremely small amounts of compounds. Thus, studying biomolecules by SERS requires a special approach and development of new SERS-active substrates which will provide label-free and non-destructive investigations of different organic compounds. In other terms, a perfect platform for SERS-based molecule detection has to be reproducible, chemically stable, biologically compatible, and should be easily fabricated in the scale production.

The silvered porous silicon (PS) is one of the very attractive materials for SERS [2] due to its reproducible and strong SERS activity as well as cost effectiveness. Nevertheless, this material is also characterized by above-mentioned drawback if low concentrations of the analytes are studied. SERS study results in fluctuations of the characteristic bands in the Raman spectra of the analyte indicating its degradation. In addition to the bioincompatibility it can be caused by a high local electric-field of the SERS substrate which leads to the photoinduced damages of the analyte molecules (especially for organic molecules) and results in photophysical desorption or photochemical reactions [3].

In this paper we propose to cover the silvered PS by graphene (Gr) to prevent degradation of the organic molecules. The hybrid structures based on the silvered PS and Gr are expected to provide an enhancement of Raman signal by the plasmonic coupling of silver nanoparticles and Gr [3]. Furthermore, Gr can be used as a fluorescence quencher during study of different dyes [5]. Moreover SERS effect in Gr-based substrates is provided by not only contribution of the electromagnetic mechanism but also chemical mechanism [3,5].

2. Experimental details

Monocrystalline n^+-type Si wafers were used for the PS fabrication. Each wafer was cleaned in a solution of NH_4OH, H_2O_2 and H_2O mixed at the volume ratio of 1:1:4 according to the procedure described elsewhere [2]. The native SiO_2 was removed from the Si wafer in diluted HF (4.5 %).

PS was fabricated by an electrochemical anodization of the Si wafer in a mixture of HF, H_2O and C_3H_7OH at the current density of 100 mA/cm^2. Then the sample was rinsed with deionized water and spun dried.

Samples of the Si wafer of a 5×7 cm size with PS layer were immersed in a water-ethanol solution of 3 mM $AgNO_3$ for 70 min resulting in a deposition of silver nanoparticles on PS. The silvered PS samples were rinsed with deionized water and then air-dried.

The CVD method was used for the Gr synthesis. Electrochemically polished copper foil was annealed for 1 h at 1050 °C in Ar and H_2 gas flow with a rate of 160 and 150 cm^3/min, respectively. Then 4 cm^3/min methane flow was introduced into the CVD reactor for 10 min. We used the wet-chemical transfer of Gr onto the SERS substrates without any polymer support.

The morphology and structure of the experimental samples were studied with the scanning electron microscope (SEM) Hitachi S4800, which provides the 1 nm resolution.

Organic dye Rhodamine 6G (Rh6G) was chosen as an analyte for the study of SERS activity of the Ag/PS substrates. The silvered PS samples were immersed in an aqueous solution of Rh6G for 2 h before SERS measurements. SERS spectra were registered with a 3D scanning confocal microscope Confotec NR500. Laser wavelengths of 473, 633 and 785 nm were used to excite the samples. Laser powers were 1.45, 0.68 and 0.86 mW, respectively. Laser spot diameters were about 300 nm (for the wavelength of 473 nm), 400 nm (for the wavelength of 633 nm) and 500 nm (for the wavelength of 785 nm).

3. Results and discussion

In the Ag deposition process, PS plays a role of a template which allows to obtain silver nanostructured film on its surface. Also, PS is a source of a great number of electrons and nucleation centers for Ag^+ ions reduction. Fig. 1a demonstrates the surface of the silvered PS before transferring Gr layers. A quasi-continuous silver film composed of particles with the sizes varying in the range of 100–500 nm is seen to be formed. In many places gaps between the particles are less than 10 nm. After the transfer step (Fig. 1b) the surface of the silvered PS was covered with Gr.

Figure 1. SEM images of the Ag/PS substrate without Gr (a) and covered with Gr (b).

Fig. 2a shows SERS spectra of Rh6G molecules adsorbed on the Gr-free silvered PS, adsorbed on the Ag/PS substrates under and above Gr and Raman image of the silvered PS with the analyte under Gr. It can be seen that the SERS intensity of Rh6G adsorbed on the silvered PS under Gr is three times higher than that for the analyte molecules located above Gr. What is more it is thirty times higher than that for the Gr-free substrate. We suppose that the Gr can provide an enhancement of the Raman signal due to more effective charge transfer from metal to analyte molecules. The other feature of the Gr-coated substrates was in slower decrease of the SERS signal intensity. Gr-free substrates provided stable signal for a few seconds while Gr-coated showed stable results up to the minutes.

Raman image (Fig. 2b) shows distribution of “hot spots” on the surface of the SERS substrate. For other types of substrates Raman images were the same but we managed to obtain clearer and more intensive Raman spectrum if the analyte was located between Gr and silvered PS.

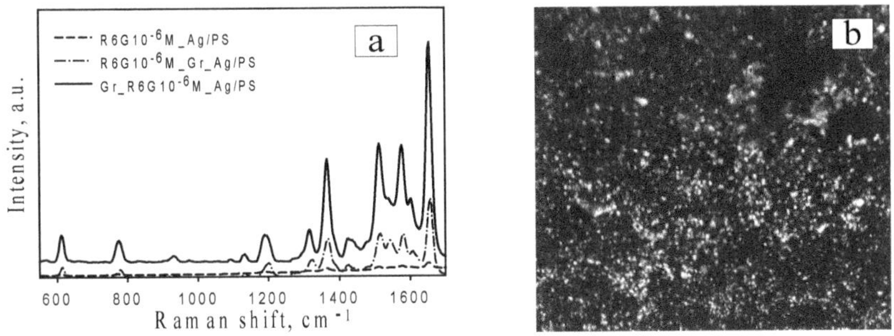

Figure 2. SERS spectra of Rh6G molecules adsorbed on different substrates (a) and Raman image of the Gr/Ag/PS surface (b).

4. Conclusion

SERS substrates based on the silvered PS covered with Gr layer were fabricated. It was revealed that intensity of the Raman signal from Rh6G adsorbed on silvered PS with Gr is much higher and stable in comparison with Gr-free substrates.

Acknowledgments

The support for the research has been provided by the Belarusian Republic Foundation for Fundamental Research (grant T16M-099) and State Research Program of Belarus "Photonics, opto- and microelectronics" (task 1.4.01).

References

1. X. Liming *et al.*, *J. Am. Chem. Soc.* **131** (2009).
2. H. Bandarenka *et al.*, *Nanoscale Research Letters* **11**, 262 (2016).
3. Y. Zhao *et al.*, *J. Phys. Chem.* C **118** (2014).
4. H. Shijiang *et al.*, *Anal. Chem.* **84**, 10 (2012).
5. S. G. Zhang *et al.*, *Applied Physics Letters* **104**, 121109 (2014).
6. W. Xu *et al.*, *Proc. Natl. Acad. Sci. U. S. A.* **109**, 24 (2012).

LUMINESCENT CARBON NANODOTS FABRICATED BY LASER ASSISTED SYNTHESIS IN LIQUIDS

N. N. Tarasenka, A. A. Nevar, A. V. Butsen, N. V. Tarasenko, V. A. Lapina
Laboratory of Laser Diagnostics of Plasma, B. I. Stepanov Institute of Physics NASB
Nezavisimosti Ave. 68, 220072 Minsk, Belarus
n.tarasenko@ifanbel.bas-net.by

The synthesis of luminescent near-spherical carbon nanoparticles by laser ablation of a graphite target in a liquid environment is reported. The particles were 1–8 nm in size with the inner structure depending on the liquid used. The exhibited strong photoluminescence (PL) which can be tuned from the near-UV to the green wavelength region by controlling experimental parameters. The highest intensities of the emission bands were detected for the excitation wavelength of 400 nm. An improvement of surface and optical characteristics was observed after additional laser treatment. It was suggested that the laser induced non-equilibrium liquid chemistry results in surface modification and in the enhancement of the photoluminescence of the synthesized nanoparticles.

1. Introduction

In recent years carbon nanoparticles (NPs) with sizes below 10 nm, so called carbon nanodots (C-dots), owing to their excellent biocompatibility, chemical inertness and tunable photoluminescence have received a great interest for the potential applications in biotechnology as more suitable alternatives to the traditional semiconductor quantum dots. Much progress has been achieved in the synthesis and characterization of C-dots [1]. Nevertheless, the origin of the photoluminescence (PL) in C-dots is not well understood yet, though many mechanisms have been proposed. However, the understanding of emission origins and control over the C-dots luminescence are of great importance for their successful applications and for the improvement of the quantum yield. PL of C-dots is not only dependent on the particle size, but on other parameters such as defects and surface states, that depend on the synthesis method. It should be noted that the particle surface passivation or surface functionalization are the most popular approaches to enhance the luminescence of C-dots. Surface groups, such as C–O, C=O and COOH, can introduce trapping states with different energy levels on NPs surfaces, that are considered to be responsible for their luminescent properties.

In this paper, we report the synthesis of carbon NPs by laser irradiation of a graphite target in a liquid media and their luminescence properties.

2. Experimental

Two approaches have been used for laser-assisted synthesis of luminescent carbon NPs in liquid. First, laser irradiation of graphite powder under liquid layer (octane, 10 mM aqueous glucose solution) was performed. The second technique was based on laser irradiation of a carbon target immersed in a liquid (water, 10 mM aqueous glucose solution, 0.008 M diethylenetriamine-pentaacetic acid (DTPA) solution).

The unfocused beam of the second harmonic of the YAG:Nd^{3+} laser (wavelength 532 nm, power density 660 mJ/cm^2 and exposure time 30 min) was used in the ablation experiments. Aiming to improve the luminescent properties and to modify the surface, the fabricated NPs were subjected to additional laser irradiation by the second harmonic of the YAG:Nd^{3+} laser (wavelength 532 nm, power density 350 mJ/cm^2 and exposure time 5 min). Laser-induced modification of the samples was carried out immediately after the end of the synthesis process.

Transmission electron microscope (TEM), atomic force microscope (AFM), X-ray photoelectron spectrometer (XPS), and Fourier transform infrared (FTIR) spectroscope were employed to characterize the microstructure of as-prepared NPs. UV-visible absorption and PL spectroscopy were applied to analyze optical properties and to confirm the synthesis of luminescent carbon nanostructures. Furthermore, the cytotoxicity of the synthesized NPs against BGM Cell Line from African green monkey kidney was evaluated.

3. Results and discussion

The TEM analysis performed showed a possibility to synthesize carbon NPs that were nearly spherical in shape and had an average diameter less than 10 nm. They can be classified as carbon quantum dots. The particles prepared by laser ablation in water were amorphous as could be concluded from the HRTEM and SAED images. But the conditions created upon laser irradiation of graphite in the DTPA solution favor the formation of crystalline NPs. It is noteworthy, that the crystal structure of the particles prepared in ethanol and DTPA solution is different. Irradiation of graphite in ethanol resulted in the formation of the orhthrhombic carbon phase, while in DTPA lonsdaleite carbon phase with the interplanar spacing of 0.22 nm was formed.

Most probably, the different inner structure of the carbon NPs produced by laser ablation in different liquids can be attributed to the solvent-dependent cooling rate. Another reason of the formation of different internal phase structures in water and organic solvents could be connected with a possible decomposition of organic solvents resulting in formation of carbon-containing by-products.

The PL measurements revealed an excitation-dependent PL behavior of NPs prepared by laser irradiation of graphite powder in octane. Similar to the colloid prepared by electrical discharge in octane, the emission maximum shifted from 330 to 515 nm when the excitation wavelength changed from 290 nm to 450 nm and the most intense luminescence was observed for the excitation wavelength of 290 nm.

However, the PL spectra of the colloids fabricated in water solutions do not show the excitation dependence that can be attributed to the different emissive sites responsible for the luminescence. The comparison of the PL spectra of colloids prepared in water and DTPA solutions allows drawing a conclusion that an addition of the stabilizing agent results in the considerable increase in the luminescence intensity and red shift of the PL bands. The colloidal solution, prepared by laser ablation in DTPA showed strong blue-green PL centered at approximately 500 nm when excited at 405 nm (Fig. 1). Therefore, surface passivation can be used for tailoring NPs luminescent properties and increase the PL intensity.

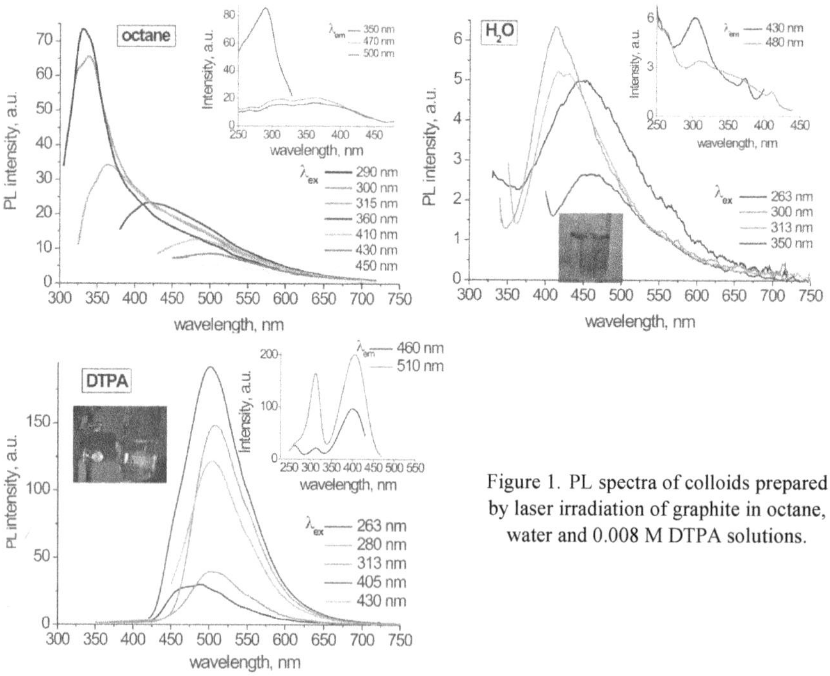

Figure 1. PL spectra of colloids prepared by laser irradiation of graphite in octane, water and 0.008 M DTPA solutions.

The PL properties of the synthesized NPs make them good candidates for nanoprobes in living cells. In this context, to evaluate the biomedical applications of the prepared NPs for drug delivery and bioimaging, it is critically important to examine their potential cytotoxicity. Our preliminary results revealed that the NPs fabricated in water had good biocompatibility and low cytotoxicity against BGM Cell Line from African green monkey kidney. We have not observed cell death under their incubation with NPs.

We also tested a suitability of carbon NPs for confocal microscopy applications by imaging BGM. The cells were incubated with NPs for 4.5 h. It was readily seen that the cells had a strong affinity for the particles, thus enhancing their image ability under the confocal microscope excitation wavelength of 405 nm. The self-luminescence of these cells was much weaker in comparison with PL of NPs. Therefore, the synthesized NPs can be used as contrast probes in bio-imaging and in other biomedical applications.

Acknowledgments

The work was supported by the National Academy of Sciences of Belarus under the project Convergence 2.4.06. The COST Action TD1208 is also acknowledged for useful interactions and exchanges.

Reference

1. Prathik Roy, Po-Cheng Chen, Arun Prakash Periasamy, Ya-Na Chen, Huan-Tsung Chang, *Materials Today* **18**, 447 (2015).

SURFACE ENERGY STATES AND PHOTOLUMINESCENCE QUENCHING IN NANOSTRUCTURES BASED ON CdSe/ZnS QUANTUM DOT AND PORPHYRIN MOLECULE: ENSEMBLE AND SINGLE OBJECT DETECTION

E. Zenkevich

Department of Information Technologies and Robotics, Belarusian National Technical University, 220013 Minsk, Belarus
zenkev@tut.by

C. Goehler, C. Krasselt, C. von Borczyskowski

Institute of Physics, Chemnitz University of Technology, 09107 Chemnitz, Germany
borczyskowski@physik.tu-chemnitz.de

It was found that the attachment of only one functionalized dye molecule on QD surface modifies not only the photoluminescence (PL) intensity but creates new energetically clearly distinguishable electronic states opening additional effective relaxation pathways stronger than Foerster resonant energy transfer (FRET). It was shown that surface properties of QDs are a complex interplay of the nature of the dye molecule and the topography of the ligand layer across the temperature range of 77–290 K.

1. Introduction

At the moment, the main obstacle on the path to the development of efficient quantum dot-based functional nanomaterials is our limited understanding of semiconductor trioctylphosphine oxide (TOPO)-capped CdSe/ZnS quantum dot (QD) surfaces, their interaction with passivating ligands, and their impact on charge and energy transfer between QDs and attached dye molecules. A one-molecule-to-one-QD base is of special interest. A large variety of functionalized organic molecules are able to modify QD optical properties which may be controlled and related to surface reconstructions [1,2].

Here, we report that upon QD-dye self-assembly, dye molecules replace a few TOPO ligand molecules, which results in QD PL energy shifts of 10–20 meV and in an increased non-radiative PL relaxation (stronger than FRET) depending on temperature. To solve several not yet well understood aspects of PL properties related to the heterogeneity of QD surfaces, we combine steady-state and picosecond time-resolved ensemble and single QD data. Our findings lie in the scope of recent calculations on detailed QD surface properties [1] and also provide new insights into the origin of PL blinking effects for QDs [3].

2. Results and discussion

QD-porphyrin nanoassemblies were formed at 293 K via one-step titration of QD solution by porphyrin molecules at a molar ratio $[C_{Porph}]/[C_{QD}] = x = 1$. A controllable self-assembly of QDs with 5,10,15,20-*meso*-*meta*-pyridyl porphyrins, H_2P and CuP [2], is realized via Zn·N-Pyridyl coordination of Zn atoms of the ZnS shell with nitrogen atoms of pyridyl substituents (see Scheme 1).

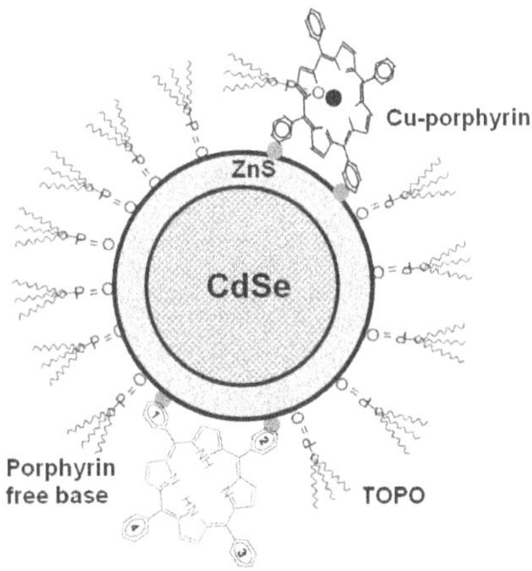

Scheme 1

Measurements at 77–300 K were carried out in methylcyclohexane/toluene mixture (6:1). Single QD samples are prepared by spin coating a toluene solution with CdSe/ZnS QDs onto a quartz substrate.

The temperature dependence of the spectroscopic observables reveals both the influence of electron-phonon coupling between 77 K and 290 K and a kind of "phase transition" probably due TOPO ligands rearrangement (Fig. 1).

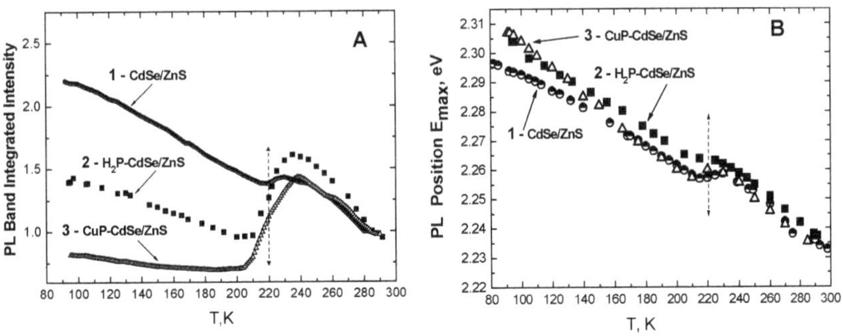

Figure 1. Manifestation of the "phase" transition of capping TOPO layer at $T_{crit} \approx 220$ K (dashed arrow) for QD PL band integrated intensities (**A**) and PL band energy position and (**B**) for TOPO-capped CdSe/ZnS (1), QDs in H_2P-CdSe/ZnS (2) and CuP-CdSe/ZnS (3) nanoassemblies at molar ratio x=1 in a methylcyclohexane-toluene (6:1) mixture for $\lambda_{exc} = 450$ nm.

Upon temperature variation the ordering and energetic position of photoluminescent states are changed at the "phase transition" of the capping

TOPO shell at $T_{crit} \approx 220$ K. According to the time resolved experiments we conclude that at least 3 basically different types of emissive states should be taken into account (Fig. 2). They respond specifically or are even created upon attachment of at most only one porphyrin molecule to the QD surface.

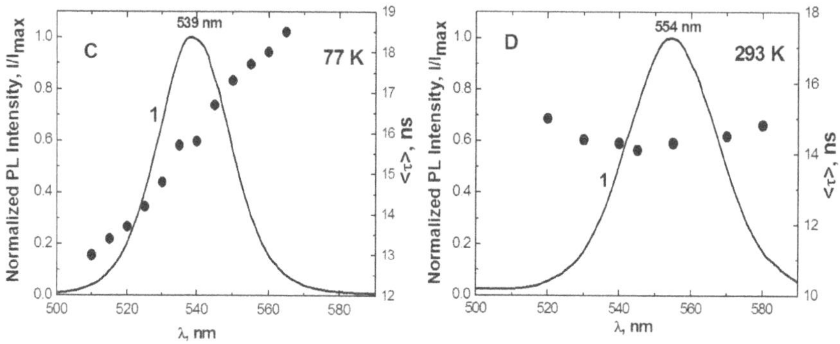

Figure 2. Normalized PL bands (1) and PL decay times $<\tau>$ at various detection wavelengths for CdSe/ZnS QDs at 77 K (**C**) and 293 K (**D**), respectively. Excitation is at λ_{exc} = 410 nm and $<\tau>$corresponds to $<\tau> = (\Sigma A_i \times \tau_i^2)/ (\Sigma A_i \times \tau_i)$.

Attachment of single porphyrin molecule to colloidal QDs renders upon nanoassembly formation PL spectra and decay of CdSe/ZnS QDs. The modifications are assigned to a dye induced generation of new distinctively dye dependent surface states. Comparison of energetically deconvoluted PL spectra of QDs reveals the appearance of new PL energies and decay times. Both, the comparison of total and deconvoluted spectral PL properties proofs that already one attached porphyrin molecule causes not only PL quenching but also changes the energy landscape of the QD PL noticeably. Lowering temperature reverses the energetic ordering of electronic states. Below the "phase" transition of TOPO ligands, PL energies depend critically on the type of the surface-attached dye molecule. This specific selectivity to the surface-attached dye molecule provides new and not yet reported insights into electronic surface properties.

The basic conclusion is that the dye molecules remove ligands from specific surface sites thus acting as "new ligands" creating a modified set of near-band edge states. This explains specifically why "non-FRET" PL quenching is often not exclusively related to FRET or charge transfer upon QD-dye nanoassembly formation. The presented results on PL emission properties of QDs and the related specific dependence on temperature and specific chemical nature of the attached dye molecule provides detailed analysis and understanding both of electronic surface properties and exciton-phonon coupling in QDs or QD-dye nanoassemblies. Only the combination of ensemble and newly designed single

QD experiments allows for a detailed and up to now not yet reported complex analysis of the PL of QDs in nanoassemblies (Fig. 3).

In both cases, namely ensemble and time averaged single QD detection, electronic states of different nature with varying energies and PL decay dynamics are subsequently explored on slow time scales typical for blinking phenomena which are buried but nevertheless present in ensemble experiments.

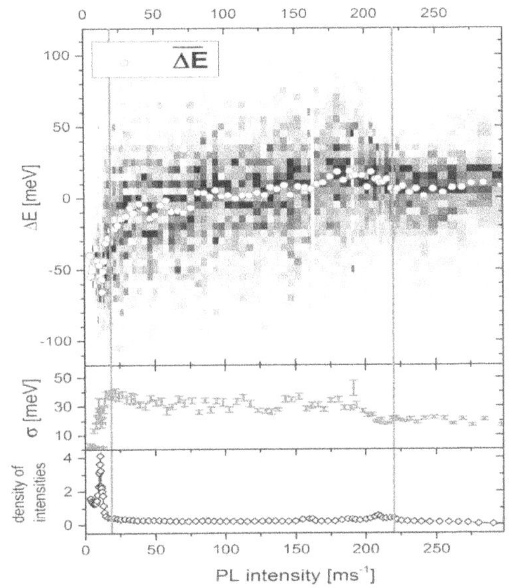

Figure 3. Top: Spectral jumps (diffusion) ΔE with respect to the dichroic beam splitter at 567 nm during a blinking time trace as a function of the PL intensity of a single CdSe/ZnS QD. Middle: σ corresponds to the Gaussian width of the spectral diffusion at a given PL intensity. Bottom: Density of intensities as a function of PL (blinking) intensity.

Acknowledgments

Financial support from Project GR 2695/8-1 "Optical Probe Semiconductor Nanoparticles" and Belarusian State Program for Scientific Research "Convergence–2020, 3.03" is gratefully acknowledged.

References

1. S. V. Kilina, P. K. Tamukong, D. S. Kilin, *Acc. Chem. Res.* **49**, 2127 (2016).
2. C. von Borczyskowski, E. Zenkevich (eds.), *Tuning Semiconducting and Metallic Quantum Dots: Spectroscopy and Dynamics* (Pan Stanford Publishing, Singapore, 2016).
3. E. Zenkevich, A. Stupak, C. Göhler, C. Krasselt, C. von Borczyskowski, *ACS Nano* **9**, 2886 (2015).

APPLICATION OF GOLD NANOPARTICLES IN TOLUENE FOR ART MATERIALS CHARACTERIZATION BY SURFACE ENHANCED RAMAN SCATTERING

E. V. Shabunya-Klyachkovskaya, E. V. Korza, L. L. Trotsiuk, A. S. Matsukovich, O. S. Kulakovich

B. I. Stepanov Institute of Physics NASB, 220072 Minsk, Belarus
elena.shabunya@gmail.com

Gold nanoparticles in toluene have been proposed as promising sensitive substrate material for Raman scattering enhancement under analysis of inorganic pigments in paint layers of real objects of cultural heritage. Gold organic sols allow to simplify the sample treatment stage and improve the quality of the analysis.

1. Introduction

A study of cultural heritage objects by means of surface enhanced Raman spectroscopy (SERS) is a great challenge for scientists [1] because of art materials heterogeneity and extremely small amount of a sample that could be taken for experiments. The huge enhancement of Raman scattering is observed in the close vicinity of noble metal nanoparticles [1]. In order to realize this effect the main task of sample treatment is providing close interaction between noble metal nanoparticles and organic or inorganic pigment chromophores. However, the binding materials in paint layers are polymer films preventing close contacts between pigments and metal nanoparticles under the SERS analysis.

In the last decades, several approaches to SERS analysis have been developed: extraction [2,3], gas-solid hydrolysis [2,4], peelable gels [5,6], *etc.* However, the most of techniques may be applicable only for organic dyes detection. Detection of inorganic pigments using SERS remains a great challenge for a long time. There were a few publications on inorganic pigments detection by SERS [7,8]. In [7], an attempt of vermilion identification in the paint layer cross-section with the help of Ag nanoparticles has been performed, but, the vermilion peaks were not observed. The ultramarine and lead white detection with the help of Ag colloids has been reported in [8]. However, the spectra obtained possess a low signal-to-noise ratio with high background luminescence and the interpretation of the results is questionable. In the recent papers [1,9,10], a systematic enhancement of Raman scattering by reference inorganic pigments has been demonstrated and the analytical protocols for colourful pastes identification in the canvas painting and icons have been proposed. These protocols are based on time-consuming procedure of pigments extraction from paint layers through washing of paint microfragments in alcohol and chloroform under slight heating. In the present paper, to avoid this stage of the sample preparation, gold organic (toluene) sol was proposed due to the ability of organic phase to dissolve the binding media in paint layers.

2. Experimental

Gold nanoparticles in toluene were synthesized by the modified Brust-Shiffrin method [11] without the stage of stabilization by alcanthiols. The optical density spectrum of the organic gold sol is characterized with the plasmon resonance maximum at 525 nm (Fig. 1a). According to the data of transmission electron microscopy (TEM), the size of particles varies from 3 to 12 nm (Fig. 1b). The SERS activity of the gold organic sol was compared with the SERS activity of the gold hydrosol. The Au hydrosol was prepared using a standard technique based on citrate reduction of $HAuCl_4$ [1,11].

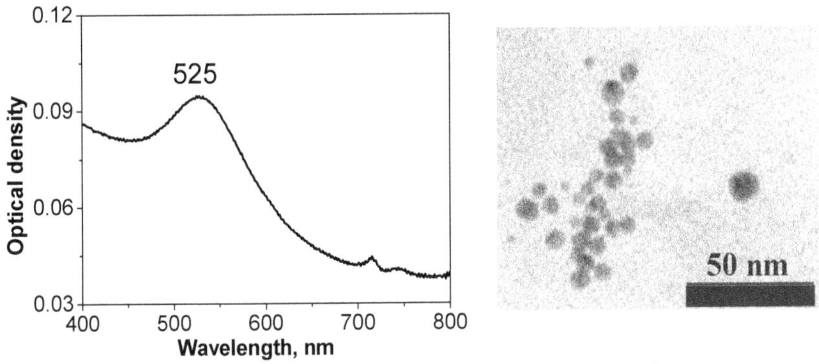

a b

Figure 1. Optical density spectrum of Au organic sols (a) and TEM image (b) of the nanoparticles.

Microfragments of the blue paint layer of the Belarusian folk icon «Madonna Hodegetria» (XIX century, The National Art Museum of the Republic of Belarus) were chosen for this study (Fig. 2a). They were divided into two parts and deposited on reflecting foil surfaces. Then 5 μl of Au organic sol and 5 μl of Au hydrosol were dripped atop the pigments and dried at the room temperature in a horizontal position.

The SERS spectra were registered in a back scattering configuration with a Nanoflex spectrometer (Solar LS) in the spectral range of 100–1000 cm^{-1} with the resolution of 1 cm^{-1}. An Ar-laser (488 nm) was used for Raman scattering excitation. The system of registration was equipped with a 1200 lines/mm diffraction grating and a CCD-detector. The acquisition time was 60 s.

3. Results and discussion

The comparison of Au hydrosol and Au organic sol efficiency is presented in Fig. 2 for analysis of the paint. Both SERS spectra demonstrate the presence of ultramarine blue in the paint layer. In the case of Au hydrosol, the chromophores peaks appear on the huge background luminescence while Au organic sol provides an efficient enhancement of Raman lines along with luminescence quenching.

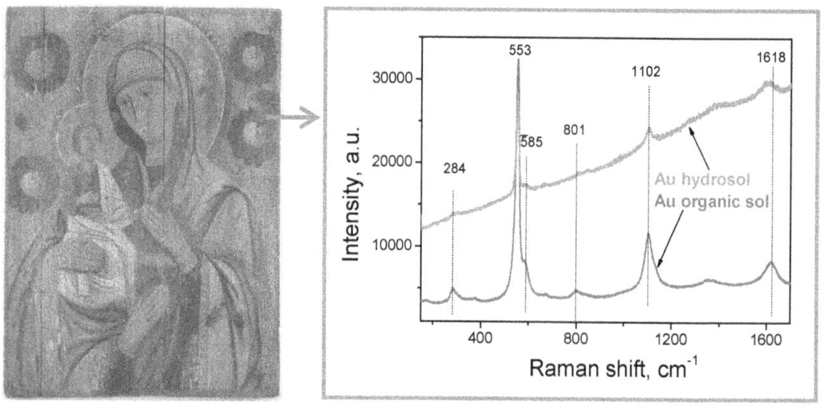

Figure 2. The comparative SERS spectra of the blue paint from the Belarusian folk icon "Madonna Hodegetria" (XIX cent.) obtained with Au hydrosol and Au organics sol.

Thus, gold nanoparticles in toluene allow to simplify the sample treatment stage and improve the quality of SERS spectra. That can be reasoned by dissolution of the binding material in the paint layers. Hence, pigments appear to be in a tight contact with Au nanoparticles. The results could be considered as a prime step of non-destructive approaches for SERS analysis which may be directly applied for art objects without a need of removing a paint specimen.

4. Conclusion

The new technique of SERS analysis with gold organic sols has been proposed and its better efficiency has been demonstrated as compared with conventional hydrosols. SERS spectra obtained are characterized by the high signal-to-noise ratio and the low background luminescence. Also, gold nanoparticles synthesized in toluene allow to avoid a preliminary stage of the sample preparation. The successful application of organic Au sol for SERS identification of inorganic art pigments in the objects of cultural heritage has been demonstrated. To the best of the authors' knowledge this result is the first case of using of organic sols in SERS experiments.

Acknowledgments

The authors thank the Belarusian Republican Foundation for Fundamental Research for financial support and The National Art Museum of the Republic of Belarus for the sample provided.

References

1. E. Shabunya-Klyachkovskaya, O. Kulakovich, S. Vaschenko, D. Guzatov, S. Gaponenko, *Eur. J. Sci. Theology* **12**, 211 (2016).
2. M. Leona, J. Steger, E. Ferloni, *J. Raman Spectrosc.* **37**, 981 (2006).
3. K. Chen, M. Leona, K. C. Vo-Dinh, F. Yan, M. B. Wabuyele, T. Vo-Dinh, *J. Raman Spectrosc.* **37**, 520 (2006).
4. M. Leona, J. R. Lombardi, *J. Raman Spectrosc.* **38**, 853 (2007).
5. B. Doherty, B. G. Brunetti, A. Sgamellotti, C. Miliani, *J. Raman Spectrosc.* **42**, 1932 (2011).
6. C. Lofrumento, M. Ricci, E. Platania, M. Becuccia, E. Castellucci, *J. Raman Spectrosc.* **44**, 47 (2013).
7. K. A. Frano, H. E. Mayhew, S. A. Svoboda, K. L. Wustholz, *Analyst* **139**, 6450 (2014).
8. O. M. Gui, A. Falamaş, L. Barbu-Tudoran, M. Aluaş, B. Giambra, S. Cinta Pinzaru, *J. Raman Spectrosc.* **44**, 277 (2013).
9. E. V. Klyachkovskaya, D. V. Guzatov, N. D. Strekal, S. V. Vaschenko, A. N. Harbachova, M. V. Belkov, S. V. Gaponenko, *J. Raman Spectrosc.* **43**, 741 (2012).
10. L. H. Oakley, D. M. Fabian, H. E. Mayhew, S. A. Svoboda, K. L. Wustholz, *Anal. Chem.* **84**, 8006 (2012).
11. M. Brust, D. Bethell, C. J. Kiely, D. J. Schiffrin, *Langmuir* **14**, 5425 (1998).
12. J. Turkevich, P. C. Stevenson, J. Hillier, *Discuss. Faraday Soc.* **11**, 55 (1951).

STRUCTURE AND SERS ACTIVITY OF GOLD NANOPARTICLES FORMED BY CHEMICAL DEPOSITION ON POROUS SILICON

S. Zavatski, N. Khinevich, V. Bondarenko

Belarusian State University of Informatics and Radioelectronics
P. Browka 6, 220013 Minsk, Belarus
h.bandarenka@gmail.com

Gold nanoparticles (NPs) have been formed by the chemical deposition of gold on porous silicon (PS) from an aqueous solution of potassium dicyanoaurate(I) and hydrofluoric acid. These Au NPs have been found to penetrate down to the bottom of PS pores as well as cover an external surface of PS. Raman measurements of rhodamine 6G (R6G) molecules adsorbed on the surface of Au/PS substrates have shown an activity of Au NPs in surface enhanced Raman scattering (SERS).We have revealed regimes of the gold deposition on PS providing optimal morphology of the Au/PS substrates favorable for the maximal SERS activity.

1. Introduction

Nanostructures of noble metals are characterized by unique optical properties associated with strong local electromagnetic fields arising due to an excitation of plasmons on the metallic surface [1,2]. This effect is widely used in many applications including SERS spectroscopy which allows to detect, identify and study trace amounts of substances in materials science [3,4], medicine, ecology, food industry, forensics and others [5,6].

Gold perfectly meets requirements for the SERS applications because of its chemical stability and strong plasmonic properties. There are known a lot of different techniques for fabrication of gold nanostructures such as deposition on a solid support by immersion in colloidal solutions [7], laser ablation [8], sputtering [9].

In this paper, we studied regularities of gold chemical deposition on PS, morphology and SERS activity of the resulting nanostructures.

2. Methods

Antimony-doped 100 mm monocrystalline Si (100) wafers of 0.01 Ω cm resistivity were used as starting substrates. Just before the PS formation each Si sample was dipped in a diluted HF solution (4.5 %) for 30 s to remove native SiO_2 layers. Immediately after the oxide removal, the Si sample was placed in an electrolytic cell made of polytetrafluoroethylene. PS layers were formed by an electrochemical anodization in an aqueous-alcoholic solution comprising of HF

(45 %), deionized water and isopropanol (C_3H_7OH) mixed at the ratio volume of 1:3:1. The anodization was performed at the current density of 100 mA/cm^2 for 85 s.

Formation of gold nanostructures on PS was carried out by the chemical deposition from the aqueous solution of 0.01 M of $KAu(CN)_2$ and 0.15 M of HF. The deposition process lasted for 30 or 70 min.

Analyte molecules for the SERS study were deposited by an immersion of the gold-plated PS samples in an aqueous solution of 10^{-6} M of R6G for 2 h.

Morphology of the gold-plated PS samples was studied with the scanning electron microscope (SEM) Hitachi-S4800. The SEM images were statistically analyzed using ImageJ software. Registration of the SERS spectra of R6G was carried out with 3D scanning confocal Raman microscope Confotec NR500 under an excitation wavelength of 633 nm that was applied for 1 s.

3. Results and discussion

Fig. 1 shows SEM images of the Au/PS samples formed for different periods of the gold deposition. It is well-seen that gold deposited both on the internal (Fig. 1a,b) and external (Fig. 1c,d) surfaces of PS as spherical particles of the nanoscaled dimensions.

Figure 1. SEM images of cross-section (a, b) and surface (c, d) PS after Au deposition for 30 (a, c) and 70 min (b, d).

Au NPs penetrated into the pores down to the depth equal to the PS thickness and their amount increased during the deposition (Fig. 1a,b). Particles in the pores have sizes varying from a few nanometers to 100 nm, *i.e.* to sizes that are comparable with the pore diameters. The deposition for 70 min led to the nearly complete filling of some pores with gold particles resulting in fabrication of gold nanowires (Fig. 1b).

Fig. 1c shows that Au NPs predominantly nucleate and grow on edges of the PS pores. We can observe that the number of Au NPs on the external PS surface increases during the deposition process (Fig. 1c,d). Therefore, there is a difference in the coverage level of the external PS surface depending on the deposition stage. This leads to the variation of the gold surface roughness.

The gold deposition for 30 min led to the formation of a quasi-continuous film which was composed of partially coalesced Au NPs (Fig. 1c). On the other hand the deposition for 70 min resulted in coalescence of Au NPs into the continuous film of 0.24 μm average thickness (Fig. 1d). Moreover, new Au NPs nucleate and grow on the surface of this film.

Visual analysis of the SEM images of the surface of Au/PS samples (Fig. 1c,d) shows that majority of the gold particles has diameters less than 100 nm. The notable difference between the top views of two Au/PS samples is in a distance between Au NPs, *i.e.* in the coverage level of the PS external surface with gold.

According to the results of the statistical analysis of the SEM images (Fig. 1c,d) the coverage levels of the PS samples formed for 30 and 70 min of the gold deposition are 79 and 98 % respectively.

Fig. 2 shows SERS spectra of the R6G molecules adsorbed on the Au/PS samples formed for different periods of the gold deposition.

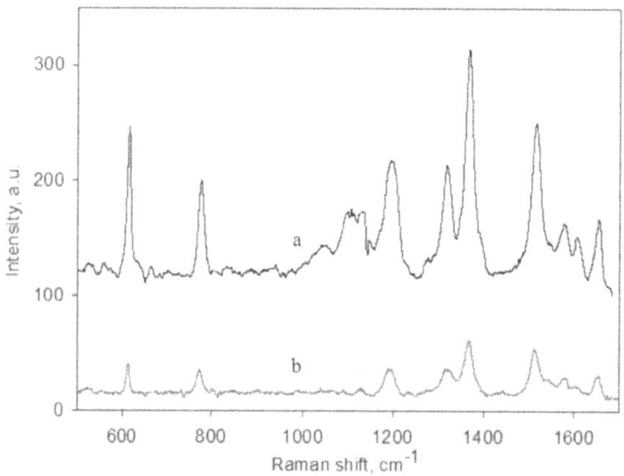

Figure 2. SERS spectra of 10^{-6} M of R6G on the Au/PS substrate formed by the gold deposition for 30 (a) and 70 min (b).

Both spectra in Fig. 2 are characterized by the typical R6G bands. Intensity of the spectrum registered on the first sample (curve a, 30 min) is about five times higher than that on the other sample (curve b, 70 min). Comparison of the results of the SEM and SERS analyzes shows dependence of the SERS activity on the morphology of the Au/PS substrates. Better activity of the sample formed by the Au deposition for 30 min takes a place as it is characterized by the lower coverage level of the PS surface.

4. Conclusion

We formed gold nanostructures by the chemical deposition of gold on PS from the aqueous solution of potassium tetracyanoaurate and hydrofluoric acid. These are gold spherical particles with diameters ranging from a few nanometers to 100 nm deposited both on the internal and external surfaces of PS. As the gold deposition time increases the filling factor of the pores and coverage level of the external surface of PS rise. The formed gold particles were found to provide great enhancement of the Raman signal under the laser excitation of 633 nm wavelength. In other words, the obtained Au/PS samples act as SERS active substrates. Dependence of the SERS-signal intensity on the morphology of the Au/PS samples was revealed. Maximal enhancement was reached for the sample characterized by 79 % coverage level of the external PS surface with gold particles due to presence of the nanoscaled gaps between them. Higher coverage lever resulted in a decrease of the SERS activity.

Acknowledgments

This work has been financially supported by the State Program of Scientific Research of Belarus "Photonics, opto- and microelectronics" (task 1.4.01).

References

1. J. R. Lombardi, R. L. Birke, T. Lu, J. Xu, *J. Chem. Phys.* **84**, 4174 (1986).
2. J. R. Lombardi, R. L. Birke, *J. Phys. Chem. C* **112**, 5605 (2008).
3. S. Lefrant, I. Baltog, B. Baibarac, *Raman Spectrosc.* **36**, 676 (2005).
4. L. Li, D. S. Zhou, G. J. Xue, *Raman Spectrosc.* **36**, 699 (2005).
5. S. C. Pinzaru, *J. Raman Spectrosc.* **35**, 338 (2004).
6. J. M. Reyes-Goddard, H. Barr, N. Stone, *Photodiag. Photodyn. Therapy.* **2**, 223 (2005).
7. Y. Jiao *et al.*, *Mater. Res. Soc. Symp. Proc.* 1:247 (2011).
8. L. Mikac *et al.*, *Croat. Chem. Acta* **88**, 437 (2015).
9. T. Ignat *et al.*, *Rom J. Inf. Sci. Tech.* **12**, 496 (2009).

GOLD NANORODS FOR SURFACE ENHANCED RAMAN SPECTROSCOPY

L. L. Trotsiuk, A. S. Matsukovich,
E. V. Shabunya-Klyachkovskaya, O. S. Kulakovich

B. I. Stepanov Institute of Physics NASB, 220072 Minsk, Belarus
l_trotsiuk@ifanbel.bas-net.by

The fabrication methods of SERS active gold nanorod films were proposed. The possibility of using gold nanorods for SERS of analytes without mercapto-groups has been shown. Long-wave excitation (785 nm) significantly increased the SERS efficiency of gold nanorod films in comparison with nanosphere films.

1. Introduction

Today surface enhanced Raman spectroscopy (SERS) is one of the most sensitive methods for detecting ultra small concentrations of matter. Rod-shaped gold nanoparticles are of significant interest in nanoplasmonics because of tunable longitudinal plasmon bands and the "lightning rod" effect on the surface electromagnetic enhancement.

Practical applications of the gold nanorods (NRs) in SERS spectroscopy are limited by the presence of adsorbed cetyltrimethylammonium bromide (CTAB) bilayer [1] used as a stabilizer. In the most of works on gold NRs for SERS, the mercapto-compounds (aminothiophenol, mercaptobenzoic acid) were used as analytes [2,3]. Mercapto-groups bind covalently to the surface of gold NRs with replacing surfactant CTAB. So, the chemical bond provides close contact between analyte molecules and metal surface. This is very important requirement for SERS since the Raman enhancement factor decreases sharply with increasing metal-analyte distance [4].

The aim of this work was to study the gold NRs formation features, to fabricate gold NR films as well as to check a possibility of using gold NRs for SERS of analytes without mercapto-groups (malachite green, mitoxantrone).

2. Experimental details

Gold nanospheres were prepared using the Turkevich method [5]. Gold NRs were prepared according to the seed-mediated growth method [2]. The size and aspect ratio of the rods were changed by varying the Au seed solution volume ($n(Au^{3+})_{seed}$:$n(Au^{3+})_{growth}$ = 1:200; 1:1700; 1:2000 for sols I, II, III, respectively) and the ratio of ascorbic acid (AA) to gold salt in the growth solution ($[Au^{3+}]$:[AA] = 1:1; 1:1.5 and 1:2 for sols I, II, III, respectively). The concentrations of the analytes were 0.8×10^{-5} M for malachite green, 1.3×10^{-5} M for mitoxantrone.

The gold NR morphology was studied with transmission electron microscope S-806TEM/SEM (Hitachi). Absorption spectra were measured using a Cary-500 spectrophotometer. (Varian).

A Nd:LSB laser (531 nm) and a diode laser (HPNIR785) emitting 785 nm line have been used for Raman spectra excitation. The registration system in the case of using Nd:LSB laser consisted of a spectrograph with a diffraction grating 1200 lines/mm (Solar TII S3801) and a cooled CCD matrix (Princeton Instruments). In the case of the diode laser the registration system consists of a Renishaw spectrograph with a diffraction grating with 1200 lines/mm and Peltier-cooled RenCam CCD detector.

3. Results and discussion

TEM analysis of the experimental samples (Fig. 1a,b,c) shows gold NRs with different size and shape depending on the synthesis condition. The average length of the rods in the sol I is 60 nm and the average diameter is 14 nm. The yield of the NRs was ~80 %. Some amount of large aggregates and gold nanospheres were also formed.

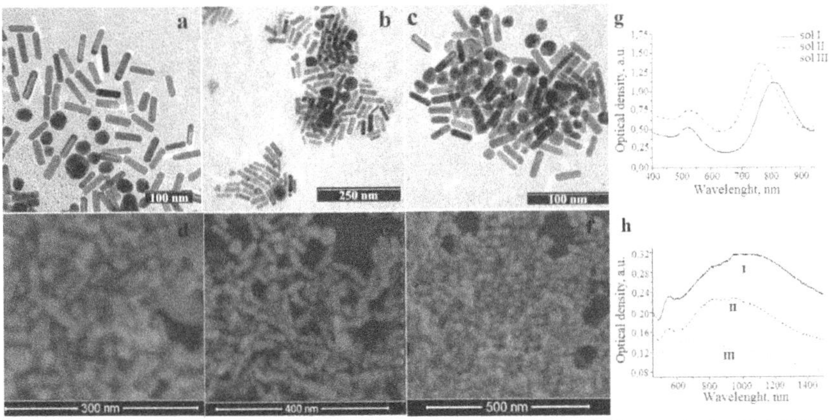

Figure 1. (a-c) TEM images of gold NRs fabricated from the sols I,II,III respectively; (d–f) SEM images of NRs films fabricated from the sols I,II,III respectively; optical density spectra of the sols (g) and films (h).

The yield of the rod-like particles in the sol II was ~84 %. However, more gold atoms were deposited on the ends than on the sides of NRs because of a slight excess of ascorbic acid in the sol II. So, the NRs of "dog bone" shape were formed [6]. The average length of the obtained particles was 43 nm and the diameter was 12 nm.

The sol III has the highest content of spherical particles. The yield of NRs was 56 % and the yield of spherical particles was ~40 %. The average size of NRs is 30 nm in length and 9 nm in diameter.

Since Au NRs have a positive charge on the surface, the layer-by-layer electrostatic deposition method was used for preparation of the films. Glass slides were treated in Piranha solution (H_2SO_4:H_2O_2=4:1), then 2 bilayers of poly(sodium-4-styrenesulfonate)/poly(diallyldimethylammonium chloride) polyelectrolytes were deposited. Gold NRs were deposited by dipping of slides in Au sols for 24 h.

Scanning electron microscopy (SEM) of gold NRs films displays noticeable difference of their uniformity (Fig. 1d-f). The film I has the most uniform structure. It contains NRs of the maximum size.

Fig. 1g shows optical absorption spectra of the sols. The 526–532 nm band corresponds to the transverse plasmon band while the long-wave bands correspond to the longitudinal one. Optical densities of the films II and III (Fig. 1h) are significantly lower than optical density of the film I. That is probably due to an excess of ascorbic acid in the sols II and III. Ascorbic acid anions may partially shield the positive charge of gold nanoparticles which prevent their electrostatic deposition on substrates.

Experimental results presented in Fig. 2 clearly indicate that the gold nanosphere films are more SERS-effective than the gold NRs films at the excitation wavelength of 532 nm.

A possible reason is the larger "analyte-metal" distance in the case of gold NRs (~5 nm) in comparison with the case of nanospheres (< 1 nm). At the excitation wavelength of 785 nm (Fig. 2b), gold NRs films became more effective. The SERS bands were better resolved with higher intensities. This is probably due to the excitation in the longitudinal plasmon mode which is more intense than the transversal one (Fig. 1). This plasmon mode is more sensitive also to the environment [7]. SERS intensity appears to be 4–5 times higher for malachite green on gold NRs films than on nanosphere films. In the case of mitoxantrone adsorbed on NRs the new bands (757 and 962 cm^{-1}) appear that were not visible on the nanosphere film.

4. Conclusion

The methods of producing of gold NRs films for SERS applications were proposed. Simultaneous increasing of the ascorbic acid concentration and the

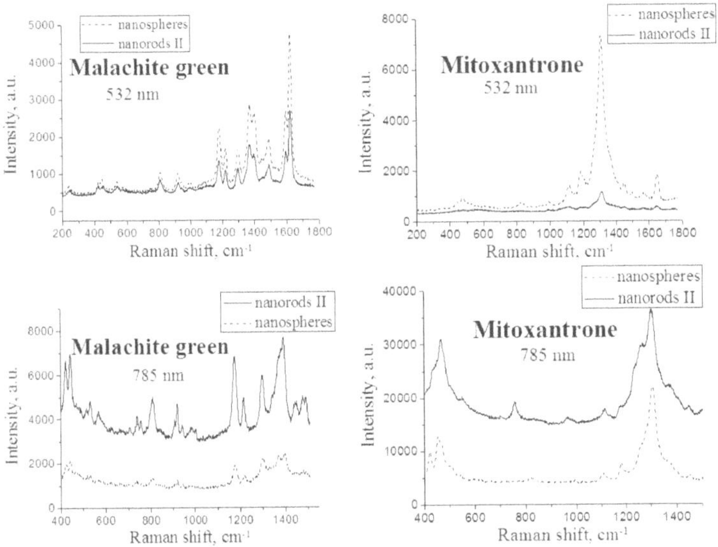

Figure 2. SERS spectra of malachite green and mitoxantrone on gold NRs II and nanosphere films: a) $\lambda_{exc.}$ = 532 nm and b) $\lambda_{exc.}$ = 785 nm.

seed amount during the synthesis reduces the size of NRs and changes their shape. Gold nanosphere films are more effective for SERS at 532 nm excitation wavelength than gold NRs. Under excitation by the longer wavelength (785 nm) the SERS efficiency of gold NRs films significantly increases as compared with nanosphere films. The developed method allows to vary positions of the plasmon band in the wide spectral range (500–1200 nm) which is important for proper control of SERS efficiency.

Acknowledgments

The work was supported by the Belarusian Republican Foundation for Basic Research.

References

1. A. Gole, C. J. Murphy, *Chem. Mater.* **16**, 3633 (2004).
2. B. Nikoobakht, M. A. El-Sayed, *J. Phys. Chem. A* **107**, 3372 (2003).
3. S. L. Smitha *et al.*, *Nanotechnology* **22**, 265705 (2011).
4. E. V. Klyachkovskaya *et al.*, *J. Raman Spectrosc.* **43**, 741 (2012).
5. J. Turkevich *et al.*, *Discuss. Faraday Soc.* **11**, 55 (1951).
6. K. A. Kozek *et al.*, *Chem. Mater.* **25**, 4537 (2013).
7. I. Ros *et al.*, *Plasmonics* **9**, 581 (2014).

CHEMISTRY OF NANOSTRUCTURES

INVITED

QUANTUM DOTS IN ROBUST MATRICES: STATE OF THE ART

F. Eichler, C. Guhrenz, A. Benad, M. Adam, N. Gaponik
Physical Chemistry, TU Dresden, 01062 Dresden, Germany
nikolai.gaponik@chemie.tu-dresden.de

Colloidal quantum dots are very attractive for optoelectronic applications, however, they need to be protected from rigid operating conditions. Robust matrices based on all-inorganic, polymer or hybrid materials have been widely investigated as protecting hosts. This paper summarizes recent advances in embedding of quantum dots into various macrocrystalline matrices and their potential applications.

1. Introduction

Colloidal semiconductor nanocrystals (NCs) also known as quantum dots (QDs) have recently gained an increasing interest as spectrally tunable and bright luminescent materials [1]. However, commercial applications as color conversion fluorophores in lighting devices (*e.g.*, white light-emitting diodes (WLEDs)) or in the displays face several challenges including photodegradation, electro-corrosion, environmental stability as well as chemical instability during the traditional processing steps (*e.g.*, embedding in silicone matrices) [2–4].

Although a variety of attempts had been made, overcoming all of these obstacles at the same time without deteriorating the optical performance of the QDs appeared to be a great challenge. As a possible solution, the approach of embedding of colloidal QDs into crystalline host matrices was recently introduced [5]. These QD-salt composites both in their monolith and powder forms demonstrate superior photostability and chemical durability, accompanied with high photoluminescence (PL) quantum yields (QYs). Their spectral properties are widely tunable (from visible to near infrared range) with the size and the composition of the embedded QDs due to the quantum confinement effect [3].

Here, the state of the art methods for embedding colloidal QDs into salt matrices are overviewed. We further discuss necessary conditions for obtaining strongly emitting and chemically stable QDs-in-macrocrystals. Subsequently, we present excellent suitability and compatibility of the resulting color-converting powders for WLEDs. Furthermore, potential future opportunities as well as challenges related to the development of these technology are summarized.

2. Experimental

All necessary experimental details on chemicals and apparatus used for the synthesis of the QDs and their composites can be found in [3,5–14]. Both QDs synthesized in aqueous media (typically thiol-capped CdTe QDs) and those prepared by the pyrolysis of precursors in high-boiling organic solvents (typically CdSe-based core-shell materials) are in the focus of the following discussion. The list of the presented here crystalline matrices includes (but not limited with) NaCl, KCl, KBr, KI, LiCl, CsI and $Na_2B_4O_7 \cdot 10H_2O$ (borax).

3. Results and discussion

Typically, embedding can be performed by the crystallization of the matrix in the presence of the well dispersed QDs [5-9]. As the crystallization can be done from the oversaturated salt solutions in water or in some other polar solvents, the QDs should possess colloidal stability in these solvents. Thus, as the most prominent candidate, very stable and well dispersed thiol-capped CdTe QDs were chosen for direct embedding from the aqueous media. Fig. 1 shows this strategy as well as typical PL spectra of the resulting composites.

Taking into account that the QDs are electrostatically stabilized, high ionic strength of the saturated salt solutions is not favorable for the QD colloids. As a result, the formation of the QD-salt composites is usually limited to relatively low QD loadings. Nevertheless, the obtained composites of CdTe QDs with NaCl, KCl, and borax are extremely robust and strongly luminescent. In case of the NaCl matrix even an enhancement of the PL QY as compared to the original QD solution was observed and carefully investigated [7]. This enhancement was explained by the formation of an additional passivating $CdCl_x$ shell around the embedded QDs.

It should be noted, that this strategy, although very straightforward, has several limitations. Indeed, the long crystallization time (up to several weeks) makes it impossible to embed CdSe-based core-shell QDs, as they have to be transferred into water before embedding and possess a limited colloidal stability after this transfer. At the same time, embedding of these type of QDs is very desired as the proper designed core-shell QDs prepared at high temperatures in organic solutions are of better quality and possess as a rule much higher (approaching unity) QYs [3]. To overcome this limitation, the following strategies were developed.

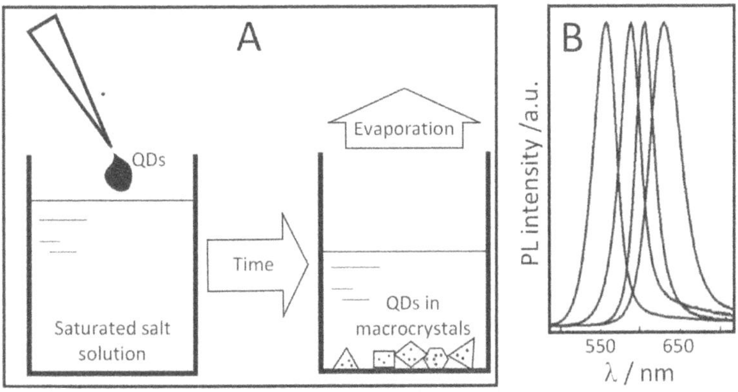

Figure 1. (A) Direct embedding of QDs into salt crystals growing from the saturated solution during its slow evaporation. The crystals formed can be separated from the solution and washed with ice cold water and "bad solvents", *e.g.*, methanol. After drying, they can be directly utilized for further processing, *e.g.*, milling and blending with silicone. (B) Typical PL spectra of the CdTe QDs embedded into NaCl crystals. Narrow emission bands and spectral tuneability are evidenced from these data.

(i) Acceleration of the evaporation process by increasing the temperature (typically to 30-40 °C). Higher temperatures are not desired as this will simultaneously increase the solubility of the salt. Although, this method may help to accelerate the crystallization, the conditions have to be individually optimized for each QD/salt pair and concentration ratio.

(ii) Embedding of CdSe-based QDs transferred into water may be performed if borax is used as a host matrix. Borax possesses relatively low solubility in water and as a result its saturated solutions show a lower ionic strength, which is favorable for a better solubilization of the QDs [10].

(iii) QDs may be quickly embedded into low soluble salts (*e.g.*, $BaSO_4$) by co-precipitating them from aqueous solutions [11].

(iv) Utilization of a polar organic solvent which can solubilize both QDs and the salt. This method was successfully applied to crystallize green-emitting composites of CdSe/CdZnSeS/ZnS QDs in LiCl from tetrahydrofuran solutions, however, its extension to other QDs and salts is still challenging [12].

(v) The embedding of the QDs into salts may be accelerated to at least one order of magnitude by the liquid–liquid diffusion-assisted crystallization (LLDC). This method relies on the diffusion and solubility change of inorganic salts (*e.g.*, NaCl) in solvents of various polarities. For example, due to the reduced solubility of NaCl in water by the interdiffusion of the worse solvent (*e.g.*, methanol), the complete crystallization finishes within less than one day! The method was additionally extended to a two-step approach (seed-mediated

LLDC) allowing the direct incorporation of organically prepared CdSe-based core-shell QDs [13].

It should be noted, that in spite of the described success, the composites obtained via the approaches (iii), (iv) and (v) appear in a form of solely fine powders and not of well-shaped bulk crystals, as it may be desired for some advanced applications.

At this stage a recently introduced approach to the embedding of QDs into salts via "cold flow" process should be discussed in more detail [14]. This procedure is illustrated in Fig. 2A. The original colloidal solution of QDs (*e.g.*, in organics) is mixed with the salt powder and exposed to the vacuum evaporation, *e.g.*, in rotor evaporation setup. After the removal of the solvent, the colored salt powder is pressed at high pressure (*e.g.* 2.2 GPa) for 5 min. Under this pressure the salt crystals "melt". The resulting transparent and highly emitting pellets (PL spectra are presented in Fig. 2B) can be directly used for optical applications, *e.g.* as color converters mounted on top of a blue LED. An exemplarily prepared, efficient WLED is demonstrated in Figs. 2C,D. Stacking of the pellet reduces reabsorption of the used QDs. In comparison to the methods (ii) and (v), a prior ligand exchange is not needed using "cold flow" and it allows to use high QD loadings. However, the method is limited to the so called "soft" salts melting under high pressure. Among the most useful soft salts KCl, KBr, KI, NaCl, CsI and AgCl should be mentioned.

Table 1 summarizes the approaches presently developed for embedding of various QDs into salt matrices. It can be seen, that not only CdTe and CdSe based QDs, but also Cd-free materials, *e.g.* those based on InP, InGaP and ZnInCuS can be embedded, which is a great step toward environment friendly processing and applications.

Table 1. Overview of the state of the art approaches to embedding of QDs into salt crystals.

Method of embedding	Examples of applicability (QD-salt pairs)	Loading	Reproducibility
Direct embedding from aqueous solution	CdTe in NaCl, KCl, KBr, KI, Borax, …	low to middle	high
	CdSe/CdS in NaCl	low	low
Direct embedding from polar solvents	CdSe/CdZnSeS/ZnS, ZnInCuS in LiCl	middle	middle
Co-precipitation with low soluble salts	CdTe in $BaSO_4$	low to middle	high
LLDC	CdTe, CdSe/ZnS, InZnP/GaP/ZnS in NaCl	low	high
Embedding via "cold flow" process	CdSe/ZnS, CdSe/CdS, InP/ZnS in KCl, KBr, KI, CsI, NaCl	high	high

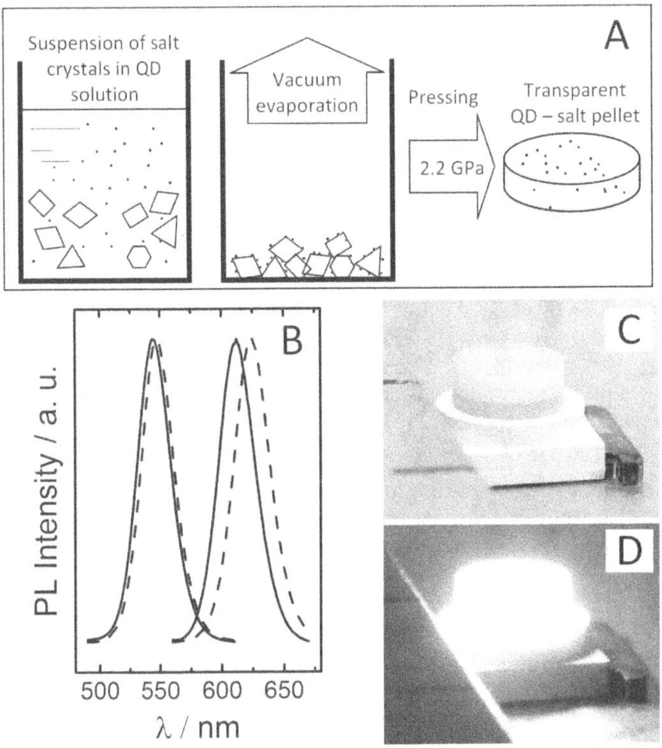

Figure 2. (A) Embedding of QDs into salt crystals via "cold flow" process. After the vacuum evaporation of the solvent from the suspension of the salt crystals in the QD solution, the resulting powder is pressed under at least 2.2 GPa. At these conditions, the cold flow transforms the mixed powder into a transparent, QD-containing crystalline pellet. (B) Typical PL spectra of the arbitrary chosen green- and red-emitting CdSe-based core-shell QDs before (solid lines) and after (dashed lines) embedding into KCl pellets. Red shifts may result from the reabsorption in the highly loaded pellets. Moreover, possible aggregation of the QDs cannot be excluded. (C) Mounting of the differently colored pellets on a blue LED and (D) the resulting WLED under operation.

4. Conclusion

Embedding of colloidal QDs into a protective matrix under preservation of long-term stability, PL QY and emission behavior is a challenging task of the current research. QD-salt composites are promising materials which show comparable or even better results than the widely used polymer matrices. Using aqueous synthesized QDs a broad ranges of incorporation processes via crystallization of the salt from its solution has been developed. To apply the same approaches for embedding of QDs synthesized in organic solutions, a ligand exchange and phase transfer to the polar media are typically necessary, which is sometimes challenging. A recently introduced approach using embedding of QDs under high pressure, when salt matrices undergo "cold flow", allowed to overcome most of the limitations. The resulting composites can be further processed in a form of pellets or powders and are promising for the applications as color converters or sources in solid-state lighting, solar cell concentrators, lasing, optical thermometry.

Acknowledgments

This work was supported by the M-ERA.NET Network within the project ICENAP, GA 1289/3-1 as well as by the DFG Project EY 16/14-3.

References

1. D. V. Talapin, J.-S. Lee, M. V. Kovalenko, E. V. Shevchenko, *Chem. Rev.* **110**, 389 (2010).
2. H. V. Demir, S. Nizamoglu, T. Erdem, E. Mutlugun, N. Gaponik, A. Eychmüller, *Nano Today* **6**, 632 (2011).
3. M. Adam, N. Gaponik, A. Eychmüller, T. Erdem, Z. Soran-Erdem, H V. Demir, *J. Phys. Chem. Lett.* **7**, 4117 (2016).
4. Z. Y. Luo, D. M. Xu, S. T. Wu, *J. Display Technol.* **10**, 526 (2014).
5. T. Otto, M. Müller, P. Mundra, V. Lesnyak, H. V. Demir, N. Gaponik, A. Eychmüller, *Nano Lett.* **12**, 5348 (2012).
6. M. Adam, R. Tietze, N. Gaponik, A. Eychmüller, *Z. Phys. Chem.* **229**, 109 (2015).
7. M. Müller, M. Kaiser, G. M. Stachowski, U. Resch-Genger, N. Gaponik, A. Eychmüller, *Chem. Mater.* **26**, 3231 (2014).
8. S. Kalytchuk, O. Zhovtiuk, A. L. Rogach, *Appl. Phys. Lett.* **103**, 103105 (2013).
9. S. Kalytchuk, O. Zhovtiuk, S. V. Kershaw, R. Zbořil, A. L. Rogach, *Small* **12**, 466 (2016).
10. M. Adam, T. Erdem, G. M. Stachowski, Z. Soran-Erdem, J. F. L. Lox, C. Bauer, J. Poppe, H. V. Demir, N. Gaponik, A. Eychmüller, *ACS Appl. Mater. Interfaces* **7**, 23364 (2015).
11. Y. Chang, X. Yao, Z. Zhang, D. Jiang, Y. Yu, L. Mi, H. Wang, G. Li, D. Yu, Y. Jiang, *J. Mater. Chem. C* **3**, 2831 (2015).
12. T. Erdem, Z. Soran-Erdem, V. K. Sharma, Y. Kelestemur, M. Adam, N. Gaponik, H. V. Demir, *Nanoscale* **7**, 17611 (2015).
13. M. Adam *et al.*, *Adv. Funct. Mater.* **25**, 2638 (2015).
14. A. Benad *et al.*, *ACS Appl. Mater. Interfaces* **8**, 21570 (2016).

EXPERIMENTAL EVIDENCE FOR CHEMO-ELECTRONIC CONVERSION OF WATER ADSORPTION ON THE SURFACE OF NANOSIZED YTTRIA-STABILIZED ZIRCONIA

A. Lyubchyk[1], H. Águas[1], E. Fortunato[1], R. Martins[1], O. Lygina[2], S. Lyubchyk[2], N. Mohammadi[3], E. Lähderanta[3], A. S. Doroshkevich[4], T. Konstantinova[5], I. Danilenko[5], O. Gorban[5], A. Shylo[5], V. K. Ksenevich[6], N. A. Poklonski[6]

[1]i3N/CENIMAT, Department of Materials Science, Faculty of Science and Technology Universidade NOVA de Lisboa and CEMOP/UNINOVA Campus de Caparica, 2829-516 Caparica, Portugal
andrey.lyubchik@campus.fct.unl.pt

[2]LAQV Requimte, Department of Chemistry, Faculty of Science and Technology Universidade NOVA de Lisboa, Campus de Caparica, 2829-516 Caparica, Portugal
o.lygina@fct.unl.pt

[3]Physics Department, Lappeenranta University of Technology 53850 Lappeenranta, Finland
Nafiseh.Mohammadi@student.lut.fi

[4]Joint Institute for Nuclear Research, 141980 Dubna, Russia
doroh@jinr.ru

[5]Material Science Department, Donetsk Institute for Physics and Engineering NAS of Ukraine, Nauki Ave. 46, 03680 Kyiv, Ukraine
matscidep@aim.com

[6]Department of Semiconductor Physics and Nanoelectronics, Faculty of Physics Belarusian State University, 220030 Minsk, Belarus
poklonski@bsu.by

Direct energy conversion techniques are now dominated by photovoltaics. But, this dominance will be soon challenged by the emergence of a new conversion methods, known as so called "chemovoltaic". Here we demonstrate a possibility of the electric energy generation through interactions of atmospheric moisture and ZrO_2 based nanopowder systems. Such electric energy generation occurs by the conversion of the energy of exothermic heterophase electrochemical reactions. Creation of electromotive force is confirmed by electrical measurements of experimental specimens at a saturation flux density gradient of atmospheric moisture.

1. Introduction

Creation of novel alternative energy generation processes is of major importance in nowadays conditions of extreme electricity consumption [1]. Water is the one of the main conventional sources of alternative energy used by mankind. Its liquid phase is traditionally used for the kinetic into electric energy conversion by so-called hydropower plants. In 2016 they generated 16.7 % of the world's total electricity and around 70 % of all produced renewable electricity [2]. However, direct conversion of water molecule energy into electricity remained unrealizable for centuries as a top scientific and practical aim.

The pioneer of the experiments on capturing electricity from the air was Nikola Tesla (1932). Since the beginning of the 21st century [3,4], the hypothesis of Nikola Tesla [5] to use atmospheric energy (humidity, as founded later) as a renewable energy source, started to be explored by the scientific community. However, these developments are on their infancy [6,7].

With the development of nanotechnology, the implementation of these ideas become potentially possible. Nowadays it is known that gas adsorption (including water vapour) can generate electric charges on metal surfaces [8]. Due to the interaction of gas molecule with a metal film of nanometer thickness < 10 nm the so-called "hot" electrons can occur and ballistically *i.e.*, without the loss of energy reach a Schottky barrier, cross it through kinetic energy, and by entering into the semiconductor material create an electric current in the external chain. However, efficiency of the such conversion method is limited by a low number of "hot" electrons (10^{-3}–10^{-5}) generated per an interaction act. Moreover, a small cross-section of free charge carriers in semiconductor crystals leads to a decrease of the conversion efficiency. Thus, the electric field (of heterojunction) decays drastically with an increase of the distance from the surface even for a few interatomic distances, therefore, only a small fraction of gas can act effectively at the heterophase process on the front surface of a semiconductor crystal. From this aspect it is of high interest to use dielectric materials with a developed surface area. We expect the electrostatic field of dielectric materials to exceed significantly the electric field of heterojunction and consequently leads to an increase of conversion efficiency.

In this paper, we show experimental verification of water induced electromotive force (EMF) generation through its adsorption on the surface of nanosized yttria-stabilized zirconia (YSZ, ZrO_2+3 mol%Y_2O_3).

2. Experimental

The compact (compressed) structures of tetragonal polycrystalline zirconia (TPZ) (ZrO_2+3 mol% Y_2O_3) nanopowder of 9–12 nm particles were used (Fig. 1). Detailed description of the nanopowder production can be found elsewhere [9]. Its general steps are the following. At first, zirconia hydroxide is produced by co-precipitation method from the chloride raw materials. After 0.4 h of dehydration in microwave oven at 120 °C, amorphous nanopowder was crystallized through annealing at 400 °C for 2 h.

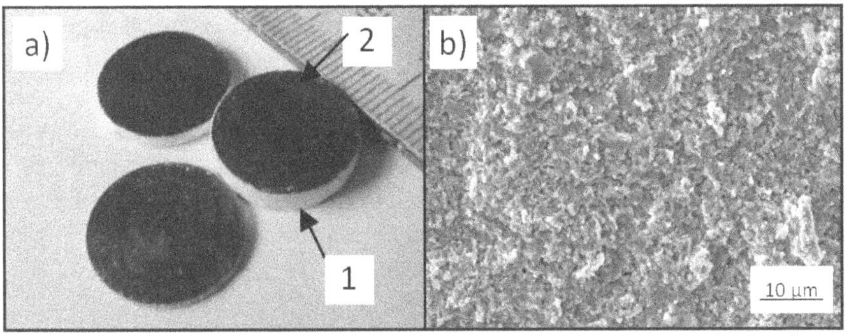

Figure 1. Sample general view (a) and TEM image of TPZ nanopowder annealed for 2 h at 400 °C (b).

Compact specimens were obtained by compression of 1 g of TPZ nanopowder using uniaxial pressure of 40 MPa until the tablets of 20 mm in diameter and of 2–3 mm height were formed. For the final formation after the compression tablets were pressed by a high hydrostatic pressure (HHP 500 MPa). Then graphite electrodes were deposited mechanically on the sides of the tablets (Fig. 2).

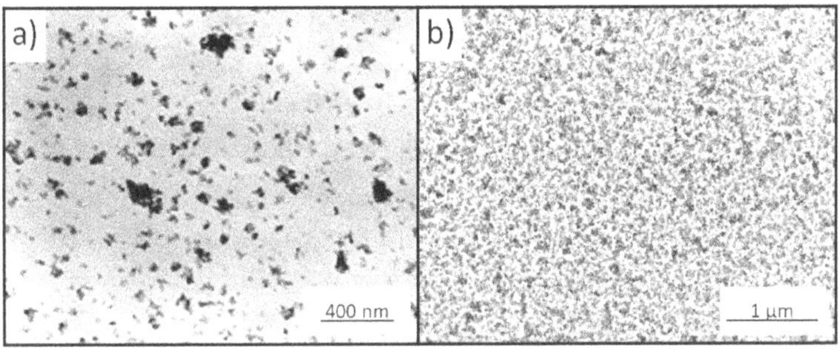

Figure 2. TEM images of TPZ nanopowder annealed for 2 h at 400 °C at different magnifications.

Morphology of the samples was investigated by transmission electron microscopy (TEM) by using the procedure described elsewhere [10] and scanning electron microscopy (SEM) using JEM 200A and JSM640LV (JEOL) instruments, respectively.

The EMF measurements and sample humidification were performed in a static mode at the fixed relative humidity (RH), barometric pressure of 760 mm Hg and temperature of 19 °C.

Saturation of the specimens by water within a certain time was measured gravimetrically using the Moisture Balances ADGS50 and calculated by the following formula: $\Delta m = f(t)$, where $\Delta m = (m - m_0)/m$, m_0 is the mass of the dehydrated tablet at the beginning of the measurement, m is the mass of the hydrated tablet.

The EMF measurements were performed under the same humidity/temperature/pressure conditions. The potential difference across the electrodes ($V = f(t)$) of the specimens was registered by the voltmeter JT-87 at the load resistor (R_L) of 1 MOhm. Voltage points were taken with an interval of 10 min for 2 h.

Prior to the experiment one end side of the tablet was covered by a waterproofing layer of paraffin in order to create a gradient water adsorption thus, allowing the movement of generated charge carries.

It should be noted that all measurements were repeated 3 times and the estimated standard deviation did not exceed 3 %.

3. Results and discussion

The typical SEM image of the tablet cross-section is shown on Fig. 2b. The microstructure of the tablets is relatively friable. The calculated density of the tables is about 3.1 g/cm^3, which corresponds to about 40–50 % from the density of the typical ZrO_2-based ceramics. Moreover, there is a branched network of relatively large (about 1 μm) pores in the bulk of the specimens. Thus, water can easily get into the bulk of a tablet. Most probably, at first water molecules draw into the pores of large size and then fill with the slower rate the remaining volume of the material spreading through smaller pores. As can be seen from the gravimetric curve (Fig. 3, circles) the amount of the adsorbed moisture increases monotonically with time until reaching a constant plateau.

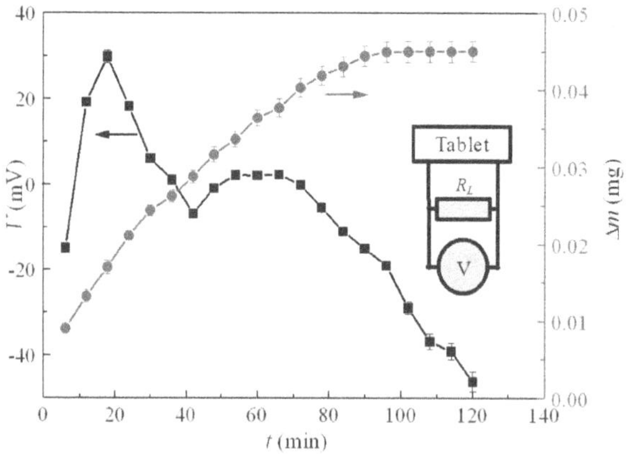

Figure 3. Variation of the weight, Δm (circles) and the EMF, V (squares) of the tablet specimen within saturation at 75 % of RH. The inset represents EMF measurement circuit at the load resistor (R_L) of 1 MOhm. The lines are given as a guide for an eye.

After 100 min, the saturation occurs where the weight of the tablet remains unchanged. The change of the specimen weight at saturation corresponded to the adsorption of 4 wt.% of moisture. From the curves represented in Fig. 3 the correspondence between the specimen saturation dynamics (circles) and its electrical properties (squares) can be established.

Herewith after 20 min (the first stage) the potential difference between the electrodes $V(t)$ increases up to 60 mV (Fig. 3 squares). The dependence of $V(t)$ in contrast to the dependence of $\Delta m(t)$ is not monotonic and during the following 20 min (the second stage) the potential difference between the electrodes decreases and the polarity of voltage gets inversed (the third stage). The amplitude of the signal of the reverse polarity is approximately 25 % from the total EMF amplitude.

According to the form of the EMF curve, it can be concluded that the electrochemical process in the sample material has two steps. Initially, the electrodes were separated by an insulating layer of a dewatered specimen material and have fixed potentials. A slight potential difference (–20 mV) observed at the beginning of the measurements ($t = 0$ min) might be induced by polarization of the sample during the compaction process (electrets effect [11]). During humidification, the potential of the water permeable electrode increases because of localization in the bulk of the adjacent space region of generated free charge carriers through the nonequilibrium chemoelectronic process (the first step) [8], while the electrical resistance of the layer between electrodes is still large (about 10^6–10^7 Ohm/mm) in accordance with impedance spectroscopy (not shown here). Meanwhile, the potential of the waterproofed electrode remained constant.

Within moisture distribution through the pores of the tablet and adsorption in the bulk, the potential equalization and the bulk conduction current occur (the second step). The bending of the potential curve is observed after 40 min of the measurements. The increase in the amplitude of the EMF and its reduction occur at a constant rate of the tabled saturation (an unchanged slope of the curve). Consequently, the inversion of the sign of the derivative $V(t)$ over the time happened not due to the change of nature of the mass transfer process, but rather, due to the specimen geometry.

4. Conclusion

We show a possibility of the electric energy generation through conversion of moisture adsorption energy on the surface of ZrO_2 based nanopowder systems into electrical charges. Such chemoelectronic coupling can be a base for novel methods of sensing, in which the sensor is based on active charge carrier production rather than on passive variations in equilibrium thermodynamic quantities, such as resistivity or capacitance-based devices or at an appropriate level of development be a novel energy conversion device "chemovoltaic". Maximum power density produced by the investigated tablets was estimated to be around 10 nW/cm^2.

Acknowledgments

The authors acknowledge EU funding by the EU H2020-MSCA-RISE-2015 through the HUNTER project (grant 691010) and EU H2020-MSCA-RISE-2015 through the Nanoguard2Ar project (grant 690968).

References

1. G. Brauner *et al.*, *Electrical Power Vision 2040 for Europe* (Brussels, February 2013), http://www.eurel.org/home/TaskForces/Documents/EUREL-PV2040-Short_Version_Web.pdf.
2. REN21, *Renewables 2016 Global Status Report* (Renewable Energy Policy Network for the 21st century, 2016), http://www.ren21.net/wp-content/uploads/2016/06/GSR_2016_Full_Report_REN21.pdf.
3. R. F. Gouveia, F. Galembeck, *J. Am. Chem. Soc.* **131**, 11381 (2009).
4. Y. Xie *et al.*, *Nature Commun.* **5**, 3575 (2014).
5. N. Tesla, *Sci. Am.* **150**, 132 (1934).
6. N. Miljkovic *et al.*, *Appl. Phys. Lett.* **105**, 013111 (2014).
7. M. Ma *et al.*, *Science* **339**, 186 (2013).
8. B. Gergen *et al.*, *Science* **294**, 2521 (2001).
9. T. E. Konstantinova *et al.*, *J. Nanoparticle Research* **13**, 4015 (2011).
10. A. S. Doroshkevich, I. A. Danilenko, T. E. Konstantinova, *Electron Microscopy and Strength of Materials* **13**, 151 (2006).
11. A. N. Gubkin, G. I. Skanavi, *Izv. AN SSSR, Ser. Fiz.* **22**, 330 (1958).

SOLVOTHERMAL SYNTHESIS OF NANOSTRUCTURED MOLYBDENUM-VANADIUM OXIDES WITH HIGH CONCENTRATION OF REDOX CENTERS

T. V. Sviridova, A. S. Logvinovich, D. V. Sviridov
Department of Chemistry, Belarusian State University
Nezavisimosti Ave. 4, 220030 Minsk, Belarus

A. I. Kokorin
Semenov Institute of Chemical Physics RAS, Kosygina 4, 119991 Moscow, Russia

Mixed V_2O_5:MoO_3 oxide with high concentration of well separated vanadyl-like groups capable to behave as redox centers was synthesized in an aqueous solution via solvothermal technique using hydrated molybdenum and vanadium oxide nanoparticles as precursors.

1. Introduction

Preparation of mixed molybdenum-vanadium oxides which are promising catalysts for selective oxidation of hydrocarbons [1,2] and desulfonation of hydrocarbon fuels [3] provides a serious and only partly solved problem due to a number of mixed oxide polymorphs radically different in catalytic activity. The solvothermal synthesis of oxide nanophases employing molybdenum and vanadium acids as precursors offers new opportunities in synthesis of mixed oxides. This synthesis can present an efficient control over the phase composition, morphology and doping of the resultant oxide products [4]. In this paper, we demonstrate that thermoinduced polycondensation of molybdic and vanadic acids enables the synthesis of lamellar molybdenum-vanadium oxide in the form of solid solution with high concentration of vanadyl centers.

2. Results and discussion

Vanadic and molybdic acids were obtained from soluble vanadates and molybdates by an ion-exchange technique. They exhibit strong tendency to formation of colloidal particles in aqueous solutions. Dynamic light scattering and ζ-potential measurements evidenced that polycondensation yields negatively-charged oxide-hydroxide aggregates *ca.* 1 nm in size in the case of molybdic acid and several tens nanometers in size in the case of vanadic acid. The structure of these aggregates can be described as: $[MeO_p{\cdot}kH_2O]\ nMe_xO_y^{z-}/(zn-m)H^+/mH^+$. The resultant colloidal particles differ by stability: whereas colloidal vanadic acid gradually converts into the gel, the further polycondensation of molybdic acid occurs only under boiling conditions. Incubation of the mixture of vanadic and molybdic acids at 100 °C results in formation and subsequent recrystallization of the hydrated oxide phase.

SEM images presented in Fig. 1 display that morphology of the oxide particles grown via the solvothermal method is very sensitive to the mixed Mo-V oxide composition. Vanadium-enriched V_2O_5:MoO_3 samples exhibit lamellae morphology and show no signs of phase segregation, while the enrichment in molybdenum results in formation of perfect prismatic crystals of hexagonal molybdenum oxide along with lamellar structures.

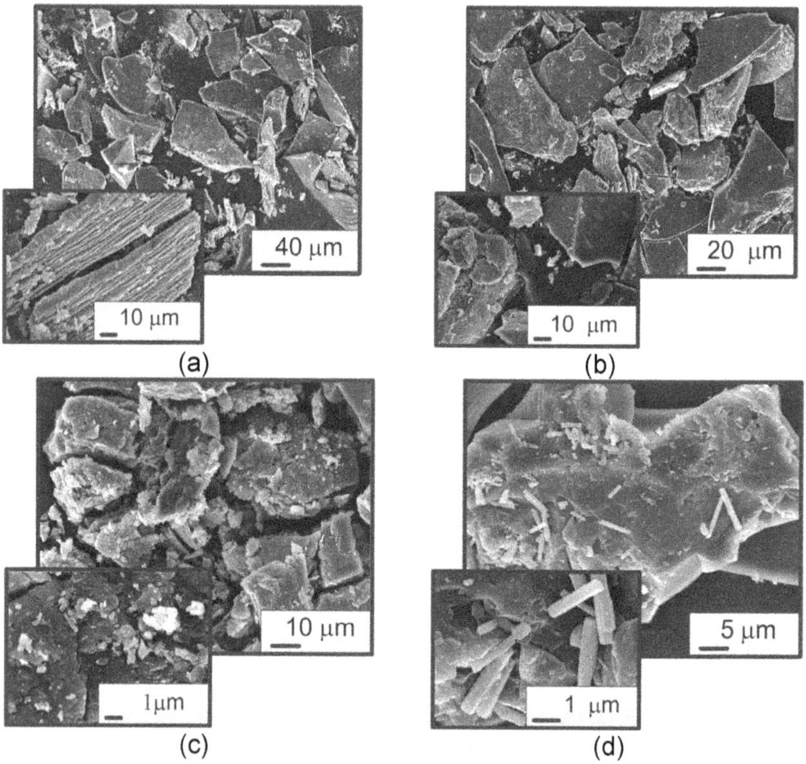

Figure 1. SEM images of mixed molybdenum-vanadium oxides with V:Mo molar ratios of (a) 3:1, (b) 3:2, (c) 2:3, and (d) 1:3.

The XRD analysis performed for the mixed oxides shows that at molybdenum concentrations below 40 mol% the solvothermally-derived oxides are constituted by xerogel of hydrated V_2O_5. At higher concentrations of molybdenum hexagonal MoO_3 becomes observable in the diffraction patterns.

In the IR spectrum of the samples prepared, the bands corresponding to the valence vibrations of V=O groups at 1009.6 cm^{-1} and V-O bending vibrations at 470 cm^{-1} are shifted to the lower frequencies as compared to pure V_2O_5. The observed weakening of V-O bonds is an indicative of partial substitution of vanadium for molybdenum in the V_2O_5 lattice. The excessive positive charge arising because of the substitution of V^{+5} by Mo^{+6} can be compensated by the transition of some vanadium centers from the pentavalent to the tetravalent state. In the IR spectra this effect appears as a broad peak with the maximum at 969.1 cm^{-1}, which is the superposition of two peaks corresponding to Mo=O and V(IV)=O stretching vibrations. These observations permit to conclude that the solvothermal synthesis of mixed-oxide molybdenum-vanadium phases produces a substitutional solid solutions isomorphic to the xerogel of hydrated V_2O_5, which contains an admixture of MoO_3 at high molybdenum concentrations.

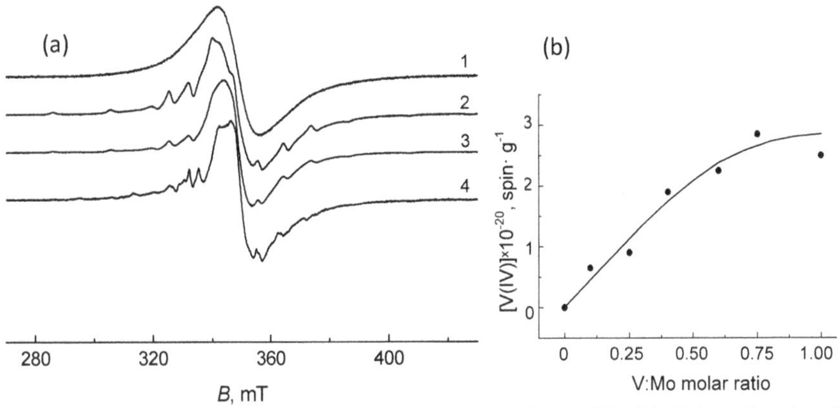

Figure 2. (a) ESR spectra at 77 K of mixed molybdenum-vanadium oxide with V:Mo mole ratios of (1) 1:0, (2) 3:1, (3) 3:2, (4) 2:3. (b) Dependence of concentration of V(IV) centers on V:Mo molar ratio.

ESR studies have evidenced that the concentration of paramagnetic Mo(V) centers in the mixed V_2O_5:MoO_3 oxide does not exceed 2 % due to interaction between molybdenum and vanadium centers. In the case of vanadium-enriched samples (at vanadium concentrations above 60 %), the ESR spectrum appears to be a superposition of the multiplet and wide anisotropic singlet line (Fig. 2) that points to formation of regions with high local concentration of paramagnetic VO^{2+} centers (*i.e.* V(IV) associates). However, even in this case the vanadyl groups remain mostly isolated in the mixed oxide (as evidenced by the dependence of concentration of V(IV) centers with V:Mo molar ratio given in Fig. 2) that permit them to behave as redox centers. From this point of view, the synthesized V_2O_5:MoO_3 solid solutions differ considerably from mixed oxides derived by the most commonly used techniques of thermal decomposition of coprecipitated molybdates and vanadates which generally yields double oxides $V_nMo_mO_p$ exhibiting relatively low concentrations of V(IV) centers [5]. These centers are extensively associated.

3. Conclusion

The results described above evidence that solvothermal synthesis yields molybdenum-vanadium oxides in the form of substituted solid solutions characterized by very high concentration of individual vanadyl centers. These materials are promising redox catalysts toward important oxidation reactions (*e.g.*, fuel desulfation).

Acknowledgments

This work was supported in part by the Russian Foundation for Basic Research (project 16-53-00136-bel-a) and by the Belarusian Foundation for Basic Research (project Kh 15-078).

References

1. W. Zhang, S. T. Oyama, *J. Phys. Chem.* **100**, 10759 (1996).
2. P. A. Vakhrushin, T. V. Sviridova, M. V. Vishnetskaya, D. V. Sviridov, A. I. Kokorin, *Russ. J. Phys. Chem. B* **6**, 711 (2012).
3. E. V. Boikov, P. A. Vakhrushin, T. V. Sviridova, M. V. Vishnetskaya, D. V. Sviridov, A. I. Kokorin, *Russ. J. Phys. Chem. B* **7**, 251 (2013).
4. T. V. Sviridova, L. I. Stepanova, D. V. Sviridov, in: *Molybdenum: Characteristics, Production and Application.* Ed. by M. Ortiz and T. Herrera (Nova Sci. Publ., New York, 2012), p. 147.
5. T. V. Sviridova, D. V. Sviridov, A. I. Kokorin, *Russ. J. Phys. Chem. B* **7**, 734 (2013).

CATALYTIC PROPERTIES OF Pd_xZn_{1-x} NANOPARTICLES PREPARED BY POLYOL METHOD IN SELECTIVE HYDROGENATION OF 2-METHYL-3-BUTYN-2-OL

L. B. Okhlopkova, M. A. Kerzhentsev

Boreskov Institute of Catalysis, Lavrentyeva Ave. 5, 630090 Novosibirsk, Russia
mila65@catalysis.ru

Z. R. Ismagilov

Boreskov Institute of Catalysis, Lavrentyeva Ave. 5, 630090 Novosibirsk, Russia
Institute of Coal Chemistry and Material Science
Sovetsky Ave. 18, 650000 Kemerovo, Russia
IsmagilovZR@iccms.sbras.ru

Highly dispersed polymer-protected Pd/Zn bimetallic nanoclusters with the composition ranging from 1/9 to 6/1 have been obtained by an polyol reduction method. High resolution transition electron microscopy results show different PdZn bimetallic structures having reproducible narrow distribution of particle sizes and average particle diameter of about 2 nm. These bimetallic nanoclusters were used as prepared and after embedding in a titania matrix for selective hydrogenation of 2-methyl-3-butyn-2-ol. Their catalytic activity and selectivity are demonstrated to be dependent on the metal composition. The effect of the particle composition is different for the colloidal and embedded particles. A monometallic Pd colloid shows the highest activity and selectivity. Among supported catalysts, the highest selectivity can be achieved for an embedded bimetallic nanocluster with a molar ratio of Pd/Zn = 2/1.

1. Introduction

In heterogeneous catalysis, materials containing two or more metals might show significantly different catalytic properties compared to the parent metals and thus are widely utilized in several reactions [1]. For selective hydrogenation of alkyne alcohols, high-performing heterogeneous catalysts mainly consist of multi-metallic systems facilitating an increased selectivity to alkene. Pd metal can be improved by adding another metal such as Zn, Cd, Pb or Ga [2]. High selectivity to alkenes caused by the charge-transfer phenomena between Pd and second metal may favorably decrease alkene adsorption. It is achieved by adding an electron-donor compound to the nanoparticles of Pd. The catalytic properties of the metal can be precisely controlled by nanoparticles size and the stoichiometry of the multi-metallic nanoparticles. The latter method is poorly understood. The immobilization of the nanoparticles in a support matrix applied for the synthesis of industrial catalysts results in the drastic changes of their properties. The considerable attention should be pay to the support and morphological and structural evolution of the nanoparticles during different stages of the synthesis, when understanding of the origin of the catalytic reactivity of supported nanoparticles is desired.

In this paper, we prepare a series of PdZn nanoparticles with different compositions by modifying load of metal precursors and deposit them by embedding to titania matrix. The prepared nanoparticles and catalysts were tested in the hydrogenation of 2-methyl-3-buten-2-ol. The effects of PdZn molar ratios on the activity and selectivity of the catalyst were investigated.

2. Experimental

The colloids of Pd_xZn_{1-x} particles were prepared by reduction within the solvent method in the presence of a stabilizer. Pd_xZn_{1-x}/TiO_2 catalysts were synthesized by sol-gel method in the presence of Pluronic F127 according to the procedure described elsewhere [3], followed by the reduction at 523 K for 2 h.

The colloids and catalysts were characterized by a high resolution transition electron microscopy, inductively coupled plasma atomic emission spectrometry (ICP-AES), X-ray fluorescence spectroscopy, X-ray photoelectron spectroscopy (XPS). Hydrogenation of 2-methyl-3-butyn-2-ol (MBY) was carried out in an autoclave reactor with the total volume of 130 ml at 323–333 K and 4–5 bar of H_2 under intensive stirring at 2000 rpm. Analysis of the reaction mixture was performed off-line with a two-dimensional gas chromatograph equipped with a DB-5MS primary capillary column. Reaction rates were expressed as moles of 2-methyl-3-butyn-2-ol (MBY) per mole of Pd metal per second, and turnover frequency (TOFs) were defined as moles of MBY consumed per mole of the Pd surface per second. MBY conversion was defined as $X=(C_{MBY,0}-C_{MBY})/C_{MBY,0}\times100$ %, and the selectivity was defined as $S_i=C_i/(C_{MBY,0}-C_{MBY})\times100$ %.

3. Results and discussion

Fig. 1a shows the relationship between catalytic activity and the metal composition of PdZn colloidal nanoclusters during the hydrogenation of MBY. The Pd colloid shows a 3-fold reactivity as compared to that of $Pd_{16}Zn_{84}$ colloids. To examine the effect of the surface area of the catalyst, the TOFs have been calculated. Dependence of the turnover frequency on the metal composition has a shape akin to that of the apparent activity. This indicates that the decline in the catalytic activity is not due to the decrease in the surface area of the particles. Other reasons such as an ensemble effect in the bimetallic nanoclusters may account for this improvement.

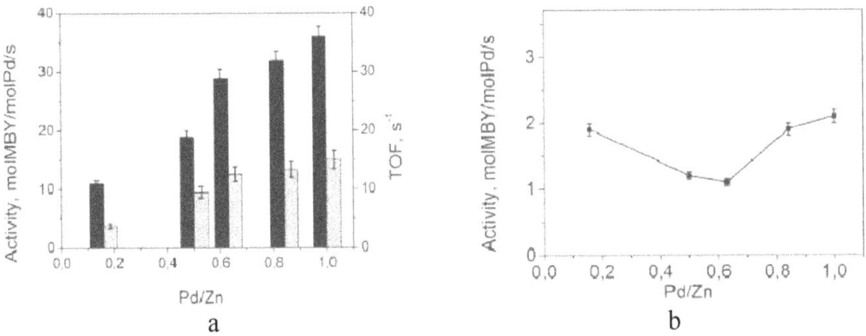

Figure 1. Effect of the metal composition of Pd_xZn_{1-x} colloids (a) and embedded nanoparticles Pd_xZn_{1-x}/TiO_2 (b) on the catalytic activity at 50 % conversion.

The ratio of surface atomic concentrations of Zn determined from XPS data is found to be larger than bulk concentration determined by ICP-AES analysis, suggesting that part of the active Pd is blocked by Zn oxide. The activity variation with the Pd/Zn ratio is mainly attributed to the dilution of Pd active sites by Zn, which is derived from the presence of PdZn solid solution and the possible partial coverage of the Pd surface by Zn oxide.

Embedding nanoparticles resulted in an order of magnitude reduction in the activity (Fig. 1b). The difficult access of the active component appears to provide low activity. As it was in the case for colloidal catalysts, monometallic catalysts were more active than bimetallic ones. $PdZn/TiO_2$ show lower activities and TOFs than Pd monometallic catalyst. In particular, $Pd_{66}Zn_{34}/TiO_2$ catalyst shows the lowest TOF, 1.5 times lower than that of the monometallic Pd/TiO_2.

For colloidal particles the selectivity to MBE at 95 % conversion decreases with increasing the Zn/Pd ratio up to 2, being of about 91 % for Pd, 82 % for $Pd_{86}Zn_{14}$, 83 % for $Pd_{63}Zn_{37}$ and $Pd_{50}Zn_{50}$, as shown in Fig. 2a. The improved selectivity to MBE of the Pd nanoparticles can be ascribed to the electronic effect in the presence of PVP stabilizer. Interestingly, the activity proved to have a regular decrease for higher Pd fractions in bimetallic Pd_xZn_{1-x} nanoparticles (Fig. 2). Meanwhile, the selectivity of $Pd_{16}Zn_{84}$ ones increases in comparison with other Pd_xZn_{1-x} nanoparticles and reaches the same values as for the Pd particles. The reason for the increased selectivity of $Pd_{16}Zn_{84}/TiO_2$ can be ascribed to the "ligand" effect rather than to the electronic effect, since the XPS analysis has demonstrated the absence of the charge transfer between Zn and Pd in colloid particles.

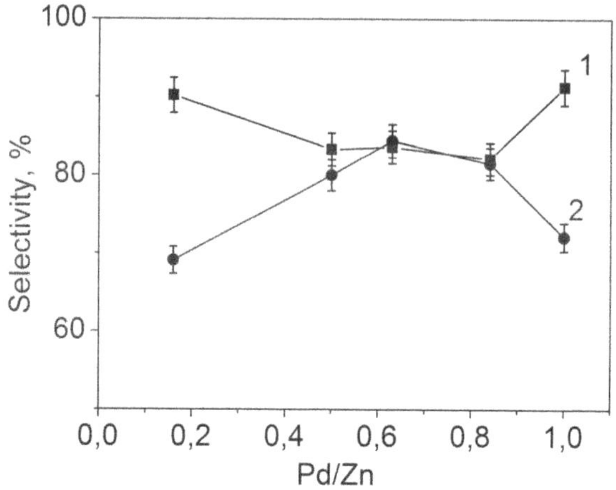

Figure 2. Effect of the metal composition of Pd_xZn_{1-x} colloids (1) and embedded bimetallic nanoclusters Pd_xZn_{1-x}/TiO_2(2) on their selectivity at 95 % conversion.

Quite different patterns for effect of Zn on selectivity were found for embedded catalysts (Fig. 2). An addition of Zn at a Pd/Zn molar ratio 2/1 ($Pd_{66}Zn_{34}/TiO_2$) improves the selectivity to MBE at high MBY conversion, while the selectivity of $Pd_{16}Zn_{84}/TiO_2$ are lower than that of $Pd_{66}Zn_{34}/TiO_2$. This can be rationalized by the presence of the PdZn alloy. The presence of charge transfer from Pd to Zn has been evidenced by XPS analysis. The changes in electronic properties from Pd to PdZn affect alkene adsorption more strongly resulting in the faster desorption of alkene and increased selectivity to alkene [4].

Acknowledgments

The support of this work by Russian Federal Agency of Scientific Organizations (project 0303-2016-0004) is gratefully acknowledged.

References

1. M. Armbruster, M. Behrens, F. Cinquini, *Chem. Cat. Chem.* **4**, 1048 (2012).
2. J. P. Boitiaux, J. Cozyns, M. Derrien, G. Leger, *Hydrocarbon Process* **64**, 51 (1985).
3. L. B. Okhlopkova, E. V. Matus, I. P. Prosvirin, I. Z. Ismagilov, M. A. Kerzhentsev, Z. R. Ismagilov, *J. Nanopart. Res.* **17**, 475 (2015).
4. E. V. Rebrov, E. A. Klinger, A. Berenguer-Murcia, E. M. Sulman, J. C. Schouten, *Org. Process Res. Dev.* **13**, 991 (2009).

PECULIARITIES OF OPTICAL, REDOX AND CATALYTIC PROPERTIES OF NANOSIZED 1D-3D COPPER PARTICLES STABILIZED IN ZSM-5

S. A. Yashnik[1], V. N. Parmon[1], R. Furiga[2]
[1]*Boreskov Institute of Catalysis SB RAS*
Akad. Lavrentieva Ave. 5, 630090 Novosibirsk, Russia
yashnik@catalysis.ru
[2]*Institute of Coal Chemistry and Material Science*
Sovetskiy Ave. 18, 650000 Kemerovo, Russia
zinfer1@mail.ru

Optical, RedOx, and catalytic activity in NO and N_2O decomposition over Cu-containing zeolites prepared through different procedures have been studied. The correlation between 1D-3D structures of the active sites and DeNOx catalytic performance of these catalysts are discussed.

1. Introduction

Nanoparticles are known to have novel optical, electronic, magnetic, and chemical properties owing to their extremely small dimensions. The limited dimensions of zeolite micropores (channels) allow stabilization of nanoscale transition metal oxide structures with unique redox and catalytic properties, *e.g.* chain-like 1D, 2D structures consisting of Cu^{2+} and Cu^{+} cations [1,2], square-plain oxide clusters [3], CuO-like nanoparticles [4] and metal nanoparticles [5]. The stabilization of 1D and 2D structures of copper ions in the zeolite channels is due to the specific properties of copper salts, in particular the cooperative Jahn-Teller effect and the predominance of copper dimer $[Cu_2(OH)_2]^{2+}$ in the solutions with a wide range of pH and concentrations.

Here we report on the study of optical, RedOx, and catalytic activity in NO and N_2O decomposition over Cu-containing zeolites prepared through different procedures aiming to find correlation between 1D-3D structures of the active sites and catalytic performance of these catalysts.

2. Experimental

The 1D-3D copper particles were introduced in ZSM-5 zeolites via the ion exchange procedure [1], the ion exchange with subsequent polycondensation [2], and the solid-state transformation (SST) [4]. The copper loadings were 0.1–3 wt.% for the ion-exchanged and SST catalysts.

All materials were characterized by the ICP, XRD, EPR, UV-Vis DR, H_2-TPR and HRTEM. ESR spectra of the samples were recorded with a Bruker

EMX EPR spectrometer operating in the microwave range. UV-Vis DR spectra of the samples were registered using a Shimadzu UV-2501 PC spectrophotometer equipped with a diffuse reflectance accessory (ISR-240 A). The temperature-programmed reduction (H_2-TPR) experiments were conducted using a flow reactor connected with a thermal conductivity detector. The sample were pretreated in the oxygen flow at 400 °C for 0.5 h, cooled to room temperature. Then, a hydrogen-argon mixture (10 vol.% of H_2) was passed through the sample at a feed rate of 30 cm^3/min. The temperature range and heating rate were 25–630 °C and 10 °C/min, respectively. The catalysts were tested in DeNOx reactions of NO and N_2O (5000 ppm in He) at 24000 h^{-1} and in the range of 200–550 °C.

3. Results and discussion

It was found that the nanosized 1D-3D copper oxide structures have special features of optical and RedOx properties that correlate with their catalytic activity in nitrogen oxides removal. The isolated Cu^{2+} ions, observed as the absorption band at 12500 cm^{-1} (d-d transition) by UV-Vis, are reduced first to Cu^+ ions and then Cu^0 at 200 and 400 °C (Fig. 1, curve 1). They have a moderate catalytic activity in DeNOx. The overall activity of Cu-containing ZSM-5 catalyst enhances with an increase of the isolated ions concentration [3].

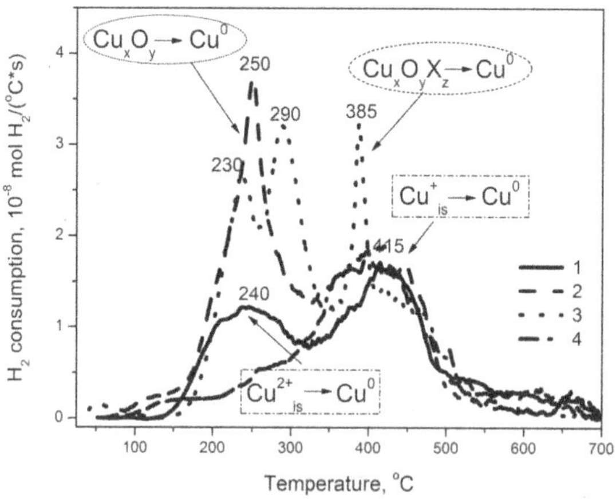

Figure 1. H_2-TPR of 1 % Cu-ZSM-5 catalysts prepared by ion-exchange from aqueous (1) and ammonia (2) $Cu(CH_3COO)_2$ solution, by ion exchange with subsequent polycondensation (3) and by solid-state transformation (4).

The chain-like 1D and square-plain 2D structures are observed by UV-Vis as charge transfer bands (CTBs) L→M 18000–23000 cm^{-1} and

28000–32000 cm^{-1}, respectively [1]. The 3D structures of Cu^{2+} ions obtained by polycondensation can be detected as the absorption at 32000–35000 cm^{-1} [2]. The 3D Cu^{2+} structures that predominate in the catalysts prepared by SST are shown by UV-Vis as the edge of fundamental absorption of CuO phase (*ca.* 11000 cm^{-1}).

The 1D-3D copper oxide structures are reduced in one-step, similar to finely dispersed CuO at 170–300 °C. When the dimension of the structures is decreased from 3D to 1D, a hydrogen consumption peak shifts to lower temperatures, for example from 290 °C to 230 °C (Fig. 1, curves 2,3). The inverse effect is observed after decoration of copper particles by a zeolite matrix. In this case the 2D copper oxide structures are reduced more hardly than 3D structures mainly at temperatures above 400 °C (Fig. 1, curve 4).

It was shown that 1D and 2D copper oxide structures inside zeolite channels were the most active in the H_2-TPR and DeNOx reactions. That is possibly caused by the high covalency of the Cu-O band, by the reversibility of reduction of Cu^{2+} ions to Cu^{+} and reoxidation of Cu^{+} to Cu^{2+} in the presence of NO (N_2O), and by an opportunity to stabilize the mixed-valent pairs Cu^{2+}/Cu^{+}. The catalysts containing 3D structures of Cu^{2+} ions possess low activity in the DeNOx of NO and N_2O. When their share increase, 3D structures of Cu^{2+} ions create steric hindrances for contact of the reactants with more active sites in the zeolite and explain the decrease in the catalyst activity.

Acknowledgments

This work was supported by Russian Science Foundation (project 14-13-01155) and project 0303-2016-0014 for Boreskov Institute of Catalysis.

References

1. S. A. Yashnik, Z. R. Ismagilov, V. F. Anufrienko, *Catal. Today* **110**, 310 (2005).
2. S. A. Yashnik, A. V. Salnikov, N. T. Vasenin, V. F. Anufrienko, Z. R. Ismagilov, *Catal. Today* **197**, 214 (2012).
3. S. Yashnik, Z. Ismagilov, *Appl. Catal. B* **170-171**, 241 (2015).
4. S. M. Ciocılteu, M. Salou, Y. Kiyozumi, Sh. Niwa, F. Mizukami, M. Haneda, *J. Mater. Chem.* **13**, 602 (2003).
5. S. A. Yashnik, V. F. Anufrienko, V. I. Zaikovskii, V. A. Rogov, S. Ph. Ruzankin, Z. R. Ismagilov, *Stud. Surf. Sci. Catal.* **174A**, 177 (2008).

TUNING THE STRUCTURE AND PHOTOCATALYTIC ACTIVITY OF TiO_2 SOL-GEL FABRICATED FILMS BY INCORPORATION OF NANO- AND SUBMICRON PARTICLES

N. M. Denisov, A. V. Baglov

Belarusian State University of Informatics and Radioelectronics
P. Browka 6, 220013 Minsk, Belarus
waefae@gmail.com

TiO_2 nanostructured films were fabricated by sol-gel technique with an addition of TiO_2 nano- or/and submicron particles into the sol. The highest photocatalytic activity along with high crystallinity predominantly in the anatase phase were achieved by adding both types of the particles. Photocatalytic activity of the films was found to get maximum after calcination at 500–550 °C corresponding to the best crystallization of the deposited material both with and without added preliminary formed nano- and submicron particles.

1. Introduction

The problem of environmental pollution by industrial wastewaters has been addressed unproportionally to rapid industrialization taking place in modern society. So far, conventional water purification technologies are ineffective against toxic organic pollutants released by chemical enterprises. Photocatalysis has been proposed to remove organic contaminations, but its efficiency requires much increase in order to be applied at a large scale.

Since the discovery of photocatalytic hydrogen production on the surface of titanium dioxide (TiO_2) [1] this material has been acknowledged as nearly best photocatalyst, which is due to its low cost, high activity, durability, chemical and thermal stability and safety for the environment. A sol-gel fabrication of TiO_2 films has advantages behind other technologies as being of low cost, technologically simple, scalable and compatible with various sizes and shapes of substrates. Turning sols into suspensions by an addition of nanoparticles has been found to yield TiO_2 films with an increased surface area and higher photovoltaic efficiency [2]. Considering the availability of commercial TiO_2 nanoparticles with various purity, particle sizes and crystalline structure, usage of the nanoparticles-doped suspensions opens new opportunities for tuning structural and photocatalytic characteristics of such films.

In this paper, we compare structure and photocatalytic properties of the sol-gel derived TiO_2 films containing nanoparticles, submicron particles and their combination. A role of the substrate and calcination temperature is also demonstrated and analyzed.

2. Materials and methods

TiO_2 films were formed on aluminum foils, silicon wafers and stainless steel grids by the sol-gel technique. All the substrates were preliminarily washed in isopropanol, sodium hydroxide water solution (100 g/L) and distilled water.

The starting sol contains sequentially mixed ethanol, 1-butanol, acetylacetone, titanium isopropoxide and distilled water with the molar ratio of 37:23:0.7:1:5. Commercial TiO_2 nanoparticles (P90, Evonik), submicron particles or their combination at the equal quantity corresponding to 7.4 wt.% of each component were added into the sol. These suspensions were deposited onto the above substrates by the dip-coating technique. The substrates were immersed in the suspension for 60 s, slowly withdrawn and dried in air at 200 °C for 5 min. The deposition/drying procedure was repeated 5 times. Then the samples were annealed for 30 min in air at a fixed temperature in the range of 450–600 °C. Particular combinations of the substrates with different suspensions and calcination temperatures used are collected in Table 1.

Table 1. General characteristics of the fabricated samples.

Substrate	Suspension used for fabrication of TiO_2 films	Calcination temperature
Aluminum foil	Without particles With nano- and/or submicron particles	500 °C
Silicon wafer	With nano- and submicron particles	450 °C, 550 °C
Stainless steel grid	With nano- and submicron particles	450–600 °C

Surface morphology of the films was analyzed with scanning electron microscopy (SEM) using a Hitachi S-4800 microscope. Their crystalline structure was determined by X-ray diffraction (XRD) recorded with a DRON-4 diffractometer operating with CuKα radiation (λ = 1.54179 Å). Photocatalytic activity of the fabricated films was assessed by degradation of Rhodamine B (RhB) organic dye in the water solution (10 mg/L) under UV irradiation at the wavelength of 365 nm. Optical absorption spectra of the RhB solutions after photocatalytic tests were recorded with a Proscan MC-122 spectrophotometer in the wavelength range of 400–600 nm.

3. Results and discussion

Fabrication of TiO_2 from the original sol without any particles has produced highly cracked films on the aluminum substrate. Adding the nanoparticles and submicron particles to the sol resulted in formation of about 3 μm thick films with rough highly developed surfaces. In these films the strain induced by unequal thermal expansion of the film and the substrate materials can be released without cracking. They are composed by 80–200 nm grains surrounded by much smaller particles of about 15 nm in size (Fig. 1). No significant structural differences were observed between the films on aluminum and stainless steel annealed at 500 °C, as well as for the stainless steel annealed at 450 °C, 500 °C and 550 °C.

Figure 1. SEM surface images of the TiO_2 film fabricated from the sol containing nano- and submicron particles on stainless steel annealed at 500 °C [3]. Two different parts of the sample at different length scales are presented for comparison.

XRD data revealed that all the formed films mostly consisted of anatase (Fig. 2). The highest intensity diffraction peaks are observed for the samples with submicron particles added into the sol. Smaller peaks from rutile phase of TiO_2 are additionally observed for the samples comprising TiO_2 P90 nanoparticles, which confirms the crystalline phase composition (90 % of anatase, 10 % of rutile) declared by their manufacturer.

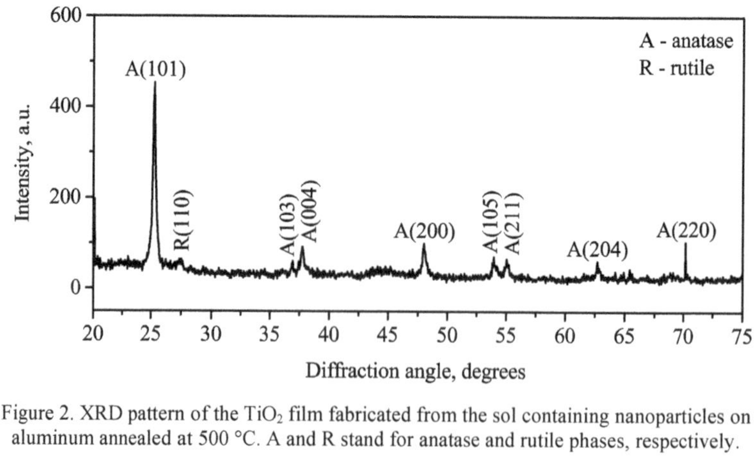

Figure 2. XRD pattern of the TiO_2 film fabricated from the sol containing nanoparticles on aluminum annealed at 500 °C. A and R stand for anatase and rutile phases, respectively.

The TiO_2 films on aluminum annealed at 500 °C were used in photocatalytic tests. The highest degradation rate of RhB was observed for the films containing both nano- and submicron particles (SMP+NP), followed by the films only with submicron particles (SMP), only nanoparticles (NP) and without particles (WP). The experimental data are illustrated in Fig. 3a. Assumingly, the microscale surface roughness formed by the submicron particles combined with high surface area of nanoparticles facilitates dye adsorption and more efficient photocatalytic degradation.

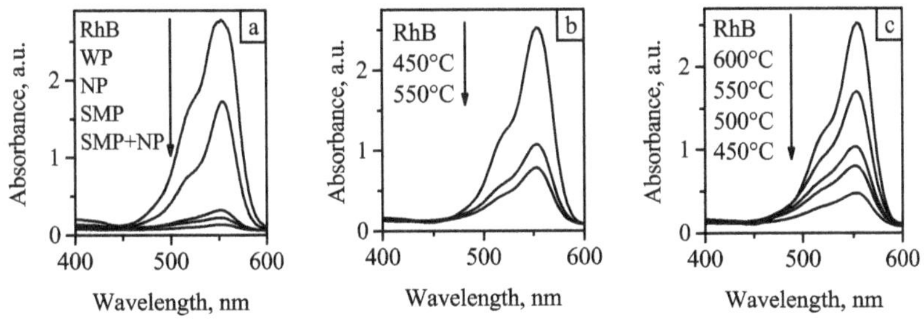

Figure 3. Absorption spectra of RhB solutions after UV-irradiation in the presence of TiO_2 films on aluminum annealed at 500 °C (a) and TiO_2 films comprising nano- and submicron particles on silicon (b) and stainless steel (c) annealed at different temperatures.

The TiO_2 films composed by both nano- and submicron particles on silicon annealed at 550 °C exhibited the higher photocatalytic activity than the films annealed at 450 °C (Fig. 3b). Evidently, annealing at the higher temperature produces TiO_2 films with better crystallinity and lower content of defects, decreasing the recombination losses of photogenerated charge carriers.

On the contrary, in the case of TiO_2 films with both nano- and submicron particles on stainless steel, the photocatalytic activity decreases with an increase of the annealing temperature in the range of 450–600 °C (Fig. 3c). Moreover, the films annealed at the temperatures above 450 °C turned yellow during UV-irradiation in both water and water solution of RhB, which indicates possible deposition of inorganic compounds on their surface.

In the previous study [3] we proposed the possible mechanism for deposition of Fe- and Cr-containing compounds on the surface of TiO_2 films during photocatalytic tests, as the result of dissolution of the substrate material (stainless steel). Here we confirm that the increase in the deposition rate of these compounds with annealing temperature can be caused by the photocatalytic activity increase of TiO_2 formed. Evidently, the TiO_2 film with the higher photocatalytic activity generates more electron-hole pairs and •OH radicals during UV-exposure, which results in faster dissolution of the substrate material (caused by electrons) and faster deposition of Fe- and Cr-compounds on the surface of the TiO_2 films (caused by holes and •OH-radicals).

4. Conclusion

Nanostructured TiO_2 sol-gel derived films with an increased photocatalytic activity can be fabricated by addition of TiO_2 nano- and submicron particles into the precursor sols. The highest photocatalytic activity along with high crystallinity predominantly in the anatase phase can be achieved by using both types of particles. While the photocatalytic activity of the films on silicon wafers increases with annealing temperature, the opposite tendency was observed for the films on the stainless steel grids, which can be ascribed to anodic dissolution of the substrate material (Fe, Cr, Ni) followed by the cathodic deposition of inactive compounds of the dissolved components on the surface of TiO_2 films during photocatalytic tests. Such electrochemical dissolution/deposition processes involving substrate material have to be accounted for in photocatalytic experiments and in a choice of the substrate material for industrial photocatalytic purifiers.

Acknowledgments

The work was financially supported by the Belarusian National Program "Convergence" (project no. 3.2.04).

The authors thank Professor V. Uglov and Mr. D. Zhigulin for XRD and SEM analysis of the experimental samples, respectively.

References

1. A. Fujishima, K. Honda, *Nature* **238**, 37 (1972).
2. C.-T. Wang, C.-F. Yen, *Surf. Coat. Technol.* **206**, 2622 (2012).
3. N. M. Denisov, A. V. Baglov, V. E. Borisenko, *Inorg. Mater.* **53**, 176 (2017).

PHOTOCATALYTIC ACTIVITY OF NANOSTRUCTURED SEMICONDUCTING METAL OXIDES FABRICATED ON MICROSTRUCTURED ALUMINUM FOIL

A. V. Baglov, L. S. Khoroshko, N. M. Denisov
Belarusian State University of Informatics and Radioelectronics
P. Browka 6, 220013 Minsk, Belarus
baglov@bsuir.by

Sol-gel derived nanostructured coatings of titanium dioxide, zinc oxide and their combinations fabricated on a microstructured aluminum foil were shown to get an increased photocatalytic activity in the process of UV decomposition of organic dye Rhodamine B.

1. Introduction

Photocatalytic degradation of organic pollutants by semiconducting metal oxides, in particular by titanium dioxide (titania, TiO_2) and zinc oxide (ZnO), has received an increasing attention during the past decade as a promising technology for water and air purification from organic compounds and bacteria [1–3]. Besides, ZnO being a direct gap semiconductor possesses light emission in the near-UV spectral region at room temperature [4–6]. Both TiO_2 and ZnO are emerged as leading candidates for green environmental management systems. That stimulates development of fabrication techniques of photocatalytic systems based on these oxides providing a most possible increase of their efficiencies combined with a low cost.

In order to meet the above requirements we developed the technology of micro- and nanostructuring of cheap aluminum foils served as substrates for sol-gel derived TiO_2 and ZnO coatings. The results of their structure characterization and photocatalytic activity tests are presented in this paper.

2. Experimental

The experiments were performed with 16×25 mm samples cut from 1 mm thick commercial foil of aluminum A0 (GOST 11069, 2001). Organic contaminations from their surfaces were removed by stirring the samples in isopropanol (C_3H_8O) for 2 min. Inorganic impurities were etched off in an aqueous solution of NaOH (111 g/l). Each chemical treatment step was finalized by washing the samples in distilled water and drying in flowing air at about 70 °C.

Treatment of the sample surfaces was performed in the etchant consisted of 31.8 % aqueous solution of hydrogen chloride (HCl) supplemented with 0.05 %

copper (II) chloride dihydrate ($CuCl_2 \cdot 2H_2O$) [7]. The etching process lasted for 2 min at room temperature. Then, the samples were washed in distilled water for 30 s and dried in flowing air at about 70 °C. The foils without treatment served as reference samples.

A sol-gel technique was used to fabricate TiO_2 and ZnO coatings. The sol for TiO_2 coatings contained 0.788 g of titanium tetraisopropoxide ($Ti(i\text{-}OC_3H_7)_4$) as a precursor, 3.7 g of ethanol (C_2H_6O) and 3.7 g of *n*-butanol ($C_4H_{10}O$) as solvents, 0.166 g of acetylacetone ($C_5H_8O_2$) as a stabilizer and 0.1 g of distilled water for hydrolysis of the precursor. The sol for ZnO coatings included 11.8 g of isopropanol (C_3H_8O) as a solvent, 1.76 g of ethanolamine (C_2H_7NO) as a stabilizer and 3.72 g of zinc acetate ($ZnC_4H_6O_4$) as a precursor.

The experimental and reference samples were immersed in corresponding sols for 1 min, withdrawn at the rate of 10 mm/s and dried at 200 °C for 3 min in air. The immersion/withdrawal/drying cycles were repeated for 3–4 times in order to get an appropriate thickness of the coatings. Then the samples were annealed at 500 °C for 30 min in air. Their composition and number of the deposited layers are summarized in Table 1.

Table 1. Characteristics of the coatings.

Sample	Coating	Number of depositions
№ 1	TiO_2	4
№ 2	ZnO	4
№ 3	TiO_2+ZnO	3 of each
№ 4	ZnO+TiO_2	3 of each

Surface morphology of the samples was examined by scanning electron microscopy (SEM). The photocatalytic activity was evaluated by the degree of decomposition of organic dye Rhodamine B in the water solution (10 mg/l) under the filtrated UV radiation of a mercury discharge lamp (8 W, I-line, $\lambda = 365$ nm). The lamp was placed at the distance of 40 mm from the surface of the samples put into the Petri cup with the dye solution. The solution was stirred during irradiation by an electromechanical shaker with the radial speed of 180 rpm. Concentration of the dye in the solution was determined by optical absorption spectra recorded in the wavelength range of 330–900 nm with a single-channel spectrophotometer Proscan MC122 at room temperature.

3. Results

Fig. 1 presents SEM images of the surface of the reference (not etched) and etched aluminum foil. The etched surface layer is evident to have a sponge like structure with 12–15 μm random pores.

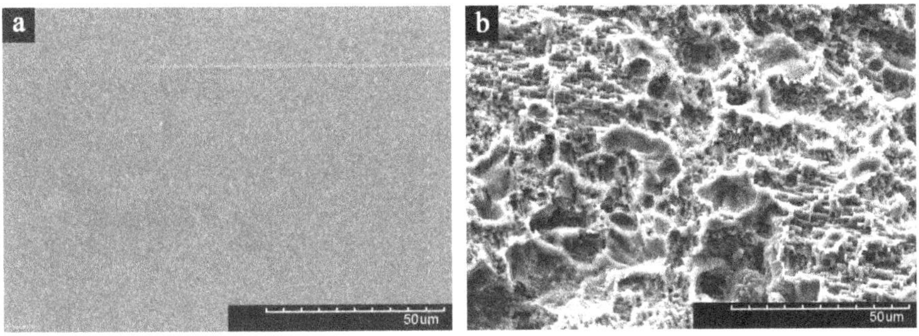

Figure 1. The surface of the aluminum foil without (a) and after (b) mirostructuring etching.

The TiO_2 and ZnO coatings formed on the reference samples are not continuous. They look like a film divided by cracks originated from the difference in the thermal expansion coefficients of the substrate (Al) and catalyst (TiO_2 or ZnO) materials. Formation of the cracks is reasonable to suppose at the cooling down step upon calcination at 500 °C. As a result of relaxation of the induced thermomechanical stresses, initially continuous film is divided into flat islands of 2–5 μm in size separated by 1–3 μm gaps.

The coatings on the microstructured samples have quite different morphology. It is illustrated by the SEM image in Fig. 2 with an example of the ZnO coating.

Figure 2. Surface morphology of the ZnO coating formed on the surface of the aluminum foil subjected to microstructuring.

The coatings on the etched aluminum surfaces are continuous and rough. They are composed by the grains with the sizes varied from hundreds to tens nanometers. As far as the calcination temperature of 500 °C has been many times experimentally confirmed to be sufficient for the sol-gel synthesis of TiO_2 and ZnO, we believe that the observed grains are crystallites of the synthesized oxides. This conclusion is also supported by the photodecomposition tests performed. The results of the tests are illustrated in Fig. 3.

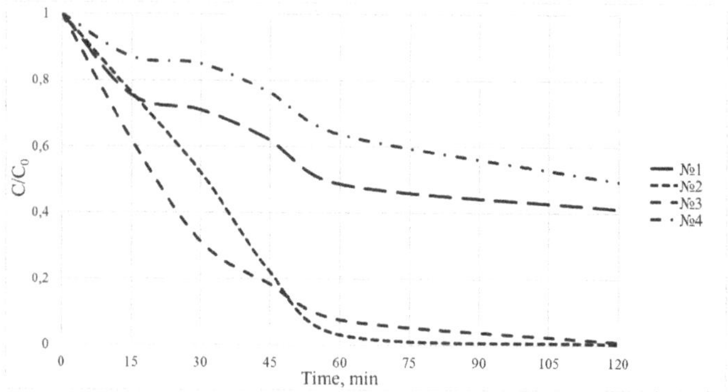

Figure 3. Kinetics of UV photocatalytic decomposition of Rhodamine B in the presence of TiO_2 (№ 1), ZnO (№ 2), TiO_2+ZnO (№ 3) and ZnO+TiO_2 (№ 4) coatings on Al foil.

In general, the coatings formed on the etched surface of the aluminum samples demonstrated approximately 1.5 fold increase of the photocatalytic activity as compared with the reference ones. The highest activity belongs to the ZnO coatings that can be explained by the most favorable well developed morphology in this case combined with electronic properties of ZnO.

4. Conclusion

TiO_2 and ZnO coatings were successfully fabricated through the simple sol-gel technique on aluminum foils. The proposed treatment helps to overcome the difference in the thermal expansion coefficients of the substrate and coating materials and fabricate the coatings with an extended surface favorable for an increase of their photocatalytic activity.

This work was financially supported by the National Program "Convergence" (project 2.1.02). The authors are grateful to V. E. Borisenko for the fruitful discussion. D. V. Zhigulin is acknowledged for SEM analysis.

References

1. S. M. Gupta, M. Tripathi, *Chin. Sci. Bull.* **56**, 1639 (2011).
2. Chien-Tsung Wang *et al.*, *Surf. Coat. Technol.* **206**, 2622 (2012).
3. N. M. Denisov *et al.*, *Inorg. Mater.* **52**, 523 (2016).
4. A. Janotti, C. G. Van de Walle, *Rep. Prog. Phys.* **72**, 126501 (2009).
5. D. C. Reynolds, *Phys. Rev. B* **60**, 2340 (1999).
6. Yefan Chen *et al.*, *J. Appl. Phys.* **84**, 3912 (1998).
7. A. V. Baglov, L. S. Khoroshko, in: *Proceedings of the VIII International Conference «Electronics and Applied Physics»* (Kyiv, Ukraine, 2016), 59.

PROSPECTIVE GLASS-CERAMICS BASED ON COMPLEX COMPOUNDS WITH GARNET STRUCTURE

Y. V. Bokshits, Y. U. Tratsiak
Research Institute for Physical Chemical Problems, Belarusian State University
Leningradskaya 14, 220030 Minsk, Belarus
slon.zhenya@gmail.com

A. Borisevich, M. V. Korjik
Research Institute for Nuclear Problems, Belarusian State University
Bobruiskaya 11, 220030 Minsk, Belarus

E. E. Trusova
Belarusian Technological State University, Sverdlova 13a, 220006 Minsk, Belarus

A method of obtaining complex substituted compounds with a garnet structure in the glass-ceramic form has been developed. It allows to produce garnets with substituting Y^{3+} and Al^{3+} ions in dodecahedral and octahedral positions into Ca^{2+}/Mg^{2+} and Ge^{4+} ions, respectively. Structural and luminescence properties of the glass-ceramics have been investigated.

1. Introduction

Nowadays compounds with a garnet structure in particular Y-Al garnets (YAG), are widely used in combination with InGaN LED chips for white light generation [1]. This is due to matching of the emitting band of InGaN with the excitation band of Ce^{3+} ions in YAG:Ce. The basic principle of such light sources consists in partial absorption of LEDs radiation followed by reemission in the yellow-green region of the spectrum. The simplest and most cheap way of producing such light sources is coating of InGaN chips by polymer layer containing the phosphor powder. The shape, size and position of the layer are chosen for providing uniformity of color spot on its whole area at minimum light losses. Despite this, LEDs based on YAG:Ce phosphors have other disadvantages: the low color rendering index and high color temperature [2]. Both disadvantages can be compensated by increasing emission of YAG:Ce in red part of spectrum. This may be achieved by using rare earth ions emitting in the red and modification of the garnet structure leading to the red shift of luminescence [3,4].

It is well known, that compounds with a garnet structure can be obtained in the form of powders, ceramics, single crystals and glass-ceramics. The latter is the best choice for using in a combination with LEDs. It is characterized by a much cheaper technology in comparison with that for ceramics or single crystals at comparable optical efficiency and it is much less prone to optical degradation as compared to coatings based on organic polymers with a phosphor.

In this paper we focus on the possibility of obtaining modified YAGs compounds ($Y_2CaAlGe(AlO_4)_3$:Ce and $Y_2MgAlGe(AlO_4)_3$:Ce) in the form of glass-ceramics and study of their structural and luminescence properties.

2. Experimental

Y_2O_3, Al_2O_3, CeO_2, GeO_2, Li_2CO_3, SiO_2, $CaCO_3$, $MgCO_3$ have been used. All reactants were qualified not less than p.a.

Glass matrix composited of 13.3 % Y_2O_3, 27.3 % Al_2O_3, 42.5 % SiO_2, 3 % Li_2O, 0.5 % CeO_2, 6.7 % Mg(Ca)O and 6.7 % GeO_2 (mol%) was synthesized in a reducing atmosphere at 1550 °C for 180 min. The glass was formed by casting the melt onto a metal plate. The obtained glasses were clear and colorless. The glass-ceramics has been obtained by crystallization of the glasses at 1500 °C for 105 min. It was translucent and had yellowish color characteristic for the YAG:Ce.

An instrument DRON 3 with CoKα radiation was used for X-ray diffraction (XRD) measurements. Photoluminescence measurements was performed using a SDL-2 spectrometer.

3. Results and discussion

XRD patterns for $Y_2CaAlGe(AlO_4)_3$:Ce and $Y_2MgAlGe(AlO_4)_3$:Ce glass ceramics are presented on Fig. 1.

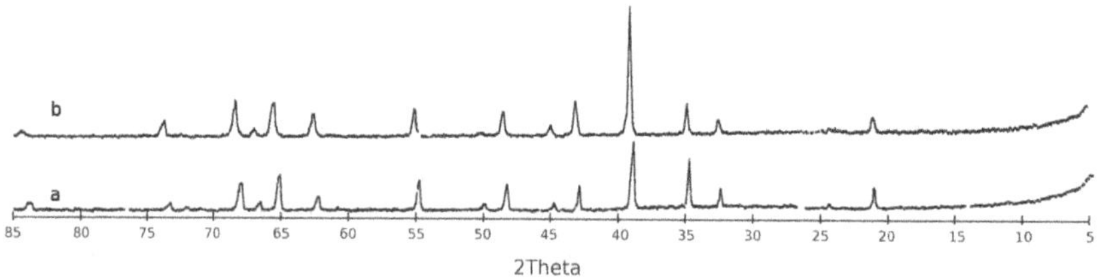

Figure 1. XRD patterns for $Y_2CaAlGe(AlO_4)_3$:Ce (a) and $Y_2MgAlGe(AlO_4)_3$:Ce (b) glass-ceramics.

The intensive reflections corresponding to the garnet phase (space group Ia3d) for both samples are observed. It should be noted that for the $Y_2CaAlGe(AlO_4)_3$:Ce sample there is a shift of principal reflections to the lower 2Θ angles (Fig. 1a) in comparison with the $Y_2MgAlGe(AlO_4)_3$:Ce sample (Fig. 1b). The reflections of the last correspond well to the reference data (PDF [33–40]). Generally, it indicates that $Y_2CaAlGe(AlO_4)_3$:Ce has a larger unit cell parameter than $Y_2MgAlGe(AlO_4)_3$:Ce. The shift can be attributed to the entering of Ca^{2+}/Mg^{2+} and Ge^{4+} ions into Y^{3+} and Al^{3+} positions in the garnet structure.

It should be noted, that the replacement of Y^{3+} and Al^{3+} ions by Mg^{2+} and Ge^{4+} ions leads to an increase (decrease) of the distance Ge^{4+}-O^{2-} (Mg^{2+}-O^{2-}), respectively. Such replacement decreases the bond distance between Ce^{3+} and O^{2-}ions. The crystal field strength increases resulting in the red shift of the emission.

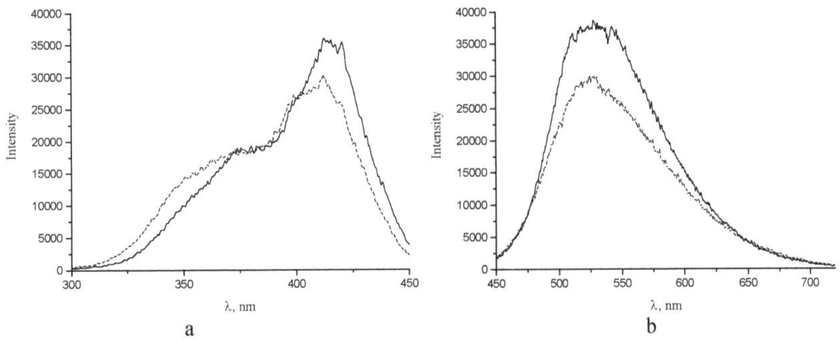

Figure 2. Excitation (a) and luminescence (b) spectra for $Y_2CaAlGe(AlO_4)_3$:Ce (solid line) and $Y_2MgAlGe(AlO_4)_3$:Ce (dashed line) glass-ceramics.

The room temperature spectra of luminescence (λ_{ex} = 410 nm) and luminescence excitation (reg. at 550 nm) are presented in Fig. 2. For both samples the excitation spectra are complex and consist of few bands unresolved at room temperature. The two observed bands in the excitation spectra located at ~350 and ~410 nm are caused by the transitions from the $^2F_{5/2}$ ground state to the Stark components of $4f^05d^1$ configuration. The band at 370 nm corresponds to the excitation of Ce^{3+} ions in the glass matrix. Glass-ceramic samples are characterized by the blue spectral shift of the first excitation band by 40 nm in comparison with unmodified YAG:Ce samples. It makes possible the effective luminescence excitation with 405 nm in both $Y_2CaAlGe(AlO_4)_3$:Ce and $Y_2MgAlGe(AlO_4)_3$:Ce samples.

The luminescence spectra of the samples under study are presented in Fig. 2b. The spectrum for Ce^{3+} ions in the garnet structure at room temperature consists of two components those may be deconvoluted from the spectra bands corresponding to transitions from the lowest radiating level of $4f^05d^1$ configuration to $^2F_{5/2}$ and $^2F_{7/2}$ states. $Y_2CaAlGe(AlO_4)_3$:Ce and $Y_2MgAlGe(AlO_4)_3$:Ce samples are characterized with wide FWHM approximately 90 and 100 nm, respectively. The maximum peak position for both samples is located at ~530 nm.

4. Conclusion

The possibility to produce complex compounds with a garnet structure in the form of glass-ceramics has been demonstrated. The observed shift of XRD reflections can be attributed to partial replacement of Y^{3+} and Al^{3+} ions in the garnet structure with Ca^{2+}/Mg^{2+} and Ge^{4+} ions, respectively. In the excitation spectra, three bands corresponding to the excitation of Ce^{3+} ions in the garnet structure are observed. The first excitation band shifts by 40 nm in comparison with unmodified YAG:Ce that makes possible the effective excitation of these compounds with LEDs emitted at 405 nm. $Y_2CaAlGe(AlO_4)_3$:Ce and $Y_2MgAlGe(AlO_4)_3$:Ce glass-ceramic samples are characterized by a wide FWHM values, approx 90 and 100 nm, respectively.

Acknowledgments

This work is supported by Belarusian Republican Foundation for Fundamental Research (grant X17L-012) and INTELUM MC RICE Project.

References

1. T. Tamura, T. Setomoto, T. Taguchi, *J. Lum.* **87-89**, 1180 (2000).
2. R. Marin *et al.*, *J. Nanopart. Res.* **14**, 1 (2012).
3. S. H. Lee *et al.*, *J. Alloys Comp.* **477**, 1 (2009).
4. H. Yang, D.-K. Lee, Y.-S. Kim, *Mater. Chem. Phys.* **114**, 665 (2009).

STRUCTURAL AND LUMINESCENT PROPERTIES OF $YAl_3(BO_3)_4$:Ce^{3+},Tb^{3+} POLYCRYSTALS SYNTHESIZED BY COLLOID CHEMICAL APPROACHES

G. P. Shevchenko, Y. V. Bokshits, Y. U. Ivlieva, G. V. Shishko

Research Institute for Physical Chemical Problems, Belarusian State University Leningradskaya 14, 220030 Minsk, Belarus
gvidonashevchenko@gmail.com

T. G. Khottchenkova, G. E. Malashkevich, I. I. Sergeev

B. I. Stepanov Institute of Physics NASB, Nezavisimosti Ave. 68, 220072 Minsk, Belarus

Polycrystalline phosphors of yttrium aluminum borates $YAl_3(BO_3)_4$ activated with Tb^{3+} ions and co-activated with Ce^{3+} ions were synthesized by colloid chemical approaches: co-precipitation of hydroxides from corresponding salts aqueous solutions in the presence of boric acid with a subsequent heat treatment. Their phase composition, morphology and luminescence were investigated. The intensity of luminescence of the Tb^{3+} under UV excitation (λ_{exc} = 310 nm) is increased by two orders of magnitude for samples co-activated with Ce^{3+} and Tb^{3+} ions. That indicates the effective sensitization of Tb^{3+} luminescence by Ce^{3+} ions.

1. Introduction

Borates with the general formula $RM_3(BO_3)_4$, where R is rare earth ions or yttrium, M is Al, Fe, Ga, Cr form crystals with a huntite structure. They have pronounced fluorescent and nonlinear optical properties as well as high thermal, chemical and mechanical stability. In addition, they exhibit an abnormally low concentration quenching of the luminescence of lanthanide ions due to ion-ion excitations transfer related to the large minimum distance between the ions equal to 0.59 nm [1]. This is due to the structure of huntite in which $[RO_6]$ polyhedra separated by $[BO_3]$ ones.

Yttrium alumoborate ($YAl_3(BO_3)_4$, (YAB)) characterized by optical transparency over the wavelength range from 200 to 7000 nm is the most promising among borates. Luminescent materials with desired parameters could be obtained by replacing of Y^{3+} ions by rare earth ions in yttrium alumoborate. For example, YAB: Sm^{3+}, Eu^{3+}, Tb^{3+}, Dy^{3+} can be used for solid-state lasers emitting in the visible range, YAB: Nd^{3+}, Yb^{3+}, Er^{3+} can be used for lasers in the near infrared and YAB co-activated with Ce^{3+} and Tb^{3+} ions can be used for high-performance UV-visualizers [2] the demand for which in biological and medical research is growing.

We have fabricated and described in this paper a series of polycrystalline samples of YAB doped with Ce^{3+}, Tb^{3+} and Ce^{3+}/Tb^{3+} ions. Colloid chemical approaches providing high homogeneity of the doped ions in host materials as compared to solid phase synthesis was used. Their phase composition, morphology and luminescence were investigated.

2. Experimental

Ultrafine yttrium alumoborate powders doped with cerium and terbium were prepared according to [3] by co-precipitation of metal hydroxides with ammonia from water solutions of appropriate salts of nitric acid with subsequent heat treatment in air. Solutions of nitrates of yttrium and aluminum were mixed in accordance with the stoichiometry of Y_2O_3:$3Al_2O_3$:$4B_2O_3$. The boric acid was added in 10 % excess to stoichiometry. Rare earth ions as activators were added as nitrate salts to a mixture of aluminum and yttrium nitrate solution at the amounts of 1 at.% Ce and 5 at.% or 10 at.% Tb of the substitutable Y^{3+} ions. The final co-precipitation of the hydroxides was allowed for a day. The precipitates formed were separated by centrifugation, washed with alcohol and dried in air at 60–80 °C. They were then heat treated in air at 600 °C for 2 h to remove structurally bound water and residual acid and at 1150 °C for 2 h to form yttrium alumoborate.

X-ray diffraction (XRD) analysis of the samples was performed with DRON-3 diffractometer using CuKα radiation. The morphology and particle sizes were determined by scanning electron microscopy (SEM) in a LEO-1420 device. The luminescence spectra (LS) and luminescence excitation spectra (LES) were recorded at 298 K with a SDL-1 spectrofluorimeter.

3. Results and discussion

The crystal structure of the samples was found to correspond well to crystalline $YAl_3(BO_3)_4$ with the huntite lattice. Owing to the small differences in ionic radii of Y^{3+}, Ce^{3+} and Tb^{3+}, the influence of the latter on the position of XRD reflections is small. According to SEM, the samples $YAl_3(BO_3)_4$ doped with Ce and Tb are composed of 0.2–2 μm particles. The average particle size of YAB sample containing 1 % Ce and 5 % Tb is about 0.5 μm and of YAB sample containing 1 % Ce and 10 % Tb is about 1 μm. The particles have a polygonal shape with a number of surfaces from 4 to 6 and the rod-like form. The size of the rod-shaped particles ranges from 0.2 to 1 μm.

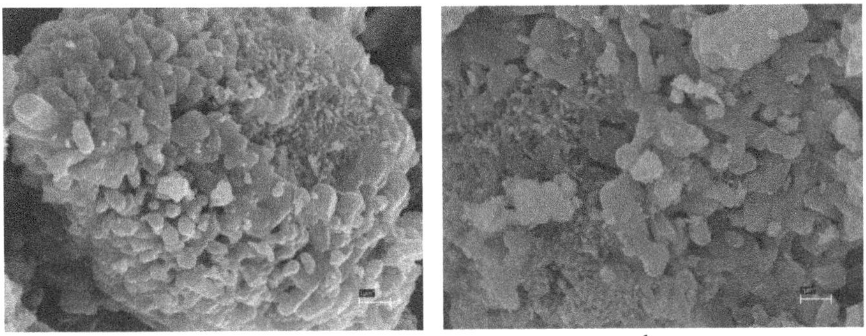

a b

Figure 1. SEM of $YAl_3(BO_3)_4$ samples containing: a – 1 % Ce + 5 % Tb, b – 1 % Ce + 10 % Tb.

The luminescence spectra of Tb^{3+} ions in mono-activated and co-activated with Ce^{3+} YAB samples at a concentration of Tb $\geq$ 5 at.% are only due to $^5D_4 \rightarrow {}^7F_j$ transitions (Fig. 2a). They practically do not depend on the concentration of doping elements. The absence of such relationship is due to sufficiently large minimum distance (Ln-Ln) and the recessed position of the $^5D_4$ state.

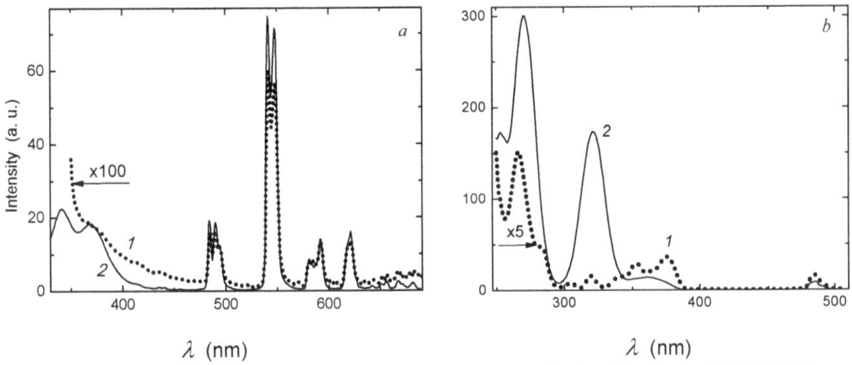

Figure 2. Corrected luminescence (*a*) and luminescence excitation (*b*) spectra of $YAl_3(BO_3)_4$ polycrystals doped with (1) Tb^{3+} (5 at.%) and co-doped with (2) Ce^{3+}(1 at.%) and Tb^{3+} (5 at.%). λ_{exc} = 310 nm (*a*), λ_{rec} = 545 nm (*b*).

Low-intense luminescence excitation bands of ions in the region of 300–400 nm (Fig. 2b, curve 1) are due to f^8-f^8 transitions. The two-humped band at 260 nm is due to $4f^8 \rightarrow 4f^7(^8S_{7/2})5d^1(E)$ transitions. At the excitation wavelength about 310 nm, the luminescence intensity of Tb^{3+} ions in the co-activated sample is of about two orders of magnitude higher than that in the mono-activated one (curves 1 and 2 in Fig. 2a). That indicates a sensitization of Tb^{3+} ions luminescence by Ce^{3+} ions. The efficiency of such sensitization according to the ratio of the partial intensity of the Tb^{3+} luminescence in the integrated, including an additional band luminescence of Ce^{3+} (two-humped band at $\lambda \leq 350$ nm), is about 50 %. Increasing of the concentration of Tb^{3+} ions up to 10 at.% with the constant concentration of Ce^{3+} ions (1 at.%) is followed by approximately 10 % increase in the efficiency of the sensitized luminescence of Tb^{3+} ions. The implementation of the described luminescence sensitization is confirmed by the luminescence excitation spectra (curves 1 and 2 in Fig. 2b) clearly demonstrating an emergence of two intense UV bands at 270 and 320 nm typical for Ce^{3+} ions in YAB.

4. Conclusion

$YAl_3(BO_3)_4:Ce^{3+},Tb^{3+}$ polycrystals synthesized by colloid chemical approaches with average sizes of polygonal particles from 0.5 to 1.0 μm are characterized by effective (50 % and higher) sensitizing of Tb^{3+} luminescence by Ce^{3+} ions. This allows application of such polycrystals for UV imaging.

Acknowledgments

This work was partially supported by the State Program (Photonics, Opto- and Microelectronics, task 1.2.03) and by the Belarusian Republican Foundation for Basic Research (grant F15D-007).

References

1. H. Y.-P. Hong, K. Dwight, *Mater. Res. Bull.* **9**, 1661 (1974).
2. G. E. Malashkevich, V. N. Sigaev, N. V. Golubev, E. Kh. Mamadzhanova, M. Z. Ziyatdinova, T. G. Khottchenkova, A. A. Sukhodola, in: *XV Int. Feofilov Symposium*. Book of Abstracts (Kazan, 2013), p. 57.
3. G. E. Malashkevich, G. P. Shevchenko, G. V. Shishko, *Bel. Patent* 16 458, 2010.

FORMATION AND PROPERTIES OF LANGMUIR-BLODGETT MONOLAYERS OF COPPER COMPLEX OF 2,4-HENEICOSANEDIONE WITH AMORPHOUS CARBON PARTICLES

A. E. Salamianski, V. E. Agabekov

Institute of Chemistry of New Materials NASB, F. Skaryna 36, 220141 Minsk, Belarus
solasy@mail.ru

Tribological properties of composite Langmuir–Blodgett monolayers of 2,4-heneicosanedione copper complex (HDCu) with particles of amorphous carbon (C) formed on silicon surfaces by horizontal precipitation method were studied. Incorporation of C (the average particles size of about 400 nm) into the HDCu Langmuir layer resulted in an increase of minimal area per molecule from 0.33 to 0.42 nm^2 and wear stability of HDCu monolayer.

1. Introduction

Currently, methods of physical or chemical vapor deposition, such as plasma ion nitriding, vacuum deposition of multilayer coatings based on molybdenum sulphides and selenides, titanium nitrides and other compounds are widely used for creation of solid lubricant coatings of precision friction units [1]. Composite Langmuir–Blodgett (LB) films of fatty acids, diketones with graphite or amorphous carbon particles also can be used as protective layers for these friction units because their formation does not require vacuum conditions and high temperatures [1,2].

The goal of this work is to create protective coatings for precision friction units based on composite LB monolayers of 2,4-heneicosanedione copper complex (HDCu) with amorphous carbon particles.

The horizontal precipitation (HP) method was used to fabricate LB films [2]. This method allows producing more defectless and uniform LB monolayers in comparison with traditional LB technology regardless of amphiphilic compounds structure and inorganic particle size [2,3].

2. Experimental

HDCu and HDCu–C monolayers were formed on silicon plates of rectangular shape with area ~ 2 cm^2. The silicon substrates were preliminary hydrophilyzed at 320 K for 15 min in the solution of H_2O:NH_4OH:H_2O_2 with the volume ratio 5:1:1. Then the silicon substrates were rinsed in distilled water several times and dried under a nitrogen flow.

A LT–201 device (Microtestmachines) [3] was used for measurements of surface pressure, area per molecule (π–A) isotherms and the deposition of HDCu and HDCu–C coatings by HP method under surface pressure of 30 mN/m. The minimal effective area per molecule in HDCu and HDCu–C Langmuir layer was calculated by extrapolation of the isotherms linear parts to zero surface pressure [2,3].

The composite HDCu–C monolayer with particles of amorphous carbon (N990, the average particles size of 400 nm) was formed from the suspension of C particles and HDCu in chloroform. The HDCu and C were taken in 1.0:0.5, 1.0:1.0 and 1.0:2.0 ratios by weight, respectively, whereafter the suspensions were sonicated for 15 min at 35 kHz in an ultrasonic bath.

A linear microtribometer RPT-02 (V.A. Belyi Metal-Polymer Research Institute NAS of Belarus) was used for the tribological tests [2,4]. Conditions of the testing included normal load of 0.5 N, a 3 mm diameter stainless steel ball as an indenter, 3 mm stroke, linear speed of 4 mm/s. The destruction of the oxide layer on the silicon substrate at the friction coefficient of 0.4 was the threshold of the experiment [2].

The morphology of the samples was studied by atomic force microscopy (AFM) using a Nanoscope 3D microscope (Veeco). Conditions of surface scanning: a contact mode, the cantilever of silicon nitride with constant rigidity *ca.* 0.32 N/m, the scanning rate of 1–5 Hz, the applied force of 1–3 nN.

3. Results

The minimal effective area per molecule (A_0) of the HDCu–C monolayers was 0.36, 0.40 and 0.42 nm^2 at the weight ratios of HDCu and C as 1.0:0.5, 1.0:1.0 and 1.0:2.0, respectively (Fig. 1).

The increase of A_0 value compared to HDCu monolayer ($A_0 \sim 0.33$ nm^2) testifies about incorporation of C particles into the structure of the HDCu film [2].

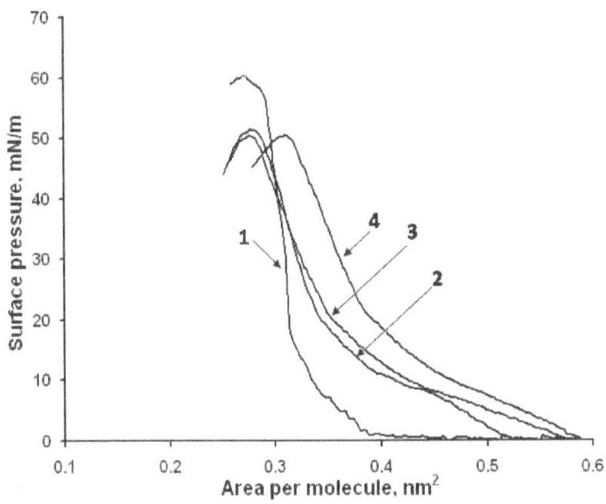

Figure 1. The surface pressure–area per molecule isotherms: (1) HDCu monolayer; (2), (3), (4) HDCu –C composite monolayers at the weight ratios of HDCu and C as 1.0:0.5, 1.0:1.0, 1.0:2.0, respectively.

The incorporation of C particles and their aggregates with sizes of ~220 nm – 6.8 μm into the HDCu layers was also confirmed by AFM data shown in Fig. 2.

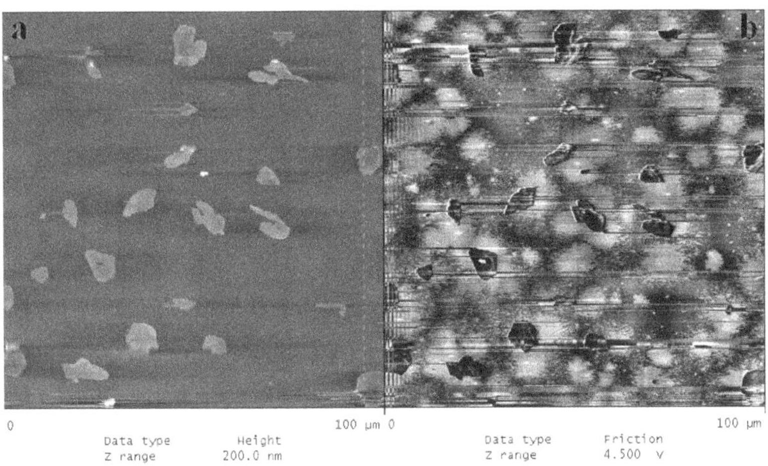

Figure 2. AFM images of morphology (a) and phase contrast (b) of the HDCu–C monolayer formed from the suspension of C particles and HDCu at the weight ratio 1.0:1.0.

An abrasion of the silicon surface modified by HDCu–C monolayers formed from the suspension of C particles and HDCu at the weight ratios of HDCu C as 1.0:0.5, 1.0:1.0, 1.0:2.0 was observed after ~ 760, 876 and 440 sliding cycles, respectively. It is illustrated in Fig. 3.

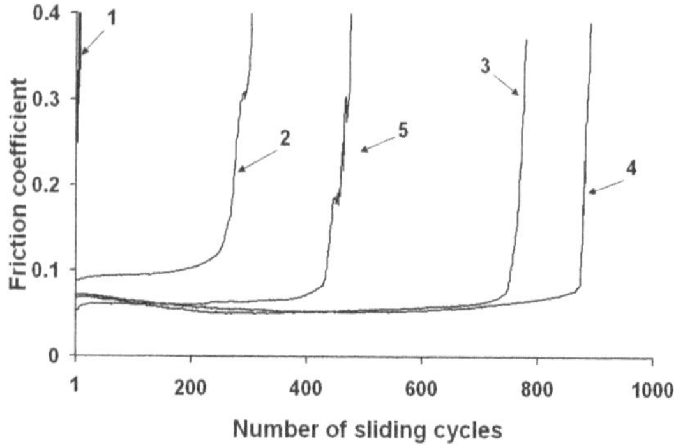

Figure 3. The friction coefficient as a function of sliding cycles number: 1 - unmodified silicon surface, 2 – HDCu monolayer, 3, 4, 5 – HDCu–C composite monolayers at the weight ratio of HDCu:C of 1.0:0.5, 1.0:1.0, 1.0:2.0, respectively.

C particles and their aggregates remain in contact area of the surfaces and protect them from wear [1,2]. That leads to an increase of the wear resistance of the HDCu–C coating compared with the wear stability in the friction process of the HDCu monolayer, which was destroyed after 260 cycles. The highest endurance against mechanical influence of the steel indenter was observed for the HDCu–C monolayer formed from the suspension of C particles and HDCu at the weight ratio of 1.0:1.0 (Fig. 3, curve 4). Augmentation of C amount in the HDCu–C coating resulted in the decrease of the wear resistance (Fig. 3, curve 5) because the main fraction of C particles and their aggregates are located outside of the HDCu matrix [2].

4. Conclusion

It was found that the incorporation of amorphous carbon particles into LB monolayer of 2,4-heneicosanedione copper complex increases its wear stability by 3.4 times. Thus, composite HDCu–C films can be used as protective coatings in precision friction units.

References

1. B. Bhushan, *Principles and Applications of Tribology* (John Wiley & Sons Ltd, Chichester 2013).
2. A. E. Salamianski, G. K. Zhavnerko, V. E. Agabekov, *Surface & Coatings Technology* **227**, 62 (2013).
3. G. K. Zhavnerko, V. E. Agabekov, A. E. Salamianski, S. A. Chizhik, A. A. Suslov, V. V. Chikunov, Device for forming mono- or multilayers of amphiphilic compounds on a solid surface, *Patent BY* № 15411, published 28.02.2012.
4. O. Yu. Komkov, *J. Friction Wear* **24**, 642 (2003).

ULTRASONICALLY PREPARED MAGNETITE COLLOIDS WITH LAYER-BY-LAYER HYDROPHILIC POLYELECTROLYTE SHELL

K. S. Livanovich[1], T. G. Shutava[1], V. V. Pankov[2]

[1]*Institute of Chemistry of New Materials NASB, F. Skaryny 36, 220141 Minsk, Belarus*
konstantin.livonovich@yandex.by
[2]*Department of Physical Chemistry, Belarusian State University*
Leningradskaya 14, 220030 Minsk, Belarus
pankov@bsu.by

Magnetite nanoparticles encapsulated by mono- and multilayered shells including modified polysaccharides were fabricated by ultrasonication of suspensions in a polyelectrolyte solution followed by layer-by-layer assembling. Hydrodynamic diameter, polydispersity, and ζ-potential of the nanocapsules depend on the structure of polyelectrolytes and number of bilayers in the shell.

1. Introduction

Magnetite nanoparticles are of great interest in biotechnology and medicine for magnetic separation, targeted drugs delivery and contrast agent-enhanced magnetic resonance imaging. They possess low toxicity and degradability in a human body [1–4]. A considerable challenge in this field is their surface functionalization, vectorization and enhanced colloidal stability.

In this paper, we optimize the preparation conditions of magnetite nanoparticles via sonication of suspensions aiming at production of monodisperse colloids with 90–250 nm particles stabilized by layer-by-layer (LbL) polyelectrolyte shells based on chitosan and dextran-grafted copolymer.

2. Experimental

Chitosan (Chit, 450 kDa), dextran (Dex, 6 kDa), methoxypoly(ethylene glycol) (PEG, 5 kDa), poly(diallyldimethylammonium chloride) (PDDA, 100–200 kDa), and dextran sulfate (DexS, 6.5–10 kDa) were purchased from Sigma-Aldrich. Chitosan of low M_w (DChit, *ca.* 60±3 kDa [5]) was synthesized by depolymerizing Chit with hydrogen peroxide [6]. Grafted copolymer of DChit and PEG (DChit-PEG) was obtained by carbodiimide crosslinking of PEG oxidized with potassium permanganate and amino groups of DChit [6]. To synthesize DChit-Dex, dextran was grafted onto glucosamine chain via reductive amination of its aldehyde group by amino group using cyanoborohydride [7]. The grafting degree (g) is *ca.* 5 and 3 for DChit-PEG and DChit-Dex, respectively [8].

Magnetite suspensions were prepared by coprecipitation with a Na_2CO_3 solution (pH 11) from the solution of $Fe(NO_3)_3$ and $FeSO_4$ mixed at the stoichiometric ratio Fe^{2+}:Fe^{3+} = 1:2 mol/mol [9]. Magnetite colloids were prepared by sonication of the suspension in the presence of positive polyelectrolyte (PE). Fe_3O_4 (0.2–0.4 mg/mL) was dispersed in an aqueous solution of PDDA, DChit, DChit-PEG, or DChit-Dex. The ratio of C_{Fe3O4}/C_{PE} varies in the range of 0.01–21 (Fig. 1). The mixture (5 mL) was subjected to strong ultrasound treatment (22 kHz, 60 W) 2 times for 45 s each. In the case of chitosan and its copolymers, pH of the mixtures before sonication was adjusted to 4.0 by adding 0.01 M HCl. Magnetite nanoparticles stabilized with PE were separated by permanent magnet and washed with distilled water (pH 6.0).

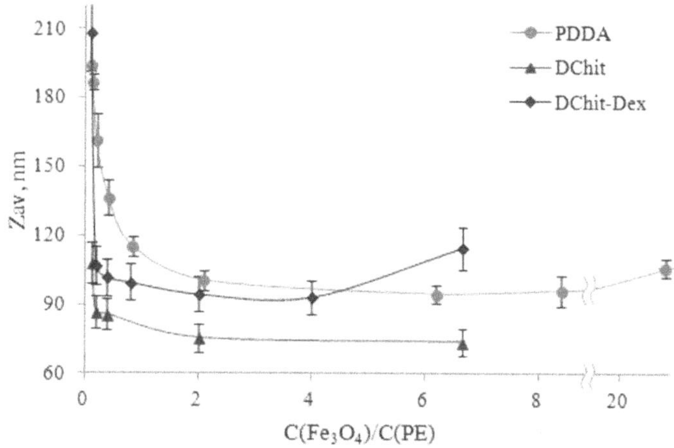

Figure 1. Average hydrodynamic diameter Z_{av} of Fe_3O_4/PE nanoparticles.

A polyelectrolyte multilayer shell was formed on the surface of nanoparticles by the layer-by-layer assembly method [10]. The PE-stabilized nanoparticles were dispersed in 1 mg/mL DexS solution and the mixture was put in an ultrasonic bath (35 kHz) for 1 min. The nanoparticles coated with one PE/DexS bilayer shell were separated from the excess of the latter polymer, washed with distilled water and redispersed in a 0.1 mg/mL PE solution following the same procedure as above. Thus, by alternating polyelectrolytes, a shell consisted of PE/DexS was formed (up to 3 bilayers). The hydrodynamic diameter by light scattering intensity (Z_{av}), polydispersity index (PDI), and ζ-potential of nanoparticles were characterized with a NanoSizer ZS, (Malvern) instrument.

3. Results and discussion

Four polycations (PDDA, DChit, DChit-Dex, and DChit-PEG) were compared as stabilizers of magnetite nanoparticles upon ultrasonication (Fig. 1). At the optimum conditions (~ 4 mg Fe_3O_4 per 1 mg PE), the average hydrodynamic diameter Z_{av} of PE-stabilized nanoparticles is lower than 100 nm with a narrow size distribution (Table 1). The PDI values in the range of 0.14–0.17 indicate formation of nanoparticles with a rather monodisperse size distribution. The presence of the polyelectrolyte during ultrasonicaton notably decreases Z_{av} as compared with magnetite suspension and stabilizes the colloid.

Table 1. Hydrodynamic diameter Z_{av}, ζ-potential, and polydispersity index (PDI) of the magnetite nanoparticles stabilized with different polycations.

PE	Z_{av}, nm	PDI	ζ-potential, mV
-	1850±560	0.68±0.23	13.1±1.0
PDDA	90±5	0.16±0.01	38.6±1.8
DChit	83±9	0.17±0.03	34.6±1.4
DChit-Dex	95±8	0.14±0.04	24.9±2.4
DChit-PEG	82±4	0.17±0.03	14.5±2.0

In all cases, the analysis of ζ-potential shows formation of the positive net charge on the nanoparticle surface, therefore, PE envelops individual magnetite crystallites. Though the nanoparticles stabilized with DChit-Dex and DChit-PEG show lower ζ-potential attributed to lower surface concentration of amino groups, the copolymers are highly effective as stabilizers and allow to produce nanoparticles of a size comparable to those stabilized with DChit and PDDA.

A formation of the LbL shells around magnetite core is proven by surface charge alternation (Fig. 2a). The value of ζ-potential depends on the structure of the polyelectrolyte adsorbed on the outer shell surface. Being adsorbed atop, both chitosan and Dex-grafted copolymer bring lower surface charge than PDDA due to a reduced number of charged groups in the former at pH 6.

The hydrodynamic diameter of the nanoparticles with a polyelectrolyte shell grows with increasing number of polyelectrolyte adsorption cycles (Fig. 2b). A minimal diameter increase per bilayer adsorption step was observed for $(PDDA/DexS)_n$ shells build on both strong polyelectrolytes. For the shells based on lower charge density of DChit and DChit-Dex, the increase of Z_{av} is more pronounced but the difference between these two PE is insignificant indicating a comparable degree of nanoparticle aggregation in the samples. Nevertheless, PDI increases insignificantly and stays in the range of 0.2–0.3 for all shell compositions.

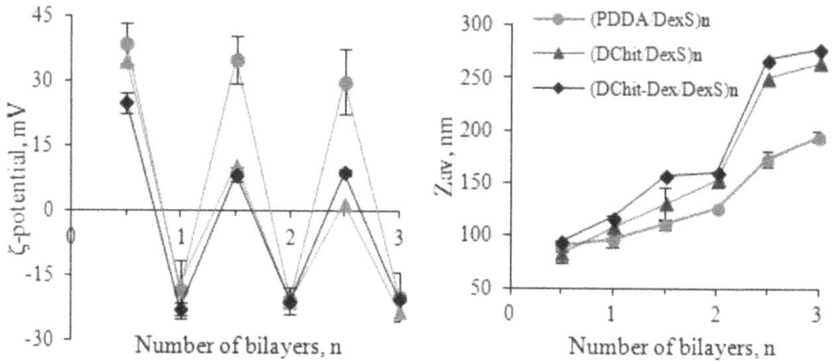

a b

Figure 2. ζ-potential (a) and hydrodynamic diameter (b) of nanoparticles coated with different polyelectrolyte pairs as a function of number of bilayers. The fractional and integer values of *n* correspond to polycation and DexS-terminated shells.

4. Conclusion

Sonication in the polycation solutions allows to produce stable colloids of magnetic nanoparticles that can be encapsulated with the LbL shell of variable composition. The approach gains control over size and surface properties of the nanoparticles by changing shell composition and number of polyelectrolyte layers. Magnetite nanoparticles coated with 2 bilayer shells retain the diameter smaller than 200 nm.

Acknowledgments

The work was supported by the Belarusian Republican Foundation for Fundamental Research (grant X015-018).

References

1. A. H. Lu *et al.*, *Angew. Chem. Int. Ed. Engl.* **43**, 4303 (2004).
2. S. Giri et al., Angew. Chem.-Ger. Edit. **177**, 5166 (2005).
3. S. Mornet *et al.*, *J. Mater. Chem.* **14**, 2161 (2004).
4. J. Gallo, N.J. Long, E.O. Aboagye, *Chem. Soc. Rev.* **42**, 7816 (2013).
5. C. Schatz *et al.*, *Langmuir* **19**, 9896 (2003).
6. T. G. Shutava, A. A. Mastich, K. S. Livonovich, *Trudy BGU* **9**, 102 (2014).
7. U. Janciauskaite *et al.*, *React. Funct. Polym.* **68**, 787 (2008).
8. H. S. Soedjak, *Anal. Chem.* **66**, 4514 (1994).
9. E. G. Petrova *et al.*, *Sviridov Readings* **11**, 113 (2011).
10. T. G. Shutava *et al.*, *Curr. Opin. Pharmacol.* **18**, 141 (2014).

INVESTIGATION OF ANTIBODY MODIFIED LIPOSOMES BINDING WITH PATTERNED FIBRINOGEN FILMS

K. I. Dubatouka, I. V. Paribok, V. E. Agabekov

Institute of Chemistry of New Materials NASB, F. Skaryny 36, 220141 Minsk, Belarus
d_katerina@tut.by

The complex of pectin conjugated with antibody and cationic liposomes containing a hydrophobic CdSe/ZnS quantum dots was synthesized. Selective binding of the liposomes modified with the antibodies and patterned fibrinogen films was studied by atomic force microscopy (AFM) and microcontact printing methods.

1. Introduction

Recently, the intensive researches of drug delivery systems based on liposomes (closed spherical particles built of lipid bilayers) were performed [1]. One methods of liposomes modification is their interaction with oppositely charged polyions (pectin, chitosan, dextran) to form polyelectrolyte complexes [2,3]. Self-assembled nanocomplexes are formed by mixing of a pectin solution with cationic liposomes, in which vesicles are arranged on the polysaccharide chains [3]. Liposomes modified by fluorescent semiconductor nanocrystals CdSe/ZnS can be used in medical diagnostics [4,5].

In this study we investigated the interaction of cationic liposomes containing hydrophobic quantum dots CdSe/ZnS and antibody conjugate modified by pectin with patterned fibrinogen films and propose a possibility to use this complex in medical diagnostics.

2. Experimental

Cationic liposomes containing hydrophobic quantum dots stabilized by trioctylphosphine oxide (CdSe/ZnS)TOPO (BSU), were obtained by the thin film hydration method [6] from the mixture (2:1:0.1 mol%) of egg phosphatidylcholine (Sigma), cholesterol (Acros Organics), hexadecylamine (Sigma), and 50 μg of quantum dots in chloroform (CLip). The solvent was removed in vacuum in a rotary evaporator (IR-1M2). Then distilled water (1 ml) was added and the mixture was stirred for 30 min and sonicated (35 kHz, Bandelin Electronic) for 2 min.

Pectin-antibody conjugate (PectAb) was obtained by the carbodiimide method [7] from the solution of citrus pectin (Sigma) and fibrinogen antibodies (monoclonal UC45, Abcam). The complex of cationic liposomes conjugated with PectAb (CLip-PectAb) was prepared by mixing these components and incubation for 1 h. The hydrodynamic diameter of the liposomes was determined by dynamic light scattering. Zeta potential was measured by electrophoretic mobility (ZetasizerNanoZS, Malvern). Fluorescence spectra were recorded with a Solar CM 2203 (Solar).

Bovine serum albumin (BSA, Sigma) patterned films were prepared via microcontact printing (mCP) according to [8]. Fibrinogen (FG), Pect-Ab and CLip-(PectAb) were subsequently deposited between the BSA bands via their adsorption. Surface morphology was investigated by AFM with a NanoScope MultiMode (Veeco) in a contact mode (silicon nitride cantilevers with the spring constant of 0.12 N/m). AFM images were analyzed using «Nanoscope 5.31r1» software.

3. Results

The conjugate formation of PectAb is based on covalent binding of carboxyl groups of pectin with amino groups of antibodies. Immobilization of the polysaccharide was proved by local surface activation with BSA microcontact printing, followed by adsorbtion of FG and PectAb conjugate. At the first stage, the periodic structure of proteins on a silicon substrate was formed by mCP (Fig. 1A,D).

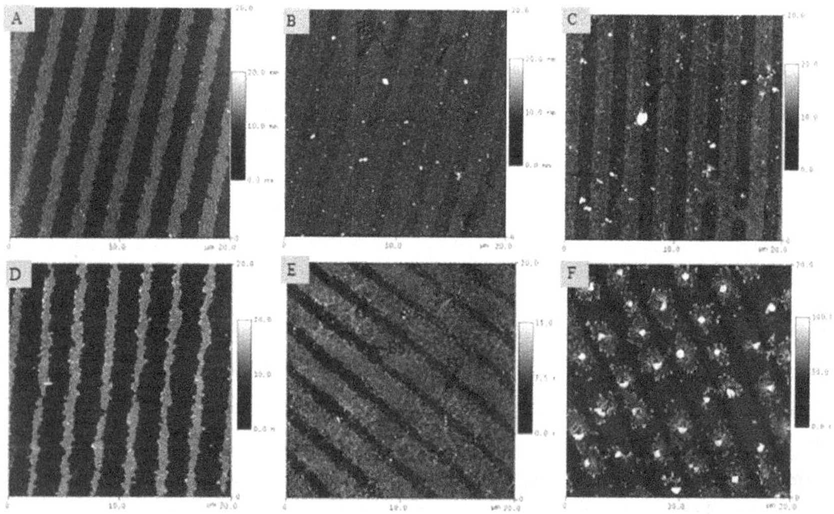

Figure 1. AFM-images of patterned BSA films (A, D), after FG adsorption (B, E), after selective adsorption of PectinAb (C), cationic liposomes and pectin-Ab complexes (F).

It is known that BSA blocks nonspecific protein adsorption [8]. So it is possible to fix a target protein between the BSA bands due to its self-assembly. Molecules of FG and BSA are different in size (4×8 and 8×45 nm) and molecular weight (350 and 67 kDa, respectively). Therefore after the FG adsorption the bands have to be equal or higher than BSA [9]. Surface section profile analysis of AFM images indicated an increase in the height of the bands after albumin mCP, FG and conjugate adsorption (Table 1).

Table 1. The bands height after adsorption of BSA, FG, PectAb, CLip-PectAb.

Layer	The total bands height, nm
BSA (mCP)	4.2±0.8
FG (non-specific adsorption)	8.1±1.0
FG/PectAb (selective adsorption)	11.5±0.8
FG/CLip-PectAb (selective adsorption)	68.0±11.0

Bovine serum albumin had the height of about 4.2 nm after its transferring onto the silicon substrate by mCP. The height of the bands after fibrinogen adsorption was ~ 8.1 nm (Fig. 1B,E). After the selective binding of PectAb conjugate with FG (Fig. 1C) the height increased to 11.5 nm, what testified about selective binding of the antibodies to the antigen (FG). Experimental data indicate the PectAb conjugate formation with the saving of antibody activity to FG.

Cationic liposomes containing hydrophobic quantum dots CdSe/ZnS with the hydrodynamic diameter of 45–55 nm and their complex with PectAb were prepared. Complex formation was possible due to the presence of amino groups on the surface of liposomes. Complex CLip-PectAb had the hydrodynamic diameter of 150–200 nm. The zeta potential of the liposomes changed during their modification from +45 to –50 mV. That indicates an interaction between the conjugate and CLip. The intensity of quantum dots luminescence entrapped into the liposomes was slightly changed: 82 ± 2 a.u. for the unmodified and 80 ± 1 a.u. for the modified liposomes (λ_{max} = 560 nm). The features of the luminescence can be used for medical diagnostics.

Binding of the CLip-PectAb complex with the patterned FG was investigated by mCP method. In this case, for correct AFM data interpretation and better visualization the narrower bands of BSA were formed (Fig. 1D). As in the previous case, FG adsorbed between the bands of BSA (Fig. 1E). Then, the CLip-PectAb complex was immobilized on the substrate. As a result (Fig. 1F) the targeted fixation of the modified liposomes occurred primarily on FG by its selective binding to the antibodies. The height in these areas was ~ 64.0 nm (Table 1).

4. Conclusion

We have developed the synthesis of the complex between pectin conjugated with the antibody and cationic liposomes containing hydrophobic CdSe/ZnS quantum dots with the size of 150–200 nm, zeta-potential of –50 mV and the luminescence emission maximum of 560 nm. The selective binding of the liposomes modified with antibodies and patterned FG films was determined by AFM and microcontact printing methods. Such antibody modified liposomes are of interest for medical diagnostics.

References

1. A. Bangham, M. Hill, N. Miller, in: *Methods in Membrane Biology, Vol.* 1 (New York: Plenum, 1974).
2. M. Antonietti, A. Wenzel, *Colloids and Surfaces A* **135**, 141 (1998).
3. P. Sriamornsak *et al.*, *Carbohydrate Polymers* **71**, 324 (2008).
4. N. V. Beloglazova, *Anal. Chem,* **85**, 7197 (2013).
5. C. Ye., *Microchim Acta* **180**, 117 (2013).
6. M. Paternostre, *Liposomes: Preparation and Membrane Protein Reconstitution, Manual on Membrane Lipids* (Springer-Verlag, Berlin Heidelberg, 1996).
7. J. P. Lopez-Alonso *et al.*, *Bioconjugate Chem.* **20**, 1459 (2009).
8. I. V. Paribok, G. K. Zhavnerko, V. E. Agabekov, Yu. A. Zmachinskaya, A. V. Yantsevich, S. A. Usanov, *Russ. J. Gen. Chem.* **77**, 363 (2007).
9. R. Green, M. Davies, C. Roberts, S. Tendler, *Biomaterials* **18**, 405 (1997).

STABLE COLLOIDS THROUGH ULTRASONICATION AND TWO HYDROPHILIC COMPONENT LAYER-BY-LAYER SHELLS

T. G. Shutava, K. S. Livanovich, A. A. Sheremet

Institute of Chemistry of New Materials NASB, F. Skaryna 36, 220141 Minsk, Belarus

shutova@ichnm.basnet.by

Nanocapsules with layer-by-layer shells consisting of poly(ethylene glycol)-grafted chitosan and poly(styrene sulfonate) and a top layer of bovine serum albumin have been fabricated. The factors influencing aggregative and storage stability of the nanocapsules in different media (water, NaCl solution, and fetal bovine serum) are discussed.

1. Introduction

High aggregative and storage stability is desirable for nanoparticles useful for fluorescent and magnetic diagnostic imaging as well as for drug delivery applications, but their conventional colloidal stability in salt solutions is low [1].

We report on high colloidal stability in water, isotonic sodium chloride, and serum and prolonged storage stability of layer-by-layer (LbL) nanocapsules containing two hydrophilic components. There are poly(ethylene glycol) (PEG)-grafted copolymer of chitosan (CH) and bovine serum albumin (BSA). The LbL assembly allows to integrate components of different classes in one thin coating, thus combining and complementing their properties [2].

2. Experimental

Chitosan (CH1, 450 kDa), methoxypoly(ethylene glycol) (PEG, 5 kDa), poly(styrene sulfonate) (PSS, 70 kDa), bovine serum albumin (BSA), fetal bovine serum (FBS) were obtained from Sigma-Aldrich. Chitosan of low M_w (CH2, 18 kDa) and PEG-grafted copolymers of CH1 and CH2 were prepared [3]. The synthesized copolymers have similar degree of amino group substitution by PEG, *ca.* 0.15. Its drafting degree **g** is *ca.* 5. Four positively charged polyelectrolytes (PE), *i.e.* CH1, CH2, and their copolymers CH1-PEG and CH2-PEG, were compared.

The polycation-stabilized ferrite cores (NP/PE) were synthesized as described elsewhere [4]. Shells consisted of 1–4 PE/PSS bilayers were formed over the cores *via* LbL assembly utilizing magnetic separation [2,3]. The colloidal stability of nanocapsules was tested in H_2O, 0.9 % NaCl, and FBS at 25 mg/L as recalculated to ferrite. The hydrodynamic diameter, polydispersity index, and ζ-potential of nanocapsules were obtained with the use of ZetaSizer

NanoZS, instrument (Malvern). The morphology of the LbL capsules was characterized by a MultiMode Nanoscope III, scanning probe microscope (Veeco).

3. Results and discussion

NP/PE colloids with narrow size distribution and a mean diameter $d_N < 100$ nm were prepared by ultrasound treatment of ferrite powder in solutions of CH1, CH2, and CH2-PEG. High molecular weight CH1-PEG (~10^6 Da) gives 220 nm particles (see Fig. 1a). Colloids of nanoparticles modified with a single layer of CH1-PEG or CH2-PEG slowly deteriorate over 5–14 days. Due to grafted PEG chains the copolymers stabilize colloids better than unmodified chitosans do. For the latter the size reaches 1–2 μm overnight. A rearrangement of polymers adsorbed on the nanoparticle surface presumably occurs upon storage. As a result their stability degrades. A layer of BSA adsorbed on NP/PE does not increase their aggregative and storage stability.

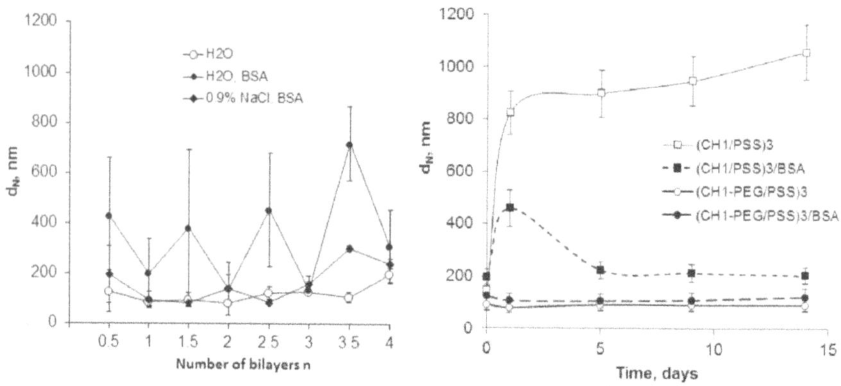

a b

Figure 1. a) Diameter of (CH2-PEG/PSS)$_n$ capsules as a function of number of bilayers; b) influence of shell architecture on d_N of (CH2/PSS)$_n$ and (CH2-PEG/PSS)$_n$ capsules at 25 °C.

Nanocapsules (CH2-PEG/PSS)$_n$ retain a low d_N value upon adsorption of both copolymer and PSS (see Fig. 1a), while the long glucosamine chain on the surface of CH1-PEG-based nanocapsules induces formation of aggregates despite a comparable degree of PEG-grafting. Upon adsorption of the next PSS layer, the surface charge changes to a high negative value and d_N reduces. The increase of d_N with increasing n is more pronounced for unmodified CH1 and related with formation of aggregates on every adsorption step.

The (CH1/PSS)$_n$ colloids aggregate completely after 24 h at room temperature forming a sediment. The (CH1-PEG/PSS)$_n$ and (CH2-PEG/PSS)$_n$

capsules with an integer number of bilayers ($n = 2,3,4$) are stable for more than 14 days (see Fig. 1b). At the same time, the diameter of PE-terminated capsules in water increases up to 0.5–1.0 μm.

The adsorption of BSA on the surface of nanocapsules on the basis of unmodified chitosans CH1 and CH2 slightly increases their diameter in water due to formation of a soft protein layer, while for the copolymer capsules alternation of d_N depends on the preceding BSA polyelectrolyte layer (Fig. 1a) due to specificity of BSA interaction with polyelectrolytes and PEG [5]. Regardless of M_w of glucosamine chain of copolymer, BSA slightly improves storage stability of copolymer-terminated capsules by slowly diffusing deeper into the LbL shell [6].

A BSA layer on the outer shell surface allows to completely avoid aggregation of CH2 and CH2-PEG-based nanocapsules in 0.9 % NaCl and keep it on the level reliable for injectable preparations, *i.e.* 200 nm (Fig. 1a). For high molecular weight chitosans CH1, independently on PEG modification, the effective diameter increases up to 450–500 nm, indicating a partial deterioration of stability. The PSS/BSA-terminated capsules are always more stable in 0.9 % NaCl than polycation/BSA-terminated ones.

In FBS, the hydrodynamic diameter estimated by light scattering intensity Z of PE-terminated nanocapsules is 50–75 nm higher than that for the PSS-terminated ones while for the nanocapsules with an additional BSA layer the difference is negligible (see Fig. 2a). It is due to specificity of adsorption of different FBS protein fractions on PEG-modified positive and negative surfaces [5]. The LbL nanocapsules are stable in FBS at 37 °C for at least 8 h and do not cause significant aggregation of FBS proteins.

4. Conclusion

Thus, the factors that influence aggregative and storage stability of nanocapsules on the basis of chitosan and its PEG-grafted copolymers in different media can be divides into two groups, *i.e.* related with polycation structure and governed by LbL shell architecture (Table 1).

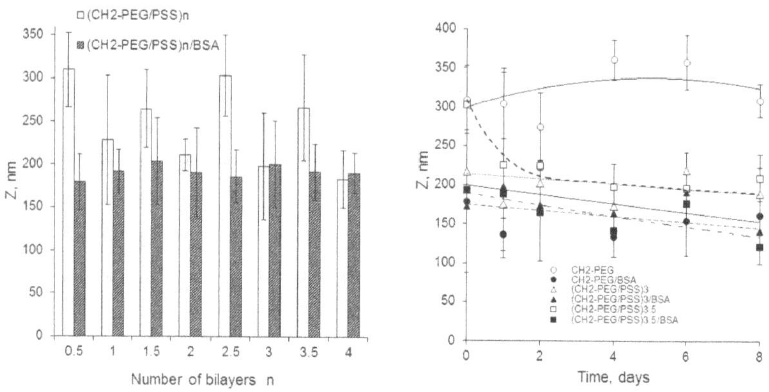

a b

Figure 2. Diameter of $(CH2\text{-}PEG/PSS)_n$ and $(CH2\text{-}PEG/PSS)_n$/BSA capsules as a function of the number of bilayers (a) and time (b). FBS, 37 °C.

Table 1. Stability evaluation of $(CH\text{-}PEG/PSS)_n$ capsules.

Factor	Medium			
	H_2O		0.9 % NaCl	FBS
	After preparation	14 days		
Shell structure				
Number of bilayers, *n*	No influence	Integer *n*, *n*>1 for copolymers	Integer *n*	Integer *n*
Surface charge	No influence	Negative	Negative	Negative
BSA layer	No influence	Increases storage stability	Requirement	Increases stability, decreases diameter
Copolymer structure				
M_w of chitosan or copolymer	Low M_w	Low M_w	Low M_w if BSA	No influence
PEG grafting	d_N increases if copolymer precedes BSA	Requirement	No influence	No influence

References

1. R. Hao *et al., Advanced Materials*. **22**, 2729 (2010).
2. T. G. Shutava *et al., Curr. Opin. Pharmacol.* **18**, 141 (2014).
3. T. G. Shutava, A. A. Mastich, K. S. Livanovich, *Proceedings of BSU* **9**, 102 (2014).
4. V. V. Pankov *et al., Proceedings of NASB, Ser. Chem. Sci.* **2** (2017).
5. G. Ladam *et al., Biomacromolecules* **1**, 674 (2000).
6. G. Schneider, G. Decher, *Langmuir* **24**, 1778 (2008).

PORPHYRIN-INDUCED AGGREGATION OF SILVER NANOPARTICLES

S. N. Terekhov, A. Yu. Panarin
B.I. Stepanov Institute of Physics NASB, Nezavisimosti Ave. 68, 220072 Minsk, Belarus
terekhov@imaph.bas-net.by

A. V. Abakshonok, A. N. Eryomin
Institute of Chemistry of New Materials NASB, Skariny 36, 220141 Minsk, Belarus

Water-soluble cationic Cu-porphyrins are able to promote aggregation of silver nanoparticles due to the effect of Coulomb interaction, displayed by the strong long-wavelength shift of the surface plasmon resonance bands in the extinction and scattering spectra. The observed effect of porphyrin-induced silver particle association can be used as a method of optical resonant light scattering detection of noble metal nanoparticles.

1. Introduction

Silver nanoparticles (Ag NPs) are related to the most widely produced nanomaterials due to their unique optical, electronic and antibacterial properties. They have various applications such as catalysis, electronics, imaging, therapy, nanophotonics, light harvesting and sensing [1]. Wider application of Ag NPs should result in their release and accumulation into the environment and organisms, leading to adverse impacts on plants and human [2]. In this regard, it is important to develop simple, reliable, and highly sensitive methods of Ag NPs analysis. Actually, perfect detection and identification of minor concentrations of metal NPs in aqueous solutions are still not advanced.

We believe that the resonance light scattering spectroscopy may be promising for the sensitive detection of Ag NPs. As is known, the scattering cross section significantly increases with the size of NPs, accompanied by decreasing the absorption efficiency [3]. Our design strategy is based on analyte-induced aggregation of Ag NPs to improve analysis of NPs by their scattering spectra. As an agglomerate promoting compound, we propose cationic metalloporphyrins which can act as electrostatic bridges between Ag NPs. For most colloids, the surface of Ag particles is negatively charged due to the presence of anions, *e.g.* citrate. Therefore, NP aggregation can be mediated by addition of positively charged porphyrin molecules which are able to attract Ag NPs of different size. It should be stressed that our approach to promote particle aggregation by porphyrins is aimed at detection of metal NPs unlike widespread concept of plasmonic sensors where the aggregation of nanoparticles is employed for sensitive and selective detection of various analytes such as DNA, proteins, and organic molecules.

This study focuses on interaction between cationic Cu(II)-tetrakis(4-N-methylpyridyl) porphyrin (CuTMPyP4) and anionic Fe(II)-5,10,15,20-tetrakis(4-sulphonatophenyl)porphin (FeTSPP) with colloidal Ag NPs. Optical properties of the hybrid system are characterized by UV-visible absorption and resonant light scattering spectroscopies.

2. Experimental

Ag NPs were formed by reduction of $AgNO_3$ with glucose (50 mM) in water solution of 10 mM $NaHCO_3$ (as a pH-regulator) and 1 mM sodium citrate (as a stabilizer) under ultrasonic treatment (3 h). The particles were purified from the synthesis medium components by centrifugation (20000 rpm, 30 min). The precipitate was washed with aqueous solution of 0.5 mM sodium citrate. The size and shape of Ag NPs were assessed by transmission electron microscopy (TEM, JEM-100CX). Water soluble porphyrins CuTMPyP4 and FeTSPP were purchased from Frontier Scientific Inc (USA).

UV-vis absorption spectra of the solutions were recorded on a Cary 500 Scan (Varian) spectrophotometer in 5x10 mm quartz cells. Resonant light scattering spectra were measured using the spectrometer reported earlier [4].

3. Results and discussion

Fig. 1 shows the extinction spectra of Ag NPs colloid and its mixture with cationic porphyrin CuTMpyP4. The absorption spectrum of the CuTMpyP4 net aqueous solution is also depicted for comparison. In the absence of porphyrin molecules, the surface plasmon resonance (SPR) band at 417 nm is observed which is typical for absorption of spherical Ag NPs with a diameter of ~30 nm. This size is in accordance with TEM image (Fig. 2a), where separately arranged quasi-spherical Ag NPs with the diameter of 25±5 nm are clearly seen.

After CuTMpyP4 addition, a significant red shift of the SPR band accompanied by the strong broadening was observed. The low intensity new peaks at 426 and 548 nm apparently correspond to B- and Q-bands of porphyrin molecules absorption (Fig. 1, spectrum 3). The strong shift of the SPR to the longer wavelengths may be due to the association of Ag NPs induced by CuTMpyP4 molecules. The TEM data (Fig. 2b) clearly demonstrate the association of Ag NPs accompanied by the tendency to form polymer chains and agglomerates.

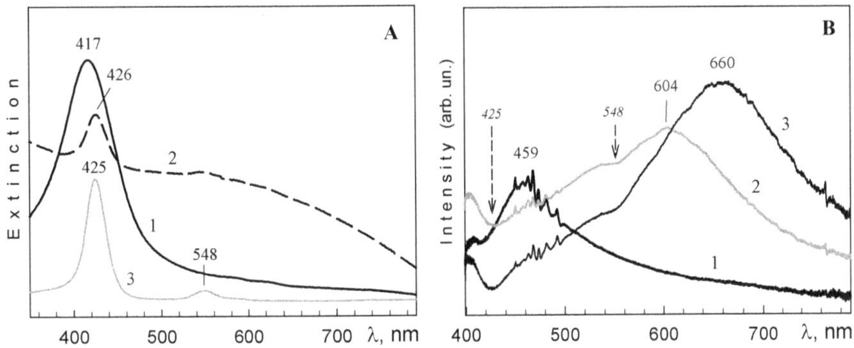

Figure 1. A: extinction spectra of Ag NPs solution (1) and Ag NPs in presence of 2 μM CuTMpyP4 (2); absorption spectrum of CuTMpyP4 (3). B: scattering spectra of Ag NPs solution (1) and Ag NPs in presence of 0.3 μM and 2 μM CuTMpyP4 (2 and 3, respectively).

Changes in the resonance light scattering spectra (Fig. 1b) upon interaction of Ag NPs with CuTMpyP4 are similar to those observed for absorption. With the increase of CuTMpyP4 concentration, low intensity SPR band at 459 nm is gradually shifted to longer wavelengths up to 660 nm. Unlike the weak extinction spectrum, the scattering peak is rather intensive that is consistent with the theoretical concepts on scattering cross-section increase with an increasing particle size.

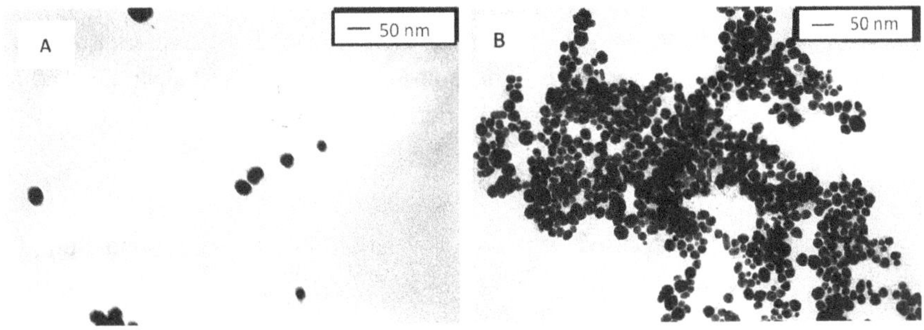

Figure 2. TEM images of Ag NPs (A) and Ag NPs - CuTMpyP4 complex (B).

Thus, absorption and scattering spectra together with TEM images clearly show that CuTMpyP4 promotes agglomeration of Ag NPs, which can be due to the Coulomb interaction between positively charged pyridyl groups of porphyrin and Ag NPs having an excess negative surface charge (due to citrate anions). Porphyrin molecules are located between individual Ag NPs forming electrostatic bridges in agglomerates (Fig. 2B).

In contrast to the case of CuTMpyP4, in the presence of anionic porphyrin, FeTSPP, no transformations in extinction and scattering spectra of Ag NPs were detected. These data confirm the electrostatic nature of the observed Ag NPs agglomeration induced by cationic porphyrin. In fact, the Coulomb interaction of repulsion type can occur between the anionic FeTSPP and negatively charged Ag NPs impeding particles association.

It is noteworthy that in the scattering spectra of Ag NPs mixture with CuTMpyP4 the pronounced dips located at wavelengths coinciding with the absorption bands of the porphyrin are present (Fig. 1b, spectra 2,3). The dips cannot be associated with the reabsorption of the scattered light by the porphyrin molecules because dilute solutions with the CuTMpyP4 concentration $< 10^{-6}$ M were used in this experiments. We suggest that the dips in the scattering counters are attributed to the plasmon resonance energy transfer. This effect occurs when the absorption band of an adsorbed molecule overlaps with the scattering band of NPs, which is realized in this case of CuTMpyP4-Ag NP complex.

4. Conclusion

From the experimental studies and discussion, we can conclude that cationic porphyrin CuTMpyP4 significantly affects the aggregation ability of Ag NPs. Interaction of CuTMpyP4 with Ag NPs results in the strong long-wavelength shift of the SPR band both in extinction and resonant light scattering spectra. The observed effect of porphyrin-induced association of Ag NPs can be used for development of optical methods of NPs detection in solution.

Acknowledgments

This work was supported by the Belarusian Republic Foundation for Fundamental Research (grant F16-096).

References

1. T. M. Tolaymat, A. M. El Badawy, A. Genaidy, K. G. Scheckel, T. P. Luxton, M. Suidan, *Sci. Total Environ.* **408**, 999 (2010).
2. D. Ye, M. N. Raghnaill, M. Bramini, E. Mahon, C. Åberg, A. Salvati, K. A. Dawson, *Nanoscale* **5**, 11153 (2013).
3. U. Kreibig, M. Vollmer, *Optical Properties of Metal Clusters* (Springer Series in Materials Science. Springer, Heidelberg, 1995), V. 25, 535 p.
4. A. Yu. Panarin, A. V. Abakshonok, V. E. Agabekov, A. N. Eryomin, S. N. Terekhov, *J. Appl. Spectrosc.* **81**, 940 (2014).

CHARACTERIZATION OF SURFACE SPECIES ON MESOPOROUS TiO_2 PREPARED BY TiC OXIDATION

E. Ovodok, M. Ivanovskaya, S. Poznyak
Research Institute for Physical Chemical Problems of the Belarusian State University
Leningradskaya 14, 220030 Minsk, Belarus
ovodok@bsu.by

N. Scharnagl
Helmholtz-Zentrum Geesthacht Centre for Materials and Coastal Research GmbH
Max-Planck-Straße 1, 21502 Geesthacht, Germany

J. Tedim
Department of Materials and Ceramics Engineering, CICECO-Aveiro Institute of Materials, University of Aveiro, 3810-193 Aveiro, Portugal

Surface species responsible for low pH values of zero charge point and high sorption capacity of mesoporous TiO_2 powder obtained by TiC oxidation with nitric acid were studied by FTIR, EPR and XPS methods. It was found that during the titania synthesis the oxide surface was modified by nitrate-nitrite, carboxylic groups, NO_x and ·C≡ radicals which determined adsorption properties of the TiO_2 powder.

1. Introduction

Titanium dioxide is used as a photocatalyst for UV decomposition of a wide variety of contaminants in water and air [1]. Mesoporosity and high surface area of TiO_2 powder enhances the efficiency of photocatalysis. Modification of the titania surface allows increasing the adsorption ability of TiO_2 powder and, thus, its activity in photodegradation processes can be improved.

This paper is focused on characterization of surface species on mesoporous TiO_2 powder obtained by TiC oxidation with nitric acid. FTIR, EPR, and XPS spectroscopies were applied for this purpose.

2. Experimental

Mesoporous TiO_2 powders were fabricated by oxidation of titanium carbide powder with nitric acid according to the method described in [2]. The samples were calcinated at 200 °C in air.

Fourier transform infrared spectra were collected on an AVATAR-330 spectrometer (Thermo Nicolet). XPS spectra were collected by Kratos DLD Ultrac spectrometer using AlK_α radiation (E = 1486.6 eV). Calibration was made relatively to C 1s line (E = 284.8±0.2 eV). EPR spectra were recorded with

VARIAN E 112 spectrometer using the resonance frequency of 9.35 GHz (X-band) at 77 K and 298 K. Zeta-potential of TiO_2 particles was measured by Malvern Zetasizer S90.

3. Results and discussion

Oxidation of black TiC powder by concentrated nitric acid gives a yellowish powder with a fine-dispersed structure. Calcination of the powder at 200 °C results in a change of color to light-brown. X-ray diffraction pattern shows the presence only anatase TiO_2 phase. The crystallite size calculated using Debye-Scherrer equation is 4 nm. According to BET analysis the powder has mesoporous structure with an average pore size of 3.4 nm. The specific surface area is found to be 353 m^2/g.

The synthesized TiO_2 particles have negative surface charge in a water suspension at pH > 3.5. Zero charge is observed at pH 3.5. This is lower than pH values commonly reported for TiO_2 formation [3].

It was found that the prepared titanium dioxide powder efficiently adsorbed positively charged molecules at neutral pH. Rhodamine B was chosen as a model pollutant for investigation of the sorption capacity of the TiO_2 powder. The maximum sorption capacity of the powder is near 40 mg/g (Fig. 1).

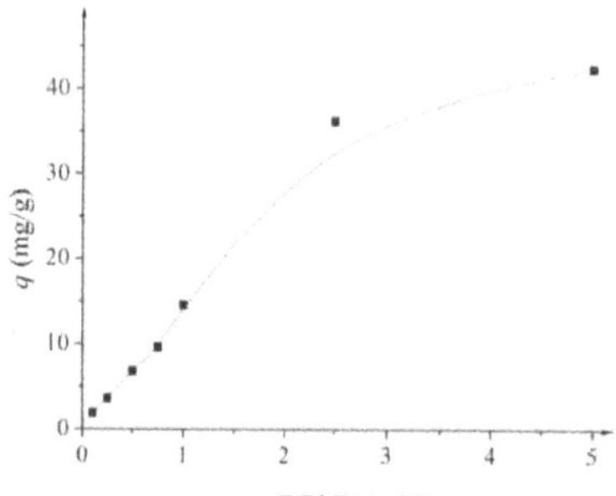

Figure 1. Sorption capacity of the synthesized TiO_2 powder to Rhodamine B.

High sorption activity of the synthesized powder is caused by large surface area, from one side, and the states at the TiO_2 surface, from the other side. Some nitrogen and carbon-containing species, which are produced during synthesis, might modify TiO_2 surface and influence adsorption properties of the TiO_2 powder. We used XPS, FTIR, and EPR spectroscopies for identification of these species.

The XPS results are collected in Table 1. Ti $2p_{3/2}$ and Ti $2p_{1/2}$ single lines of symmetric shape are registered at the energies, which correspond to Ti^{4+} (E_{bond} Ti 2p) in TiO_2. O 1s, C 1s and N 1s peaks indicate that OH-, CO_x-, and NO_x-groups are present on the oxide surface. C 1s line has a complex shape which can be approximated by lines assigned to sp^3- and sp^2- carbon states and oxidized –CO and COO^- species. XPS data show that $Ti^{\delta+}$ and $C^{\delta-}$ states, which are characteristic for titanium carbide, are absent in the synthesized titania samples, indicating that TiC is completely oxidized during the synthesis.

Table 1. XPS data of TiO_2 obtained by oxidation of TiC with nitric acid.

Lines	E_{bond}, eV	Element species
O 1*s*	530.1	O^{2-} (TiO_2)
	531.6	HO^- (TiO_2)
C 1*s*	284.8	C–C, C–H
	286	>C=O
	288.8	$-COO^-$, CO_3^-
Ti $2p_{(3/2)}$	458.7	Ti^{4+} (TiO_2)
N 1*s*	400.2	NO_x

The FTIR spectrum of the sample is presented in Fig. 2. The dominating wide band at 400–850 cm^{-1} is attributed to vibrations in octahedrons [TiO_6]. It is a superposition of several normal ν(Ti–O) vibrations characteristic of TiO_2. The spectrum shows two intensive bands related to deformation δ(H–O–H) vibrations at 1629 cm^{-1} and asymmetric ν_{as}(O–H) vibrations at 3200–3400 cm^{-1} of water molecules. The region from 800 to 1800 cm^{-1} shows vibrations of nitro, nitrates and carboxylate groups. In this region there are vibrations from unbound nitrate groups (at 1680 cm^{-1}), monodentate groups (at 1665 and 1680 cm^{-1}) and bidentate ones (at 1644 and 1710 cm^{-1}). The lines at 1407 and 1107 cm^{-1} are the symmetric and asymmetric stretching vibrations of nitro groups, respectively. The broad absorption band at 2000 cm^{-1} may be due to an appearance of complexes with CO adsorbed on the TiO_2 surface.

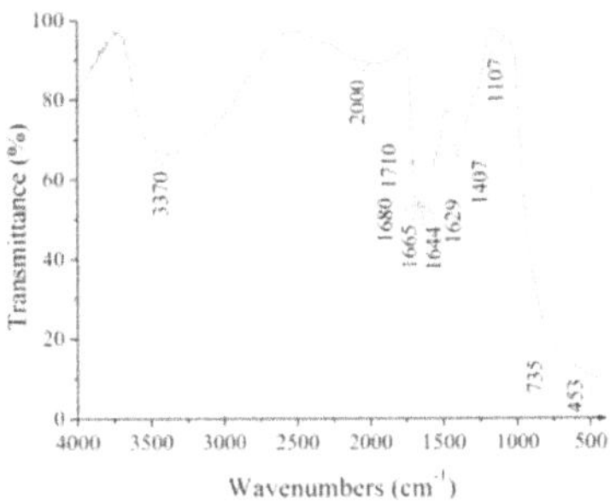

Figure 2. FTIR spectrum of TiO_2 obtained by oxidation of TiC.

The EPR spectrum shows the central line with *g*-factor of 2.0025 and weakly resolved broadened neighboring lines (Fig. 3). The half width of the central line (ΔB) is 0.5 mT. The spectrum represents a superposition of two triplet signals of NO_x radicals. The radicals include one unpaired electron (S=1/2) and ^{14}N atom (nuclear spin I = 1, ^{14}N isotope abundance is 99.6 %). These two signals

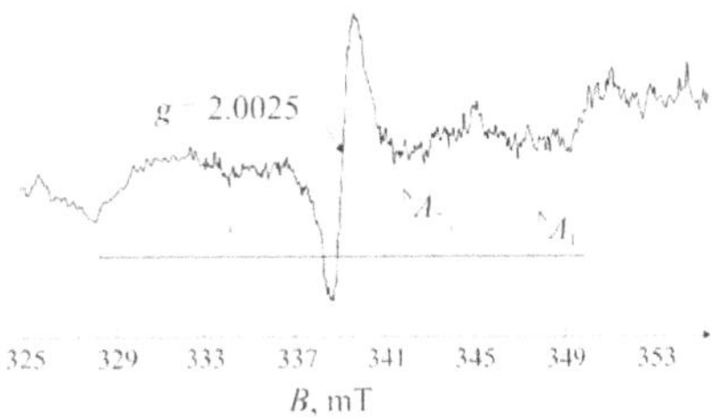

Figure 3. EPR spectrum of TiO_2 obtained by oxidation of TiC.

have a central line with $g = g_e$ and different hyperfine structure (HFS) constants (A^N). This constant characterizes magnetic interaction of the unpaired electron with ^{14}N nuclear. The parameters of the EPR signals and corresponding paramagnetic centers (PCs) are presented in Table 2.

Table 2. Parameters of the EPR signals and nature of the corresponding PCs, which are observed in the titanium dioxide obtained by oxidation of TiC with nitric acid.

Parameters of the EPR signals	Paramagnetic centers
$g = 2.002$; $^{N}A_1 = 5.6$ mT	$\cdot NO_2$
$g = 2.002$; $^{N}A_2 = 2.4$ mT	
$g_{iso} = 2.002$	·C≡

Intensity of the central EPR line exceeds the total intensity of the neighboring $\cdot^{14}N$ HFS lines, which can be due to an existence of additional signals at $g = g_e$ without hyperfine splitting. ·C≡ radicals in the amorphous carbon phase have such type of EPR signal. They can be formed as a byproduct of TiC oxidation by nitric acid under heating. The total concentration of the PCs (NO_x + ·C≡) is 4×10^{16} spin/g.

4. Conclusion

Using XPS, FTIR and EPR methods the presence of hydroxyl, carboxylic, nitrate-nitrite groups, $(Ti)_2$>CO, Ti–CO complex, and NO_x and ·C≡ radicals in mesoporous TiO_2 powder prepared by TiC oxidation with HNO_3 are confirmed. The nitrogen and carbon-containing species modify the TiO_2 surface, change the charge state of the surface and provide high sorption capacity of the TiO_2 powder to a positively charged dye.

Acknowledgments

The work was supported by the project (645662 – SMARCOAT – H2020-MSCA-RISE-2014) of «HORIZON-2020» EU program.

References

1. X. Chen, S.S. Mao, *Chem. Rev.* **107**, 2891 (2007).
2. S. Poznyak, E. Ovodok, H. Maltanava, M. Ivanovskaya, A. Kudlash, N. Scharnagl, J. Tedim, in: *Proc. 7th International Conference on Advanced Nanomaterials (ANM 2016)* (Aveiro, Portugal, 2016).
3. M. Kosmulski, *J. Colloid Interface Sci.* **253**, 77 (2002).

PHOTOCATALYTICALLY PREPARED Au NANOPARTICLES ON TiO_2 NANOTUBES FOR ELECTROCATALYTIC APPLICATION

H. M. Maltanava, S. K. Poznyak, M. I. Ivanovskaya, T. V. Gaevskaya

Research Institute for Physical Chemical Problems, Belarusian State University
220030 Minsk, Belarus
poznyak@bsu.by

Composites containing Au nanoparticles on TiO_2 nanotubes have been produced using photocatalytic deposition. Electrocatalytic activity of the prepared system in oxygen electroreduction reaction was found to depend strongly on the annealing temperature of the TiO_2 support. This effect was explained in the frame of a model supposing the formation of Schottky barrier at the TiO_2-Au interface.

1. Introduction

In fuel cells, effective performance of both anode for oxidation of fuel molecules (hydrogen, methanol, formic acid, *etc.*) and cathode, where oxygen electroreduction (ORR) occurs, is necessary to achieve the current density needed for wide-spread applications [1]. Despite extensive research progress in recent years, commercialization of fuel cells has been impeded by slow kinetics of cathodic reaction and by high cost and limited supply of platinum proved as the most efficient electrocatalyst for ORR. Therefore, significant efforts have been devoted to development of non-platinum electrocatalysts [2].

Compared to the platinum-group metals, gold has attracted less attention in electrocatalysis mainly because of its low catalytic activity. However, when the dimension of Au particles is reduced to the nanoscale, the material properties exhibit a dramatic deviation from those of bulk material [1]. Usually, to evaluate the activity of Au nanoparticles (NPs) they are dispersed on a solid surface. Thus, properties of both active phase and support are important. The support has to combine high conductivity, corrosion stability and low cost. In searching for support materials, considerable interest was given to titania nanotubular layers (TNTs) due to their unique features such as huge surface area, low cost, high stability, and nontoxicity [3]. Recently, efficient ORR in acidic media has been reported on TiO_2 nanotubes loaded with Au NPs [4].

The purpose of the present work is to define the role of semiconducting TiO_2 support in electrocatatalytic activity of Au NPs-loaded TNTs in alkaline medium. Since the efficiency of electrodes in ORR can be strongly affected by the metal-support interaction, Au NPs were deposited onto TNT layers using photocatalytic deposition. This technique allows obtaining more tight contact between NPs and TiO_2 surface without any additional post-treatment.

2. Experimental

Self-organized highly ordered nanotubular TiO_2 layers were produced on commercially pure Ti sheets by the two-steps anodization in the ethylene glycol electrolyte containing 0.75 % NH_4F and 2 vol.% H_2O. The electrochemical anodization for both steps consisted of a potential ramp (200 mV s^{-1}) from 0 to 40 V followed by holding the potential constant for 1 h. The oxide film formed during the first step of anodization was removed by detachment in an ultrasound bath with deionized water. After the second anodization step, the samples were washed with ethanol and then their surface was cleaned from debris by treatment in an ultrasound bath with distilled water during 30 s. To study the influence of semiconductor doping level on the activity of Au NPs-TNT system in ORR, TiO_2 samples were annealed at different temperatures ranging from 350 °C to 550 °C for 3 h in air.

Au NPs were loaded onto annealed TNT by photocatalytic reduction of Au(III). The solution for Au deposition contained 1.25×10^{-4} M $HAuCl_4$ as an Au source and 0.05 % $Na_5P_3O_{10}$ as a stabilizer. The TNT electrodes were emerged into the solution and then were UV-irradiated during 3 min. Scanning and transmission electron microscopy (SEM, TEM), UV-vis and X-ray diffraction (XRD) methods were used for characterization of the fabricated samples.

The electrocatalytic activity of the electrodes in ORR was examined by a cyclic voltammetry (CV) in a 0.1 M KOH solution saturated with oxygen during 1 h using an Autolab PGSTAT 302N potentiostat. Mott-Schottky measurements were performed under *dc* polarization applied by successive displacements of 50 or 100 mV in the cathodic direction from 0.6 V to –0.8 V with superimposing a sinusoidal perturbation (100 Hz with a modulation of 10 mV) onto *dc* bias voltage.

3. Results and discussion

The annealed TNT samples have well-aligned nanotubular structure with a relatively narrow distribution of the inner diameters (60±5 nm) and the wall thicknesses (12±2 nm) (Fig. 1a). The thickness of the nanotubular layers was 10±1 μm. According to the results of XRD measurements, all annealed TNT samples contain anatase phase of TiO_2.

Fig. 1b shows SEM images of the Au-loaded TiO_2 nanotubular layers. Gold nanoparticles can be found inside and on the surface of the TNT layers. The NPs are well separated and distributed inside nanotubes without aggregation. The size of the deposited Au NPs is 20–25 nm. Optical reflectance spectra of Au-TNT samples exhibits the band peaked at 540 nm.

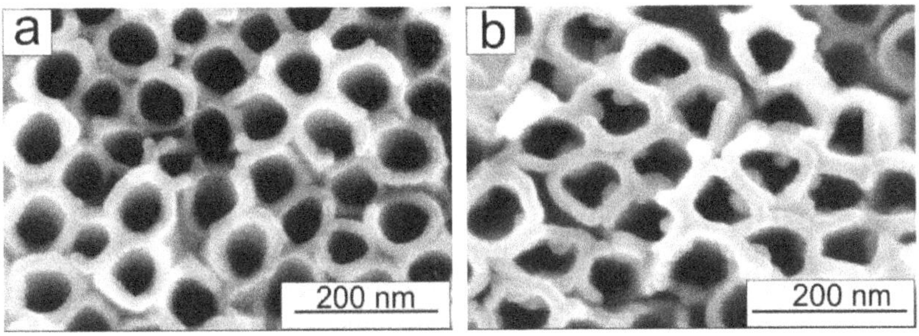

Figure 1. SEM images of bare (a) and gold-loaded (b) TiO_2 nanotubular layers.

Cyclic voltammograms for TiO_2 samples without Au NPs have one wave of ORR at potentials more negative than –0.6 V. For the TNT electrodes loaded with Au NPs, an additional wave, which can be related to oxygen electroreduction on the gold surface, appears at less negative potentials (Fig. 2a). The position of the wave corresponding to ORR on Au NPs shifts significantly (by *ca.* 0.25 V) to more negative potentials with increasing the annealing temperature of TiO_2 support from 350 to 550 °C.

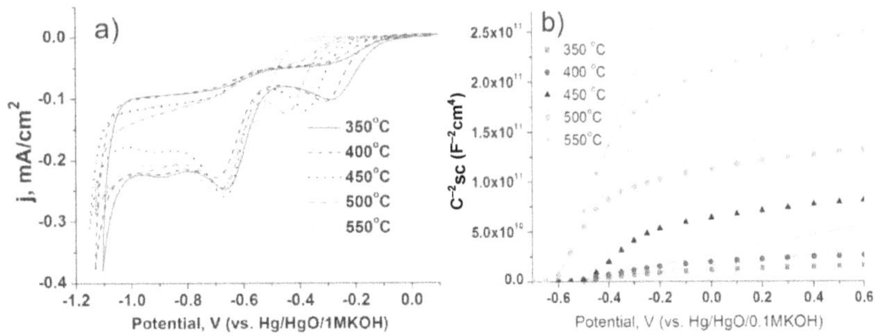

Figure 2. CV curves of ORR (a) and Mott-Schottky plots (b) for TNT-Au NPs electrodes annealed at different temperatures.

This strong difference in the overvoltage for ORR observed for Au NPs deposited on the TiO_2 crystalline nanotubular layers with the same nature but only annealed at different temperatures can be rationalized by peculiarities of electron transport through the metal/semiconductor interface. At the contact between Au nanoparticle and TiO_2 support, the Schottky barrier is formed locally as a result of the equilibration of the Fermi level in metal and semiconductor. Under cathodic polarization, electrons on the path from the TiO_2 bulk to Au NPs must come through this barrier to participate in ORR on the Au surface. The efficiency of the charge transfer *via* the metal/semiconductor interface depends mainly on the Schottky barrier width and can be realized *via* two paths: thermionic emission and electron tunneling. When the width becomes narrower, the probability of electron tunneling through the space charge region (SCR) increases sharply, providing an additional way for electron transport through the Schottky barrier.

Mott-Schottky plots were obtained for characterization of electronic properties of the TNT layers annealed at different temperatures (Fig. 2b). From the slope of the linear part (at negative potentials) of the C_{sc}^{-2} *vs.* E plots, the doping density and the Schottky barrier width can be determined. Increasing the annealing temperature leads to the growth of the curve slope. These data prove the widening of SCR in TiO_2 annealed at higher temperatures and changing the mechanism of electron transport through SCR from tunneling to thermionic emission. As a result, the ORR overvoltage of the Au-TNT electrodes increases with the increase of annealing temperature of the TNT support.

4. Conclusion

UV-assisted photocatalytic deposition was used for fabrication the Au NPs-titania nanotube systems. The overvoltage of O_2 electroreduction on TNT-supported Au NPs grows as the annealing temperature of the TNT matrix increases. The observed effect can be explained by peculiarities of the electron transport through the Schottky barrier at the Au NPs/TiO_2 interface.

References

1. W. Chen, S. Chen, *Angew. Chem. Int. Ed.* **48**, 4386 (2009).
2. M. Shao (ed.), *Electrocatalysis in Fuel Cells: A Non- and Low- Platinum Approach* (Springer-Verlag, London, 2013).
3. P. Roy, S. Berger, P. Schmuki, *Angew. Chem. Int. Ed.* **50**, 2904 (2011).
4. J. M. Macak, F. Schmidt-Stein, P. Schmuki, *Electrochem. Commun.* **9**, 1783 (2007).

DICHROIC POLARIZING FILMS BASED ON POLY(VINYL ALCOHOL)-TUNGSTOPHOSPHORIC ACID NANOCOMPOSITES

T. V. Shevchenko, N. I. Sushko, O. N. Tretinnikov

B. I. Stepanov Institute of Physics NASB, Nezavisimosti Ave. 68, 220072 Minsk, Belarus
o.tretinnikov@ifanbel.bas-net.by

We report on a new type of dichroic polarizing films based on the acid-catalyzed thermal dehydrated of poly(vinyl alcohol). The films are produced by stretching and subsequent annealing of poly(vinyl alcohol)-tungstophosphoric acid film nanocomposites. These polarizing films are superior in moist heat resistance to conventional poly(vinyl alcohol)-based films. They possess a high degree of optical polarization and transmittance and can be obtained with a thickness of the order of micrometers.

1. Introduction

The most common dichroic polarizing films are designed as stretched films of poly(vinyl alcohol) (PVA) containing dichroic iodine complexes. The PVA/iodine-type polarizing films are used in virtually all liquid crystal displays (LCDs) [1]. However, these films are not resistant to heat and water. This fact strongly limits their applicability in LCDs under harsh environmental conditions. There are also PVA/polyene-type dichroic polarizing films. Their visible dichroism occurs due to features of long-chain linear polyenes, $–(CH=CH)_n–$, formed in the acid-catalysed dehydration of PVA and chemically incorporated into the PVA backbone [2]. PVA-polyene polarizing films have the higher resistance to heat and water than the films based on PVA-iodine complexes. Also, they can be relatively thin. In spite of these advantages, PVA-polyene polarizing films have not found wide commercial application because their production necessarily involves a usage of large quantities of hazardous and highly corrosive mineral liquid acids to catalyze the dehydration of PVA.

In this paper, we present the approach to overcome deficiencies of traditional PVA-polyene polarizing films avoiding mineral liquid acids in the catalysis of PVA dehydration. We employ a solid catalyst, the Keggin-type heteropolyacid (HPA), 12-tungstophosphoric acid (TPA, $H_3PW_{12}O_{40}$). By casting aqueous solutions of PVA and TPA we obtain nanocomposite films consisting of the polymer matrix embedded with small (40–50 nm in diameter) spherical nanoparticles of the solid acid catalyst. Subsequent stretching and annealing of this nanocomposite film produce polarizing films of the PVA-polyene type.

2. Experimental

PVA (>99 % hydrolyzed, weight-average molecular weight of 145000 g/mol, Moviol 28-99) was purchased from Sigma-Aldrich (Germany). TPA hydrate (analytical grade) was supplied by Vekton (Russia). The materials were used as received without purification.

PVA films containing TPA nanoparticles were obtained from a mixed solution of PVA and TPA in water. For this, one part of TPA solution (4.0 wt.%) was slowly added to ten parts by weight of PVA solution (4.0 wt.%) under continuous stirring. PVA-TPA nanocomposite films with TPA concentration of 10 wt.% (0.15 mol%) relative to PVA were prepared by casting the solution onto plastic culture dishes followed by drying. The dry films were peeled off the substrates and stored under ambient conditions until used. For preparation of polarizing PVA-polyene-TPA film, a sample of PVA-TPA film was fixed on a mechanically-driven stretching device, equilibrated at relative humidity of 90 % and room temperature for 30–60 min, and uniaxially stretched at the extension rate of 50 mm/min to the draw ratio of 4. The stretched film fixed on the stretching device was air dried at 50 °C for 1 h and annealed in air at 140 °C for 5 to 20 min, depending on the film thickness.

The optical absorption of the films was recorded on a UV-Vis-NIR Cary 500 spectrophotometer (Varian) at a scan rate of 120 nm/min and a resolution of 2 nm (slit bandwidth). Polarized spectra were obtained with a dichroic sheet polarizer.

3. Results and discussion

PVA-TPA films fabricated were colourless, homogeneous and had the same high transparency as pure PVA films on visual observation. This implies that TPA is present in this polymer matrix in the form of nanosized particles. Indeed, as can be seen from TEM image of ultrathin PVA-TPA film (Fig. 1), TPA particles have spherical shape and size near 40 nm. Clustering of the particles is likely caused by the very low concentration of PVA-TPA solution used for the TEM sample preparation.

Upon uniaxial stretching and subsequent annealing, the nanocomposite PVA-TPA film acquires polarizing properties [3]. Polyenes that are formed during thermal treatment of the stretched film are predominantly oriented in the direction of stretching. Therefore, they transmit light polarized perpendicular to the stretching direction and strongly absorb light with the parallel polarization as demonstrated by the polarized absorbance spectra shown in Fig. 2. For the parallel polarization, the spectral contour is bell-shaped and contains absorption band maxima at 370, 392, 418, 443, 466, and 487 nm assigned to polyenes with conjugated bonds, $n = 6–11$. The shoulders at 503, 521, and 539 nm are indicative of polyenes with $n = 12$, 13, and 14, respectively [4]. Because of small size, low concentration (0.15 mol%) and low absorbance in the visible region, TPA nanoparticles do not have any adverse effect on the polarization efficiency.

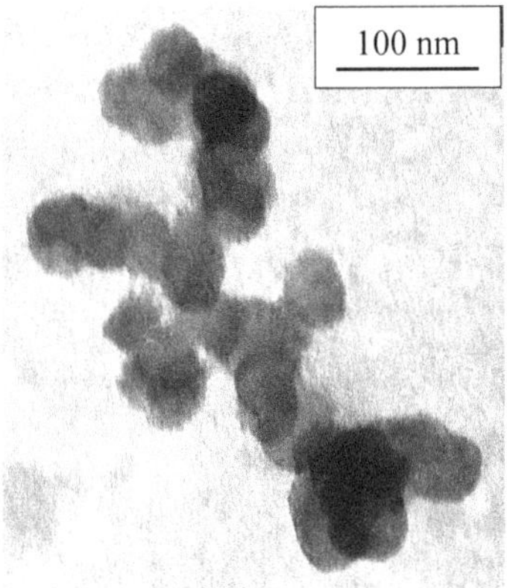

Figure 1. TEM image of ultrathin PVA-TPA film.

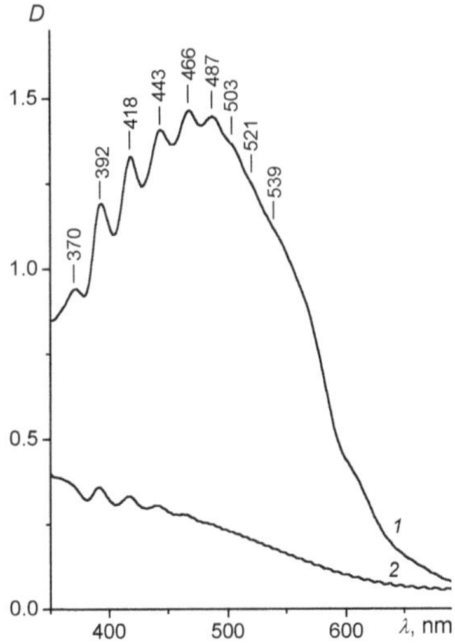

Figure 2. Optical absorption spectra of uniaxially stretched and subsequently annealed at 140 °C PVA-TPA nanocomposite film, measured for parallel (1) and perpendicular (2) light polarization relative to the stretching direction. The film is of 14.3 μm thick.

Unlike PVA-iodine polarizing films, PVA-polyene polarizing films based on the PVA-PTA nanocomposites can be readily obtained with the thickness less than 10 μm. As an example, Fig. 3a shows the polarized transmittance ($T_\perp$, $T_\parallel$) spectra of PVA-polyene-TPA polarizing film with the thickness of 8.8 μm. In the wavelength range of 380–570 nm, the film is not transparent to the light polarized parallel to the stretching direction while it transmits 40 to 70 % of the light with the orthogonal polarization. The degree of polarization of light passed through the film, designed as $P(\%) = 100\times(T_\perp - T_\parallel)/(T_\perp + T_\parallel)$, is as high as 95–98 % in the middle spectral range (420–510 nm) and decreases to 80 % at the edges.

The film described above was produced by stretching the PVA-TPA film four times its original length. Apparently, polarizing films with even better optical properties could be obtained by increasing the degree of film stretching and optimizing the stretching conditions. Fig. 3b shows the optical transmittance of the film as a function of its residence time in hot humid air (60 °C, 90 % RH). Its polarizing properties do not deteriorate even upon 50 h exposure to these harsh environmental conditions.

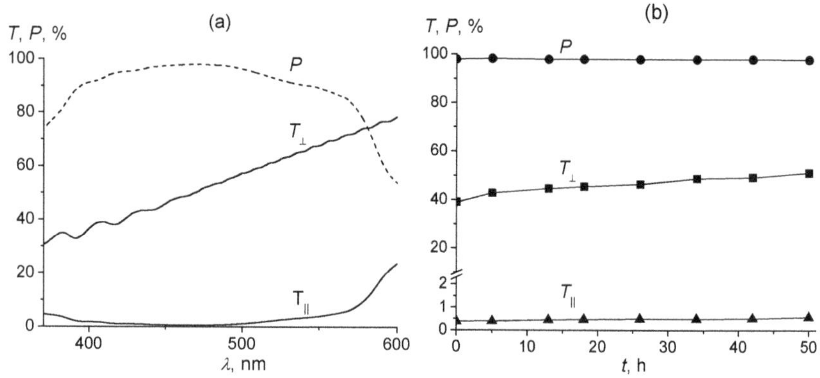

Figure 3. (a) Polarized transmittance ($T_\perp$, $T_\parallel$) spectra and the degree of polarization (P) of 8.8-μm-thick polarizing film obtained by stretching and subsequent annealing of PVA-TPA nanocomposite film. (b) Transmittance of this film at $\lambda = 466$ nm as a function of its residence time in hot humid air (60 °C, 90 % RH).

4. Conclusion

Dichroic polarizing films based on PVA-TPA nanocomposites were prepared and investigated. These nanocomposite films are superior in moist heat resistance to conventional PVA-based polarizing films. They have high transmittance and polarization degree and can be readily fabricated with the thickness less than 10 μm.

References

1. J. Ma, X. Ye, B. Jin, *Displays* **32**, 49 (2011).
2. J. W. Gilman, D. L. VanderHart, T. Kashiwagi, *ACS Symp. Ser.* **599**, 161 (1994).
3. O. N. Tretinnikov, N. I. Sushko, *Patent RU No.* 2520938 (2014).
4. K. Maruyama, H. Akahoshi, M. Kobayashi, Y. Tanizaki, *Bull. Chem. Soc. Jap.* **58**, 2923 (1985).

TITANIUM OXIDE AND SILICA-BASED DELIVERY SYSTEMS FOR NUCLEIC ACID FRAGMENTS

N. V. Shikina, E. V. Bessudnova, D. V. Korneev
SB RAS, Boreskov Institute of Catalysis, 630090 Novosibirsk, Russia
shikina@catalysis.ru

A. S. Levina, M. N. Repkova, V. F. Zarytova
SB RAS, Institute of Chemical Biology and Fundamental Medicine
630090 Novosibirsk, Russia
levina@niboch.nsc.ru

O. C. Efimova, Z. R. Ismagilov
SB RAS, Institute of Coal Chemistry and Materials Science
650000 Kemerovo, Russia
zinfer1@mail.ru

Titanium dioxide (TiO_2) nanoparticles in the anatase form and a hydrolytic product of (3-aminopropylntriethoxy)silane ($Si{\sim}NH_2$) were used to design nanocomposites bearing DNA fragments. The properties of the nanoparticles and nanocomposites were studied. It was shown that the nanoparticles are promising vehicles for the delivery of DNA fragments into cells.

1. Introduction

A rational approach to design efficient agents to treat viral, oncological, and genetic diseases is a use of nucleic acid fragments (NAF), which can interfere in the functioning of genetic material in cells, thus suppressing an expression of viral genes. However, generally poor penetration of NAF into cells necessitates designing special delivery systems to facilitate nucleic acid internalization. The use of nanoparticles is a new approach to deliver nucleic acid fragments into cells [1]. Due to high biocompatibility and ability to various functionalization, TiO_2 and SiO_2 nanoparticles are of great interest for the use in medicine [2,3]. Nanoparticles provide penetration of NAF into cells, and NAF of a special structure can interact with target nucleic acid inside the cells.

The goal of the work is the development of a method for fabrication of stable nanoparticles based on oxides of Ti or Si, which is adapted to the immobilization of NAF (oligonucleotides and their analogs), the comparative analysis of the properties of the prepared nanoparticles and nanocomposites, and the evaluation of the nanocomposites to penetrate into cells.

2. Experimental

TiO_2 nanoparticles were synthesized by hydrolysis of titanium tetraisopropoxide ($Ti(O\text{-}i\text{-}C_3H_7)_4$) in the presence of acidic catalyst, followed by the modification of external surface of the nanoparticles with glycidylisopropyl ether and the addition of LiOH up to neutral pH value. The sample of the nanoparticles is an opalescent sol.

Tetraethoxysilane is usually utilized to prepare SiO_2-based nanoparticles. We used aminosubstituted silane because the presence of the amino groups attached to silica enhances functional ability of the material for the immobilization of various molecules, particularly, nucleic acid fragments, due to strong electrostatic interaction between positively charged amino groups and negatively charged internucleotide phosphate groups in oligonucleotides. Hydrolysis of (3-aminopropyl)silane was carried out at pH 10 and 70 °C, followed by neutralization of 50 % of the amino groups by acetic acid. The resulting aminosubstituted silanol $Si(OH)_3(C_3H_6NH_2)$ ($Si\sim NH_2$) is a clear solution.

The TiO_2-PL•NAF nanocomposites were synthesized in two stages. The TiO_2-containing sol was mixed with an aqueous solution of polylysine, and the mixture was incubated in a solution of NaCl. The solution of a NAF at a certain concentration was added to polylysine-containing TiO_2 nanoparticles (TiO_2-PL), and the mixture was incubated in NaCl solution at room temperature for 20–30 min.

The synthesis of the $Si\sim NH_2$•NFA nanocomposites was performed by incubating solutions of $Si\sim NH_2$ with a certain amount of NFA in a solution of NaCl.

3. Results and discussion

X-ray and TEM analyses showed that the main phase of titanium dioxide nanoparticles was anatase. The TEM, SAXS, DLR, and AFM methods showed almost the same results. The predominant size of TiO_2 nanoparticles was 5–6 nm. Aminosubstituted silane appeared to be amorphous material, with the particle size of 0.8–1.2 nm according to the DLR and AFM data.

The analysis of TiO_2 nanoparticles and nanocomposites on their basis by the SAXS method showed the broadening of the bands of the samples containing oligonucleotides in the size distribution diagrams compared to the samples of unbound TiO_2 nanoparticles (Fig. 1). This indicates the formation of complexes between the nanoparticles and oligonucleotides. An important parameter, which characterizes the capacity of the nanoparticles for NAF is the P/Ti ratio (the calculated weight ratio is 1/60 for the case of 10 nM NAF/12.5 mM TiO_2). According to the EDX analysis, an average P/Ti ratio from seven samples was 1/62, which indicated the almost quantitative immobilization of NFA to the nanoparticles.

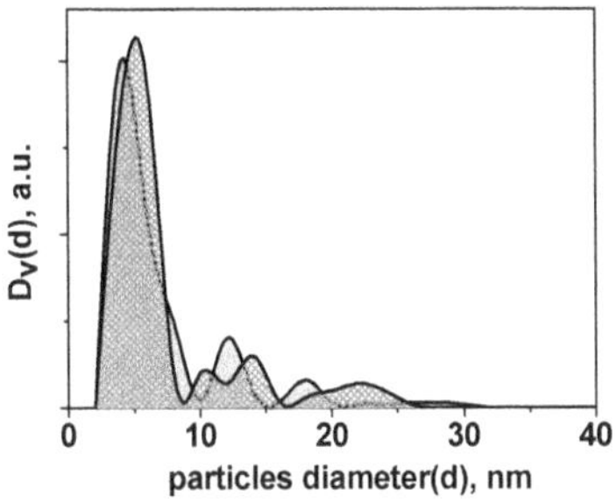

Figure 1. SAXS size distribution diagrams: solid color, TiO_2-nanoparticles; grid, TiO_2-PL•NFA nanocomposite.

The formation of Si~NH_2•NAF nanocomposite was demonstrated by the AFM data (Figs. 2). The height of the Si~NH_2 nanoparticles did not exceed 1.2 nm (Fig. 2a). The Si~NH_2•NAF nanocomposites were visualized in Fig. 2b. It should be noted that the objects in the silanol-containing systems can be only conditionally called nanoparticles because their size is on the border of the adopted gradation between true solutions and nanoparticles.

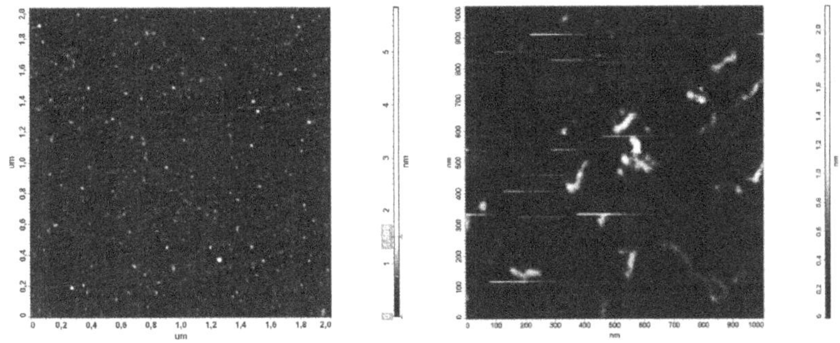

a b

Figure 2. AFM images of the Si~NH_2 (a) and Si~NH_2•FNA (b) samples.

The proposed nanocomposites were shown by confocal laser microscopy to penetrate into eukaryotic cells (HeLa, MDCK, and KST). For this purpose, we synthesized fluorescently labeled TiO_2-PL•NAF^{Flu} and Si~NH_2•NAF^{Flu} nanocomposites. The nanocomposites were observed in a partially agglometarted form on membranes and inside the cells (Fig. 3).

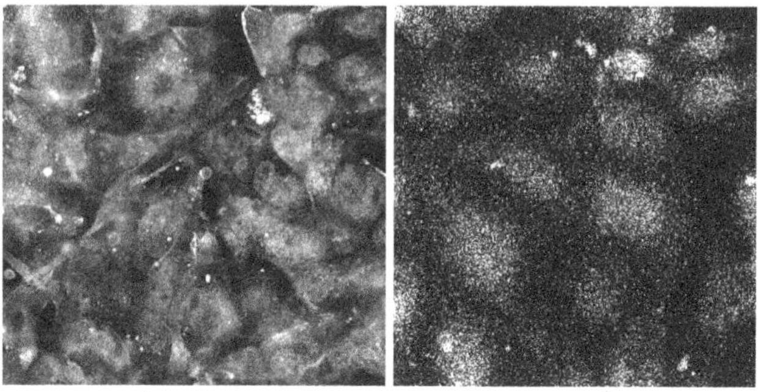

Figure 3. Images of MDCK cells after their incubation with TiO_2-PL•NAF^{Flu} (a) and Si~NH_2•NAF^{Flu} (b) nanocomposites (fluorescently labeled nanocomposites are observed as clusters of bright points).

4. Conclusion

We developed methods for synthesis of the TiO_2- and Si-based nanoparticles and the nanocomposites consisting of these nanoparticles and nucleic acid fragments. It was shown that the nanocomposites penetrate into eukaryotic cells without additional use of transfection agents and physical impacts.

Acknowledgments

The work was supported within the framework of budget project No. 0303-2016-0004 for Boreskov Institute of Catalysis.

References

1. T. Paunesku, S. Vogt, B. Lai, J. Maser, N. Stojicevic, K. T. Thurn, C. Osipo, H. Liu, D. Legnini, Z. Wang, C. Lee, G. E. Woloschak, *NanoLett.* **7**, 596 (2007).
2. Z. R. Ismagilov, L. T. Tsikoza, N. V. Shikina, V. F. Zarytova, V. V. Zinoviev, S. N. Zagrebelny, *Russian Chemical Reviews* **78**, 873 (2009).
3. L. Tang, J. Cheng, *Nano Today* **8**, 290 (2013).

FLUORESCENT SILVER NANOCLUSTERS IRREVERSIBLY BOUND TO PLASTIC SUBSTRATES

A. A. Gorbachev, T. M. Sheypak, P. P. Pershukevich, O. N. Tretinnikov
B. I. Stepanov Institute of Physics NASB, Nezavisimosti Ave. 68, Minsk 220072, Belarus
o.tretinnikov@ifanbel.bas-net.by

A method for the preparation of fluorescent Ag nanoclusters (NCs) irreversibly bound to the surface of a plastic substrate is described. It consists in surface functionalization of the plastic substrate with chemically end-grafted poly(acrylic acid) (PAA) followed by photo-activated template synthesis of Ag NCs in the grafted polymer. The surface-immobilized Ag NCs show bright visible fluorescence with intensity and wavelength dependent on the surface density of PAA template and on the time of NCs synthesis.

1. Introduction

Nanoparticles (NPs) of noble metals consisting of several to some tens of atoms exhibit properties different from those of the atoms and larger NPs. One such property is visible fluorescence [1]. These NPs have sizes below 2 nm and are conventionally referred to as nanoclusters (NCs). The fluorescent metal NCs have recently attracted considerable attention because of their high potential in biological labelling, imaging, and sensing [2]. Silver NCs have been particularly widely studied due to simple fabrication methods and strong fluorescence [3].

Ag nanoclusters, unlike Ag NPs, cannot be obtained simply by dissolving a silver salt in water and reducing Ag^+ ions, because Ag atoms and NCs have a strong tendency to aggregate into large NPs. Template synthesis with the use of certain water-soluble polymers containing ionizable functional groups has proven to be an efficient method to immobilize Ag^+ ions, restrict the size of growing species, and stabilize them in water. In particular, poly(acrylic acid) (PAA) homopolymers and block copolymers have been identified as ideal templates for synthesizing stable Ag NCs with bright fluorescence [4].

Most of the published papers are concerned with the synthesis and fluorescence of metal NCs dissolved in water [5]. On the other hand, the development of fluorescent metal NCs immobilized on solid supports, especially plastic ones, may open new possibilities in sensors, optical data storage and electroluminescent devices. Zhang *et al.* [6] fabricated fluorescent Ag NCs on glass substrates by photo-reduction of Ag^+ ions loaded in LbL films of poly(vinyl pyrrolidone) and PAA. Su *et al.* [7] obtained surface-immobilized BSA-protected Au NCs by complexing them with polyelectrolytes of opposite charge in water, followed by casting on a glass slide.

The surface immobilization of fluorescent metal NCs by the above methods is not irreversible because the NCs-loaded polymer matrices are linked to the surface by non-covalent (physical) bonds which can be disrupted by water and other solvents. Here we address this issue by the covalent attachment of the template polymer to the surface of plastic substrate by means of surface graft polymerization and subsequent *in-situ* synthesis of fluorescent Ag NCs in the grafted polymer layer. Polypropylene (PP) film was selected as the substrate, and PAA graft-polymerized on the surface of PP film served as the template.

2. Experimental

PP film of 40 μm thickness was purified by Soxhlet extraction with acetone. Acrylic acid (AA) (99 %, Aldrich), benzophenone (BP) (99 %, Sigma-Aldrich), and silver nitrate (analytical grade, Reakhim) were used as received.

The surface functionalization of PP films with PAA was achieved by UV-induced surface graft-polymerization. PP film was immersed in 5 % BP in acetone for 5 min, dried in air, and placed on a quartz plate. A droplet of 20 % AA aqueous solution was deposited on the film and covered with the second quartz plate. This assembly was fixed with clumps and exposed at 35–40 °C to UV light (365 nm, 70 mW/cm^2) to effect graft polymerization. The grafted PP film was washed in tap water and then in DD water for 8 h under stirring. The film was then washed in HCl solution of pH 5 for 5 min, rinsed with DI water and dried.

The PP film, with the PAA-grafted surface pointing up, was placed on the bottom of a plastic dish, fixed by a glass ring, and $AgNO_3$ solution (0.1 M) in DI water was added. The dish was covered with a lead, stored in the dark for 20 min, and exposed to a UV light (365 nm, 50 mW/cm^2). After irradiation, the film was washed in DI water for 1 min with constant shaking, rinsed with a new portion of water, and dried in air.

UV-Vis absorption and fluorescence spectra were recorded with a Cary 500 UV-Vis-NIR spectrophotometer and a SDL-2 spectrofluorimeter, respectively.

3. Results and discussion

Fig. 1 shows electronic absorption spectra of PAA-grafted PP films irradiated in $AgNO_3$ solution for 3, 8, and 12 min. Irradiation for 3 min results in strong absorption in the range of 200–300 nm that quickly rises with decreasing λ and reveals weak shoulders indicative of bands at about 250 and 225 nm (Fig. 1a). The absorption strengthens further without change in the shape of the spectra for

8 min of irradiation. However, when the exposure time increases to 12 min, the absorption weakens and the shoulders at 250 and 225 nm almost disappear.

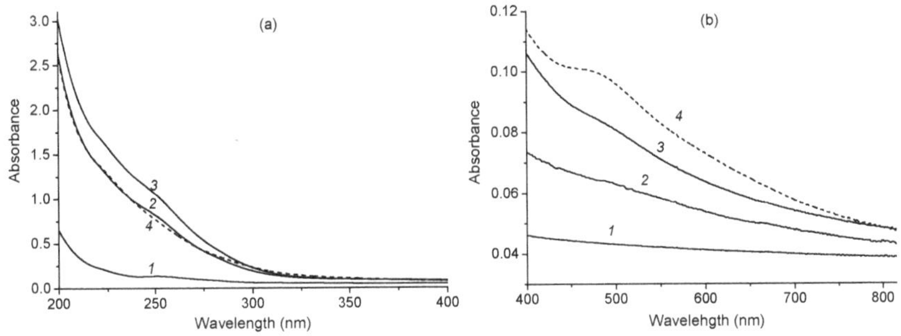

Figure 1. UV (a) and Vis electronic absorption spectra (b) of PAA-grafted PP films measured before (1) and after UV irradiation of the films in an $AgNO_3$ solution for 3 (2), 8 (3), and 12 min (4).

Strong UV absorption increasing with decreasing λ is characteristic of small Ag NCs (less than 10–20 Ag atoms) giving discrete bands and larger NCs producing continuous absorption. Therefore, the observed changes in the spectra suggest that both small and relatively large Ag NCs formed at the start of irradiation and that their concentrations increased as the irradiation time was increased to 8 min. Nucleation of new NCs slowed during longer irradiations and eventually stopped. At the same time, the already formed clusters continued to grow. As a result, small NCs grew to large sizes and large NCs converted into NPs of >2 nm, which do not absorb in the UV region but give a surface plasmon resonance band in the visible region. Indeed, a characteristic plasmon band of Ag NPs with a maximum at 470–480 nm is clearly visible in the spectrum of the film irradiated for 12 min (Fig. 1b).

The formation of Ag NCs on the surface of PAA-grafted PP films upon their exposure to UV radiation in an aqueous $AgNO_3$ solution was evident from the fact that the films become highly fluorescent. The orange emission was easily visible to the naked eye under excitation with $\lambda_{ex} = 365$ nm, even with 10 mW/cm^2 excitation power. Spectral data indicated that the fluorescence lies in the range of 600–900 nm and is observed over a broad range of λ_{ex} (300–600 nm). Fig. 2 shows the fluorescence spectra ($\lambda_{ex} = 485$ nm) of Ag NCs synthesized on the surface of PAA-grafted PP films by their UV irradiation in an aqueous $AgNO_3$ solution for 1–12 min. Irradiation for 1–8 min leads to a complex fluorescence band with a maximum at 720–740 nm and a shoulder at 625–650 nm. The complex character of the band should be explained by the presence of populations of Ag NCs of various sizes with the larger clusters emitting at longer wavelengths. On the increase of irradiation time from 1 to

8 min, the fluorescence increases in intensity equally over all wavelengths so that the positions of the maximum and shoulder stay nearly constant.

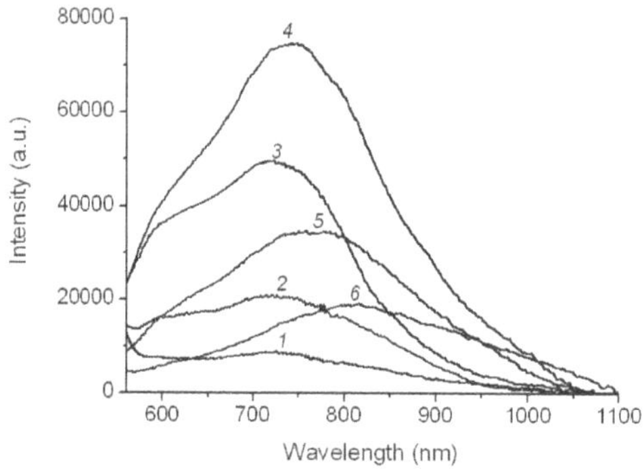

Figure 2. Fluorescence spectra (λ_{ex}= 485 nm) of Ag NCs synthesized on the surface of PAA-grafted PP films by their UV irradiation in an aqueous $AgNO_3$ solution for 1 (1), 3 (2), 5 (3), 8 (4), 10 (5), and 12 min.

This means that the number of Ag NCs continuously increases while their size distribution remains unaltered. The intensity of fluorescence weakens considerably, the maximum shifts to longer wavelength, and the short-wavelength shoulder disappears if the irradiation time is increased to 10 and 12 min. This indicates that the number of fluorescing Ag NCs decreases with a concurrent increase in their sizes. The reason for this behaviour was presented above in the analysis of electronic spectra. Under prolonged UV exposure, nucleation of new NCs slows down and eventually stops while already formed clusters continue to grow and transform into nonfluorescent NPs.

4. Conclusion

Highly fluorescent Ag NCs irreversibly bound to plastic substrates have been obtained for the first time by the covalent attachment of polymer template to the substrate surface by means of surface graft polymerization and subsequent template synthesis of the NCs in the surface-grafted polymer layer.

References

1. L. A. Peyser, A. E. Vinson *et al.*, *Science* **291**, 103 (2001).
2. L. Shang, S. Dong, G. U. Nienhaus, *Nano Today* **6**, 401 (2011).
3. H. Xu, K. S. Suslick, *Adv. Mater.* **22**, 1078 (2010).
4. Z. Shen, H. Duan, H. Frey, *Adv. Mater.* **19**, 349 (2007).
5. Z. Luo, K. Zheng, J. Xie, *Chem. Commun.* **50**, 5143 (2014).
6. W. Zhang, J. Song, W. Liao *et al.*, *J. Mater. Chem. C* **1**, 2036 (2013).
7. L. Su, T. Shu, Z. Wang *et al.*, *Biosens. Bioelectron.* **44**, 16 (2013).

NANOTECHNOLOGY

INVITED

CORRELATED EVOLUTION OF SURFACE MORPHOLOGY, STRUCTURE AND MAGNETIC PROPERTIES OF NANOPOROUS Co/Pd FILMS WITH PERPENDICULAR MAGNETIC ANISOTROPY

J. Fedotova, J. Kasiuk, V. Bayev

Institute for Nuclear Problems, Belarusian State University, 220030 Minsk, Belarus
Julia@hep.by

O. V. Kupreeva

Belarusian State University of Informatics and Radioelectronics, 220013 Minsk, Belarus

T. N. Anh Nguyen, H. M. Do, D. L. Vu

Institute of Materials Science, Vietnam Academy of Science and Technology
Hanoi, Vietnam

In this study, we consider the technological route to conserve perpendicular magnetic anisotropy in porous Co/Pd multilayer thin films deposited over porous anodized templates of TiO_2 and Al_2O_3 with smoothed surface morphology. It is proved that these templates with large contribution on flat-surface interpore areas promote conservation of high magnetic anisotropy constants K_{eff} up to 1.9×10^6 erg/cm^3) and M_r/M_S (up to 0.99) as well as magnetic hardening with twice increase of the coercive force H_C (up to 2.7 kOe) indicating an excellent potential of the fabricated materials for various magnetic and spintronic devices.

1. Introduction

Heterogeneous and patterned nanostructured films with a perpendicular magnetic anisotropy are considered as promising materials for high density recording media [1,2], systems for visualization of magnetic fields distribution and magnetic tunneling junction sensors [3]. Among the most convenient methods to produce magnetically oriented planar structures there are template-assisted synthesis of magnetic nanoarrays with a shape anisotropy [4–6] and formation of Pt, Pd-3*d* metal compounds with L_{10} structure that exhibits high magnetocrystalline anisotropy [7].

Nowadays the attempts to fabricate such media for data storage with density as high as 1 Tbit/in^2, and the bit size around 20 nm^2 face strong limitation because of superparamagnetic (SP) effect which is undesirable thermal fluctuations of magnetic moments when grain size falls down to a few nm [2]. The formation of percolated perpendicular magnetic media containing exchange-coupled magnetic films with densely distributed artificially engineered pinning centers (antidots) [5] enables to overcome this restriction.

In this context, fabrication of antidot arrays in [3*d* (Co, Fe) – 5*d* (Pt, Pd)]n films by deposition onto porous substrates like anodized Al_2O_3, TiO_2 or ZrO_2 can be an effective method to tailor the magnetization reversal mechanism with simultaneous stabilization of the orientation of magnetization at room temperature due to the strong domain-wall pinning [5]. Magnetic anisotropy of such nanoporous films is very sensitive to topographic features of templates used as a substrate, especially at length scales close to domain-wall width and exchange length. Complex morphology of nanoporous films often results in the deterioration of intrinsic perpendicular anisotropy due to the local misalignment of easy magnetization axis at the proximity of nanopore rims. This becomes well apparent in the decrease of the ratio between remanent magnetization M_r and saturation magnetization M_S in porous films. At the same time, an application of anodized porous templates with flattened interpore areas could promote the conservation of perpendicular anisotropy in the films despite of their porous structure. For that purpose, we applied two types of anodized templates – thick-walled nanotubes of TiO_2 and flat-surface Al_2O_3 anodized and ion-plasma polished in specifically selected regimes in order to increase the contribution of smoothed areas between homogeneously distributed nanopores. Co/Pd multilayers (MLs) deposited over such flat-surface templates should combine the advantage of exchange-couple films possessing perpendicular magnetic anisotropy with magnetic hardening due to the pinning effects.

Present paper summarizes our latest results related to the correlation between morphological features, structure and perpendicular magnetic anisotropy observed in Co/Pd nanoporous multilayered (ML) films fabricated on Al_2O_3 and TiO_2 templates with smoothed surface morphology.

2. Experimental

The fabrication of nanoporous films is a two-step process consisting of the preparation of porous Al_2O_3 and TiO_2 templates and subsequent deposition of Co/Pd and Co/Pt MLs [8,9]. Multilayers of Co/Pd were deposited on the surface of porous TiO_2 and Si/SiO_2 wafers by two techniques, namely, thermal evaporation in ultra-high vacuum with pressure of 7.5×10^{-10} Torr and by *dc* magnetron sputtering in the AJA system with a base pressure lower than 3×10^{-8} Torr. Simultaneous deposition on Si/SiO_2 wafers and templates of porous TiO_2 permits to compare the effect of porous TiO_2 morphology on magnetic properties of the films with respect to flat multilayers.

The surface morphology of TiO_2 and Al_2O_3 templates as well as the deposited MLs was checked by scanning electron microscope (SEM) HITACHI S-4800. X-ray diffraction (XRD) of MLs were performed with two-circle laboratory diffractometer (Panalytical X'Pert Pro) equipped with PW3050/60 vertical goniometer and PW3383/00 X-ray tube (2.2 kW, Cu = 1.54056 Å). Magnetization (M) of Co/Pd MLs was measured at room temperature in the magnetic fields H up to 20 kOe using MPMS-XL SQUID (Quantum Design).

3. Results and discussion

3.1. *Co/Pd porous MLs fabricated on nanotubular thick-walled TiO_2 templates*

SEM images of Co/Pd MLs deposited onto TiO_2 templates and the cross-sectional image of the template are shown in Fig. 1. Images illustrate that TiO_2 possesses tubular structure with quite homogeneous distribution of nanopores over its surface. It is proved also that Ar plasma etching of the template's surface led to the efficient smoothing of surface relief as compared to the initial relief observed immediately after anodization.

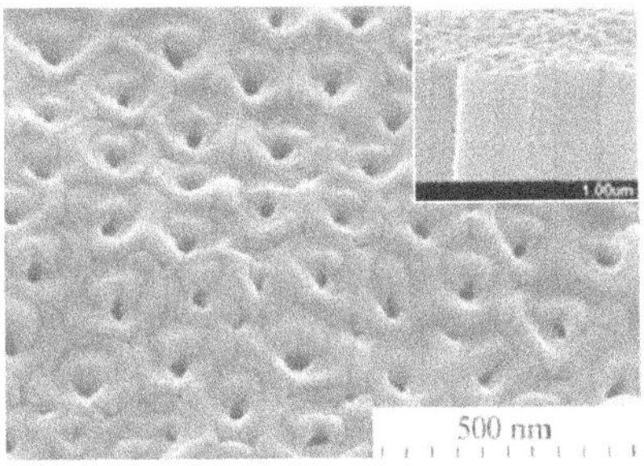

Figure 1. SEM image of Pd/[$Co_{0.3\,nm}$/$Pd_{0.55\,nm}$]/Pd film deposited on porous TiO_2 taken at the angle of 45 deg with respect to the sample surface. The inset illustrates nanotubular structure of the initial TiO_2 template.

Fig. 1 proves that the surface of Co/Pd MLs possesses nanoporous structure inherited from the initial template with medium pores diameters $d_E \approx 30$ nm and interpore distance ≈ 153 nm. Deconvolution of XRD patterns with pseudo-Voigt functions reveals a combination of peaks at $2\theta \approx 40$ deg and 41 deg associated with (111) Pd buffer/cap layer with the lattice parameter $a \approx 3.9$ Å and (111) peak of $Co_{100-x}Pd_x$ quasi-alloy with $a = 3.8$ Å [10].

Fig. 2 shows magnetization loops $M(H)$ recorded for continuous and porous Pd/[$Co_{0.3\,nm}$/$Pd_{0.55\,nm}$]/Pd films. The continuous film shows high $H_C = 2.2$ kOe in perpendicular direction ($H_\perp$) and $M_r/M_S \approx 0.96$, evidencing its prominent perpendicular magnetic anisotropy. However, the ($H_{||}$) loop for the film shows S-shape hysteretic behavior which is the signature of in-plane anisotropy contribution typically originating from the shape anisotropy of the film. As demonstrated in Fig. 2b, the perpendicular magnetic anisotropy is conserved in the nanoporous film, but with somehow reduced value of M_r/M_S=0.71 and enhanced coercive field H_C = 2.7 kOe for $H_\perp$ loop associated with the pinning of magnetic moments on nanopore edges [1,4]. It is noteworthy that porous MLs of similar composition deposited onto Al_2O_3 templates with U-shaped surface relief demonstrate drastic fall of M_r/M_S to 0.5 just because of an advanced morphology [10].

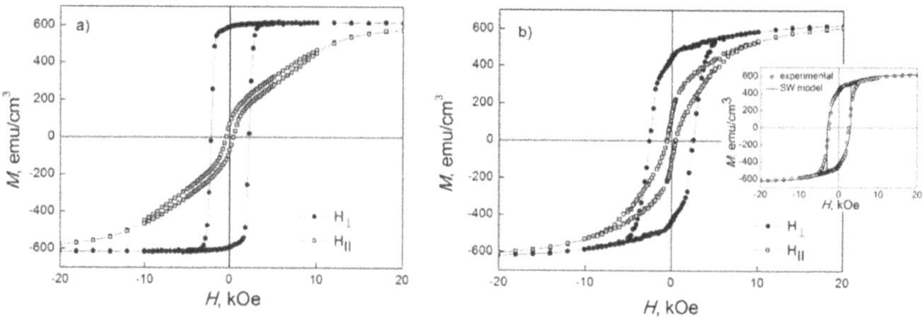

Figure 2. Magnetization loops $M(H)$ measured in perpendicular ($H_\perp$) and parallel ($H_{||}$) geometries for the continuous (a) and porous (b) Pd/[$Co_{0.3\ nm}$/$Pd_{0.55\ nm}$]/Pd films. The inset shows approximation of $M(H_\perp)$ curve with the Stoner-Wohlfarth model.

In the case of a continuous film with well-defined hard and easy magnetization axis, the effective anisotropy constant K_{eff} could be calculated as $K_{eff} = H_{eff} \cdot M_S/2$. The effective perpendicular anisotropy field H_{eff} could be estimated directly from $M(H_{//})$ loop as the field H required for saturation of magnetization along hard magnetization axis. However, Fig. 2 clearly reveals that $M(H_{//})$ curves for the studied continuous and porous Co/Pd MLs are not ideally linear typical for magnetic orientation along hard magnetic axis thus drawing to some uncertainty in the estimation of H_{eff} from $M(H_{//})$ curve. For that reason effective constant in both films should be more accurately characterized as the difference of areas under $M(H_\perp)$ and $M(H_{//})$ loops. Calculated K_{eff} and H_{eff} are summarized in the Table 1.

It should be noted that effective anisotropy field $H_{eff} = H_a - H_d$, where H_d is the magnetostatic shape anisotropy (demagnetizing field), H_a is the magnetocrystalline anisotropy. For continuous infinite thin films $H_d = -4\pi M_S^{Vol}$, where M_S^{Vol} is the bulk magnetization of the film. Evaluated H_d and H_a (for continuous film only) equal to 7.7 kOe and 18.1 kOe, correspondingly. After that the magnetocrystalline anisotropy constant K_U for the continuous film was calculated as $K_U = M_S^{Vol}(H_{eff} + 4\pi M_S^{Vol})/2$, accounting that $H_a = H_{eff} + H_d$. Obtained $K_U \approx 5.6\times10^6$ erg/cm^3 is consistent with the previously reported one for continuous Co/Pd ML films [6,7]. It should be emphasized that H_d for porous film cannot be calculated reliably basing on available experimental data thus making impossible determination of K_U.

Table 1. Effective anisotropy field H_{eff}, effective anisotropy constant K_{eff} and uniaxial (magnetocrystalline) anisotropy constant K_U for continuous and porous Co/Pd films.

Film	*Area difference method*		*Stoner-Wohlfarth model*		K_U, $\times 10^{-6}$, erg/cm^3
	H_{eff}, kOe	K_{eff}, $\times 10^{-6}$, erg/cm^3	H_{eff}, kOe	K_{eff}, $\times 10^{-6}$, erg/cm^3	
Continuous	10.4	3.2	—	—	5.6
Porous	4.8	1.5	6.0	1.9	—

In order to clarify possible mechanisms of the magnetization reversal in the continuous and porous films, the dependencies of the coercive field H_C on the angle θ between the direction of easy magnetization and the applied field H are presented in Fig. 3.

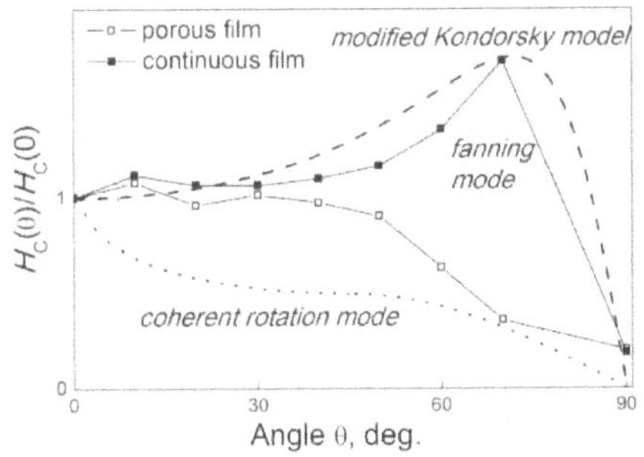

Figure 3. Coercive fields H_C as a function of the angle θ between the applied field H and the easy magnetization axis for porous and continuous films. Theoretical $H_C(\theta)$ dependencies were calculated according to the modified Kondorsky model [20] (dashed curve) and the rotational mechanisms [21] (dotted curve).

For the continuous film we observed the maximum for $H_C(\theta)$ at $\theta \approx 70$–75 deg. Such shape is typical for the magnetization reversal considered in the modified Kondorsky model, which combines the magnetization reversal by domain wall propagation with the rotational mechanism [11]. The coercive field H_C for the porous film is constant in the angle θ range of 0–50 deg, and then it slowly decreases. Such dependence is similar to that for the magnetization reversal by the incoherent/coherent rotation of magnetic moments in the nanostructures [11]. Alternatively, experimentally obtained curves could be interpreted as a combination of a domain-wall motion and the rotational mechanisms describing the magnetization reversal in the interpore areas of the film and areas around the nanopore edges, respectively.

Basing on the analysis of $H_C(\theta)$ dependencies, magnetization curve $M(H_\perp)$ for porous Co/Pd MLs was approximated in terms of the Stoner-Wohlfarth model for coherent rotation of magnetic moments (see the inset in Fig. 2s) [11]. The theoretical curve $\cos\varphi(h)$ is a solution of the equation

$$h \cdot \sin\varphi + \frac{1}{2} \cdot \sin 2(\varphi - \theta) = 0, \qquad (1)$$

where $h = H/H_{eff}$, φ is the angle between the magnetic moment and external magnetic field H. The approximation of the experimental curve $M(H_{\perp})$ by Eq. (1) makes it possible to estimate parameter $\theta = \arccos(M_r/M_S)$, which determines the orientation of the easy-magnetization axis of the films with respect to their normals and H_{eff}. The effective magnetic anisotropy of the films are given in Table 1. The field H_{eff} obtained by the approximation is ~6 kOe, which corresponds to the effective magnetic anisotropy constant $H_{eff} \times M_S/2 = 1.9 \times 10^6$ erg/cm^3 (Table 1). This value correlates with those obtained for a similar multilayer system [6,7].

3.2. *CoPd porous MLs fabricated on flat-surface Al_2O_3 templates*

SEM image of $Ta_{5\,nm}/Pd_{15\,nm}/[Co_{0.5\,nm}/Pd_{1\,nm}]_{x5}/Pd_{3\,nm}/Ta_{5\,nm}$ film deposited on porous Al_2O_3 is shown in Fig. 4a. This image clearly illustrates nanoporous morphology of Co/Pd MLs where nanopores with the average diameter D of about 20 nm are homogeneously distributed over the perfectly smoothed surface of anodized Al_2O_3 (shown in the inset). Such morphology is very different from typical U-shaped relief of Al_2O_3 templates usually applied for fabrication of magnetic nanoperforated MLs [10].

Experimental XRD patterns of porous Co/Pd film and correspondent continuous (reference sample) Co/Pd one presented in Fig. 4b demonstrate two clear intense diffraction lines at ≈ 40 deg and ≈ 41 deg corresponding well to the position of the main (111) peaks of *fcc* structure of Pd ($2\theta \sim 40$ deg) mainly from the buffer layer and of Co_xPd_{1-x} quasi-alloy ($2\theta \sim 40.7$–41.1 deg) [10]. Two additional reflections at $2\theta = 40.3$ deg and 41.1 deg indicate possible formation of Co_xPd_{1-x} quasi-alloy with a variable composition and, consequently, with a variable lattice parameter a.

Fig. 5 shows experimental field dependences of the normalized magnetization $M/M_S(H)$ for nanoporous and continuous Co/Pd films measured in perpendicular geometry. Magnetization curves of continuous film demonstrate high values of $M_r/M_S = 0.96$. This observation together with high coercive field $H_C = 1.1$ kOe, indicates pronounced PMA in the films.

Remarkably, Co/Pd porous MLs also possess perfect $M_r/M_S = 0.99$, evidencing conservation of PMA despite of the engraved nanoporous structure of their surface. This value is much better that in the case of Co/Pd MLs deposited onto smoothed-surface TiO_2 templates considered above. Additionally, the nanoporous film exhibits H_C increased up to ~1.9 kOe as

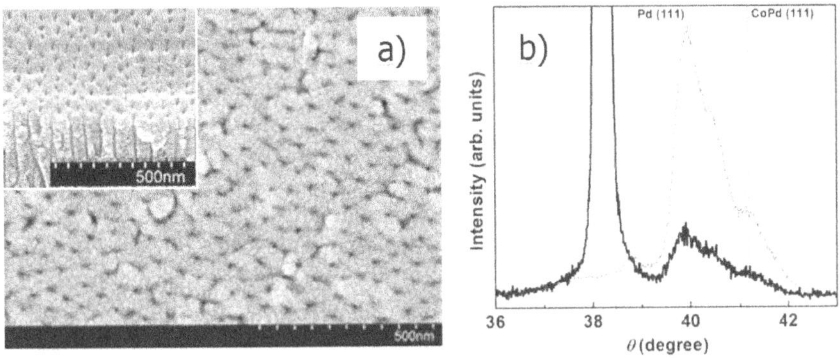

Figure 4. SEM image (a) of $Ta_{5\ nm}/Pd_{15\ nm}/[Co_{0.5\ nm}/Pd_{1\ nm}]_{x5}/Pd_{3\ nm}/Ta_{5\ nm}$ film deposited on porous Al_2O_3. The inset illustrates the morphology of initial Al_2O_3 template. XRD patterns of $Ta_{5\ nm}/Pd_{15\ nm}/[Co_{0.5\ nm}/Pd_{1\ nm}]_{x5}/Pd_{3\ nm}/Ta_{5\ nm}$ on porous Al_2O_3 template (black line) and flat Si/SiO_2 wafer (dashed line) (b).

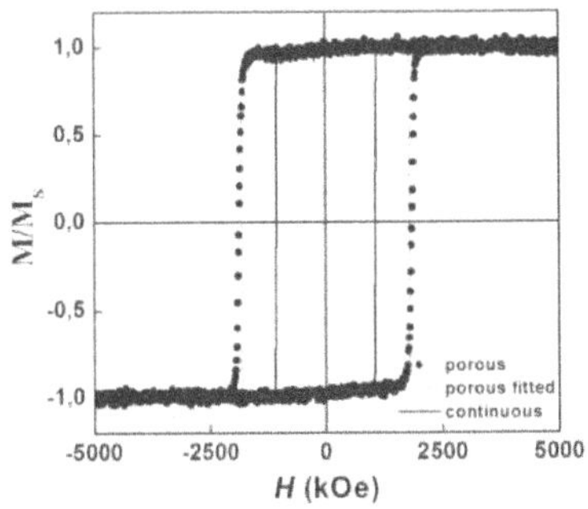

Figure 5. Normalized magnetization dependences of $Ta_{5\ nm}/Pd_{15\ nm}/[Co_{0.5\ nm}/Pd_{1\ nm}]_{x5}/Pd_{3\ nm}/Ta_{5\ nm}$ MLs on nanoporous Al_2O_3 template (black dots) and flat Si/SiO_2 wafer (black line). The dotted line presents the approximation of experimental $M/M_S(H)$ curves for the porous films in accordance with the Stoner-Wohlfarth model [33].

compared to the reference film. Such a 2 times increase of H_C is supposed to be governed by *i*) hindering of the domain wall motion and *ii*) change of the magnetization reversal process, both induced by the films porosity. It is generally accepted that the increase of H_C in nanoporous films is associated with pinning of the magnetic moments at the pores edges during the magnetization reversal [1–5]. At the same time, the change of the magnetization reversal mechanism is often observed in the films characterized with nanosized pores and small distances between them [2,3]. In accordance with the detailed analysis of $H_C(\theta)$ curves recorded on porous MLs fabricated by deposition onto TiO_2 templates, we presume that rotational mechanisms are dominating also for Co/Pd MLs deposited onto flat-surface Al_2O_3 template and applied the Stoner-Wohlfarth model for simulation of correspondent $M(H_\perp)$ loop. The results of the approximation are presented in Fig. 3. The calculations lead to $H_a = 2.6$ kOe for Co/Pd porous MLs. The average angle θ characterizing deviation of easy magnetization axis of the porous films from their normal appears to be 8 deg, according to the approximation results. The origin of such deviations is in the local morphology inhomogeneities of the porous films related to pores edges. Thus, a quite low value of θ extracted from the modeled curves confirms a perfect conservation of high PMA effect in the studied porous MLs.

4. Conclusion

Templates of thick-walled TiO_2 nanotubes and flat-surface porous Al_2O_3 were successfully used for fabrication of porous Co/Pd MLs characterized with a well-conserved perpendicular magnetic anisotropy. Their SEM and XRD characterization confirms smoothed relief and flattened interpore areas both in porous Co/Pd//TiO_2 and Co/Pd//Al_2O_3 MLs containing predominantly Co_xPd_{100-x} quasi-alloy. Smoothed surface morphology of porous MLs favors perfect conservation of perpendicular magnetic anisotropy that is reflected in extremely high M_r/M_S (up to 0.99), high H_a (up to 6.0 kOe) and small deviation angles θ (up to 8 deg) of easy magnetization axis from normal to the film surface when magnetization curve is recorded in the perpendicular geometry. Values of K_{eff} estimated by $M(H_\perp)$ approximation within the Stoner-Wohlfarth model reached up to 1.9×10^6 erg/cm^3.

The authors acknowledge the support from the Vietnam Academy of Science and Technology (project VAST.HTQT.BELARUS.03/16-17), the National Foundation for Science and Technology Development of Vietnam (project 103.99-2015.83), from the State Research Program “Physical materials science, new materials and technologies” (task 2.44) and the Belarusian Republic Foundation for Fundamental Research (project F16V2-004).

References

1. M. Grobis *et al.*, *Appl. Phys. Lett* **98**, 192504 (2011).
2. M. T. Rahman, N. Shams, C. H. Lai, *Nanotechnology* **19**, 325302 (2008).
3. M. T. Rahman *et al.*, *Phys. Rev. B* **81**, 014418 (2010).
4. V. Neu *et al.*, *Nanotechnology* **24**, 145702 (2013).
5. C. Schulze *et al.*, *Nanotechnology.* **21**, 495701 (2010).
6. S. K. Kim, S. C. Shindoi, *J. Appl. Phys.* **89**, 3055 (2001).
7. O. Hellwig *et al.*, *Appl. Phys. Lett.* **95**, 232505 (2009).
8. S. K. Lazarouk *et al.*, *Thin Solid Films* **526**, 41 (2012).
9. S. K. Lazarouk *et al.*, *J. Appl. Phys.* **107**, 033527 (2010).
10. A. Maximenko *et al.*, *J. Magn. Magn. Mater.* **400**, 200 (2016).
11. J. V. Kasiuk *et al.*, *Phys. Solid State* **58**, 2312 (2016).

NANOSTRUCTURE ENGINEERING TO TAILOR InAs/InAsSb SUPERLATTICE FOR MIDWAVE INFRARED BROADBAND DETECTION

J. P. Perez, Q. Durlin, P. Christol

University Montpellier, IES, UMR 5214, F- 34000, Montpellier, France

CNRS, IES, UMR 5214, F- 34000, Montpellier, France

We studied the Ga-free InAs/InAsSb type-II superlattice (T2SL) period, thickness and antimony composition, in order to define an optimized structure suitable for detection of the full mid-wavelength infrared domain (MWIR). The superlattice (SL) structures were fabricated by MBE on n-type GaSb substrate and exhibited cut-off wavelength between 5 μm and 5.5 μm at 150 K. Growth procedure to achieve strain-balanced structures are reported and first structural and optical results are presented and analyzed.

1. Introduction

Recently, new Sb-based technologies such as diffusion-limited InAsSb barrier [1] photodetector has reached impressively low dark current allowing temperature operation as high as 150 K [2] for a cutoff wavelength around 4.2 μm. Nevertheless, only the blue-MWIR domain is addressed in that case.

According to the Planck's law and considering a blackbody at 300 K, without any IR system or transparency windows considerations, the power emitted per unit area at the surface of the blackbody in the blue-MWIR range 3–4.2 μm represents only 18 % of the total power in the range of 3–5 μm. Consequently, there is an obvious need to extend the operational wavelength of the InAsSb barrier photodetector.

An alternative to the widely investigated type-II InAs/GaSb superlattices could be the Ga-free InAs/InAsSb (T2SL) highlighting the carrier lifetime as long as 10 μs at 80 K in the MWIR [3].

In this paper, strain balanced InAs/InAsSb T2SL grown by molecular beam epitaxy (MBE) is studied. Choices of the superlattice period and antimony composition of the InAsSb ternary alloy in order to obtain high absorption in the full MWIR domain are first presented. Next, MBE growth conditions to achieve strain-balanced InAs/InAsSb T2SL structure on GaSb substrate are detailed. Finally, photoluminescence characterizations performed on such Ga-free T2SL are reported and analyzed.

2. Choice of the Ga-free InAs/InAsSb superlattice period

$InAs/InAs_{1-x}Sb_x$ T2SLs can be strained balanced on GaSb by setting the average lattice parameter of one period of the SL equal to the lattice parameter of GaSb. It follows that InAsSb and InAs layer thickness (t_{InAsSb} and t_{InAs}) as a function of the antimony composition (x_{Sb}) and SL period (P) can be calculated by using:

$$t_{InAsSb} = \left(\frac{a_{GaSb} - a_{InAs}}{a_{InSb} - a_{InAs}}\right)\left(\frac{P}{x_{Sb}}\right) = 0.09\left(\frac{P}{x_{Sb}}\right), \qquad (1)$$
$$t_{InAs} + t_{InAsSb} = P,$$

where $a_{GaSb} = 6.0954$ Å; $a_{InAs} = 6.0584$ Å and $a_{InSb} = 6.4794$ Å are the lattice parameters of the binary compounds.

Assuming a type II-b InAs/InAsSb band offset [4] with electrons confined in the InAs layer and holes confined in the InAsSb one, the quantized miniband energies of the strain balanced $InAs/InAs_{1-x}Sb_x$ T2SL were calculated with next**nano**³ commercial software [4]. At $T = 150$ K, for x_{Sb} composition varying from 0.25 to 0.4 and SL period P varying from 4 nm to 8 nm, the cut-off wavelength corresponding to ground heavy hole VH1 to conduction C1 interminiband energetic transition is plotted in Fig. 1. In addition, $|\langle\Phi_{C1}|\Phi_{VH1}\rangle|^2$ wavefunction overlap values calculated for each fundamental VH1-C1 transition are specified.

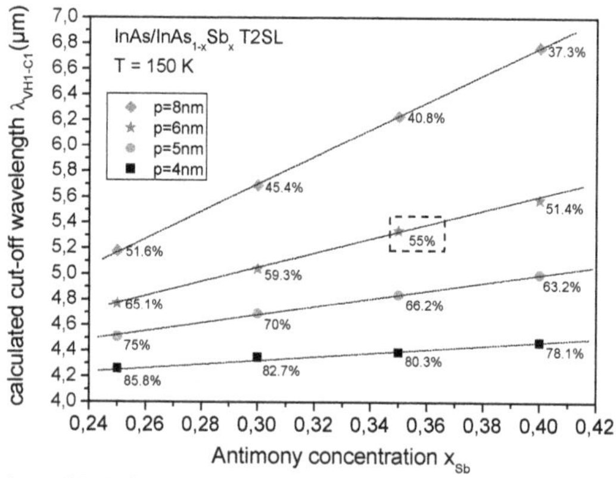

Figure 1. Strain balanced $InAs/InAs_{1-x}Sb_x$ T2SL on GaSb substrate. Calculated cut-off wavelength at $T = 150$ K and associated wavefunction overlap of the ground VH1-C1 miniband transition as a function of Sb concentration (x_{Sb}) and for different period thicknesses (P).

In order to reach the Red-MWIR spectral band, $InAs/InAs_{1-x}Sb_x$ SL structure with $x_{Sb} = 0.35$ and SL $P = 6$ nm could be of interest since it exhibits cut-off wavelength higher than 5 μm at 150 K for a wavefunction overlap higher than 50 % (Fig. 1).

3. MBE growth of Ga-free InAs/InAsSb superlattice structures

InAs/InAsSb strain balanced superlattices lattice matched to GaSb have been grown on n-type GaSb:Te (100) substrates by solid source MBE equipped with valved crackers set up to produce As_2 and Sb_2 species. Following the thermal oxide desorption at 530 °C, a 220 nm-thick GaSb buffer layer was grown at 500 °C. The substrate was then cooled to 435 °C for the growth of the structure occurring at the indium speed of the growth $v_{In} = 0.3$ ML/s. Except the buffer, the samples consist of a 500 nm to 3 μm thick superlattice region composed of alternating InAs/InAsSb layers sandwiched between two 20 nm thick AlSb barriers and finally capped by a 10 nm thick GaSb layer. Due to the competition for incorporation between arsenic (As) and antimony (Sb) species in the InAsSb layer, several test samples have been grown to determine the couple speed growth (v_{As} and v_{Sb}) leading to a lattice-matched structure. The corresponding X-ray diffraction spectrum shown in Fig. 2a exhibits many intense satellite peaks with a full-width at half-maximum (FWHM) of the first-order peak equal to 44 arcsec, attesting the good crystalline quality of the layers.

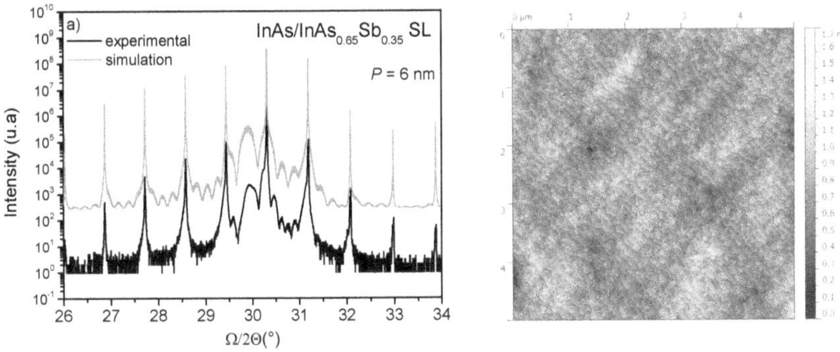

Figure 2. Left: High-resolution Ω/2Θ X-ray diffraction patterns for the (004) reflection of InAs/$InAs_{0.65}Sb_{0.35}$ T2SL sample (P = 6 nm) lattice-matched to the GaSb substrate ($v_{In} = 0.3$ ML/s, $v_{As} = 0.34$ ML/s, $v_{Sb} = 0.27$ ML/s). Right: AFM scan of InAs/In $As_{0.65}Sb_{0.35}$ T2SL sample. Clear monolayer steps can be observed.

Moreover, atomic force microscopy (AFM) on a 5×5 μm^2 scan area highlights atomic steps and measured root-mean-square (RMS) surface roughness only equals to 0.15 nm (Fig. 2b).

4. Photoluminescence of the Ga-free InAs/InAsSb superlattices

The samples are placed in a cryostat allowing accurate control of the temperature from 10 K to 300 K and are optically excited at 50 W/cm^2 with a

784 nm laser diode modulated at 133 kHz through a CaF_2 window. The luminescence signal is analyzed with a Nexus 870 FT-IR system equipped with a MCT detector (12 μm cut-off wavelength). Fig. 3 shows the normalized PL spectra at 80 K of two $InAs/InAs_{0.65}Sb_{0.35}$ T2SL structures having periods equal to 8 nm and 6 nm and compared to the PL spectrum of bulk $InAs_{0.91}Sb_{0.09}$ lattice matched to GaSb. These measurements are in agreement with the calculated fundamental valence to conduction interminiband transition (Fig. 1) and confirm the choice made on the P = 6 nm T2SL since its PL peak is observed at the wavelength equal to 5 μm at 80 K while the P = 8 nm T2SL sample exhibits a PL peak at a wavelength higher than 6 μm.

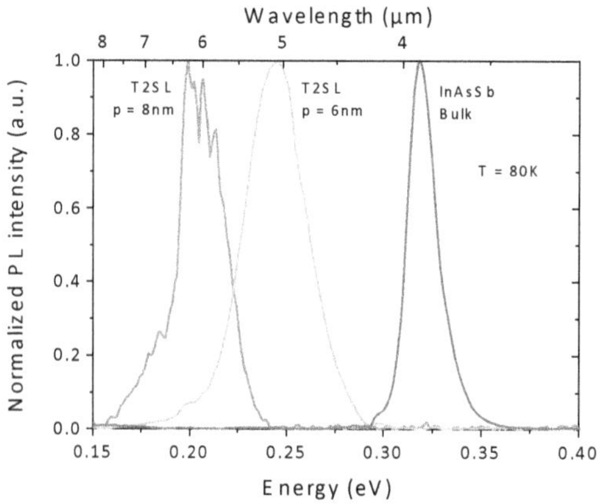

Figure 3. PL spectra at T = 80 K for T2SL samples (with P = 6 nm and 8 nm) compared with $InAs_{0.91}Sb_{0.09}$ bulk.

5. Conclusion

A 6 nm period $InAs/InAs_{0.65}Sb_{0.35}$ SL has been investigated by numerical simulations and grown by MBE as a good candidate for full MWIR bandwidth. Despite basic characterizations (AFM, X-ray diffraction, PL) highlighting good material quality, lifetime measurements (not presented here) revealed a high density of recombination centers, requiring extended studies on growth conditions of such Ga-free T2SL.

References

1. S. Maimon *et al.*, *Appl. Phys. Lett.* **89**, 151109 (2006).
2. P. C. Klipstein *et al.*, *Inf. Phys. Technol.* **59**, 172 (2013).
3. Y. Aytac *et al.*, *Appl. Phys. Let.* **105**, 022107 (2014).
4. D. Lackner *et al.*, *J. Appl. Phys.* **111**, 034507 (2012).
5. See http://www.nextnano.de/nextnano3/ for more information about the next**nano**^{3} software.

FORMATION OF MONO- AND MULTI-LAYERED FILMS OF LATERALLY ORIENTED SEMICONDUCTOR COLLOIDAL NANOPLATELETS

A. Mikhailov[1], G. Isic[2], S. Askrabic[2], A. Antanovich[1], A. Prudnikau[1], M. Artemyev[1]

[1]Research Institute for Physical Chemical Problems, Belarusian State University
Leningradskaya 14, 220006 Minsk, Belarus
[2]Institute of Physics Belgrade, University of Belgrade
Pregrevica 118, 11080 Belgrade, Serbia

We demonstrate successful utilization of the Langmuir technique for deposition of thin films of laterally oriented CdSe nanoplatelets from their colloidal solutions. We studied the morphology of the deposited films by TEM and AFM techniques, as well as their optical absorption and photoluminescence.

1. Introduction

Semiconductor nanoplatelets (NPls) are 2D nanostructures with physical properties similar to classic quantum wells, that makes them of a great fundamental interest as quantum objects with anisotropic optical properties [1]. Along with luminescent properties semiconductor NPls are perspective for practical applications, like photovoltaic or laser-gain active media. In order to investigate anisotropy-caused properties it is necessary to obtain a flat film with uniform lateral orientation of NPls. In the present paper, we utilized the Langmuir-like technique to fabricate thin films comprised of monolayer or multilayers of laterally oriented CdSe NPls and studied their morphology and optical properties.

2. Experimental

Colloidal 4.5 ML CdSe NPls were synthesized as described in [2]. NPls in the colloidal solution in chloroform were additionally treated with cadmium acetate and oleic acid according to the procedure described in [3] in order to avoid stacking of nanoplatelets into large aggregates. Afterwards, CdSe NPls were purified by precipitation with acetonitrile, washed with isopropanol and then redispersed in chloroform. Thus, the final solution contained colloidal CdSe NPls terminated by Cd atoms and capped with oleic acid. The solution was drop-cast on the water surface and dried until thin solid film of NPls formed on the water surface. These films were transferred onto quartz substrates and polymer-coated grids and dried in ambient conditions, and subsequently subjected to optical, AFM and TEM investigations. Multilayered thin films of

CdSe NPls were prepared by subsequent deposition of several monolayers by the same technique.

Room temperature optical absorption and photoluminescence (PL) spectra of thin films of CdSe NPls were measured with an OceanOptics HR-2000 spectrometer and a Jobin-Yvon Fluoromag spectrofluorimeter, respectively. Spatial arrangement of CdSe NPls in thin films was studied using a Leo 906E transmission electron microscope. AFM examination of thin NPLs films was performed using a NT-MDT Ntegra Prima SPM system in the tapping mode.

3. Results and discussion

The TEM image in Fig. 1 shows that the film consists of domains of laterally oriented CdSe NPls.

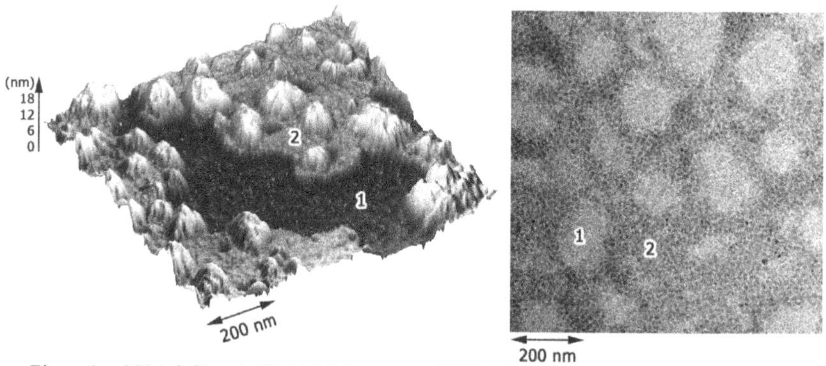

Figure 1. AFM (left) and TEM (right) images of CdSe NPls film prepared by the Langmuir technique (1 – empty area containing no NPls, 2 – area covered with a monolayer of NPls).

The NPls domains in the films are separated by empty areas free from NPls. Such domain structure forms when the thin liquid film of NPls in chloroform quickly spreads off the water surface and dries. The resulting solid film of laterally oriented NPls is rigid and cannot be compressed into a homogeneous monolayer without applying an external force. AFM measurements show a consistent height difference of around 6 nm between areas denoted by 1 and 2 in Fig. 1 indicating the substrate and the NPl monolayer, respectively. The inferred thickness of 6 nm is consistent with the thickness of a monolayer of 4.5 ML CdSe NPls (≈ 2 nm) and two monolayers of oleic acid molecules on the top and bottom surfaces of each NPl. Some hillock exceeding 10 nm in the AFM image and invisible in this TEM image (Fig. 1) are probably related to residual organic molecules crystallized from the colloidal solution during the Langmuir film formation.

Fig. 2 shows a series of optical absorption spectra of a monolayer and multilayered films of 4.5 ML CdSe NPls deposited onto quartz substrate.

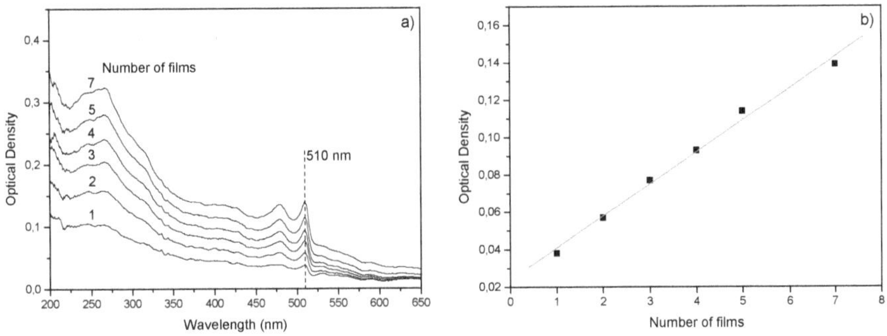

Figure 2. a) Absorption spectra of multilayered films of 4.5 ML CdSe NPls on quartz. b) Linear relationship between the first exciton (λ = 510 nm) optical density and the number of monolayers.

All spectra contain the characteristic first exciton peak at λ = 510 nm similar to that in the colloidal solution of NPls in chloroform. The optical absorption increases monotonously with the number of monolayers over the entire investigated spectral range demonstrating that the Langmuir technique can be successfully used for a reliable control of the precise amount of NPls deposited onto the substrate. Moreover, the linear relationship between the optical density at first exciton and the number of subsequently deposited layers indicates that each new layer contains a nearly constant amount of NPls.

Fig. 3 shows absorption and PL spectra of 4.5 ML CdSe NPls in the colloidal solution and in the monolayer film.

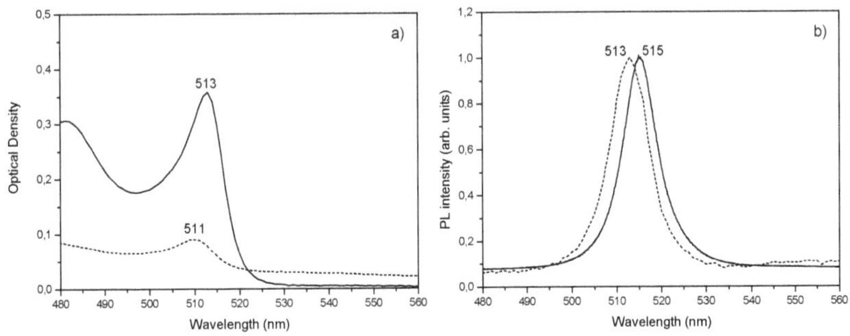

Figure 3. Absorption (a) and normalized luminescence spectra (b) of 4.5 ML CdSe NPls in the colloidal solution in chloroform (solid lines) and in the monolayer film (dashed lines).

The first exciton in the monolayer is blue-shifted by 2 nm both in the absorption and PL spectra relative to the colloidal solution spectra due to the changes in dielectric surrounding. The magnitude of the optical absorption of a

CdSe NPls monolayer is much higher as compared to that of quantum dots or nanorods of the same size, *i.e.*, the same energy of optical transitions due to the effect of giant optical oscillator (GOST) [4]. We did not observe any remarkable spectral shifts with increasing the number of monolayers which pointed to the absence of collective effects.

4. Conclusion

Uniform monolayers and multilayers of colloidal CdSe NPls were fabricated by Langmuir technique. The films consist of monolayers of laterally oriented NPls forming domains. The first exciton absorption peak of multilayered films is linearly proportional to the number of NPLs monolayers. Such structures can be utilized to study anisotropy of optical properties of oriented 2D quantum confined nanostructures.

Acknowledgements

We acknowledge financial support from BRFFR X16SRBG-001 project. This work was supported by the Science and Technology Committee of Belarus and the Ministry of Education, Science and Technological Development of Serbia through the bilateral collaboration program 2016-2018 (#451-03-00293/01).

References

1. M. Nasilowski, B. Mahler, E. Lhuillier, S. Ithurria, B. Dubertret, *Chem. Rev.* **116**, 10934 (2016).
2. A. W. Achtstein *et al.*, *ACS Nano* **8**, 7678 (2014).
3. A. Antanovich, A. Prudnikau, A. Matsukovich, A. Achtstein, M. Artemyev, *J. Phys. Chem. C* **120**, 5764 (2016).
4. A. W. Achtstein, A. Antanovich, A. Prudnikau, R. Scott, U. Woggon, M. Artemyev, *J. Phys. Chem. C* **119**, 20156 (2015).

CdTe QUANTUM DOTS NANOCOMPOSITE FILMS OBTAINED BY THERMAL DECOMPOSITION OF PRECURSORS EMBEDDED IN POLYMERIC MATRIX

F. Antolini

Department of Fusion and Technologies for Nuclear Safety, Photonics Micro and Nanostructures Laboratory ENEA Frascati, Frascati, 00044 Rome, Italy
francesco.antolini@enea.it

L. Ortolani

CNR-IMM, Laboratory, P.le Gobetti 101, 40129 Bologna, Italy
ortolani@bo.imm.cnr.it

The synthesis of nanoparticles of CdTe within a polymer film has been illustrated. The CdTe QDs have been obtained by thermal decomposition of two precursors embedded within the polymeric matrix before the thermal annealing. The formation of the CdTe QDs is studied by optical and structural techniques. The optical analysis showed the formation of CdTe QDs of 3–3.6 nm and with an emission photoluminescence included in the 550–650 nm region. The X-ray diffraction confirms the presence of nanocrystals of CdTe.

1. Introduction

Colloidal photoluminescent semiconductor nanocrystals, also referred to as quantum dots (QDs) [1] have received considerable interest due to the unique optical and electronic properties of QDs including narrow emission, broad absorption and high photostability compared with the macrocrystalline materials.

The generation of QDs directly in a polymeric matrix can be done in a space-selective way by using an external laser source [2] on a polymer doped with a metallorganic single source precursor (SSP). This kind of strategy lead to the synthesis of several SSPs [3] that have to be soluble within the polymers and with relatively low decomposition temperature to avoid the polymer decomposition.

In this paper, we present the formation of CdTe QDs starting from a simple precursor embedded in a polymer [4]. The CdTe QDs have been synthesized in poly(methyl methacrylate) (PMMA) by starting from a solution of Te complexed with trioctylphosphine (TOP) and cadmium dodecanoate. The formation of CdTe QDs has been studied both by optical and structural measurements directly on films. The UV-Vis and photoluminescence (PL) spectra show formation of the typical CdTe shoulder in the region between 550 and 600 nm and the PL emission peak at 580 nm. The X-ray diffraction (XRD) shows the presence of the typical broad peaks of the cubic nanophase of the CdTe QDs.

2. Materials and methods

The solution of the Te precursor has been prepared by mixing in a pyrex tube 0.3 mmol of Te and 300 μl of TOP. The mixture is heated under stirring at 100 °C in N_2 until Te dissolution. This solution is then combined with 600 μl of PMMA solution (50 mg/ml in chloroform), then 2.4 ml of chloroform is added (final volume of the mix is 3 ml).

The solution of the Cd precursor has been prepared adding 0.6 mmol of CdO and 2.4 mmol of DA in a pyrex tube. Then 800 μl of ODE are added and the solution is stirred at 250 °C in N_2 until the complete dissolution of the reagents.

Finally, 300 μl of TOP-Te/PMMA solution is added to 100 μl of the Cd precursor and this solution is immediately used to deposit a film by spin coating at 1200 rpm for 45 s on a quartz substrate.

The film annealing has been carried out at 170 °C under vacuum for 10 min.

The backed films were immediately characterized by means of UV-Vis spectrophotometer (Cary 100 Scan). The PL signal was measured by using an spectrofluorimeter (Edinburgh FLS920) equipped with an Edinburgh Xe900 450W xenon arc lamp (exc 350 nm, emission interval 450–800 nm). All the PL spectra were corrected for the detector response.

The XRD analysis was performed with a Philips conventional Bragg-Brentano vertical diffractometer with CuKα radiation.

3. Results and discussion

The preparation of the cadmium and tellurium precursors has been carried out to form complex of these atoms that are soluble in organic solvents to allow their mixing with the PMMA polymer. Cadmium atoms become soluble in organic solvent when are bound with a fatty acid to form a cadmium carboxylate. On the other side the Te solubilisation has been obtained by dissolving the Te in TOP. In the film all the reagents can react under the thermal treatment to give the expected CdTe product.

The optical characterization shows the formation of the CdTe indicated by the shoulder in the region between 550 and 600 nm [5] (Fig. 1). The first excitonic absorption has its maximum at 544 nm (2.28 eV) that is blue shifted with respect to the CdTe bulk energy gap situated at 1.51 eV. The PL signal from the annealed sample is shown in Fig. 1b. It has a maximum at 585 nm and with a FWHM of 50 nm corresponding to the first excitonic level.

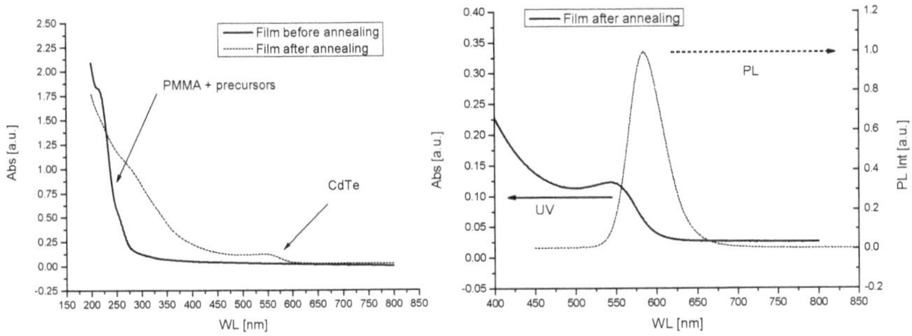

Figure 1. a) UV-Vis spectra of the precursors' film before (black line) and after annealing (gray line); b) the absorption (black line) and the PL signal (gray line) of the film after the annealing. The annealing has been carried out at 170 °C 10 min under vacuum.

The evaluation of the particle size has been carried out using the so called "sizing curve" as reported in [5] and resulted to be 3.4 nm at 170 °C for 10 min. The structural characterization of the new material by XRD is illustrated in Fig. 2.

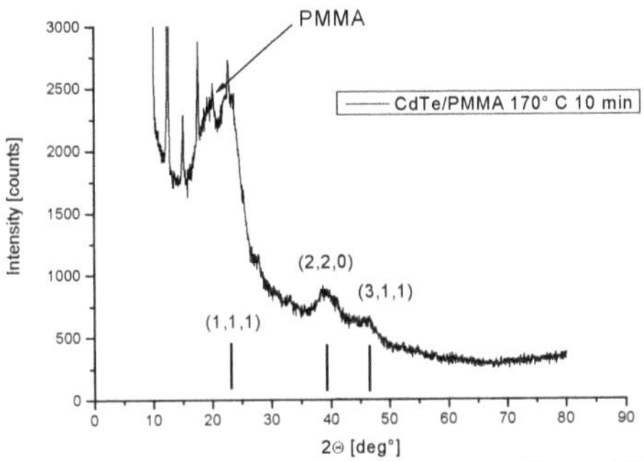

Figure 2. XRD of the CdTe/PMMA film after the annealing at 170 °C 10 min.

The XRD pattern shows the presence of PMMA at about 20 deg and the diffraction peaks of the crystallographic planes (1,1,1), (2,2,0) and (3,1,1) of zinc blend (cubic phase) phased CdTe (PDF # 75-2086). The peaks are broad confirming the nanostructuration of the CdTe crystals.

4. Conclusion

CdTe QDs were grown in a polymeric film. They were *in situ* synthesized directly in the polymer starting from soluble precursors by thermal treatment.

The CdTe QDs grown with these precursors in the solid state show a lower decomposition temperature with respect to similar process carried out in a solution [5].

Acknowledgments

We acknowledge financial support from FP7 project "Laser Induced Synthesis of Polymeric Nanocomposite Materials and Development of Micro-patterned Hybrid Light Emitting Diodes (LED) and Transistors (LET)"-LAMP project (G. A.247928).

References

1. M. V. Kovalenco *et al.*, *ACS Nano* **9**, 1012 (2015).
2. A. K. Bansal *et al.*, *Nanoscale* **7**, 11163 (2015).
3. M. Z. Malik, M. Afzaal, P. O'Brien, *Chem. Rev.* **110**, 4417 (2010).
4. F. Antolini, L. Ortolani, *Method to Prepare a Nanocomposite Inside a Polymer Matrix*, It. Patent RM2014A000259 (2014).
5. L. Algieri *et al.*, *Phys. Status Solidi (c)* **12**, 147 (2015).

XPS STUDY OF GRAPHENE GROWN BY ATMOSPHERIC PRESSURE CVD FROM N-DECANE PRECURSOR WITH NITROGEN AS A CARRIER GAS

N. G. Kovalchuk, I. V. Komissarov, S. L. Prischepa

Belarusian State University of Informatics and Radioelectronics
P. Browka 6, 220013 Minsk, Belarus

M. Andrulevicius, A. Lazauskas, S. Tamulevicius

Institute of Materials Science, Kaunas University of Technology
K. Baršausko Str. 59, 51423 Kaunas, Lithuania

We present X-ray photoelectron spectroscopy study (XPS) of graphene films grown on copper foil by atmospheric pressure CVD with n-decane as a precursor, a mixture of nitrogen and hydrogen as a carrier gas, under different hydrogen flow rates. The XPS study revealed the characteristics of incorporated nitrogen, which was found to have a binding energy around 402 eV, the atomic concentration of incorporated nitrogen is higher for the sample synthesized with higher hydrogen flow rate.

1. Introduction

Nitrogen can be incorporated into graphene sheet (i) in situ, using ammonia as a component of gas carrier mixture or nitrogen containing precursors and (ii) by post treatment [1]. There are not many reports on a topic of using N_2 gas as a nitrogen source during the CVD growth. In our opinion, the main difficulties in using N_2 gas as a nitrogen source arises from the fact that nitrogen molecule possesses one of the strongest bond with the energy of 226 kcal/mol, which means that the temperature of the CVD process is not enough to decompose significant amount of nitrogen molecules into atoms. We report our results of XPS study of graphene grown by atmospheric pressure CVD from n-decane precursor with nitrogen as a carrier gas.

2. Experimental details

Two graphene films (samples A and B) were grown by atmospheric pressure CVD from n-decane precursor. Sample A was prepared at 1050 °C in N_2 and H_2 gas flow with the rates of 100 and 60 cm^3/min, respectively. Sample B was prepared under similar conditions, except the H_2 gas flow rate was reduced to 6 cm^3/min. The feeding rate of n-decane into the reactor was 4 μL/min. The graphene samples were transferred onto SiO_2/Si by the wet chemical method. More detailed information about graphene growth and transfer procedure can be found elsewhere [2]. Thermo Scientific ESCALAB 250Xi spectrometer with a monochromatic Al Kα radiation ($h\nu = 1486.6$ eV) was used for X-ray photoelectron spectroscopy (XPS) measurements. Base pressure in the analytical chamber was better than 2×10^{-7} Pa. Original ESCALAB 250Xi Avantage software was used for peaks deconvolution and fitting procedure using a sum of Lorentzian-Gaussian (70:30) functions. The samples were analyzed as received, no surface cleaning procedure was applied.

3. Results

In the inset of Fig. 1a we show the XPS survey spectrum of sample A transferred on SiO_2/Si substrate. The main core level peaks for carbon, nitrogen, silicon and oxygen are indicated. It is important to note, that due to small thickness of the samples the contribution from the substrate with strong silicon and oxygen signals was detected. The presence of nitrogen 1*s* core level in the

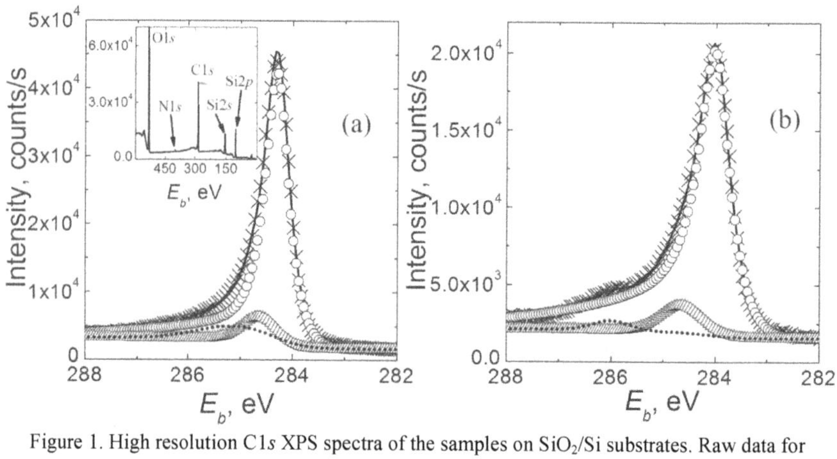

Figure 1. High resolution C1*s* XPS spectra of the samples on SiO_2/Si substrates. Raw data for sample A (a) and sample B (b). Open circle, triangle and dot lines are the result of the fitting procedure. Solid line corresponds to the envelope of the fitted peaks. Inset: XPS survey spectrum.

XPS survey spectrum could be caused either by adsorption or/and incorporation of nitrogen into graphene with the formation of C-N bonds. These bonds affect the shape and position of the carbon 1*s* response. Figs. 1a and 1b present the C1*s* core level for samples A and B, correspondingly. The experimental data can be satisfactorily deconvoluted into three peaks. The first, dominant, peak has the position of the maximum at 284.3 eV and 284.08 eV and FWHM of 0.54 eV and 0.68 eV for samples A and B, respectively. We attribute this line to the sp^2 C bonds. The dominance of this peak confirms the fact that most of carbon atoms are arranged into the honeycomb lattice. The sp^2 C peak of graphene could be observed at various values of energy depending on the material of the substrate on which graphene was deposited.

The peak position varies from 283.97 eV for graphene on Pt (111) [3] to 284.83 eV for graphene on SiC [3]. This variety of the sp^2 C peak position usually is explained by the charge transfer phenomena that take place in the substrate-graphene system [3]. Since both samples were deposited on the same substrate, we expect the same impact of the charge transfer on the peak position. The difference in the binding energy E_b can be related to the difference in concentration of sp^2 C-H bonds which cause the down shift of the E_b [4].

The second peak of the C1*s* core level, centered at 284.63 eV and 284.65 eV for samples A and B, respectively, is slightly up-shifted compare to the dominant one, and could be attributed to the existence of sp^3 C bonds. Finally, the third peak, with the characteristic energy of 285.12 eV and 286.02 eV for samples A and B, correspondingly, can be attributed to the presence of sp^2 C-N bonds. The performed analysis of the C1*s* core level data confirms the fact that the interpretation of the C1s data for graphene is not a trivial task. A number of factors, such as oxygen based functional groups, charge compensation effects and doping can significantly affect the carbon 1*s* response [5]. Therefore, analysis of the N1*s* core level is necessary to verify the status of nitrogen. Figs. 2a and 2b present the high resolution XPS for N1*s* spectrum of samples A and B, respectively, in which the

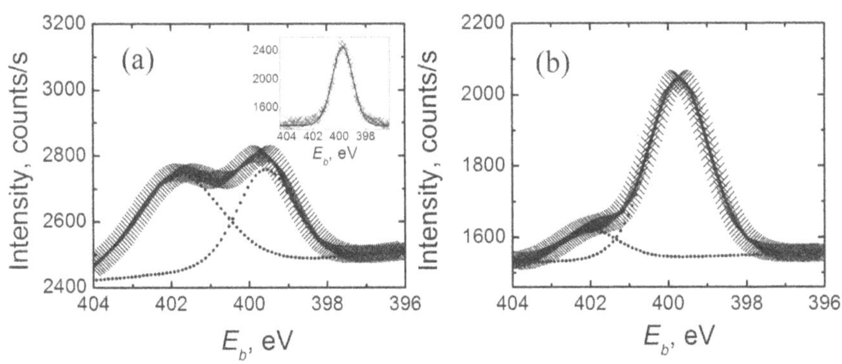

Figure 2. High resolution N1*s* XPS spectra of samples on SiO_2/Si substrate. (a) sample A (b) sample B. Crosses are raw data. Doted lines are the result of the fitting procedure. Solid line corresponds to the envelope of the fitted peaks. Inset: N1*s* XPS spectrum of a bare substrate (symbols) together with the fitting curve (solid line).

spectra could be fitted to two components of the binding energy, namely at $E_b = 399.7$ eV and 402.1 eV (sample B) and $E_b = 401.8$ eV (sample A). The peak at 399.7 eV observed for both samples is related to the adsorbed nitrogen.

Its position was quantified by the measurement on the bare substrate area, see inset of Fig. 2a, indicating that nitrogen dope both samples with a single status.

It is rather difficult to associate the binding energy with a specific configuration of nitrogen and usually it requires additional structural measurements. Indeed, the energy range of 398−404 eV can be related to the different nitrogen configurations, such as pyridine-type, pyrrolic-type, graphite-type, and pyridine N-oxide. Based only on the XPS data we cannot make unambiguous assignment concerning nitrogen configuration. For example, the binding energy for graphitic nitrogen is varied between 400.0 eV [5] and 402.7 eV [6]. The quantified surface atomic concentrations determined by XPS for samples A and B are summarized in Table 1.

Table 1. Quantified surface atomic concentrations for samples A and B on SiO_2/Si substrate as obtained from the XPS study.

Sample	Atomic concentration, %		
	C	N (399.74 eV)	N (~ 402 eV)
A	98.4	0.6	1.00
B	97.6	1.99	0.41

4. Conclusion

Based on the obtained XPS data, we may conclude that (i) nitrogen is incorporated into graphene in a single status; (ii) the concentration of nitrogen associated with the energy of about $E_b \approx 402$ eV is greater for sample A.

Acknowledgments

This work has been prepared under support of Lithuanian-Belarus Partnership Program for Bilateral Cooperation in the Field of Science and Technologies, projects TAP-LB-14-001 (MA, AL, ST) and F15LIT-009 (NGK, IVK, SLP).

References

1. H. Wang, T. Maiyalagan, X. Wang, *ACS Catal.* **2**, 781 (2012).
2. I. V. Komissarov *et al.*, *Physics Procedia* **72**, 450 (2015).
3. T. Susi, T. Pichler, P. Ayala, J. Beilstein, *Nanotechnol.* **6**, 177 (2015).
4. Y. Yamada, J. Kim, S. Matsuo, S. Sato, *Carbon* **70**, 59 (2014).
5. T. Schiros *et al.*, *Nano Lett.* **12**, 4025 (2012).
6. A. L. M. Reddy *et al.*, *ACS Nano* **4**, 6337 (2010).

SYNTHESIS OF SILICON CARBIDE NANOPARTICLES BY PLASMA-ASSISTED TECHNIQUES IN LIQUIDS

N. N. Tarasenka, A. A. Nevar, A. V. Butsen, N. V. Tarasenko, N. V. Rzheutski, E. V. Lutsenko

Laboratory of Laser Diagnostics of Plasma, B. I. Stepanov Institute of Physics NASB Nezavisimosti Ave. 68, 220072 Minsk, Belarus
natalie.tarasenka@dragon.bas-net.by

T. Velusamy, S. Chakrabarti, D. Mariotti

Nanotechnology & Integrated Bio-Engineering Centre (NIBEC), Ulster University Shore Road, BT37 0QB, Newtownabbey, UK

The capabilities of the liquid-assisted electrical discharge and laser ablation techniques with additional laser irradiation for synthesis of SiC nanocrystals have been studied. To determine optimal conditions for the binary compound nanoparticles formation, the characterization of structure, phase composition and morphology of the synthesized nanoparticles was performed by means of HRTEM, SAED and XPS techniques. The results proved the formation of near-spherical SiC nanoparticles with hexagonal crystalline structure.

1. Introduction

Silicon carbide (SiC) is a wide bandgap semiconductor which possesses high thermal stability that makes SiC-based devices well-suited for high temperature operation conditions in which conventional semiconductors (*e.g.* Si) cannot perform adequately and reliably [1]. Besides, due to high hardness and strength, chemical and thermal stability, high melting point, oxidation resistance, high erosion resistance and high thermal conductivity, SiC is promising material for use as heterogeneous catalyst support, for high power, high temperature electronic devices as well as abrasion and cutting applications [2].

The traditional Acheson process for SiC powders production is based on a conventional carbothermal reduction and involves numerous steps. It has an excessive demand for energy and provides rather poor quality materials. Also, SiC structures are produced by physical and chemical vapor deposition, sol-gel, liquid phase sintering or mechanical alloying methods [1], and laser ablation technique [3].

In this paper, the capabilities of the plasma processes for fabrication of SiC nanoparticles (NPs) are evaluated. Plasma-chemical processes have proved to be attractive for NPs synthesis due to high production rates, versatility and possibility of control over the particle size and size distribution function during the synthesis. Nonthermal plasmas have been known for their ability to produce

NPs for 20 years. Different approaches have been developed to adopt plasma processes to NPs production depending on required composition and structure [4]. They include NPs formation in electrical discharge plasma ignited inside liquid between two electrodes, plasma in contact with liquid, and laser-induced plasma generated by focusing laser beam on a target surface in laser ablation experiments.

Ignition of plasma favors initiation of electron-induced non-equilibrium reactions that provide a possibility to produce metastable phases that are difficult to get in other ways. SiC NPs were prepared both by electrical discharge as well as by laser induced plasma in liquid. The goal of producing these particles is to use them as building blocks for nanoparticle-based electronic devices, in catalysis or in biomedicine. Particles were found to be non-agglomerated and their size can be tuned via additional plasma and laser treatment. Different crystal structures of SiC NPs were observed depending on the preparation procedure.

2. Experimental

SiC NPs were prepared by the two alternative techniques.

First, by electrical discharge between Si and C electrodes immersed in ethanol. The alternating current spark discharge was initiated by applying the high voltage of 11 kV. The peak current of the high-voltage spark was about 17 A with the single discharge pulse of 50 μs.

Second, by ns and fs laser irradiation of the mixture of preliminarily prepared Si and C colloids in ethanol taken in the stoichiometric ratio. The unfocused beam of the second harmonic of Nd:YAG laser (532 nm) with pulse duration of 10 ns, repetition rate of 10 Hz and fluences in the range of 230–400 mJ/cm^2 was used for ns laser modification and 140 fs, 52 kHz, 0.7 W for fs laser irradiation. Visible images of the discharge in the solution were taken by a charge-coupled device (CCD) camera. The pH value of the mixture was measured by using a pH meter (Hanna Instruments HI 83141). The formation of SiC nanocrystals was confirmed by HRTEM, XPS and Raman studies.

3. Results and discussion

The characterization results of NPs prepared by the laser irradiation of the mixture of the Si and C colloids by nanosecond laser are shown in Fig. 1. As can be seen from the HRTEM image, non-spherical polycrystalline NPs of about 10 nm are formed due to co-melting of initial Si and C NPs. The FFT processing of the HRTEM image allowed obtaining the lattice parameters of NPs.

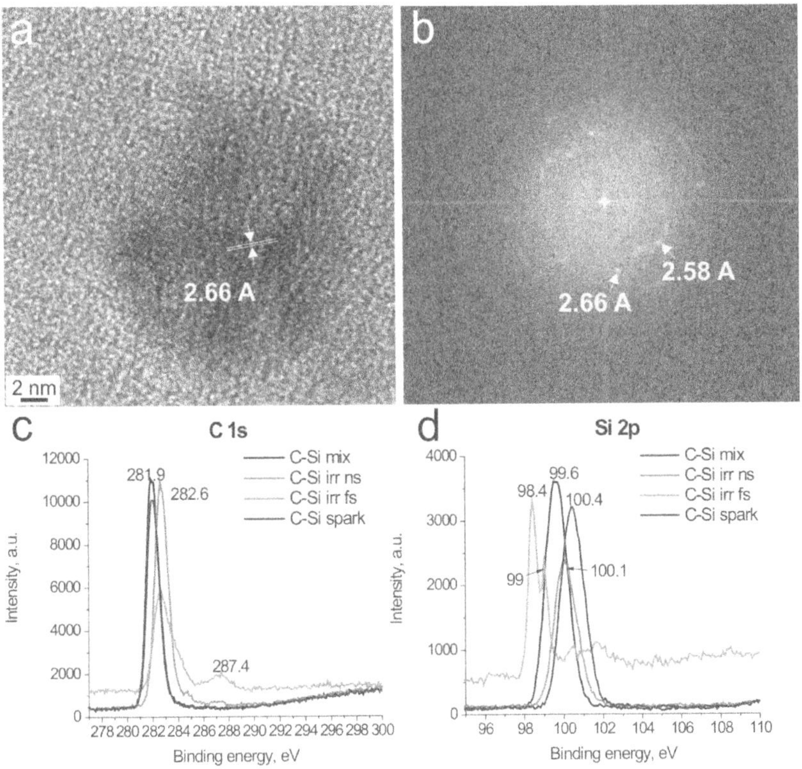

Figure 1. SiC NPs fabricated by plasma-assisted techniques: a – HRTEM image, b – FFT of the HRTEM image, the interplanar spacings found from the FFT image corresponds to the reflections from (100) and (002) planes of the hexagonal SiC phase. XPS spectra of C 1*s* (c) and Si 2*p* (d) for the samples prepared by mixing of the C and Si colloids followed by the ns and fs laser irradiation and electrical discharge between the Si and C electrodes in ethanol.

In addition to the spots due to the cubic silicon and carbon phases, the spots corresponding to the (100) and (002) reflections of the hexagonal SiC phase were found in the center of the formed nanocomposite. This observation can argue on the assumption that the main mechanism of SiC NPs formation is co-melting of initial NPs in the formed mixture with subsequent chemical interaction in the melted liquid droplet.

Analysis of the C 1*s* and Si 2*p* XPS spectra (Fig. 1c, 2d) additionally confirms formation of SiC nanocrystals as the result of ns laser irradiation of the C and Si colloid mixture. The spectra show the Si–C peak with the binding energy of 100.1–100.4 eV. Si is not oxidized since the chemical shift component in the Si 2*p* core-level spectra at around 103.3–104.0 eV was not observed [5]. In the C 1*s* XPS spectra of NPs (Fig. 1c) C–Si bonding at the binding energy of

282–282.6 eV position occurs while the peaks of the C–O and C–C groups are not present in all the samples. However, in the sample prepared by fs laser irradiation, oxidation can be clearly detected since there is the additional peak corresponding to C=O groups located at 287.4 eV.

In conclusion, the experimental results confirmed the formation of SiC phase as a result of electrical discharge in ethanol and ns laser treatment. These reactions can proceed through different mechanisms of laser-particle interaction for the fs and ns laser pulses.

Acknowledgments

The work was partially financed by the National Academy of Sciences of Belarus under the project Convergence 2.4.06 and by the Belarusian Foundation for Fundamental Researches under the grant F 15CO-013. Part of the work was carried out during an STSM supported by the COST Action TD1208, which is also acknowledged for useful interactions and exchanges. This work was also partially supported by EPSRC (award n. EP/M024938/1).

References

1. Houyem Abderrazak and Emna Selmane Bel Hadj Hmida, *Silicon Carbide: Synthesis and Properties* in *Properties and Applications of Silicon Carbide,* ed. R. Gerhardt, (InTech, 2011), pp. 361-388.
2. J.-M. Nhut, R. Vieira, L. Pesant, J.-P. Tessonnier, N. Keller, G. Ehret, C. Pham-Huu, M. J. Ledoux, *Catal. Today* **76**, 11 (2002).
3. S. Yang, W. Cai, H. Zeng, X. Xu, *J. Mater. Chem.* **19**, 7119 (2009).
4. S. Askari, A. Ul Haq, M. Macias-Montero, I. Levchenko, F. Yu, W. Zhou, K. Ostrikov, P. Maguire, V. Svrcek, D. Mariotti, *Nanoscale* **8**, 17141 (2016).
5. T. Cao, Y. Cheng, H. Zhang, B. Yan, Yi Cheng, *J. Mater. Chem C* **3**, 4876 (2015).

SYNTHESIS OF ZINC OXIDE NANORODS AND PLATES BY SOLVOTHERMAL METHOD

S. K. Lim, S.–H. Hwang
Division of Nano end Energy Convergence Research, DGIST
711-873 Daegu, Korea
limsk@dgist.ac.kr, hsungho@dgist.ac.kr

J. Jung
Korea Textile Development Institute, 41842 Daegu, Korea
jhjung@textile.or.kr

M. G. Jeong
UL Chemical Inc., 120-10 Daegu, Korea
ulchemical@naver.com

ZnO particles with rod and plate configurations were synthesized using a solvothermal method using zinc acetate and zinc chloride, respectively. The surface of the as-synthesized ZnO rods and plates were characterized using various analysis tools (XRD, XPS, photoluminescence, FE-SEM, HR-TEM, BET, and UV-Vis). The surface areas and bandgaps of the two ZnO samples were nearly identical; however, XPS and photoluminescence (PL) studies showed that the rods and the plates have relatively pronounced oxygen vacancy and oxygen interstitial contributions, respectively.

1. Introduction

Nanostructured metal oxides have been the subject of great attention in the field of nanotechnology both from a fundamental and industrial points of view. For example, their peculiar electrical properties make them suitable as active sensing materials in resistive sensors, with enhanced performance with respect to bulk materials. Pure ZnO is an n-type semiconductor with a wide bandgap (3.37 eV). It is one of the most promising materials due to its use in a wide range applications in various fields, including short wavelength light-emitting diode and room temperature ultraviolet (UV) lasing diode, solar cell, UV-absorber, transparent conductor, gas sensor, *etc.* ZnO nanostructures, such as nanowires, nanobelts, nanoflowers and other special nanostructures have been successfully synthesized using various approaches.

In this study, we synthesized rod and plate-like ZnO by solvothermal process with precursors such as zinc chloride and zinc acetate dehydrate and discusses an importance of morphological effect.

2. Experimental

A solvothermal method was employed to synthesize the crystalline ZnO rods and plates. For ZnO rods, 0.1 M zinc acetate dihydrate ($Zn(CH_3COO)_2{\cdot}2H_2O$, Junsei) was dissolved in pure ethanol (200 mL) and stirred for 10 min. Subsequently, 1 M NaOH was added to the Zn precursor solution and stirred for 30 min. This mixed solution was transferred to a teflon-lined stainless steel autoclave and was thermally treated at 150 °C for 2 h. After cooling to room temperature, the precipitate was washed with deionized water several times and dried at 70 °C in a vacuum oven to remove the solvent. For ZnO plates, zinc chloride ($ZnCl_2$, Daejung) was used as a precursor and the synthetic procedure was identical to that for the ZnO rods.

The Brunauer-Emmet-Teller (BET) surface areas of the rods and plates were 10.2 and 11.3 $m^2{\cdot}g^{-1}$, respectively. The X-ray diffraction (XRD, Rigaku D/max-2500) patterns of the samples were obtained using an X-ray diffractometer employing Cu-Kα radiation (γ = 1.5405 Å) operating at 40 kV and 100 mA with a scan rate of 0.02 deg/s. To identify the shape and structure of the samples, field emission scanning electron microscopy (FE-SEM, Hitachi S-4800) and high-resolution transmission electron microscopy (HR-TEM, JEM-2200FS) were employed. X-ray photoelectron spectroscopy (XPS, ESCALAB 250 XPS spectrometer, VG Scientifics) was used to examine the binding energies of Zn and O atoms. Photoluminescence (PL) emission spectroscopy (SPEX 1403) was employed to study charge recombination and charge trapping pathways using a He-Cd laser excitation source (325 nm).

3. Result and discussion

Fig. 1 shows the XRD patterns of ZnO samples synthesized using zinc acetate and zinc chloride precursors. All the diffraction peaks observed were indexed to the hexagonal ZnO wurtzite structure (JCPDS card no. 36-1451) and no trace of other phases was observed, indicating high phase purity [1,2]. The ZnO synthesized from zinc acetate exhibited a strong (101) peak and a relatively weak (002) peak, suggesting preferential growth orientation (*i.e.* rod type). The ZnO synthesized from zinc chloride displayed the same peak distributions, yet the peak intensities (*e.g.* (101) peak) were weak. Particularly, the relative intensity of the (002) peak with respect to the (101) peak increased from ~0.47 (zinc acetate) to ~0.64 (zinc chloride), indicating a decrease in the aspect ratio and shape change from a rod type to a plate type. This shape change with different precursors might be attributed to the different degree of basicity of the precursors. Acetate is a much stronger conjugate base (pK_b ~ 9.2) than chloride, and the concentration of OH^- in the solution is increased when acetate is used. Accordingly, the molar ratio of $[OH^-]/[Zn^{2+}]$ is higher in the zinc acetate solution compared with the zinc chloride one. It is typical that ZnO grows more anisotropically as this molar ratio increases because of the large grain sizes obtained via the Ostwald ripening process [3].

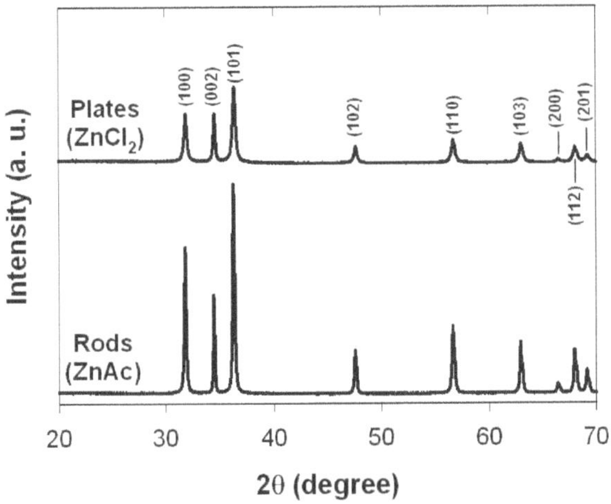

Figure 1. XRD patterns for ZnO particles synthesized from zinc acetate and zinc chloride.

The different ZnO shapes were further confirmed by SEM and TEM analyses (Fig. 2). The ZnO sample synthesized from zinc acetate was composed of a number of uniformly sized rods with an average diameter of ~60 nm and a length of ~500 nm (Fig. 2a). An HR-TEM image for a single ZnO rod further showed a uniform lattice structure and crystal defects were not found (Fig. 2b).

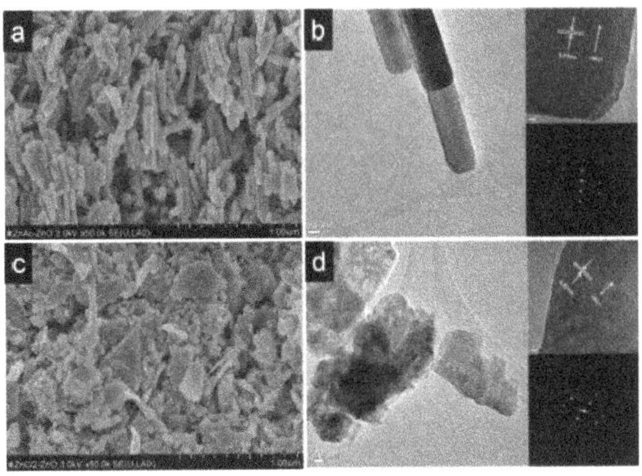

Figure 2. FE-SEM (a, c) and HR-TEM (b, d) images of the synthesized ZnO rods (a, b) and plates (c, d).

The inter-lattice fringe distance was 0.52 nm, which is close to the *d* spacing of the (001) plane. On the other hand, the ZnO synthesized from zinc chloride displayed a disc- or plate-like configurations with the widths up to submicrons and a well-crystallized lattice structure of the same (001) plane (Fig. 2c, d).

4. Conclusion

ZnO particles with rod and plate configurations were synthesized simply by varying the precursor. Surface analyses with XRD, SEM, and TEM showed a decrease in the aspect ratio when zinc acetate was replaced with zinc chloride due to the different degree of basicity. XPS and PL studies revealed that the rods have more pronounced oxygen vacancies ($V_O^{\times}$, $V_O^{\bullet}$, and $V_O^{\bullet\bullet}$) close to the conduction band energy level, which function as electron trap sites. Accordingly, the hole transfer reactions were enhanced in the rods. On the other hand, the plates have relatively pronounced oxygen interstitial (O_i, O_i', and O_i'') contributions, which are close to the valence band energy level.

Acknowledgments

This research was supported by a grant from the Fundamental R&D Program for Core Technology of Materials funded by the Ministry of Knowledge Economy (10050946), Ministry of Science, ICT and Future Planning, and the DGIST R&D Program of the MEST, also supported by the A SME R&D project for the Start-ups & Growing stage companies funded by the Small and Medium Business Administration, Korea.

References

1. S. K. Lim, S.-H. Hwang, S. Kim, H. Park, *Sens. Actuators B* **160**, 94 (2011).
2. S. K. Lim, S. H. Hong, S.-H. Hwang, S. Kim, H. Park, *J. Mater. Sci. Technol.* **29**, 39 (2013).
3. Y. H. Zheng, C. Q. Chen, Y. Y. Zhan, X. Y. Lin, Q. Zheng, K. M. Wei, J. F. Zhu, Y. J. Zhu, *Inorg. Chem.* **6**, 6675 (2007).

SOL-GEL FABRICATION OF ZINC OXIDE NANOSTRUCTURED COATINGS ON A FIBERGLASS TISSUE

T. A. Filippova, V. E. Borisenko

Belarusian State University of Informatics and Radioelectronics
P. Browka 6, 220013 Minsk, Belarus
filippova.tatyana.a@gmail.com

V. V. Uglov[1,2], S. V. Zlotski[1]

[1]Belarusian State University, Nezavisimosti Ave. 4, 220030 Minsk, Belarus
[2]Tomsk Polytechnic University, Lenina Ave. 2a, 634028 Tomsk, Russia

Nanostructured zinc oxide coatings were for the first time fabricated by sol-gel technique on a fiberglass tissue. The coatings have demonstrated high catalytic activity in UV light assisted decomposition of Rhodamine B.

1. Introduction

Zinc oxide (ZnO) is a direct gap semiconductor with the band gap of 3.4 eV which makes it useful in a wide range of applications in electronics, optoelectronics, gas sensing, photocatalysis, environmental engineering [1]. Nanostructuring of this material during fabrication brings additional advantages for applications due to an extension of the surface-to-volume ratio, which is indeed important in catalysis and sensing. In order to maximize the effect we propose to synthesize ZnO nanostructured coatings on a fiberglass tissue using a sol-gel technology. The first results obtained are presented in this paper.

2. Experimental

The fiberglass tissue Je3-200 [2] of 0.191 mm thickness was cut into 3×3 cm samples, which were used as substrates for ZnO coatings. They were washed in acetone and dried in air.

Two type of sols were composed to fabricate the coatings. Analytic grade compounds were used. The first one, referred to herein after as a basic sol, contained zinc acetate anhydrate ($(CH_3COO)_2Zn$) dissolved in the mixture of 2-aminoethanol (NH_2-CH_2CH_2-OH) and isopropanol alcohol (C_3H_7OH) with their molar ratio of 5:7:49. The second one was prepared by adding 3.5 wt.% of $ZnCl_2$ into the first sol. Both sols were kept 1 day at room temperature before the use.

The coatings were fabricated by dipping the fiberglass tissue samples into one of the sols for 60 s, drying them in air at 200 °C for 3 min and subsequent annealing at 500 °C for 30 min in air.

Surface morphology of the coatings and their chemical composition were analyzed with scanning electron microscopy (SEM) and energy-dispersive X-ray (EDX) spectroscopy, respectively.

Crystal phase composition of the samples was analyzed by X-ray diffraction (XRD) using CuKα radiation ($\lambda = 0.154179$ nm) in Ultima IV diffractometer. Crystallite sizes were calculated by Scherrer equation.

We have also estimated a photocatalytic activity of the fabricated ZnO coatings. The coatings assisted decomposition of the organic dye Rhodamine B in its water solution (100 mg/l) was performed under UV light from a 8 W mercury lamp. The solution was stirred continuously and open to air during the decomposition. A concentration decrease of Rhodamine B in the solution was monitored by optical absorption spectroscopy of the solution in the wavelength range of 400–700 nm.

3. Results and discussion

The structure of the fiberglass tissue used as a substrate is shown in Fig. 1. It has been woven by glass fibers of about 10 μm in diameter. The EDX analysis performed has shown it to contain 5.43 % of carbon, 48.77 % of oxygen, 0.79 % of fluorine, 0.23 % of sodium, 1.35 % of magnesium, 6.94 % of aluminum, 25.28 % of silicon and 15.06 % of calcium.

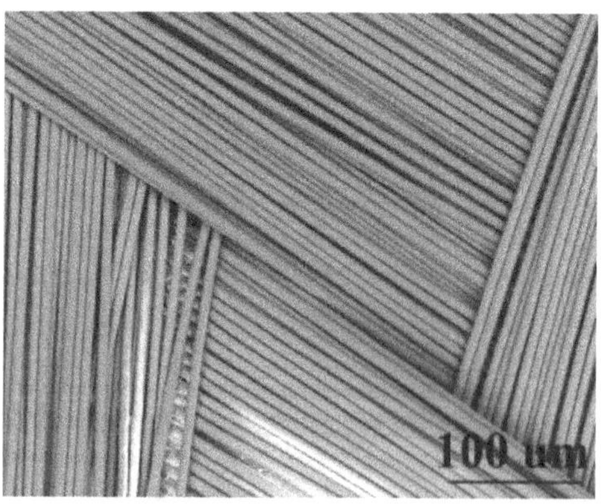

Figure 1. SEM image of the fiberglass tissue used as a substrate material for ZnO coatings.

The used sols and calcination regime (500 °C, 30 min) is known from our previous experiments [3] to guarantee formation of nanostructured ZnO films on flat glass substrates. Fig. 2 presents the surface morphology of the fiberglass tissue with the fabricated coatings.

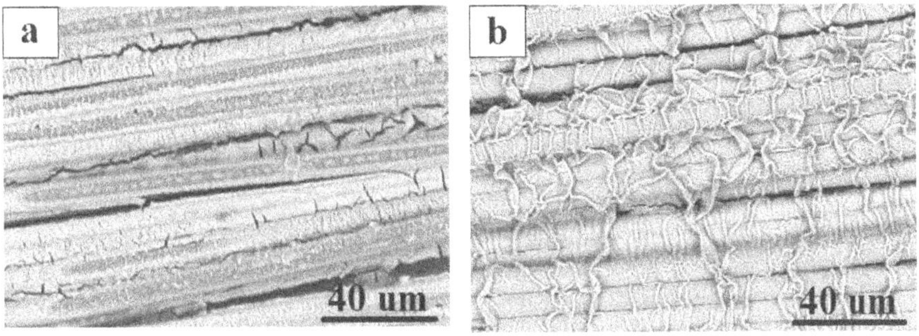

Figure 2. SEM images of the fiberglass tissue covered by ZnO derived from the basic sol without $ZnCl_2$ (a) and the sol containing $ZnCl_2$ (b).

The synthesized ZnO covered individual fibers filling small interfiber gaps. The coatings fabricated from the basic sol have cracks on the fibers separated from the nearest neighbors by not filling gaps. In contrast, no cracks are observed if $ZnCl_2$-contained sol was used. Moreover, there is a net of nanometer thick filaments covering the whole surface of the samples. In both cases XRD analysis indicates the ZnO coatings to be mainly composed of crystallites as large as 24–50 nm. No preferable orientation of the crystallites has been registered. Cl-contained products have not been identified in the coatings.

The observed morphology and composition peculiarities of the coatings fabricated from different sols can be explained by melting of $ZnCl_2$ (at 290 °C) during calcination at 500 °C, its catalyzing effect on the sol-gel processes involved [4] and subsequent decomposition yielding evaporation of Cl and Cl-contained products.

The photocatalytic tests of the ZnO coatings on the fiberglass tissue have shown their efficiency in the UV light assisted decomposition of Rhodamine B in its water solution. Upon UV irradiation for 2 h in the presence of the fabricated samples concentration of the dye estimated by the intensity of the light absorption at 554 nm was about 22 times (~96 %) decreased. Optical absorption spectra in Fig. 3 present a comparison of the efficiency of the catalysts derived from the $ZnCl_2$-free and $ZnCl_2$-contained sols. The coating derived from the $ZnCl_2$-contained sol demonstrates a little bit higher photocatalytic activity than that derived from the basic $ZnCl_2$-free sol. It can be explained by an extended surface area of the catalyst with the net of nanometer thick filaments.

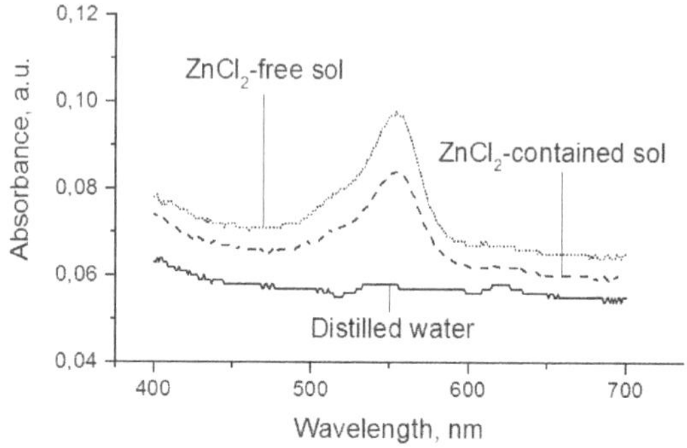

Figure 3. Optical absorption spectra of the Rhodamine B water solution subjected to UV light irradiation in the presence of the catalysts sol-gel synthesized onto the fiberglass tissue from $ZnCl_2$-free and $ZnCl_2$-contained sols.

The second run of the photocatalytic tests with the already used photocatalysts has in general reproduced the results of the first run.

4. Conclusion

Sol-gel derived ZnO coatings on the fiberglass tissue have been demonstrated to be facile in fabrication and efficient in photocatalytic applications materials. The addition of $ZnCl_2$ into the sol used for their synthesis favors better surface morphology and catalytic activity of the coatings, while its particular role should be additionally studied.

Acknowledgments

This work was financially supported the National Program "Convergence" (project 3.2.04). The authors are grateful to D. V. Zhigulin for the SEM and EDX analysis of the samples.

References

1. C. Jagadish, S. J. Pearton, *Zinc Oxide Bulk, Thin Films and Nanostructures* (Elsevier Sci., 2006).
2. Russian Federation State Standard 19907-83 *Dielectric Fabrics Made of Glass. Twister complex threads. Specifications.*
3. N. M. Denisov, E. B. Chubenko, V. P. Bondarenko, V. E. Borisenko, *Applied Catalysis B* (2017) – submitted.
4. A. Corma, H. Garcia, *Chem. Rev.* **103**, 4307 (2003).

SYNTHESIS, STRUCTURE AND OPTICAL PROPERTIES OF GLASS-CERAMICS BASED ON ZnO NANOCRYSTALS AND CODOPED WITH Er^{3+} AND Yb^{3+} IONS

I. Alekseeva, O. Dymshits*, A. Zhilin, D. Shemchuk, S. Zapalova, M. Tsenter, A. Vasilevskaya

NITIOM S. I. Vavilov State Optical Institute
Babushkina Str. 36, 192171 Saint-Petersburg, Russia
**vodym1959@gmail.com*

A. Volokitina**, P. Loiko, K. Bogdanov, A. Baranov

ITMO University, Kronverkskiy Pr. 49, 197101 Saint-Petersburg, Russia
***anna.itmo@gmail.com*

G. Arzumanyan, E. Kuznetsov

Joint Institute for Nuclear Research, Joliot-Curie Str. 6, 141980 Dubna, Russia;
Dubna State University, Universitetskaya Str. 19, 141982 Dubna, Russia

Josep Maria Serres, Xavier Mateos

Física i Cristal·lografia de Materials i Nanomaterials, Universitat Rovira i Virgili, Campus Sescelades, c/ Marcel·lí Domingo, s/n., Tarragona, Spain E-43007

Transparent glass-ceramics (GCs) based on ZnO and β-Zn_2SiO_4 nanocrystals and doped with Er^{3+} and Yb^{3+} ions are synthesized on the basis of a zinc potassium alimunosilicate glass by secondary heat treatments. Their structure is studied with XRD, TEM and Raman spectroscopy. The concentration of RE^{3+} ions determines the character of phase separation of the initial glass. Green Stokes and anti-Stokes emissions of Er^{3+} ions are detected from GCs. These GCs are promising for development of ZnO → RE^{3+} energy-transfer materials, for white light generation, as well as for polyfunctional materials codoped with Co^{2+} and Er^{3+} ions for self-Q-switched lasers.

1. Introduction

Zinc oxide nanostructures are nowadays attracting a lot of attention for various photonics applications. Transparent GCs with ZnO nanocrystals were previously intended for doping with transition-metal ions, *e.g.* Co^{2+}, and used as saturable absorbers of lasers [1], and for doping with rare-earth ions to combine intrinsic defect-related emissions of ZnO with RE^{3+} ions [2] and to reach ZnO → RE^{3+} energy-transfer [3]. In the present paper, we report on synthesis, structure and optical properties of ZnO and β-willemite based transparent GCs doped with Er^{3+} and Yb^{3+} ions.

2. Synthesis of the glass and glass-ceramics

The glass with the composition (mol%) 14 K_2O, 32 ZnO, 14 Al_2O_3, 40 SiO_2 was co-doped with 0.3 mol% Er_2O_3 and 0.3 mol% Yb_2O_3. The glass batches were prepared using reagent grade raw materials. The components were carefully mixed for 1 h in a ball mill and placed into a platinum crucible. Batches of 200 g in weight were melted in a laboratory electric furnace at 1590 °C for 3 h with stirring, poured onto a cold metal plate and annealed at 550 °C in an electric furnace. The parent glass was transparent and rose colored due to the Er^{3+} dopant. GCs were produced from the parent glass by secondary heat treatments. The temperature was ranged from 680 to 900 °C, and the duration was in the range of 2–84 h.

3. Structural study

X-ray diffraction (XRD) patterns of the initial glass and transparent GCs (heat-treated at 680–750 °C) are shown in Fig. 1a. The initial glass was X-ray amorphous. After the heat treatment in the range of 680–720 °C, the precipitation of hexagonal (wurtzite-type structure) ZnO nanocrystals was detected. Their mean size (Scherrer diameter) was increasing with the temperature from 10.5 nm to 13 nm. For the samples heat-treated at 750 °C and above, hexagonal β-willemite (β-Zn_2SiO_4) was crystallized additionally. GCs became translucent and opaque after the heat treatments at >800 °C and at >900 °C, respectively. For the temperatures >950 °C, the presence of α-willemite was clearly detected.

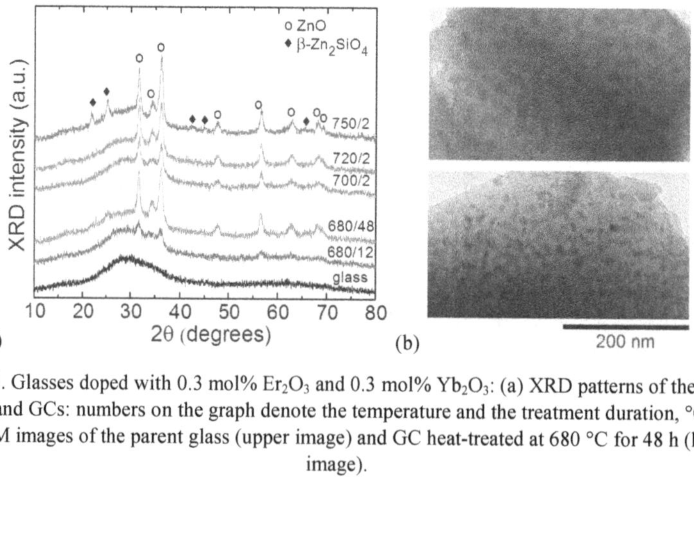

(a) (b)

Figure 1. Glasses doped with 0.3 mol% Er_2O_3 and 0.3 mol% Yb_2O_3: (a) XRD patterns of the initial glass and GCs: numbers on the graph denote the temperature and the treatment duration, °C/h; (b) TEM images of the parent glass (upper image) and GC heat-treated at 680 °C for 48 h (lower image).

The parent glass and GCs prepared by the heat treatment at 680 °C for 48 h were further characterized by transmission electron microscopy (TEM). The TEM studies (Fig. 1b) revealed a phase separation even for the initial glass, with a characteristic droplet structure (mean size 12–15 nm). For GCs the ZnO nanocrystals precipitation is evident as it is confirmed by the XRD data.

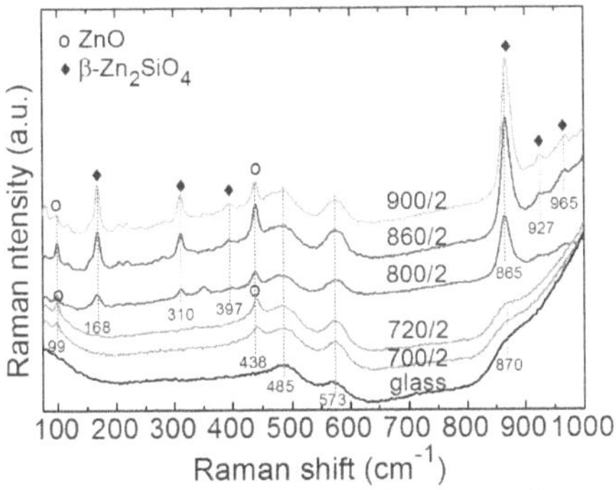

Figure 2. Raman spectra of the parent glass and GCs: the numbers denote the temperature and the treatment duration, °C/h; λ_{exc} = 488 nm.

We involved Raman spectroscopy (Fig. 2) to confirm the above observations of structural features. For the samples heat treated at < 750 °C, two bands at 99 and 438 cm^{-1} are observed and assigned to ZnO. For the higher temperatures, multiple sharp bands at high (865, 927, 965 cm^{-1}) and low energies (168, 310, 397 cm^{-1}) appear. These bands are attributed to β-Zn_2SiO_4. The fraction of β-Zn_2SiO_4 is increased with the temperature as the intensity of these mentioned bands increases.

4. Absorption and luminescence

Absorption spectra of the parent glass and GCs are shown in Fig. 3a. The spectra of GCs contain three features. First, the UV absorption edge of GCs (λ_g = 430 nm) is shifted with respect to the initial glass (λ_g = 370 nm) due to the precipitation of the ZnO nanocrystals. Second, the bands of Er^{3+} and Yb^{3+} ions remain almost unchanged with the heat treatment which indicates that these ions do not enter the ZnO nanocrystals. Third, a light scattering at >1 μm is detected and attributed to free charge carriers in ZnO.

Luminescence and upconversion luminescence spectra of GC heat-treated at 700 °C for 2 h are shown in Fig. 3b indicating the intense green emission due to the $^2H_{11/2}$+$^4S_{3/2} \rightarrow {}^4I_{15/2}$ transition.

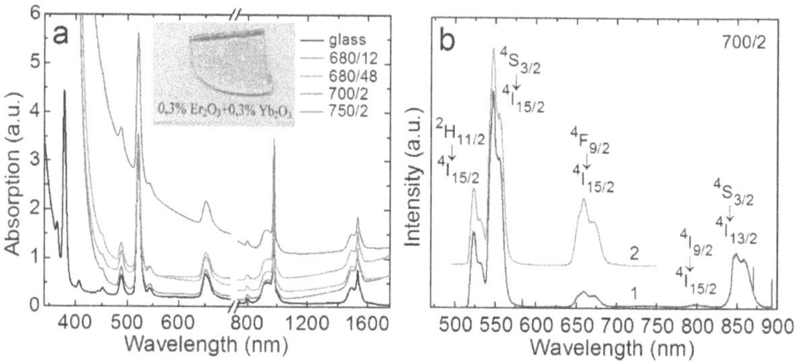

Figure 3. (a) Absorption spectra of the parent glass and GCs. The numbers denote the temperature and the treatment duration, °C/h. The spectra are shifted for the convenience of their observation, inset – photograph of the parent glass; (b) GC heat-treated at 700 °C for 2 h: 1 – luminescence spectrum recorded at λ_{exc} = 488 nm, 2 – upconversion luminescence spectrum recorded at λ_{exc} = 980 nm.

5. Conclusion

Transparent GCs containing ZnO and β-Zn_2SiO_4 nanocrystals (mean sizes of 10–13 nm) and doped with Er^{3+} and Yb^{3+} ions were synthesized. The RE^{3+} ions do not enter the zinc-containing nanophases while they assist the phase separation of the parent glass. Further work will be focused on the study of ZnO → RE^{3+} energy transfer for a broadband light emission and development of Co^{2+}/Er^{3+} codoped GCs for self-Q-switched lasers.

Acknowledgments

This work was partly supported by the RFBR (grant 16-03-01130). P.L. acknowledges financial support from the Government of the Russian Federation (grant 074-U01) through ITMO Post-Doctoral Fellowship scheme.

References

1. P. A. Loiko *et al.*, *Appl. Opt.* **55**, 5505 (2016).
2. G. M. Arzumanyan *et al.*, *Opt. Mater.* **62**, 666 (2016).
3. Q. Luo *et al.*, *Opt. Lett.* **36**, 2767 (2011).

GLASS-CERAMICS WITH Ho^{3+}:$YNbO_4$ NANOCRYSTALS: NUCLEATION, STRUCTURE, AND SPECTRAL PROPERTIES

I. Alekseeva, O. Dymshits*, M. Tsenter, A. Vasilevskaya, S. Zapalova, A. Zhilin
NITIOM S. I. Vavilov State Optical Institute
Babushkina Str. 36, 192171 Saint-Petersburg, Russia
**vodym1959@gmail.com*

A. Volokitina**, P. Loiko, K. Bogdanov, A. Baranov
ITMO University, Kronverkskiy Pr. 49, 197101 Saint-Petersburg, Russia
***anna.itmo@gmail.com*

Josep Maria Serres, Xavier Mateos
Física i Cristal·lografia de Materials i Nanomaterials, Universitat Rovira i Virgili, Campus Sescelades, c/ Marcel·lí Domingo, s/n., Tarragona, Spain E-43007

Transparent lithium alimunosilicate glass-ceramics (GCs) containing tetragonal and monoclinic nanocrystals (7–15 nm) of rare-earth niobates Ho^{3+}:$YNbO_4$ and β-quartz solid solution are synthesized. The role of RE^{3+} ions in the phase separation of the initial glass is proved with XRD, TEM, Raman and optical spectroscopy. GCs are featured by intense green luminescence and promising for development of Ho^{3+},Yb^{3+} co-doped green upconversion phosphors.

1. Introduction

Transparent glass-ceramics containing nanocrystals that incorporate rare-earth ions (RE^{3+}) are of interest for development of phosphors, bulk and fiber optical amplifiers and lasers. Rare-earth orthoniobates, $RENbO_4$, are promising host matrices for doping with other RE^{3+} ions due to low phonon frequencies, high luminescence quantum yield and possibility of high dopant concentrations. Recently, GCs based on rare-earth niobates were prepared [1]. In the present paper, we report on synthesis, structural and optical study of novel lithium aliminosilicate GCs based on Ho^{3+}:$YNbO_4$ nanocrystals.

2. Synthesis of glass and glass-ceramics

The glass with the composition of 18 Li_2O, 27 Al_2O_3 and 55 SiO_2 (mol%) was chosen for this study. It was doped with oxides: Y_2O_3 (2.7 mol%), Ho_2O_3 (0.5 mol%), and Nb_2O_5 (3.2 mol%). A batch to produce 300 g of the glass was melted in a quartz ceramic crucible in a laboratory electric furnace at 1580 °C for 4 h with stirring, then the glass melt was bubbled by oxygen for 0.5 h to remove the hydroxyl groups from the melt. The melt was poured onto a metal plate and the initial glass was annealed at 620 °C. To produce GCs, the initial glass was heat-treated in isothermal conditions in the temperature range of 680–1250 °C for 6–24 h.

3. Structural study

The structure of the initial glass and GCs was studied by X-ray diffraction (XRD), transmission electron microscopy (TEM) and Raman spectroscopy.

XRD patterns of the initial glass and GCs are shown in Fig. 1. The initial glass and the glass heat-treated at 720 °C for 6 h were transparent and X-ray amorphous. The prolongation of this heat treatment for 24 h leads to crystallization of tetragonal $RENbO_4$ crystals. After the heat treatment at 800–1250 °C, the GCs contain two crystalline phases: $RENbO_4$, and lithium aluminosilicates. Thus, $RENbO_4$ serves as a nucleating agent of the main crystalline phase. Monoclinic (fergusonite-type) $RENbO_4$ crystals appear additionally to the tetragonal ones after the heat treatment at 1000 °C and above and the GCs become translucent. For $T > 1100$ °C, lithium aluminosilicate with the β-quartz ss transforms to β-spodumene ss (keatite structure) and the GCs are opaque. After the heat treatments at $T > 1200$ °C, the stable phases of monoclinic $RENbO_4$ and β-spodumene ss are registered on the XRD patterns.

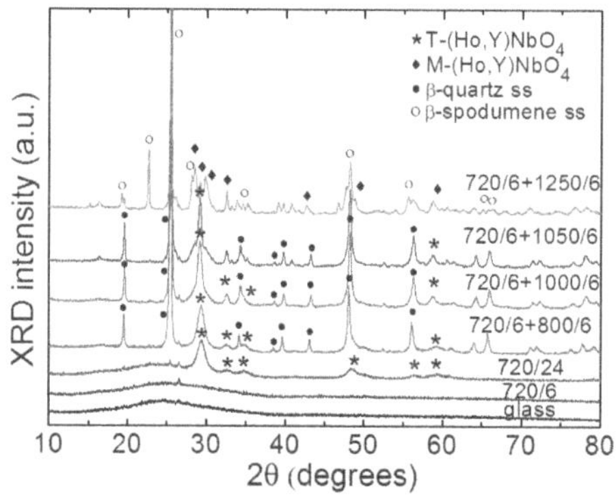

Figure 1. XRD patterns of the initial glass and GCs: numbers denote temperature and duration of the heat treatment, °C/h.

The TEM images of the initial glass and GCs are presented in Fig. 2. The initial glass is homogeneous. After the heat treatment at 720 °C for 6 h whilst it remains X-ray amorphous the GCs appears as liquid phase-separated. A well-developed droplet structure with the size of inhomogeneous regions of 5–7 nm is observed. The growth of $RENbO_4$ crystals is evident in Fig. 2c. Their size is 9–12 nm after processing at $T = 800$ °C and 12–15 nm after processing at $T = 1000$ °C.

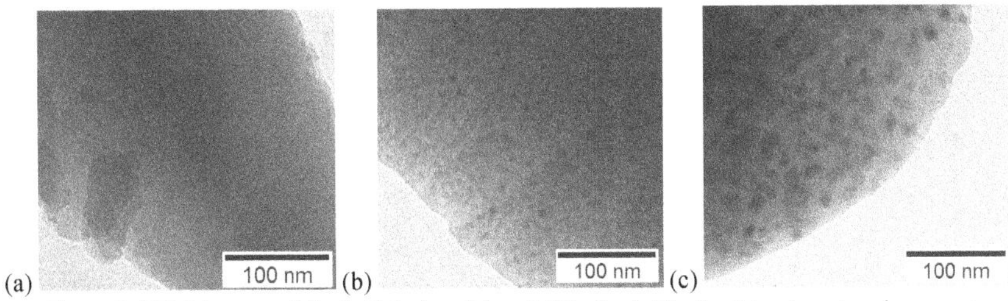

Figure 2. TEM images of the initial glass (a) and GCs (b,c). The heat treatment regimes are: 720 °C/6 h (b) and 720 °C/6 h + 800 °C/6 h (c).

The Raman spectra of the initial glass and GCs are shown in Fig. 3. They confirm the phase separation of the glass heat-treated at 720 °C for 6 h (note the shift of the high energy band at ~810 cm^{-1} being characteristic for the localized amorphous niobates) and further precipitation of tetragonal $RENbO_4$ nanocrystals for the heat treatment for 24 h. For $T > 800$ °C, the bands at 479 cm^{-1} and 1083 cm^{-1} are detected and assigned to the β-quartz ss. For $T > 1000$ °C, the crystallization of monoclinic form of $RENbO_4$ is detected.

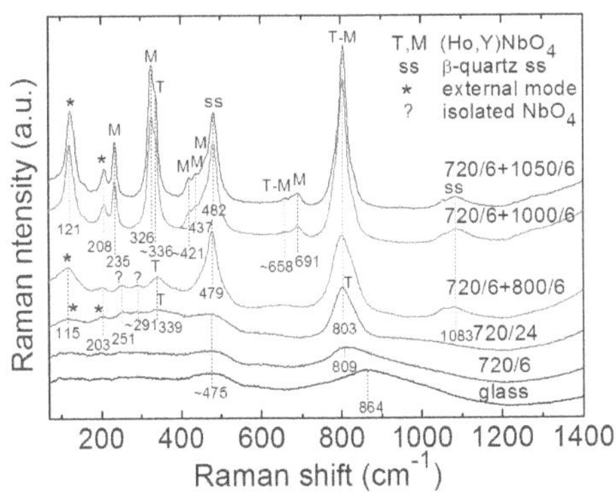

Figure 3. Raman spectra of the initial glass and GCs. Numbers on the graphs denote temperature and duration of the heat treatment, °C/h. The spectra are combined from the data for $\lambda_{exc} = 514$ nm (< 600 cm^{-1}) and 488 nm (> 600 cm^{-1}) to avoid luminescence.

4. Spectroscopic properties

To characterize optical properties of the initial glass and GCs, we have studied absorption and luminescence spectra of the Ho^{3+} ions in the visible range (Fig. 4). The spectrum of the initial glass contains broad structureless bands typical for Ho^{3+} ions in alimunosilicate glasses. After the heat treatment at 720 °C for 6 h, the spectrum slightly changes confirming that the Ho^{3+} ions enter the droplet phase, *cf.* Fig. 2a. The absorption bands become strongly structured for $T = 800$ °C and further for $T = 1000$ °C representing an increased Stark splitting and varied local field for the Ho^{3+} ions in the tetragonal and monoclinic Ho^{3+}:$YNbO_4$ nanocrystals.

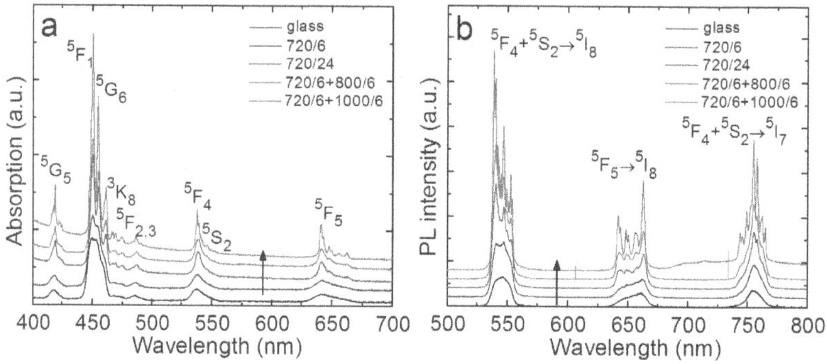

Figure 4. Spectroscopy of the initial glass and GCs: (a) absorption spectra of Ho^{3+} ions; (b) photoluminescence spectra of Ho^{3+} ions, λ_{exc} = 488 nm. Numbers denote temperature and duration of the heat treatment, °C/h.

When excited at 488 nm, the GCs exhibited an intense green luminescence (the main band in the range of 535–555 nm is related to the $^5F_4$+$^5S_2$ → $^5I_8$ transition in the Ho^{3+} ions) which intensity was enhanced with the heat treatment.

5. Conclusion

We present the first evidence of the crystallization process of lithium alimunosilicate GCs nucleated with rare-earth niobates via the liquid phase separation with subsequent crystallization. By the control of the heat-treatment regime, we have synthesized transparent GCs containing tetragonal and monoclinic Ho^{3+}:$YNbO_4$ nanocrystals with tunable optical properties. Further work will be focused on Ho^{3+}, Yb^{3+} co-doped GCs for green upconversion phosphors.

Acknowledgments

This work was partly supported by the RFBR (grant 16-03-01130). P.L. acknowledges financial support from the Government of the Russian Federation (grant 074-U01) through ITMO Post-Doctoral Fellowship scheme.

References

1. O. S. Dymshits *et al.*, *J. Lumin.* **160**, 337 (2015).
2. P. A. Loiko *et al.*, *J. Lumin.* **179**, 64 (2016).

INFLUENCE OF Y_2O_3 AND Gd_2O_3 ADDITIVES ON LUMINESCENCE PROPERTIES OF CaO-2SiO$_2$:Ce GLASSES

Y. U. Tratsiak

Research Institution for Physical Chemical Problems, Belarusian State University
Leningradskaya 14, 220030 Minsk, Belarus
slon.zhenya@gmail.com

M. V. Korjik

Research Institution for Nuclear Problems, Belarusian State University
Bobruiskaya 11, 220030 Minsk, Belarus

E. E. Trusova

Belarusian Technological State University, Sverdlova 13a, 220006 Minsk, Belarus

M. Fasoli, F. Moretti, A. Vedda

Dipartimento di Scienza dei Materiali, Università di Milano-Bicocca
via Roberto Cozzi 55, 20125 Milano, Italy

Alkali-earth glasses with Y_2O_3 and Gd_2O_3 additives have been obtained using the traditional melt technology. An influence of Y_2O_3 and Gd_2O_3 on structure, photoluminescence and radioluminescence of the glasses has been investigated and discussed.

1. Introduction

Nowadays, scintillation materials are widely used in various fields, such as high-energy physics, medicine, geology, homeland safety. In this regard, the search and development of new efficient and inexpensive scintillation materials are important and actual [1,2]. Particular attention can be paid to glasses as potential scintillators. They are characterized by a low cost, simplicity of processing, optical transparency, easy composition tuning. Moreover, silicate scintillators in the form of single crystal like rare-earth ortho- and pyrosilicates are well known. They have a high light yield (LY) (up to 40000 ph/MeV for gadolinium pyrosilicate, GPS) and energy resolution (up to 6 % for GPS) [3]. Their silicate-based composition allows to obtain similar compounds in a glass or glass-ceramic form. The combination of rare-earth ortho- and pyrosilicates with alkali-earth silica glasses looks especially attractive in order to improve LY and reduce the afterglow of scintillators [4].

This paper is focused on preparation of calcium silicate glasses with addition of Gd_2O_3 and Y_2O_3 and study their photoluminescence (PL) and radioluminescence (RL) properties.

2. Materials and methods

Glass samples with compositions of $3CaO\text{-}2SiO_2$:Ce (CaSi), $(CaO\text{-}2SiO_2)\cdot 0.34Y_2O_3$:Ce (CaY) and $(CaO\text{-}2SiO_2)\cdot 0.34Gd_2O_3$:Ce (CaGd) have been prepared using the traditional solid-state melt technology.

$CaCO_3$, SiO_2, Y_2O_3 and Gd_2O_3 were used as raw materials. The initial compounds were mixed in a mortar by grinding in the required quantities. CeO_2 was introduced for substitution of 1 at.% of Ca^{2+} ions in all glasses. The resulting mixtures were transferred in Al_2O_3/ZrO_2 crucibles, placed in a FALORNI gas furnace and annealed in the reducing atmosphere at 1550 °C for 3 h. The glass melts were casted on steel plates and placed into an electrical furnace at 600 °C for 4 h. Finally for optical measurements, glass specimens were cut on plates with ~1 mm thickness and polished.

X-ray excited RL measurements were carried out at room temperature with a home-made apparatus featuring a CCD detector (Jobin-Yvon Spectrum One 3000) coupled to a monochromator (Jobin-Yvon Triax 180) with a 300 grooves/mm grating. X-ray irradiation was performed by a Philips X-ray tube operating at 20 kV. The irradiation of all samples was carried out in the same conditions varying only the exposure time.

Steady state PL spectra were measured at room temperature with a xenon lamp as an excitation source, together with a double monochromator (Jobin-Yvon Gemini 180 with a 1200 grooves/mm grating), and recorded through a nitrogen cooled CCD detector coupled with a monochromator (Jobin-Yvon Micro HR).

Raman measurements were performed in backscattering configuration with a frequency doubled Nd:YAG laser as an excitation souce (wavelength $\lambda = 532$ nm) on a Jobin-Yvon T64000 triple monochromator equipped with a liquid nitrogen cooled CCD detector (Jobin-Yvon Symphony). The detection range was 100–3500 cm^{-1}.

3. Results and discussion

PL and RL spectra of the glasses under study are presented in Fig. 1 in comparison with PL of calcium silicate glass. All glasses reveal the broad and unresolved emission band corresponding to $5d_1\text{-}^2F_{5/2}$ and $5d_1\text{-}^2F_{7/2}$ radiative transitions of Ce^{3+} ions (Fig. 1a). The peak position (405 nm) is independent of the composition. However, CaY and CaGd samples are characterized by much higher PL intensity than CaSi.

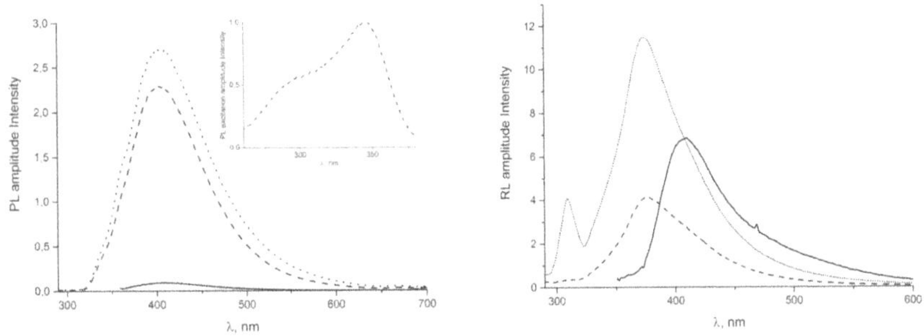

a b

Figure 1. PL (a) and RL (b) spectra for CaSi (solid line), CaY (dashed line) and CaGd (dotted line) samples. λ_{ex} for PL is 345 nm. The luminescence excitation spectrum for CaY is shown in the insert.

All samples are characterized also by broad RL spectra (Fig. 1b). An additional band at 320 nm (corresponding to the $^6P_j \rightarrow ^8S_{7/2}$ transition of Gd^{3+} ions) is observed for CaGd. The highest RL intensity belongs to this glass and can be a result of formation of nano-crystalline ortho- and pyrosilicate structures in this case. The different spectral shape of RL and PL of glasses with Gd_2O_3 and Y_2O_3 suggests the presence of Ce^{3+} ions in different surroundings and/or the occurrence of defect emissions. That deserves further investigations in future.

Raman spectroscopy was used to reveal more information on structure of the glasses (Fig. 2). Around 1000 cm^{-1} the Raman structures are evidenced, corresponding to anti-symmetric stretching vibrations of SiO_4 tetrahedra and possibly to vibrations of units with two non-bridging oxygens those are characteristic of metasilicate glass compositions. At the lower wavenumbers, the spectra are also clearly dependent on composition revealing new bands for CaY and CaGd glasses that may be assigned to formation of nanoceramic phases within glass matrices [2].

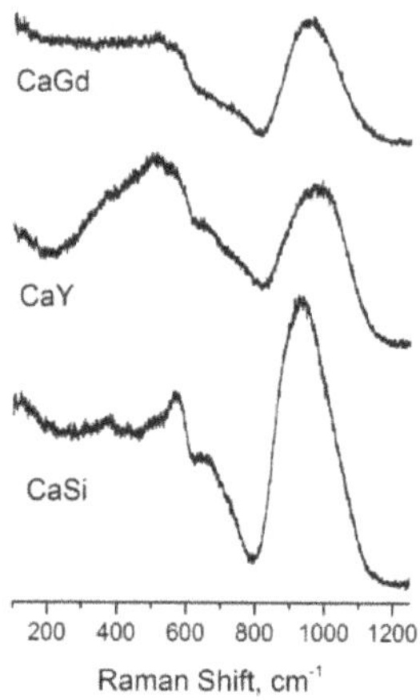

Figure 2. Raman spectra for a series of investigated samples.

4. Conclusion

We have demonstrated the possibility to prepare Ce-doped silicate glasses with mixed Y and Gd composition. Presence of Y and Gd strongly enhances PL of Ce^{3+} ions. The difference of PL and RL spectra suggests different surroundings of Ce^{3+} ions and/or an occurrence of defect emission in RL. Clear differences there are also in Raman spectra. Future work will be devoted to deeper insight on optical properties of such materials in relation to their structural features.

Acknowledgments

This work was supported by H2020 M. Curie Rise “INTELUM” Project.

References

1. G. Blasse, *Chem. Mater.* **6**, 1465 (1994).
2. D. Faoite *et al.*, *J. Phys.: Conf. Series* **620**, 012002 (2015).
3. http://scintillator.lbl.gov/
4. M. Koschan *et al.*, *J. Cryst. Growth* **352**, 133 (2012).

UNIQUE POROUS ALUMINA FORMED IN PHOSPHORIC ACID ELECTROLYTES AT HIGH ANODIC VOLTAGES

X. Huang, W. Su, L. Sun
Hangzhou Dianzi University, 310018 Hangzhou, China

A. A. Maximenko, M. Marszałek
Institute of Nuclear Physics Polish Academy of Sciences
Radzikowskiego Str. 152, PL 31-342 Krakow, Poland
Alexey.Maximenko@ifj.edu.pl

J. A. Fedotova
Research Institute for Nuclear Problems of Belarusian State University
Bobruiskaya 11, 220030 Minsk, Belarus
julia@hep.by

S. K. Lazarouk
Belarusian State University of Informatics and Radioelectronics
P. Browka 6, 220013 Minsk, Belarus
serg@nano.bsuir.edu.by

Anodic formation of porous alumina has been carried out at the anodic voltages up to 230 V in phosphoric acid electrolytes. The high anodic voltages have been achieved leading to new parameters of porous alumina such as interpore distance up to 420 nm, forming cell factor up to 1.7 nm/V, volume expansion factor up to 2.4 and porosity down to 1 %. Formation of porous alumina with multilayer walls has been demonstrated.

1. Introduction

Porous alumina is used as a template with predefined morphology for synthesis of functional nanomaterials [1]. Self-ordering of alumina hexagonal cells during an anodization process is sizably improved at high forming voltages [2]. Such electrochemical process was called the high field anodization and provided fabrication of ordered porous alumina with tubular structures [3,4].

In this paper, the high voltage anodization of aluminium in phosphoric acid electrolytes is considered. The porous alumina parameters obtained at different anodic voltages (U_a) up to 230 V have been studied.

2. Experimental

Aluminium foil (purity 99.9 %) with the thickness of 100 μm and 1.0 μm thick aluminium films deposited on silicon wafers were used as supports. A 0.2 μm niobium film was deposited on Al surface to fabricate a mask. Then the niobium mask was configurated by standard photolithography operations and plasma-chemical etching in gaseous SF_6. Anodization of aluminum was carried out in 0.2 M – 4 M H_3PO_4 aqueous solutions. The anodization voltage was linearly increased from 0 to 100–240 V with a rate of 1 V/s.

The volume expansion factor of the anodic oxide was calculated from measurements of the metal and oxide thickness. The structure parameters of the anodic films were analysed by a scanning electron microscopy (SEM). Porosity of the films was estimated by using SEM measurements of pore diameters and interpore distances. Elemental compositions of the samples were studied by EDX analysis.

3. Results and discussion

Aluminum anodization with a mask provides additional ways to the heat dissipation. The heat sink through the niobium mask prevents overheating effects in the alumina barrier layer [5]. This approach allows us to reach the maximum anodic voltage of 230 V in phosphoric acid electrolytes, whereas an ordinary full-area anodization process admits the maximum anodic voltage of 195 V [6]. Further increase of the anodic voltage is limited by local breakdown effects.

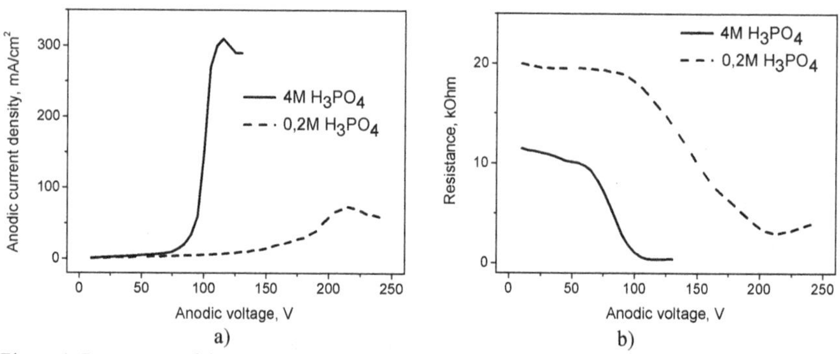

Figure 1. Parameters of the anodization processes in phosphoric acid electrolytes: a) current-voltage characteristics of the anodic processes; b) resistance of anodic films at different anodic voltages.

Fig. 1a shows the current-voltage characteristics of the aluminium anodization processes in phosphoric acid electrolytes. It should be noted that the negative differential resistance occurs at high anodic voltages. The decrease of the anodic film resistance can be a result of qualitative changes in the process of its formation at high voltages.

Fig. 2 displays the SEM images of porous alumina formed at different anodic voltages in 0.2 M H_3PO_4. The anodic film formed at 150 V has a conventional "honey comb" structure (Fig. 2a), while at 210 V it has a tubular structure (Fig. 2b). The low resistance zone in Fig. 1b corresponds to this growth mode. A special attention should be paid to the transition between high resistance and low resistance zones (Fig. 1b). The films formed at anodic voltage corresponding to this transition (195 V for 0.2 M H_3PO_4 electrolyte) have multilayer walls between pores (Fig. 2c,d). Similar multilayer wall structures have been also reported in [6,7]. It should be noted that the multilayer wall structures have been found only upon anodization phosphoric acid electrolytes while the tubular anodic films can be formed in sulfuric, oxalic and other electrolytes.

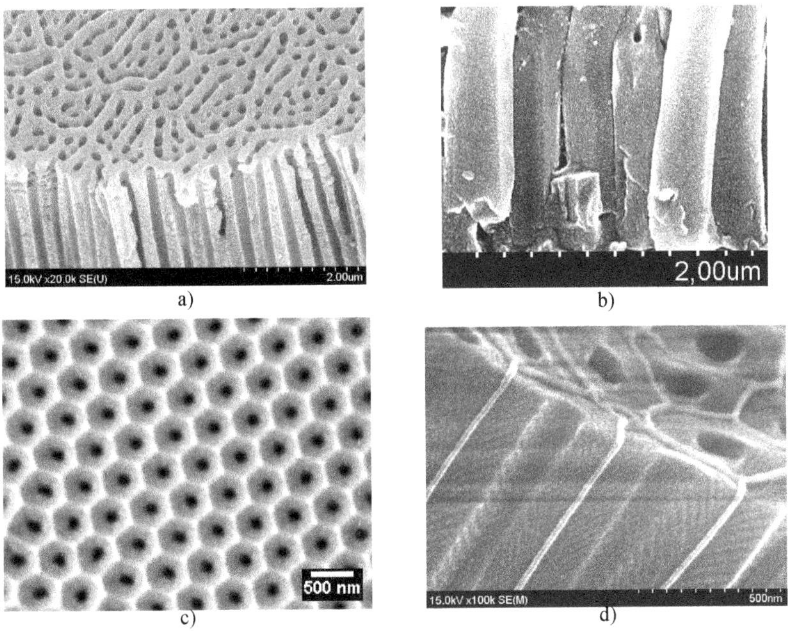

Figure 2. SEM images of porous alumina formed in 0.2 M H_3PO_4 at different anodic voltages: a) 150 V, b) 210 V, c) 195 V (planar view), d) 195 V (cross-sectional view).

Fig. 3 demonstrates structure parameters of porous alumina *vs* anodic voltage. Volume expansion factor increases up to 2.25 whereas porosity decreases down to 1 % with the anodic voltage growth. The volume expansion factor more than 2 corresponds to the tubular alumina structure while the honey comb porous alumina has the expansion factor in the range of 1.1–1.7. The low porosity (1–2 %) could be observed only in the tubular structure.

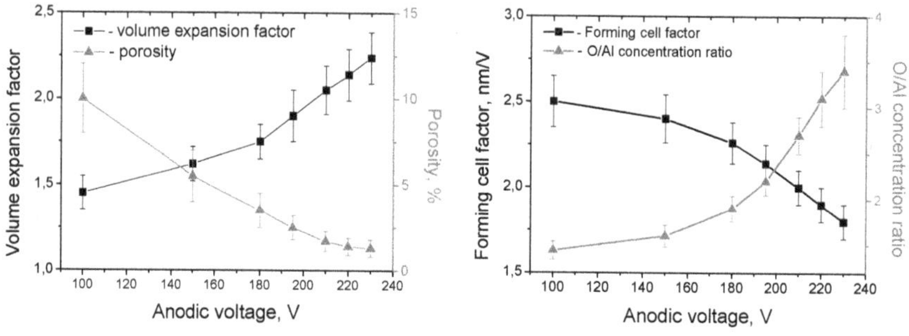

a) b)

Figure 3. Porous alumina structure parameters *vs* anodic voltage: a) volume expansion factor and porosity; b) forming cell factor and O/Al concentration ratio.

The anodic voltage growth results in the increase of the distance between pore centers (interpore distance) up to 420 nm fabricated in the 0.2 M H_3PO_4 electrolyte (Fig. 3b). The formation cell factor (the ratio between interpore distance and anodic voltage) decreases from 2.5 nm/V at $U_a = 100$ V to 1.7 nm/V at $U_a = 230$ V for the 0.2 M H_3PO_4 electrolyte.

EDX analysis results for O/Al elemental composition are shown in Fig. 3b. The O/Al ratio is 1.5 for porous alumina formed at low anodic voltages while it reaches 3.5 for the films formed at 230 V. It means that porous alumina films formed at low anodic voltages consist of stoichiometric Al_2O_3 while at high anodic voltages the films consist of $Al(OH)_3 + Al_2O_3$, as it was proposed in [3]. The volume expansion factor data validate this assumption because the value of 2.25 corresponds to the composition of $Al(OH)_3 + Al_2O_3$.

Porous alumina formed in the transition zone regimes ($U_a = 180$–200 V) has intermediate parameters. These regimes allow to fabricate self-ordering hexagonal alumina cells (Fig. 2c,d) which can be used for a template synthesis of advanced functional nanomaterials [8]. Appearance of the different multilayer wall structures of porous alumina cells can be explained by an occurrence of borders between the pure Al_2O_3 layer and the composite $Al_2O_3 + Al(OH)_3$ layer [9]. The pure Al_2O_3 layers have high dielectric properties resulting in lighter areas in the SEM images in comparison with the grey areas of the composite $Al(OH)_3 + Al_2O_3$ layer (Fig. 2c,d).

4. Conclusion

High electric field anodization of aluminum with the photolithographic mask has been carried out in phosphoric acid electrolytes at the highest forming voltages reaching 230 V. This procedure allows to get unique parameters of the anodic films fabricated. New features of the films can be explained by anodization at high electric field strength inside alumina barrier layers. They open new directions for porous alumina nanotechnology applications.

Acknowledgements

We acknowledge the support from the National Science Centre (NCN, Poland) grant 2014/13/N/ST8/00731.

References

1. W. Lee, S. Park, *Chem. Rev.* **114**, 7487 (2014).
2. S. Chu, K. Wada, S. Inoue, M. Isogai, A. Yasumori, *Adv. Mater.* **17**, 2115 (2005).
3. L. Yi, L. Zhiyuan, H. Xing *et al.*, *J. Mater. Chem.* **21**, 9661 (2011).
4. S. K. Lazarouk, D. A. Sasinovich, V. E. Borisenko *et al.*, *J. Appl. Phys.* **107**, 033527 (2010).
5. S. K. Lazarouk, in: *Physics, Chemistry and Application of Nanostructures*, ed. by V. E. Borisenko, S. V. Gaponenko, V. S. Gurin, C. H. Kam (World Scientific, Singapore, 2013), p. 355.
6. Y. Li, M. Zheng, L. Ma, W. Shen, *Nanotechnology* **17**, 5101 (2006).
7. A. F. Feil, M. V. Costa, P. Migowski, J. Dupont, S. R. Teixeira, L. Amaral, *J. Nanosci. Nanotech.* **11**, 2330 (2011).
8. J. V. Kasiuk, A. A. Maksimenko, J. A. Fedotova, M. Marszałek, S. K. Lazaruk, O. V. Kupreeva, *Phys. Solid State* **58**, 2312 (2016).
9. G. E. Thompson, G. C. Wood, *Nature* **290**, 230 (1981).

HEAT TRANSFER COEFFICIENT FROM THE ANODIC ALUMINUM OXIDE TO THE ELECTROLYTE DURING ANODIZATION PROCESS

D. A. Tsirkunov, F. Alvarado, G. G. Rabatuev

Belarusian State University of Informatics and Radioelectronics
P. Browka 6, 220013 Minsk, Belarus
tsirkunov@bsuir.by

Joule heating significantly affects the process of growth and dissolution of the emerging porous anodic alumina. Heat transfer coefficient from the anodic aluminum oxide to the electrolyte during anodization process determines the temperature increase depending on the amount of transferred heat. The paper presents the calculation of the heat transfer coefficient from the growing aluminum oxide to various organic and aqueous electrolytes. The calculation results showed that this parameter is in the range of 400–800 W/(m^2K) for organic electrolytes and in the range of 800–2 000 W/(m^2K) for aqueous electrolytes. We have proposed the method of heat transfer coefficient increasing up to 10 000 by a decrease of the characteristic size of the anodization surface.

1. Introduction

Electrochemical anodic oxidation of valve metals has been intensively investigated in recent years, because this process allows to create the nanoscale structures [1]. The uniqueness of electrochemical anodization is simplicity of implementation and a wide range of possibilities. This method allows to form nanostructures with a size from units to hundreds nanometers by choosing the electrolyte composition, electrical mode of anodization and temperature of the emerging oxide film.

The electrolyte composition is an important parameter, that affects the transferring of heat from the growing oxide. In this paper we calculate the heat transfer coefficient (HTC) from porous anodic alumina (PAA) to the electrolyte during anodization process in different electrolytes.

2. Calculation details

The convection HTC according to the similarity theory is defined by the following expression [3]:

$$h = Nu \cdot Km / L, \tag{1}$$

where Nu is the Nusselt number; Km is the liquid thermal conductivity, W/(m·K); L is the characteristic linear size of the heat exchange surface (m), defined as the ratio of the area of the sample to its perimeter. This characteristic size for a round anodizing cell is a quarter of the cell diameter.

Nusselt number characterizes the ratio between intensity of the heat exchange by convection and thermal conductivity (always more or equal to 1) [3]. The Nusselt number of a flat sample (the aluminum foil in our case) placed vertically relative to the outer surface of the electrolyte is defined by the following formula [4]:

$$Nu = 0.68 + 0.67 \cdot \left(gL^3 \beta \cdot \frac{(t_c - t_0)}{\nu \cdot \delta} \right)^{1/4} \cdot \left[1 + \left(\frac{0.492\delta}{\nu} \right)^{9/16} \right]^{-4/9}, \quad (2)$$

where g is the acceleration of free fall on Earth surface (9.81 m/s²); β is the temperature coefficient of volume expansion of the coolant, K^{-1}; t_c is the temperature of heat exchange surface, K; t_0 is the coolant temperature, 20 °C; ν is the coefficient of kinematic viscosity, m²/s; δ is the thermal diffusivity, m²/s.

3. Results and discussion

We calculated the Nusselt numbers of some liquids, that were included in the electrolyte composition. Also, there were calculated HTC depending on the temperature of heat exchange surface for different liquids.

Graphically the values of HTC are shown in Fig. 1. Water has the highest HTC. The calculation results showed that this parameters is in the range of 400–800 W/(m²K) for organic electrolytes and in the range of 800–2000 W/(m²K) for aqueous electrolytes. Therefore, the water-based electrolytes provide the best heat sink, and the temperature in the zone of growth of PAA will be less. The temperature in the zone of growth of PAA in cases of organic electrolytes will be 2–3 times higher [2].

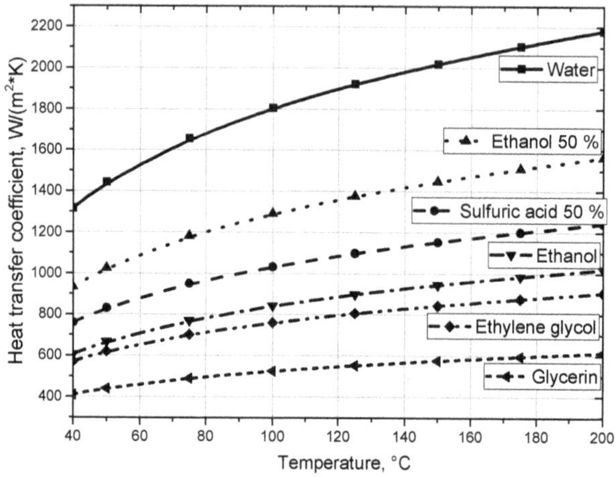

Figure 1. The HTC versus the temperature of the heat exchange surface (the temperature of the anodic film) during anodization process in different electrolytes.

A form of the anodizing surface is taken into account by its characteristic size. Round surfaces have maximal characteristic sizes. This form is typical for wafers being used in microelectronics. An influence of the diameter of the round anodizing surface on HTC was calculated. The calculation was performed for two water-based electrolytes, containing 1 % and 50 % of sulfuric acid.

The results of the calculations are presented in Fig. 2. HTC achieves its maximum at small cell diameters. Therefore, PAA will grow at lower temperatures in the anodization cells with a diameter less than 10 mm, in comparison with cells of higher diameter. It is also seen that an increase of the temperature of anodic films results in the growth of HTC. It means, that there is a self-regulation of the temperature during the anodization process. The self-regulation of temperature is very important during anodization at high anodic voltages [5]. In addition, changing of the concentration of sulfuric acid in the electrolyte from 1 % to 50 % leads to decreasing of HTC. Thus HTC from the PAA to the electrolyte during anodization process could be increased up to 10000 W/(m^{2}K) by the decrease of the characteristic size of the anodization surface.

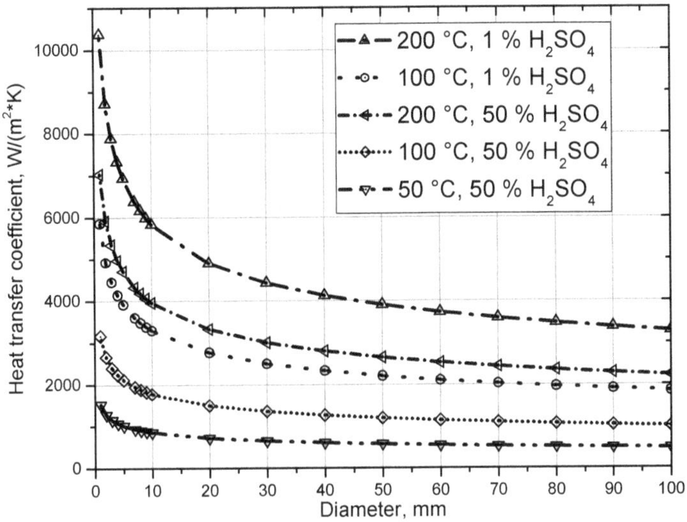

Figure 2. The dependence of HTC on the electrochemical cell diameter for aqueous electrolytes containing from 1 % to 50 % of sulfuric acid and for different temperatures of the heat exchange surface.

4. Conclusion

HTC determines the temperature in the zone of growth of PAA. Choosing the electrolyte composition and the diameter of the cell for anodizing, it is possible to control this temperature. Consideration of heat generation will allow to provide stability of the anodizing process. It is also possible to conduct controlled anodizing under extremely high temperatures.

Acknowledgments

Authors thank S. K. Lazarouk for his help and discussion of the results presented.

References

1. C. T. Sousa *et al.*, *Appl. Phys. Rev.* **1**, 031102 (2014).
2. S. K. Lazarouk *et al.*, *Doklady BSUIR* **100**, 5 (2016).
3. S. W. Churchill, H. H. S. Chu, *Int. J. Heat and Mass Transfer* **18**, 1323 (1975).
4. S. P. Sukhatme, *A Textbook on Heat Transfer* (4th ed.) (Universities Press, India, 2005), pp. 257–258.
5. S. Chu *et al.*, *Adv. Mater.* **17**, 2115 (2005).

MORPHOLOGY OF ANODIC POROUS ALUMINA MEMBRANES FABRICATED ON SUBSTRATES WITH DIFFERENT ELECTRIC CONDUCTIVITY

A. Maximenko[1,2], M. Marszałek[1], Y. Zabila[1], A. Zarzycki[1]
[1]*Institute of Nuclear Physics Polish Academy of Sciences*
Radzikowskiego Str. 152, PL 31-342 Krakow, Poland
Alexey.Maximenko@ifj.edu.pl
[2]*Research Institute for Nuclear Problems of Belarusian State University*
Bobruiskaya 11, 220030 Minsk, Belarus

This work presents variations of morphology of porous anodic aluminum oxide (AAO) membranes prepared by anodization of Al foil and Al films deposited on semiconducting or dielectric substrates working as an additional electrical junction between anode and the anodized film. The pore diameter increases when electrical conductivity of the substrate increases. However, the cell diameter depends on electrolyte type and anodization voltage rather than on conductivity of the substrate.

1. Introduction

Nanoporous anodic oxide membranes possess the larger surface area with more number of reaction sites and more developed surface morphology in comparison with conventional bulk and micro-size materials. These unique properties of the membranes are used to address technological issues of using valve metal (Al, Ti, Zr, *etc.*) oxides for applications in water splitting devices, supercapacitors, solar cells and photocatalysis. On the other hand, such nanostructured membranes can be used as biocompatible materials in biomedical applications, support matrices for nanoparticles (catalysis) and templates for organized nanostructures in particular, magnetic nanostructures (nanorods, nanodots and antidot arrays). Porous structure of anodic oxides can be easily tuned by the type of electrolytes and the applied voltage under anodization [1]. In this work, we investigated the influence of electric conductivity of substrate upon morphology of porous AAO.

2. Experimental

The anodization was performed on three types of substrates with different electrical properties: Al foil (metallic, conductive), Al films deposited on Si wafer (semiconducting) and Al films deposited on Si wafer pre-coated with 100 nm of amorphous SiO_2 layer (dielectric, insulating). Preparation of the conductive samples was performed on the high-purity aluminium foil (99.999 %, Laurand Associates, Inc.). Prior to the anodization process, the Al foil was annealed in air at 560 °C for 24 h and chemically polished in a solution of 62 % nitric acid and 85 % phosphoric acid. The Al films were deposited in high vacuum (~10^{-6} mbar) by thermal evaporation on appropriate semiconducting or dielectric substrates.

The outline of the cell used for anodization is presented in Fig. 1. The anodizing substrate is located at the bottom of the cell and pressed by the O-ring sealing to ensure tightness. The upper substrate part is in contact with the electrolyte while the bottom part is in contact with the aluminum anode. The system maintaining a constant temperature on the border between the electrolyte and anodized substrate is provided by a Peltier element. A platinum plate (99.99 %) is used as a cathode.

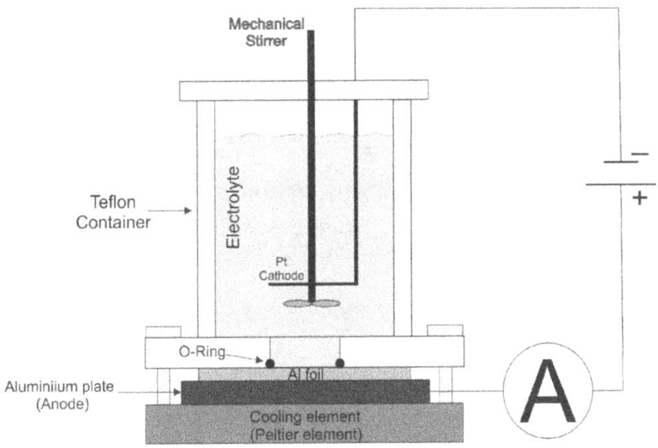

Figure 1. Outline of the anodization cell.

The cell allows to use different contacts from anode to the anodized Al (Fig. 2). We have used four types of connections between the anode and Al. Conductive, if the anodized Al film directly contacted anode, semiconducting or insulating if Al film was deposited on Si or SiO_2, respectively, and direct when Al film was connected externally to the anode regardless of type of substrate.

To ensure better pore periodicity in the anodized samples, the two-step anodization process was performed. Three types of electrolyte based on 0.3 M sulfuric, 0.3 M oxalic, and 0.2 M phosphoric acids at constant anodization voltages (U_A) 25 V, 40 V and 195 V, respectively, were used to form porous structures with different features (*i.e.* pore (D_P) and cell (D_C) diameters). The temperature at the border between the electrolyte and anodized Al was kept constant at 13 °C, 17 °C and 0 °C for sulfuric, oxalic and phosphoric acid electrolytes, respectively. The time of the second anodization was 5 min.

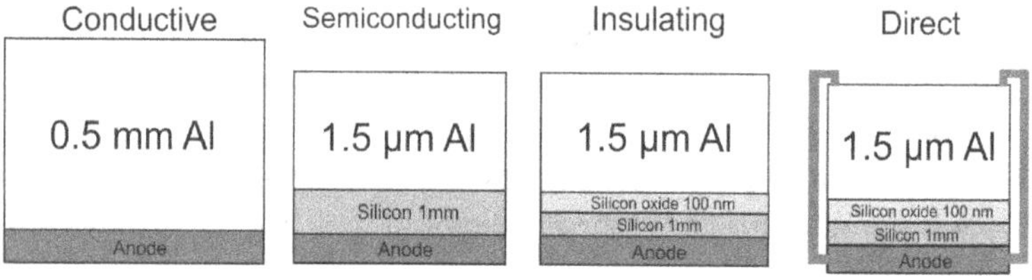

Figure 2. The different types of electrical contacts used for the anodization process.

Morphology of AAO was investigated with a scanning electron microscope (SEM) FEI Inspect S50. Morphological features of the produced templates (D_C and D_P) were estimated from top-view SEM images by a public ImageJ platform. The analysis was based on the contrast thresholding of images with further segmentation into objects of interest (pores or cells) and background.

3. Results and discussion

The dependence of morphology parameters of porous AAO for sulfuric, oxalic and phosphoric acid electrolytes on the substrate electrical conductivity is summarized in Fig. 3.

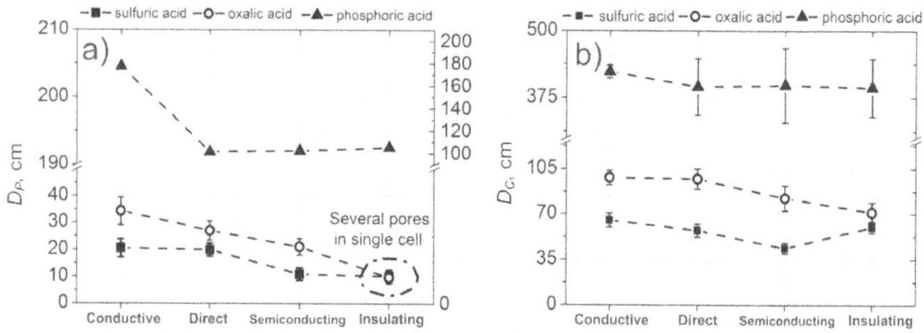

Figure 3. The pore (D_P) and cell (D_C) diameters of porous AAO formed in sulfuric, oxalic and phosphoric acid electrolytes for different types of electrical contacts between anode and anodized film.

The cell and pore diameters depend on the type of the electrolyte and the anodizing voltage. The cell diameter weekly depends on the substrate type. However, the pore diameter increases with an increase of the substrate electrical conductivity. It is interesting to mention that cells of AAO formed in oxalic and sulfuric acids with contacts provided from the anode to the anodizing film trough Si/SiO_2 substrate contain several pores.

The observed changes of AAO morphology can be explained by the fact that geometrical dimensions of the cell depends on the voltage applied between the anode and the cathode in the process of anodization [2] ($D_C \sim U_A$). On the other hand, the diameter of pores has the complicated dependence on the anodizing conditions known as [3]:

$$n \cdot D_p = 2.6 \times 10^{-17} \cdot j^{-0.32} \cdot T^{6.7}, \tag{1}$$

where n is the number of pores per cm^2 which depends inversely proportional on the current density of anodization ($n \sim j^{-\alpha}$), D_P is the pore diameter (nm), j is the current density (mA/cm^2), and T is the electrolyte temperature (K). From (1) one can see that the number of pores n increases with a decrease of the current density j. Since j has a minimum value for the insulating type of contact, a few pores can be observed in one cell.

Comparison of the data obtained for different substrates demonstrates that the anodizing process on the aluminum foil results in porous templates with the better ordering than templates produced on aluminium films deposited on silicon. The better ordering can be due to the longer anodization time during the first step of the anodization for foils. This step affects the ordering of primary oxide layer.

Acknowledgements

Authors acknowledge the support from the National Science Centre (Poland) grant 2014/13/N/ST8/00731. Authors are also grateful to F. Krok and B. R. Jany from Jagellonian University for SEM imaging.

References

1. Y. Lei, W. Cai, G. Wilde, *Progress in Materials Science* **52**, 465 (2007).
2. A. Li, F. Muller, A. Bimer, K. Nielsch, U. Gösele, *J. Appl. Phys.* **84**, 6023 (1998).
3. G. Patermarakis, K. Moussoutzanis, *Corrosion Science* **44**,1737 (2002).

MEMBRANE STRUCTURES BASED ON FREE TWO-LAYER POROUS ALUMINA FILMS

D. L. Shimanovich, V. A. Yakovtseva, S. A. Volchek, I. A. Skumin, A. D. Shimanovich
Belarusian State University of Informatics and Radioelectronics
P. Browka 6, 220013 Minsk, Belarus
ShDL@tut.by

The 73–216 μm thick nanostructured porous alumina membranes based on free-standing two-layer films of nanoporous anodic Al_2O_3 with pore diameters of about 55 nm were fabricated by the two-side anodization of Al in the oxalic acid electrolyte and further bipolar anodization. The volume growth coefficient in the conversion of Al to Al_2O_3 was 1.44–1.46. The membranes obtained demonstrated the high resistance to cracking and the stability of form at high temperature exposures.

1. Introduction

Membranes based on free-standing anodic alumina (Al_2O_3) films obtained by the two-step one-side anodization of aluminum (Al) foils followed by chemical removal of residual aluminum have been widely reported [1–3]. However, this method has disadvantages of the need for masking one side of the Al foil during the anodization and chemical removal of residual non-anodized Al. The resulting alumina membranes are not planar due to mechanical stresses at the Al/Al_2O_3 interface. They show the lack of dimensional stability at high-temperature testing and operation. There is also a variation in their thickness for the large-format membranes.

This paper discusses the method for producing alumina membranes by the two-step two-side anodization towards the complete end-to-end oxidation of initial aluminum foils. The main problem of this approach is associated with the need to ensure high surface quality of Al substrates. The surface roughness causes the effect of the supply potential cutoff at the final stage of the two-side anodization resulting in the formation of local non-anodized inclusions inside alumina membranes at the interface of the two opposing barrier layers junction. We showed that these Al inclusions can be eliminated by bipolar anodizing after the main stage of the deep end-to-end anodization.

2. Experimental

The 60, 110, and 160 μm thick aluminum foils of the purity of 99.99 % were used as initial materials. After repeated rolling using polished platens, the foils were subjected to the thermo-straightening under the pressure of ~10^7 Pa at 350 °C for 1 h to relieve mechanical stresses and increase plasticity. Then, the samples of 60×48 mm in size were formed by stamping. They were chemically cleaned in CrO_3:H_2SO_4 (1:100) for 3 min. In order to eliminate the microroughness, the electrochemical polishing of aluminum was carried out in the electrolyte based on perchloric and acetic acids (22 %:78 %) at the temperature of 7–9 °C and voltage of 25–27 V for 1 min. Thereafter the thickness of the Al samples became of about 50, 100, 150 μm, respectively.

The two-side anodization was made through two steps in the 7 % electrolyte of $H_2C_2O_4$ at the constant voltage of 55 V. The first anodization step was performed for 10 min. Then, the porous alumina film fabricated was selectively dissolved in CrO_3:H_3PO_4:H_2O solution at 85 °C for 5 min. Patterns that are replicas of the hexagonal pore arrays are preserved on the fresh aluminum surface. This allows the preparation of high regularity pores by the subsequent second anodization under the same conditions as the first anodization. The deep two-side end-to-end porous anodization was performed until the anodization current decreased to almost zero value when the two oxide layers growing towards each other are connected.

3. Results and discussion

The two-side two-step anodization of aluminum foils resulted in the fabrication of nanostructured membranes of 73, 145, and 216 μm in thickness. Fig. 1 presents SEM images of the alumina membrane obtained by the two-side two-step anodization.

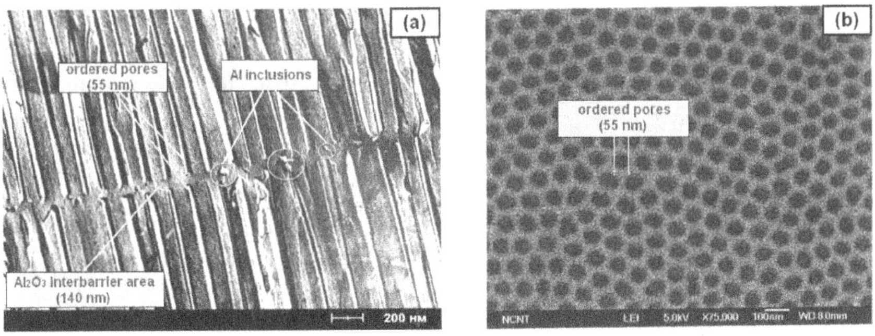

Figure 1. SEM images of alumina membrane fabricated by the two-side two-step anodization: cross-section (a) and top view (b).

The membranes consisted of two-layer nanoporous alumina with pores of about 55 nm in diameter symmetrically located on two sides of the membranes. The total thickness of the barrier layers was ~140 nm. The volume growth coefficient in the conversion of Al to Al_2O_3 was 1.44–1.46. As seen in Fig. 1, local non-anodized Al inclusions are observed inside alumina membranes at the interface of junction of the two opposing barrier layers.

The basic idea of the bipolar anodization is to use a two-chamber electrolytic bath (Fig. 2), in which the alumina membrane with Al inclusions was placed as an insulating partition. The first chamber is filled with the anodization electrolyte (7 % $H_2C_2O_4$), while the buffer electrolyte (10 % $CuSO_4$) is used in the second chamber. The cathode electrode is placed in the first chamber and the anode electrode is connected to the second one.

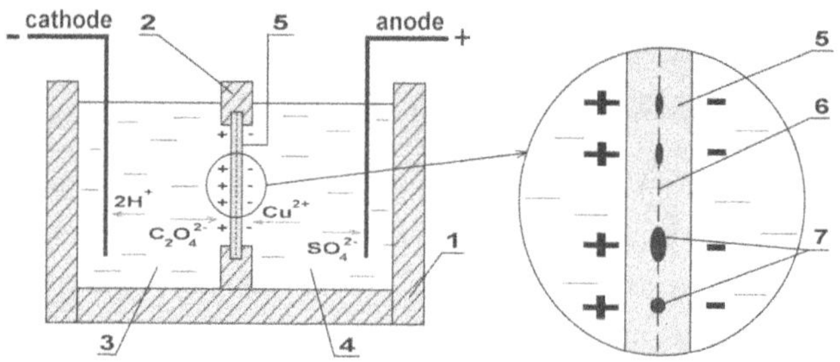

Figure 2. Schematic diagram of bipolar anodization: (1) two-chamber electrolytic bath; (2) insulating partition; (3) anodizing electrolyte in the cathode chamber; (4) buffer electrolyte in the anode chamber; (5) anodized alumina membrane; (6) interface of junction of the two opposing barrier layers; (7) non-anodized Al inclusions.

When the current was switched on (55 V), a positive charge appeared on one side of the membrane opposite the Al inclusions. The membrane became an anode, and anodization of Al inclusions took place. The opposite side of the membrane was charged negatively, it became a cathode. The recovery of buffer electrolyte cations (Cu^{2+}) on the cathode side opposite the Al inclusions was observed with guaranteed absence of sparks and burn-through of the oxidized layer in these zones.

Fig. 3 shows comparative images of the membranes fabricated to demonstrate the evolution of the bipolar anodization process during different times. The removal of copper deposits for the final membrane cleanup was carried out in 60 % HNO_3 for 2–3 min.

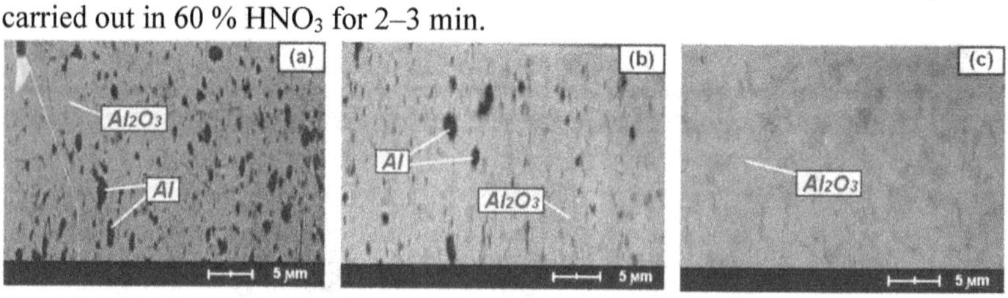

Figure 3. Nanoporous alumina membranes before and after the bipolar anodization for: 0 min (a); 15 min (b); 30 min (c).

4. Conclusion

The two-step two-side anodization of aluminum foils followed by the bipolar anodization and subsequent etching allows to fabricate nanostructured alumina membranes. The highly ordered nanostructured nature of their cellular porous morphology can be controlled by electrochemical and temperature modes of the anodization process. The technique appears to have considerable promise.

Acknowledgments

This work was partially supported by the Belarus State Program of the Scientific Research "Physical Science, New Materials and Technologies" - "Nanomaterials and Nanotechnologies" (grants 2.36 and 2.45).

References

1. O. Jessensky, F. Muller, U. Gosele. *Appl. Phys. Lett.* **72**, 1173 (1998).
2. P. C. Staira *et al.*, *Topics in Catalysis* **39**, 181 (2006).
3. V. A. Sokol, V. A. Yakovtseva, D. L. Shimanovich, *Doklady BGUIR* **64**, 21 (2012).

SOL-GEL COATINGS FOR PHOTOLITHOGRAPHY ON NANOPOROUS ANODIC ALUMINA AND ALUMINUM

L. S. Khoroshko, T. I. Orekhovskaya, M. V. Meledina
Belarusian State University of Informatics and Radioelectronics
P. Browka 6, 220013 Minsk, Belarus
l_khoroshko@bsuir.by

The results of the patterned microstructures fabrication through electrochemical anodizing, sol-gel synthesis, photolithography and chemical etching are presented. The continuous xerogel films were used as a protective mask. They are good alternative to metal in photolithography. Perspectives of application of these structures in planar optoelectronics are discussed.

1. Introduction

Microstructures based on porous anodic alumina (PAA) can be fabricated using photolithography method with various configurations according to the photomask patterning [1]. Previously, we have investigated different xerogel materials such as titania (TiO_2), strontium titanate ($SrTiO_3$) and yttrium aluminum composites as an alternative to metal protective masks, preventing penetration of photoresists into channels of the pores of anodic alumina [2,3]. Xerogels doped with lanthanides revealing luminescence in a wide spectral range expand the areas of possible applications of these structures [4]. Furthermore, sufficient adhesion of xerogel films such as TiO_2 and $SrTiO_3$ to various types of substrates like silica, silicon, aluminum and nanotextured aluminum also received significant attention for fabrication of patterned sol-gel derived coatings [5].

In this paper, we describe the route of the film microstructures fabrication obtained through electrochemical anodizing of aluminium, sol-gel synthesis of titania, photolithography and chemical etching. The possibility of using titania xerogels as a protective mask in photolithography on PAA with the pores of 20–50 nm is shown. Also, we propose to utilize the sol-gel method for fabrication the titania patterned film on an aluminum foil using a lift-off technology.

2. Experimental

The following steps were performed in the fabrication procedure for microstructures based on PAA:

(i) The synthesis of PAA with required morphology by the two-step process of anodic oxidation in various electrolytes improving the pores ordering.
(ii) The xerogel formation on the surface of PAA by spinning of titania sol (2700–3000 rev/min) with intermediate drying of each layer at 100–200 °C in air (8–10 xerogel layers are needed to complete the pores mouths resulting in the surface planarization).
(iii) The photolithographic procedure.
(iv) The anisotropic etching of titania xerogel/PAA through the windows of lithographic mask in the selective etchant comprising ammonium fluoride solution in hydrochloric acid [2].

PAA was fabricated by the two-step anodizing of aluminum foil (99.5 % of Al) or aluminum layers deposited on silicon wafers. Anodic alumina formed by the first anodizing step was removed in the mixture of orthophosphoric acid and chromium (IV) oxide at 60 °C. The second anodizing step was performed under the same conditions. The following anodizing modes were used: phosphoric acid, 10–12 °C, 100 V; oxalic acid, 12–15 °C, 40 V, combined electrolyte (oxalic and sulfuric acids), 12–15 °C, 20 V. The fabricated PAA films were dried at 100 °C for 1 h in air. Titania sol with europium was prepared from titanium tetraisopropoxide, acetyl acetone, ethylene glycol monomethyl ether, and europium nitrate dissolved in distilled water. The xerogel coating was formed by spinning of the sol followed by drying of each gel layer.

The lift-off photolithography on aluminum foil was performed by the following sequence:

(i) The cleaning treatment of aluminum foil surface.
(ii) The photoresist mask deposition and prebaking. (fragments that form the pattern remain uncovered by the photoresist).
(iii) The sol-gel film formation on the free areas (europium-doped titania).
(iv) The photoresist mask removing by the "explosion".

The photoresist SP 1813 of 1.8 μm thickness was used for the lift-off photolithography. The fragments to be patterned with the xerogel remained opened while the residual parts of a sample were coated with photoresist/xerogel. Unlike the standard lift-off photolithography, prior to sol deposition the resistive mask was prebaked at 160 °C for 30 min, preventing its dissolution during deposition of the sol containing ethanol. The film consisting of 10 xerogel layers of TiO_2:Eu was formed on the photoresist and the opened fragments of the sample. Then, the photoresist was removed in the mixture of dimethylformamide and monoethanolamine (5:1) at 80 °C. Finally, only fragments of xerogel corresponding to the photolithographic mask remained at the surface of the sample. Morphology of the structures fabricated was investigated using a HITACHI S-4800 scanning electron microscope (SEM).

3. Results and discussion

According to SEM data, quality of the walls of the etched grooves in the xerogel/PAA structure depends on the pore size of anodic alumina. Pores about 70–100 nm (orthophosphoric electrolyte) allow to fabricate inhomogeneous titania xerogel coating which do not cover the PAA surface entirely. Thus the photoresist penetrate into the pores resulting in rough edge of the walls after photolithography and etching procedures (Fig. 1a,b). On the contrary, PAA with the pores of 20–50 nm (oxalate and combined electrolyte) allows formation of the continuous titania xerogel coating on the surface preventing penetration of photoresist through the pore channels and remaining sharp edges of the walls after etching (Fig. 1c).

a b c

Figure 1. SEM images of the PAA based structures after the photolithography and etching in fluoride-containing mixture: (a) PAA formed in 1 M H_3PO_4; (b) the wall of the etched groove at PAA formed in 1 M H_3PO_4; (c) the grooves on PAA formed in the combined electrolyte.

Fig. 2 shows images formed on Al foil by the lift-off photolithography.

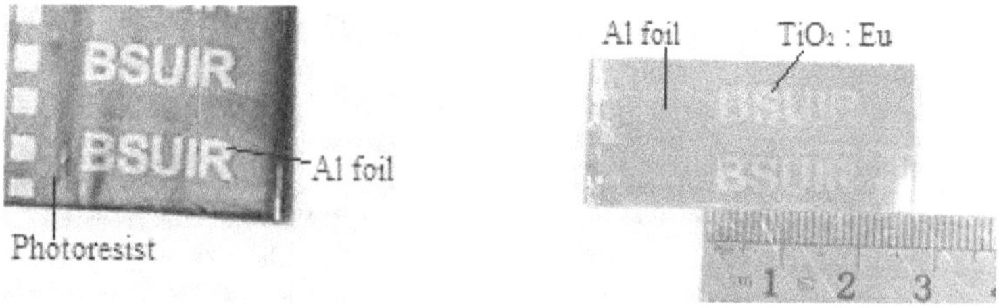

a b

Figure 2. Inscriptions "BSUIR" on Al foil: (a) photoresist mask on Al foil; (b) TiO_2:Eu xerogel on Al foil after removing the photoresist mask.

Thus, the above experiment demonstrated that the proposed method combined the sol-gel technology and lift-off lithography allowed to exclude etching of xerogels in strong acid solutions. The xerogels doped with luminescent ions (Eu^{2+}) expand the range of possible applications of the described structures. The etched groove configuration is of interest in a planar waveguide technology.

4. Conclusion

We have fabricated and studied microstructures formed on aluminum using electrochemical anodizing, sol-gel technology, photolithography and chemical etching of xerogel/PAA without metal masks, and obtained with the lift-off photolithography. The technology will be further developed for application in planar optoelectronics, for creation of security marks and new photocatalytic structures.

Acknowledgments

The authors wish to thank Prof. N.V. Gaponenko for stimulating discussion.

References

1. N. V. Gaponenko *et al.*, *J. Electrochem. Soc.* **149 (2)**, H49 (2002).
2. L. S. Khoroshko, *J. Phys.: Conf. Ser.* **643**, 012110 (2015).
3. L. Khoroshko *et al.*, in: *Physics, Chemistry and Application of Nanostructures*, eds. V. E. Borisenko, S. V. Gaponenko, V. S. Gurin, C. H. Kam, (Singapore, World Scientific, 2015), pp. 377–380.
4. N. V. Gaponenko *et al.*, patent BY 15052 (2011).
5. N. V. Gaponenko *et al.*, *Microelectron. Eng.* **90**, 131 (2012).

FEATURES OF PLASMA-ASSISTED NANOSTRUCTURING OF $(CuInSe_2)_x$–$(2ZnSe)_{1-x}$ FILMS WITH DIFFERENT CRYSTALLOGRAPHIC POLYTYPES

S. P. Zimin, D. A. Mokrov

Yaroslavl State University, Sovetskaya 14, 150003 Yaroslavl, Russia
zimin@uniyar.ac.ru

E. S. Gorlachev, I. I. Amirov, V. V. Naumov

Yaroslavl Branch of the Institute of Physics and Technology RAS
Universitetskaya 21, 150007 Yaroslavl, Russia

V. F. Gremenok

Scientific-Practical Materials Research Center NASB
P. Browka 19, 220072 Minsk, Belarus

A. N. Pyatlitski, T. W. Petliskaya

Belmicroanalysis State Center, RDC "Belmicrosystems" RPC "Integral"
220108 Minsk, Belarus

We report a study of $(CuInSe_2)_x$–$(2ZnSe)_{1-x}$ film surface after low-energy inductively coupled argon plasma treatment. The films were grown on glass and steel substrates using selenization process. They had similar polycrystalline morphology but different crystallographic polytypes. The films on the glass substrates had the chalcopyrite structure and initial metallic nanoinclusions. In the process of plasma treatment they showed self-formation of high-density vertical nanorods. The films on the steel substrates had an abnormal sphalerite structure without nanoinclusions while during the plasma treatment they reveal low-density nanocones.

1. Introduction

$CuIn_{1-x}Ga_xSe_2$ (CIGS) and $(CuInSe_2)_x$–$(2ZnSe)_{1-x}$ (CIZS) are promising absorber materials for photovoltaic applications. They allow an optical band gap variation in the wide range of ~1.0–2.7 eV and show high conversion efficiency of ~22.6 % [1,2]. However, the CIZS films have an advantage from economical point of view, since incorporation of Zn is less expensive than In. An improvement of functional characteristics of CIGS and CIZS film device structures can be achieved through nanostructuring of the film surface. Previously [3], we have developed a new approach to the formation of nanostructures on the surface of polycrystalline CIGS films using inductively coupled plasma treatment [4]. In this paper, we report an application of this plasma-assisted nanostructuring method to polycrystalline CIZS films on glass and steel substrates with different crystallographic polytypes.

2. Experimental

The CIZS films were grown using the two-stage precursor selenization process [1] on glass substrates with a Cr sublayer and steel substrates with a Ti sublayer. The films on the glass substrates (In 25.0 at.%, Zn 1.5 at.%, Cu 25.3 at.%) had the chalcopyrite crystalline structure typical for CIZS and CIGS films, but the films on the steel substrates (In 29.4 at.%, Zn 2.0 at.%, Cu 20.3 at.%) had an abnormal structure of sphalerite polytype. Plasma treatment of the films was carried out in inductively coupled argon plasma [3,4]. The treatment duration was 60 s, the RF power was 300 W, which corresponded to the low average bombarding Ar^+ ion energy of 200 eV. The surface morphology studies were carried out with scanning electron microscopy (SEM) using a Supra 40 (Carl Zeiss) microscope.

3. Results and discussion

The as-grown films of both types were polycrystalline with irregularly shaped grains of 1.0–3.5 μm. The chalcopyrite polytype film surface had nanoinclusions with the density of 4×10^{11} cm^{-2} and the size of 10 nm (Fig. 1), which are probably metallic alloy (Cu-In-Zn) nanoparticles like in CIGS films [5]. The sphalerite polytype films, however, did not have any nanoinclusions.

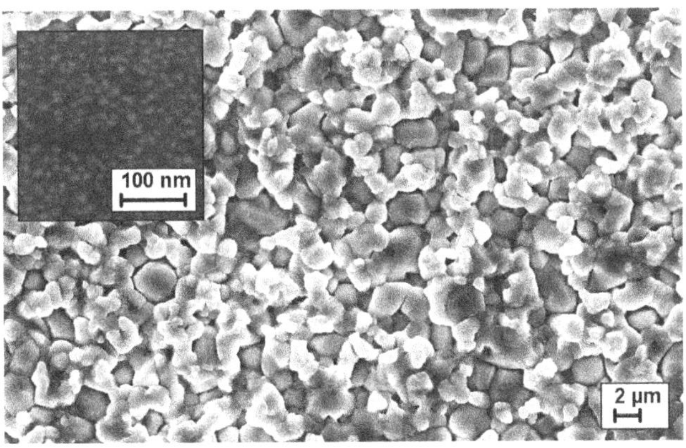

Figure 1. Typical SEM image of the initial chalcopyrite CIZS film grown on the glass substrate. The inset presents a high-resolution SEM image of nanoinclusions on the flat area surface.

For the chalcopyrite polytype films the plasma treatment resulted in formation of nanorods (height up to 120 nm, diameter of 40 nm and density of 6×10^{10} cm^{-2}, Fig. 2). For the sphalerite films this treatment leads to formation of nanocones (height up to 250 nm, base diameter of 160 nm and density of 1×10^{9} cm^{-2}, Fig. 3).

Figure 2. Typical SEM images of nanorods on the surface of the chalcopyrite CIZS film on the glass substrate (70 deg tilt view). The inset shows a surface view of the nanorods at the same length scale.

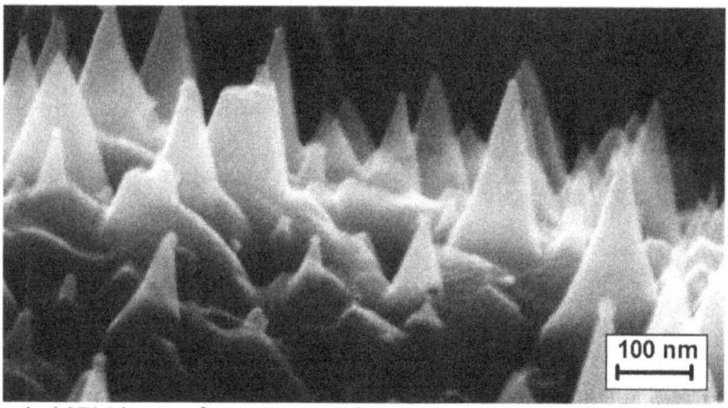

Figure 3. Typical SEM image of nanocones on the surface of the sphalerite CIZS film on the steel substrate formed during the plasma treatment (70 deg tilt view).

We can explain these results with the model of plasma-assisted self-formation of nanostructures. We postulate that the starting point for the nanorod growth is the presence of metallic nanoinclusions. The plasma treatment induces physical sputtering and surface heating, which result in the formation of the sputtered CIZS elements in a vapor phase and in the transition of nanoinclusions into a liquid phase. These processes result in the implementation of the vapor-liquid-solid mechanism, with initial nanoinclusions serving both as droplets and micromasks. As a consequence, the formation of nanorods can take place. When the initial high-density nanoinclusions are absent, only a micromasking process (presumably with random metallic or oxide particles) during plasma treatment tends to the formation of nanocones.

4. Conclusion

The study of the argon plasma treatment of the $(CuInSe_2)_x$–$(2ZnSe)_{1-x}$ film surface with the structure of chalcopyrite and sphalerite crystallographic polytypes grown on glass and steel substrates was carried out. The chalcopyrite polytype films with nanoinclusions showed self-formation of high-density vertical uniform nanorod arrays, while the sphalerite type films demonstrated formation of low-density nanocones. The results show an importance of the nanoinclusions and crystallographic polytype in the plasma-assisted nanostructure self-formation processes.

Acknowledgments

These investigations were carried out at the Facilities Sharing Center "Diagnostics of Micro- and Nanostructures" with the support of the Ministry of Education and Science of Russian Federation.

References

1. V. A. Ivanov, V. F. Gremenok, V. M. Siarheyeva, E. P. Zaretskaya, V. B. Zalesski, V. I. Kovalevski, K. Bente, *Moldavian J. Phys. Sci.* **5**, 355 (2006).
2. M. A. Green, K. Emery, Y. Hishikawa, W. Warta, E. D. Dunlop, *Prog. Photovolt.: Res. Appl.* **24**, 905 (2016).
3. S. P. Zimin, D. A. Mokrov, E. S. Gorlachev, I. I. Amirov, V. V. Naumov, V. F. Gremenok, S. A. Bashkirov, in: *Physics, Chemistry and Application of Nanostructures*, eds. V. E. Borisenko, S. V. Gaponenko, V. S. Gurin, C. H. Kam (Singapore, World Scientific, 2015), pp. 388–390.
4. S. P. Zimin, E. S. Gorlachev, I. I. Amirov, V. V. Naumov, G. A. Dubov, V. F. Gremenok, S. A. Bashkirov, *Semicond. Sci. Technol.* **29**, 015009 (2014).
5. S. P. Zimin, E. S. Gorlachev, V. F. Gremenok, K. Bente, D. A. Mokrov, I. I. Amirov, V. V. Naumov, W. Y. Kim, *Phys. Status Solidi C* (in press).

AFM INVESTIGATION OF CZTS AND CZTSe THIN FILMS

K. A. Urazov
Center of Chemical Engineering, Kazakh-British Technical University
Tole bi Str. 59, 050000 Almaty, Kazakhstan
u_kazhm@mail.ru

M. B. Dergacheva
Electrochemistry Technology Department, Sokolsky Institute of Fuel, Catalysis and Electrochemistry, Kunaev Str. 142, 050010 Almaty, Kazakhstan
m_dergacheva@mail.ru

V. F. Gremenok
Laboratory of Solid State Physics, Scientific-Practical Materials Research Centre NASB
P. Browka Str. 19, 220072 Minsk, Belarus
gremenok@ifttp.bas-net.by

K. A. Mit
Institute of Physics and Technology, Ibragimov Str. 11, 050032 Almaty, Kazakhstan
mit@sci.kz

The morphology of electrodeposited Cu_2ZnSnS_4 (CZTS) and $Cu_2ZnSnSe_4$ (CZTSe) thin films was investigated by atomic force microscopy (AFM). The thin films were prepared by the single step electrodeposition method from aqueous solution in the potentiostatic mode. As-deposited thin films were annealed at 450 °C in Ar atmosphere. The CZTSe thin films tend to have a more homogeneous particle size distribution.

1. Introduction

$Cu_2ZnSnS(Se)_4$ (CZTS(Se)) compound has the band gap of 1–1.5 eV and the high absorption coefficient (above 10^4 cm^{-1}), which are favorable for the conversion of solar radiation [1,2]. Environment friendly CZTS(Se) is a promising material, which is intensively studied at present and can be used as an absorbing layer of thin film solar cells [3–5]. A number of techniques are known for fabrication of CZTS(Se), such as co-evaporation [6], reactive magnetron co-sputtering [7], soft-chemistry method [8], sol-gel precipitation followed by sulfurization [9] and electrodeposition [10–13]. However, especially noteworthy is the electrochemical deposition, which makes it possible to perform the process at low temperatures with all four elements (Cu, Zn, Sn, and S or Se) simultaneously deposited.

We carried out a single step electrochemical deposition of CZTS and CZTSe thin films in the potentiostatic mode at room temperature. The influence of anion compositions (S,Se) on surface morphology was investigated. The results are presented in this paper.

2. Experimental

Electrochemical deposition of CZTS and CZTSe thin films was performed in a three-electrode electrochemical cell using a potentiostat/galvanostat Gill AC (ACM Instruments). A Mo/glass substrate was used as a working electrode, platinum as an auxiliary electrode and silver chloride electrode as a reference electrode.

The CZTS thin films were deposited onto the Mo/glass substrates from the electrolyte based on 0.2 M tri-sodium citrate, which contains the following salts: 0.01 M $CuSO_4 \cdot 5H_2O$, 0.01 M $ZnSO_4 \cdot 7H_2O$, 0.005 M $SnSO_4$ and 0.05 M $Na_2S_2O_3$, all of analytically pure grade. To adjust pH of the electrolyte to 4.6, 0.1 M tartaric acid was used as described in [14]. The electrolysis was performed at the constant potential of –1.0 V for 30 min at room temperature.

The CZTSe thin films were electrodeposited onto the Mo/glass substrates from the electrolyte with pH 1.5, containing 0.1 M tartaric acid, 0.005 M $CuSO_4 \cdot 5H_2O$, 0.01 M $ZnSO_4 \cdot 7H_2O$, 0.01 M $SnCl_4$ and 0.005 M $NaHSeO_3$. All reagents were of analytically pure grade. The electrolysis was performed at the constant potential of –0.6 V for 30 min at room temperature [15].

The as-deposited CZTS and CZTSe thin films were annealed at 450 °C for 30 min in Ar atmosphere for crystallization. Their chemical composition was examined by EDAX with JSM-6610 (Japan) scanning electron microscope (SEM). The morphological properties of the films were studied by AFM (JSPM-5200).

3. Results and discussion

The results of compositional analysis of the CZTS and CZTSe thin films are collected in Table 1.

Table 1. The chemical composition of CZTS and CZTSe thin films.

Composition, at.%	CZTS	CZTSe
Cu	25.7	19.5
Zn	11.4	10.6
Sn	12.4	14.6
S	50.3	–
Se	–	55.2

The study of the surface morphology of the electrodeposited CZTS film with quaternary composition Cu_2ZnSnS_4 by AFM allowed to establish that their surface was formed uniformly distributed particles (Fig. 1a). The surface is irregular and poorly homogeneous. The films consist of large and small grains. The particles grow in one direction. Sizes of larger particles are about 0.5–0.7 μm. The large particles consist of small particles.

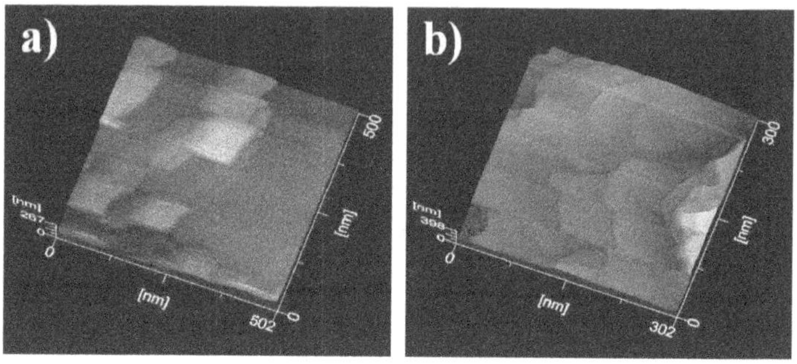

Figure 1. AFM images of CZTS (a) and CZTSe (b) thin films.

The CZTSe thin films have a smaller roughness within the range of a few tens nanometers. The diameter of smaller particles averages to 25 nm. These particles form close assembles with large grains of different forms.

SEM cross section images of CZTSe thin film are shown in Fig. 2. The as-deposited CZTSe film has cavities (Fig. 2a). They can be a consequence of poor hydrogen elimination under electrodeposition from the aqueous solution. The cavity disappeared after annealing at 450 °C (Fig. 2b). The film thickness decreases from 2.4 μm to 1.9 μm and the optimal thickness of CZTS and CZTSe films reached 1.9–2.5 μm notice that CZTSe films tend to have a more homogeneous distribution of particles size.

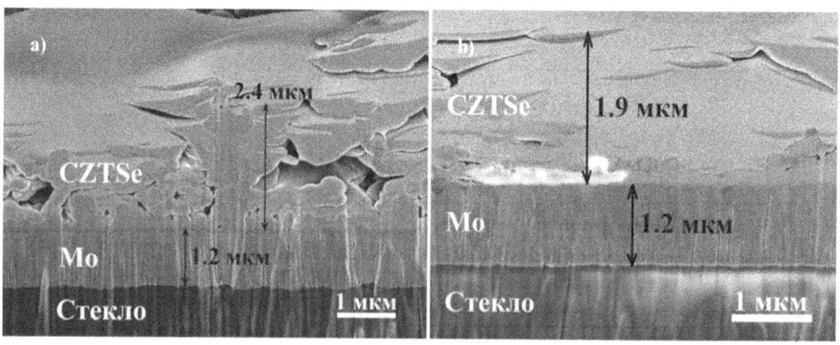

Figure 2. Cross-sectional SEM images of CZTSe thin film before (a) and after (b) annealing.

Acknowledgments

The work has been supported by the Ministry of Education and Science of the Republic of Kazakhstan (grant 0200/GF4) and Belarusian Republican Foundation for Fundamental Research (grant F15KOR-004).

References

1. T. Tanaka, T. Nagatomo, T. Kawasaki, *J. Phys. Chem. Solids* **66**, 1978 (2004).
2. K. Ito, T. Nakazawa, *Jpn. J. Appl. Phys.* **27**, 2094 (1988).
3. T. K. Todorov, K. B. Reuter, D. B. Mitzi, *Adv. Mater.* **22**, E156 (2010).
4. A. Walsh, S. Chen, S.-H. Wei, *Adv. Energy Mater.* **2**, 400 (2012).
5. J. Paier, R. Asahi, A. Nagoya, *Phys. Rev.* **B79**, 115126 (2009).
6. T. Tanaka, D. Kawasaki, M. Nishio, *Phys. Status Solidi* **3**, 2844 (2006).
7. F. Liu, Y. Li and K. Zhang, *Sol. Energy Mat. Sol. C* **94**, 2431 (2010).
8. T. K. Todorov, M. Kita, J. Carda, *Thin Solid Films* **517**, 2541 (2009).
9. K. Tanaka, N. Moritake, H. Uchiki, *Sol. Energy Mat. Sol. C* **91**, 1199 (2007).
10. S. M. Pawar, B. S. Pawar, A. V. Moholkar, *Electrochim. Acta* **55**, 4057 (2010).
11. M. Lux-Steiner, A. Weber, D. Abou-Ras, *Thin Solid Films* **517**, 2511 (2009).
12. Y. Cui, S. Zuo, J. Jiang, *Sol. Energy Mat. Sol. C.* **94**, 2136 (2011).
13. C. P. Chan, H. Lam, C. Surya, *Sol. Energy Mat. Sol. C.* **94**, 207 (2010).
14. M. B. Dergacheva, K. A. Urazov, A. E. Nurtazina, *Russ. J. Electrochem.* **53**, 1 (2017).
15. M. B. Dergacheva, K. A. Urazov, N. N. Gudeleva, K. A. Mit', *Chem. J. Kazakhstan* **4(56)**, 129 (2016).

SURFACE MORPHOLOGY OF GERMANIUM LAYERS ON SILICON SURFACES AT HIGH TEMPERATURES

A. A. Shklyaev
A. V. Rzhanov Institute of Semiconductor Physics, SB RAS, 630090 Novosibirsk, Russia
Novosibirsk State University, Pirogov Str. 2, 630090 Novosibirsk, Russia

The growth of Ge on Si(100) surfaces at 750–950 °C occurs close to the thermodynamic equilibrium conditions due to high rates of the surface and bulk diffusion and Si-Ge intermixing. Experimental results show a tendency in reordering of the surface morphology controlled by the dynamic equilibrium between the island growth and their decay by atomic diffusion. Dewetting of SiGe structures from Si surfaces takes place in the conditions far from the dynamic equilibrium.

1. Introduction

A self-organization of surface morphology during the growth of heterostructures is a consequence of total energy minimization of growing structures. The main contribution into this process comes from the elastic strain relaxation. It can occur completely in the absence of kinetic restrictions in surface reactions providing the growth in conditions close to thermodynamic equilibrium. For the growth of Si/Ge heterostructures, such conditions can be realized at high temperatures, 800–950 °C. Our studies show that the use of such temperatures leads to two types of substantially different surface morphologies depending on the amount of deposited Ge on Si(100) surfaces and on the Ge deposition rate.

At high temperatures, along with the surface diffusion of deposited adatoms and their attachment to energetically favorable sites on the surface, additional processes can occur: Si-Ge intermixing and Ge diffusion into the surface layer. Our results show that there is a boundary Ge deposition rate and temperature, along which the dynamic equilibrium between the island growth and their decay takes place. The boundary can be referred to as critical conditions for the island growth at high temperatures [1]. This means that if the Ge deposition rate is not high enough, Ge atoms accumulated on the surface are insufficient to form islands. We have shown that three-dimensional (3D) islands, formed in the conditions close to the dynamical equilibrium have an ordering tendency [2]. They get an identical shape and are characterized by a uniform strain distribution along the surface. If the Ge deposition rate is relatively high, islands of submicrometric sizes appear on the surface. In this case, the effect of SiGe dewetting from the Si(100) surface takes place producing SiGe islands with a large aspect ratio.

2. Experimental

The growth of Si/Ge heterostructures was carried out by means of molecular beam epitaxy. The vacuum system was equipped with a scanning tunneling microscope (STM) manufactured by Omicron. A Knudsen cell with a BN crucible was used for Ge deposition at the rate of 0.06–1.2 nm/min. The sample temperature was measured with an IMPAC IGA 12 pyrometer. After the samples removal from the growth chamber, scanning electron microscopy (SEM) images were obtained using a Pioneer microscope manufactured by Raith.

3. Critical conditions for SiGe island formation during Ge deposition on Si(100) at high temperatures

The island nucleation usually occurs spontaneously and results in an island random distribution along the surface. There is a specific feature of the island formation in the conditions close to critical, which describe the dynamic equilibrium between the island formation and decay due to Ge diffusion into Si substrate (Fig. 1a). The islands appear on the surface without a nucleation step through a gradual evolution of the surface morphology. The elastic strain minimization is reached by means of the formation of ordered islands with a smooth surface morphology (Fig. 1b,c).

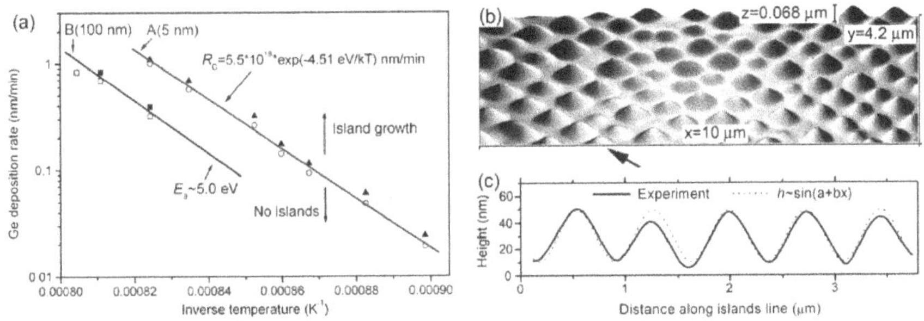

Figure 1. (a) Critical conditions for the island formation during 5 (data A) and 100 nm (data B) Ge depositions on Si(100) at high temperatures. The solid and open symbols represent the conditions at which the islands appear or not on the surface, respectively. (b) STM image of the surfaces obtained by 20 nm Ge deposition on Si(100) at 925 °C with the deposition rate of 0.39. (c) The solid line represents the height profile along the island array indicated by the arrow in (b). The dotted line is the experimental height profile approximation by the sinusoidal function.

The lateral island ordering is governed by the lattice strain anisotropy. The results obtained clearly reflect the substrate crystal symmetry. It would be suggested that ordering of 3D structures can be reached by long time annealing of the islands initially grown at relatively low temperatures. However, this was not realized experimentally, due to the necessity of rather high annealing temperature, since the ordering is required for intensive mass transport along the surface. Since the high-temperature annealing also initiates atomic diffusion from epitaxial layers into substrates, this diffusion significantly restricts the temperature and duration of annealing. Our study shows that limitations on the annealing temperature and duration can be overcome if the epitaxial layer composition is restored during annealing by the material deposition on its surface.

4. Dewetting of SiGe layers from Si(100) surfaces at high temperatures

A study of submicron dielectric particles, which scatter the light according to the Mie theory, has become attractive for the last years because they can generate well-defined resonant electromagnetic modes in good agreement with theoretical predictions for particles with large refractive index ($n > 3$) [3]. However, due to their lower optical losses and better compatibility with requirements of modern silicon technology than metal particles, they can find wider application for devices operating in the visible and near-infrared spectral regions.

The simplest and most competitive method of submicron particle formation is based on the dewetting effect of Si and Ge from the surface of SiO_2 films. The dewetting occurs during annealing of Si and Ge layers initially deposited on SiO_2 films at relatively low temperatures close to the room temperature. We have found that a bare Si surface becomes dewettable with respect to Ge at high temperatures, resulting in formation of the surface morphology with ridges in the case of Si(111) substrates [4,5] or the porous layer on Si(100) substrates [6]. We have found that the Si(100) surface exhibits dewetting features with respect to Ge which produce the formation of particle arrays (Fig. 2), when Ge is deposited on Si(100) surfaces directly at 750–950 °C. The dewetting occurs when the Ge deposition rate is relatively high producing the growth conditions that are far from the dynamic equilibrium between islands growth and decay. From the viewpoint of light scattering the use of dewetting effects are important for fabrication of particles with the aspect ratio of 0.5–0.7.

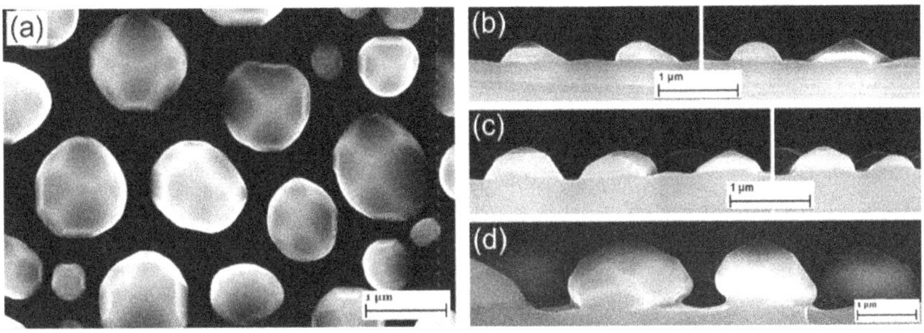

Figure 2. SEM images of (a) Si(100) surface after 30 nm Ge deposition and (b) – (d) cross-sections of the samples obtained by 20, 30 and 100 nm Ge deposition on Si(100), respectively, at 850 °C.

5. Conclusion

High-temperature Ge deposition on Si(100) surfaces allows the formation of two types of surface morphologies. One type is composed of ordered SiGe island arrays with a uniform shape and smooth surface morphology. The other type consists of submicron particles with a high aspect ratio. The features of these structures are favorable for wide practical applications, in particular, for design of sensors and in photovoltaic devices.

Acknowledgments

The partial financial support by the Russian Science Foundation (grant 14-12-01037) is gratefully acknowledged.

References

1. A. A. Shklyaev, A. E. Budazhapova, *Mater. Sci. Semicond. Proc.* **57**, 18 (2017).
2. A. A. Shklyaev, A. E. Budazhapova, *Appl. Surf. Sci.* **360**, 1023 (2016).
3. M. Naffouti, T. David, A. Benkouider, L. Farve, A. Ronda, I. Derbezier, S. Bidault, N. Bonod, V. Abbarchi, *Nanoscale* **8**, 2844 (2016).
4. A. A. Shklyaev, K. E. Ponomarev, *J. Cryst. Growth* **413**, 94 (2015).
5. A. Shklyaev, L. Bolotov, V. Poborchii, T. Tada, *J. Appl. Phys.* **117**, 205303 (2015).
6. A. A. Shklyaev, A. V. Latyshev, *Nanoscale Res. Lett.* **11**, 366 (2016).

FORMATION OF SILICON-GERMANIUM NANOCLUSTERS ON RARE EARTH OXIDES

A. S. Strogova, A. A. Kovalevsky

Belarusian State University of Informatics and Radioelectronics
P. Browka 6, 220013 Minsk, Belarus
strogova@bsuir.by

Self-organization and growth of nanoclusters containing silicon, germanium and a silicon-germanium solid solution on the surface of Dy_2O_3 and Y_2O_3 were investigated by atomic force microscopy and Raman spectroscopy.

1. Introduction

Nanostructures of silicon (Si), germanium (Ge) and silicon-germanium (Si-Ge) solid solutions have found widespread applications in fabrication of semiconductor devices, integrated circuits, and other products of micro- and nanoelectronics [1,2]. In this paper, CVD fabrication of Si-Ge nanostructures on the surface of rare earth oxides (Dy_2O_3 and Y_2O_3) is presented.

2. Experimental

Silicon (100) wafers of 100 mm in diameter covered by Dy_2O_3 and Y_2O_3 as thick as 20 nm and 40 nm, respectively, were used as substrates. Simultaneous deposition of Si and Ge was performed in the isothermal zone of "Isotron 4-150" horizontal reactor at a reduced pressure (10–140 Pa) in the temperature range of 520–630 °C [2]. GeH_4 and SiH_4 were used as precursors with their ratio changing from 1:1000 to 5:1000.

The layers thickness was determined by cleavage in a S-4800 (Hitachi, Japan) scanning electron microscope. Lateral size and height of nanoclusters were determined with Ntegra Prima scanning probe microscope operating in the AFM mode.

The composition of the nanoclusters was analyzed on the basis of Raman spectroscopy data. Raman spectra were registered on a DFS-52 spectrometer with Ar laser excitation (λ = 514.5 nm).

3. Results and discussion

Results of AFM study have demonstrated that the process of nucleation and growth of nanoclusters is predetermined by sorption of silicon and germanium atoms at the virgin silicon or Dy_2O_3 and Y_2O_3 covered substrates. They coalesce with the formation of silicon, germanium or Si-Ge nuclei, as evidenced by the Raman measurements, which are subsequently growing (Fig. 1).

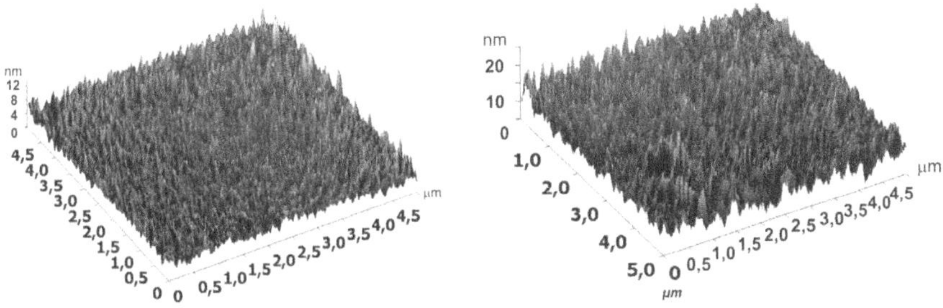

a b

Figure 1. AFM images of nanoclusters on Dy_2O_3 (a) and Y_2O_3 (b) layers on (100) silicon substrate.

All stages of the nanocluster nucleation and growth depend on topography and purity of the substrate, its orientation and chemical composition of the gas mixture used in the process of the CVD deposition. In fact, the deposited material is a mixture of polynanocrystalline silicon doped by germanium and Si-Ge solid solutions. Their internal structure can be exceedingly varied on any substrate tested and depend on such factors as gas pressure and temperature controlling the deposition rate. They determine the content of germanium atoms in the deposited layer, which in turn influences the lattice parameter of the Si-Ge solid solution and related surface reconstruction.

Fig. 2 shows the surface topography of the nanoclusters fabricated at different gas pressure in the reactor.

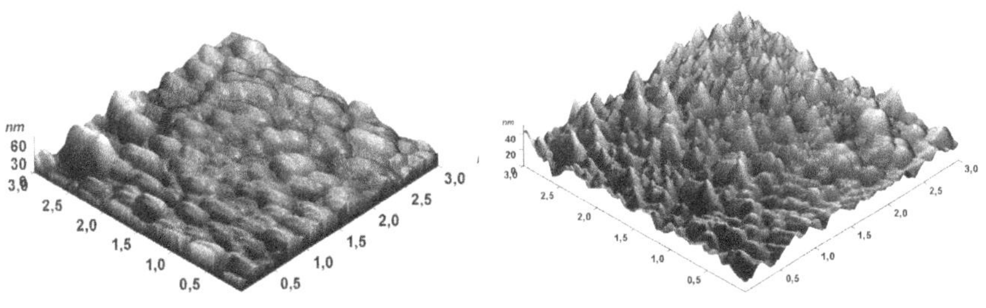

a b

Figure 2. AFM images of nanoclusters formed at the pressure of 45 Pa (a) and 35 Pa (b).

The observed increase of the height of the growing nanoclusters during material deposition occurs due to emission of germanium atoms from low to upper levels of the deposition plane, because upper parts of the clusters get insufficient amount of germanium atoms to ensure a consistent formation of the second and third layers.

The cluster growth is also dependent on the temperature in the reaction zone. At high temperatures (560–620 °C) the clusters usually begin to grow in the form of prisms (Fig. 3). Analysis of AFM images has shown that in this case the cluster growth takes place only up to a certain critical volume. Then they transform into domes due to the appearance of new lateral faces forming a larger angle with the substrate as compared to prisms.

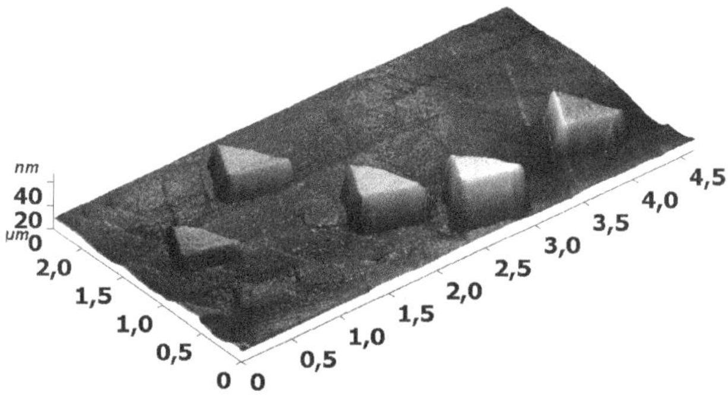

Figure 3. Fragment of the substrate with prismatic nanoclusters.

Similar changes of the cluster shape were explained in [1,2] by the more effective relaxation of the strains in the dome-shaped clusters due to the large ratio of its height to the lateral dimensions. The same can be said about truncated pyramid clusters. Probably, the truncated pyramids are formed by Si-Ge solid solutions. An increase of the deposition temperature results in an enhanced surface diffusion of germanium and silicon atoms to the already nucleated clusters. As a consequence, one may observe a decrease of the surface density of the clusters, an increase of their size and amount of silicon in the clusters, *i.e.* formation of Si-Ge solid solution. The temperature of the deposition is far from the melting point of silicon and germanium, hence the equilibrium form of nanocrystals should contain facets with low surface energy. Usually, facets with low Miller indices are formed.

The difference between the size and structure of silicon, germanium and Si-Ge nanoclusters may be conditioned by various factors related to physical and technological parameters of their formation. Also, there are effects due to the small characteristic size of defects on initial surface and its crystallographic orientation. In the first case, new phases emerging in the clusters are to be non-equilibrium, and in the second case, they may be stable at a given size and temperature.

4. Conclusion

The experimental data obtained certify a possibility to fabricate self-organized nanoclusters of Si-Ge solid solutions employing CVD deposition from a mixture of GeH_4 and SiH_4. Nanolayers of Dy_2O_3 and Y_2O_3 on a silicon substrate, gas pressure and temperature in the reaction chamber have been found to determine the surface density and the form of the clusters.

References

1. A. A. Kovalevskii, N. V. Babushkina, D. V. Plyakin, A. S. Strogova, *Russian Microelectronics* **39**, 190 (2010).
2. A. A. Kovalevskii, A. S. Strogova, D. V. Plyakin, *Russian Materials. Technologies. Tools.* **15**, 32 (2010).

ANALYSIS OF NANOWIRE DIAMETER VARIATION DURING MBE SELF-CATALYZED GROWTH: A MONTE CARLO SIMULATION

A. Nastovjak

A. V. Rzhanov Institute of Semiconductor Physics SB RAS, 630090 Novosibirsk, Russia

A. Suprunets

Novosibirsk State Technical University, 630073 Novosibirsk, Russia

N. Shwartz

A. V. Rzhanov Institute of Semiconductor Physics SB RAS, 630090 Novosibirsk, Russia
Novosibirsk State Technical University, 630073 Novosibirsk, Russia

The features of nanowire diameter self-equilibration during self-catalyzed MBE growth of III-V nanowires were studied using the Monte Carlo (MC) simulation. It was demonstrated that for dense nanowire ensembles of stationary sizes of metal droplets on the wire tip commonly depend on the initial droplet diameter. Reasons for the deviation from theoretically predicted single stationary droplet size were analyzed. The self-alignment effect of droplet sizes depends both on growth conditions and catalytic properties of the liquid metal. The growth of InAs nanowire array was simulated. Influence of indium and arsenic readsorption on the self-equilibration effect was examined.

1. Introduction

Interest to self-catalyzed grown III-V nanowires (NWs) is due to their promising applications in nano- and optoelectronic devices. A self-catalyzed growth when group-III liquid metal is used as a catalyst allows avoiding foreign metal contamination but is less stable since the liquid drop tends to swell or shrink during the growth. Recently, it was demonstrated that the self-catalyzed growth of III-V NWs can be dynamically stable over a range of III/V ratio pressures [1]. For the given growth conditions, there is a unique stationary value of the wire diameter and droplet volume. It was demonstrated that during the self-catalyzed growth of GaAs NWs from an array of gallium drops with different sizes the equalization of drop sizes occurred [2]. Since the droplet diameter for vapor-liquid-solid (VLS) grown NW determines its size, this means that the drop size equilibration can result in NWs diameter alignment. The analytical expression for a stationary drop diameter was obtained in [2]. The stationary value of the diameter depends on atomic flux of group III elements, area of their diffusion gathering and axial growth rate which in turn depends on the arsenic flux. Along with analytical models, predictions of nanowire morphology can be obtained using MC simulation. This method is the relevant technique for details of atom migration and kinetics of the nanostructure growth. In the present paper, details of a very dense InAs NW array grown by MBE according to the VLS mechanism are carefully analyzed.

2. The MC model and simulation results

The simulation was carried out using a new variant of lattice MC model suggested for examination of nanostructure growth according to VLS mechanism and embedded in the software package SilSim3D [3]. Description of liquid phase realization in the lattice model can be found elsewhere [4]. Possible elementary events in the model are adsorption, diffusion hop, desorption, and chemical reaction. To simulate InAs NW growth, the five-component system was considered: As_2, As, In(s), In(l), and $(M)_f$ – molecular arsenic, atomic arsenic, indium in solid and liquid states and a film-mask material, respectively. In this model liquid arsenic was excluded from consideration and the liquid phase was taken into account only for indium. Arsenic dissolution into the drop and arsenic diffusion through the drop in the new model are imitated by the exchange reaction: As + In(l) → In(l) + As. The difference between these two processes is determined by neighborhood of reactant atom of the reaction. The choice of energy parameters of the model is described in [3].

The self-catalyzed growth of InAs nanowires was analyzed using this model. We simulate the last stage of NW growth from a droplet array after a hole array pattering in the film-mask material and In drops gathering in the holes. We suppose that in the MBE process In and As_2 fluxes arrive normal to the substrate. Several variants of the initial model system were considered: substrates with different sized droplets and a set of substrates with identical droplets. One droplet on the model surface with the cyclic boundary condition in lateral directions is equivalent to the array of equal droplets on the substrate.

Results of self-catalyzed NW growth from droplets with different initial sizes are shown in Fig. 1. The images demonstrate the trend to droplet equalization during the growth. However, due to too much spread in the initial droplet size possible redistribution of the deposited material between the growing NWs can prevent the alignment effect. Such situation can happen when the distance between the droplets is comparable with their diameters. In the example presented in Fig. 1d diameters of the droplets on NWs tips are different. Moreover, the presence of large drop near the small one (Fig. 1c) results in small drop disappearance, since amount of In coming to small drop is less than amount assumed for NW elongation. The increase of the substrate size in two times without changing the number and size of initial droplets averts small drop collapse, but stationary values of the drops diameter remains different. For small model surface, the stationary radius of the drops, according to the analytical model predictions [2], was obtained only for special cases.

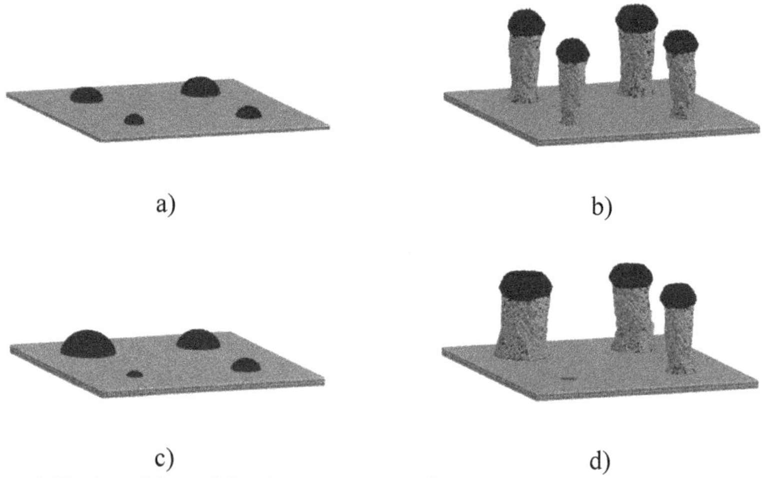

Figure 1. 3D view of the model surface ($S = 86 \times 74$ nm^2) before growth (a,c) and after NW growth (c,d) at $T = 800$ K, $F_{In} = 0.075$ ML/s, $F_{As2} = 0.5$ ML/s. Initial droplet volumes $V_0 = 75$ nm^3, $V_0 = 135$ nm^3, $V_0 = 272$ nm^3, $V_0 = 446$ nm^3 (a,b) and $V_0 = 75$ nm^3, $V_0 = 272$ nm^3, $V_0 = 565$ nm^3, $V_0 = 2254$ nm^3 (c,d).

Fig. 2a demonstrates the dependence of the droplet volume on the growth time for identical initial drops. Different curves correspond to a set of substrates with identical initial droplets of various sizes. In evaporation was forbidden in this case and readsorption of indium and arsenic was not taken into account. If initial drop size is less than the stable size, drop starts to grow, in an opposite case the drop volume decreases with NW growth time. Sizes of initially small drops achieve the single stable size. While for NW growth from droplets with sizes exceeding stationary value, the drop size continuously decreases. In this case, NW tapering is observed, so arsenic flux impinges on NW side walls. Since in this model no surface reconstruction and appearance of (110) facets with a large indium diffusion length were taken into account, In incorporation into NW side walls in the presence of arsenic becomes considerable. That is the reason of continuous drop size decrease (curves 3,4 in Fig. 2a).

An influence of readsorption on the self-equilibration effect was analyzed. At the initial stage of NW growth, the number of readsorbed In atoms increases with time due to NW elongation. At later stages it achieves some stationary value. Thus, readsorption not only changes the stable droplet size, but also leads to different additional In contribution to droplets of different size.

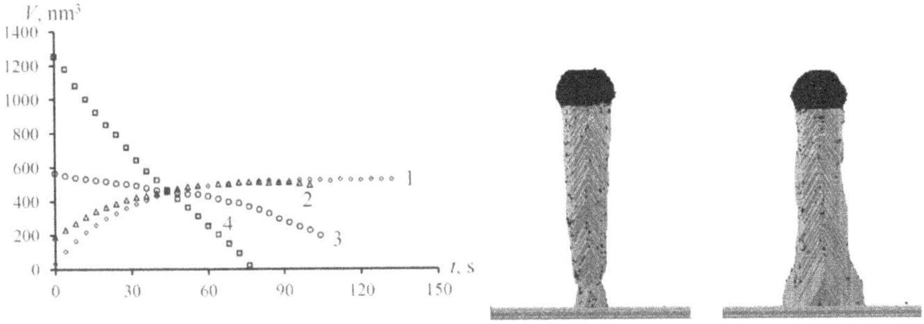

a) b) c)

Figure 2. Self-alignment effect of droplet sizes: a) In droplet volume V versus NW growth time for different sizes of initial droplets: 1 – $V_0 = 32$ nm^3, 2 – $V_0 = 189$ nm^3, 3 – $V_0 = 566$ nm^3; 4 – $V_0 = 1254$ nm^3; $T = 800$ K, $F_{In} = 0.075$ ML/s, $F_{As2} = 0.5$ ML/s, $S = 43\times37$ nm^2; b,c) 3D view of NW after 60 s growth: $V_0 = 32$ nm^3 (b) and $V_0 = 566$ nm^3 (c).

At the next point, the influence of catalytic properties on self-equilibration effect was examined. In the model, catalytic properties of the drop depends on the energy barrier of As_2 molecules dissociation at the liquid metal surface. According to the simulation results, sizes of all droplets can achieve the same value only for definite barrier of As_2 dissociation. That means that self-equilibration of the drop sizes depends both on growth conditions and the catalytic properties of liquid metal.

Presented Monte Carlo simulation results may be useful for controlling diameter uniformity of self-catalyzed grown NWs.

Acknowledgments

This work has been supported by RFBR (grant 16-31-00120 - a) and by the Russian Academy of Sciences Programs.

References

1. J. Tersoff, *NanoLett.* **15**, 6609 (2015).
2. V. G. Dubrovskii *et al.*, *NanoLett.* **15**, 5580 (2015).
3. A. G. Suprunets, M. A. Vasilenko. N. L. Shwartz, *J. Phys.: Conf. Ser.* **690**, 012011 (2016).
4. M. A. Vasilenko, I. G. Neizvestny, N. L. Shwartz, *Comput. Mater. Sci.* **102**, 286 (2015).

MODELING OF FORMATION AND GROWTH OF NANODROPLETS AT HIGH NUCLEATION RATES

S. P. Fisenko
A. V. Luikov Heat &Mass Transfer Institute NASB, 220072 Minsk, Belarus
fsp@hmti.ac.by

A. A. Rostami, D. B. Kane, Y. B. Pithawalla
Research, Development & Regulatory Affairs, Altria Client Services LLC
Richmond, Virginia, USA

M. S. El-Shall
Department of Chemistry, Virginia Commonwealth University, Richmond, Virginia, USA

A model of formation and growth of aerosol nanodroplets is developed in which main channel of nanodroplets growth is coagulation for very high nucleation rates. Results of the stochastic coagulation process are described by means of the size distribution function of droplets approximated by the Weibull function of special kind. Procedure for determining of Weibull function is proposed. Comparison with experimental data is presented.

1. Introduction

A process of aerosol condensation in which superheated vapors are rapidly mixed with a colder ambient air results in extremely high vapor supersaturations. This is common phenomenon in atmospheric precipitation, fog formation and many aerosol generators, including electronic nicotine delivery systems (ENDS). To characterize the resulting aerosol, it is important to estimate homogeneous condensation nucleation rate, droplet size distribution, droplet composition and vapor-liquid partitioning. At conditions of high supersaturation and when the critical cluster size is as small as dimer, the classical nucleation theory is not applicable.

2. Model

To overcome limitations of the classical nucleation theory we have developed the computational model that uses the kinetic theory of gases to calculate the rate of dimer and cluster formations through coagulation and condensation for multi-component vapors. The droplet growth rate is calculated in the free molecular regime of growth [1]. Non-isothermal conditions caused by the condensation latent heat release are accounted for in our model. The typical nucleation rate for these systems is on the order of 10^{30} clusters/(m^3s).

Condensation is stopped when an instable steady-state condition is reached between small droplets and vapor due to the Kelvin effect. At small droplet radii, the value of supersaturation needed for condensation to occur is so large that it does not happen naturally.

As cluster growth by molecular condensation slows down, the Brownian coagulation becomes dominant for continued growth. Coagulation appears as an effect with a characteristic time scale on the order of 10^{-8} s at the early stages. Such coagulation is stochastic but thermodynamically favorable because it decreases the surface free energy of this system.

3. Basic results

High nucleation and coagulation rates are very typical for a nanoparticle formation. Simulation results for early stages of aerosol evolution from supersaturated glycerol vapor are presented in Figs. 1, 2.

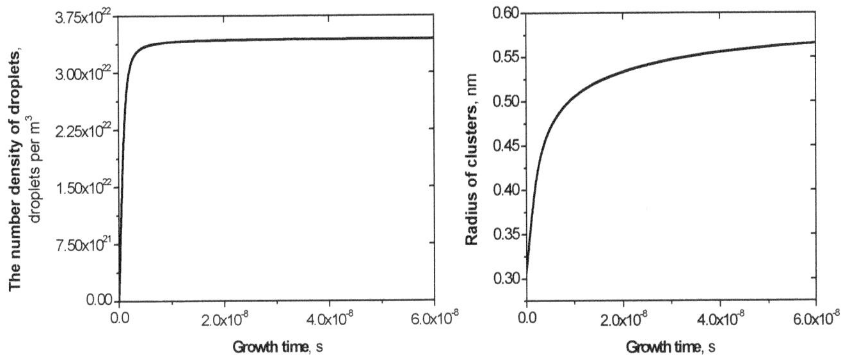

Figure 1. Nucleation and fast growth of glycerol nanodroplets.

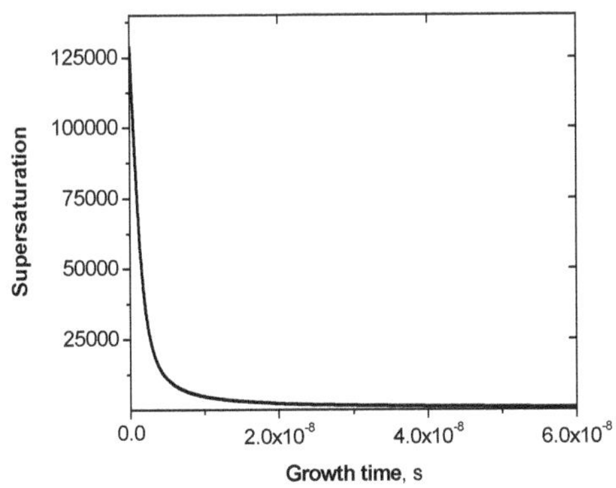

Figure 2. Supersaturation versus time for the process of glycerol nucleation.

These simulation results demonstrate the mean radius of droplets at a given time. The minimum stable droplet size during aerosol formation and growth is determined by coalescence phenomenon [2]. The minimum radius for the conditions of our model is about 60 nm. Due to the Kelvin effect, particles with sizes below this value are not stable and will evaporate, transferring their mass to the larger nearest droplets.

The droplet size distribution function is determined using the special Weibull distribution function, which minimizes the entropy for a total given droplet mass and averaged cube of radii. The example of such function at time of 0.01 s for a droplet with the mean cube of droplet radii value of 8×10^{-21} m^3 is shown in Fig. 3.

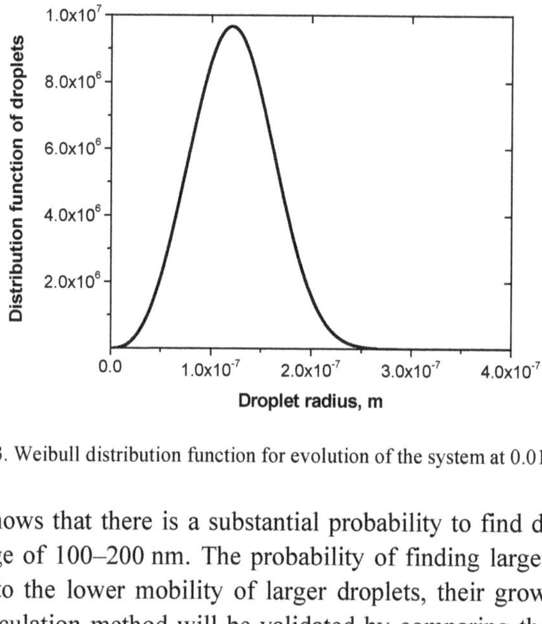

Figure 3. Weibull distribution function for evolution of the system at 0.01 s.

This plot shows that there is a substantial probability to find droplets with radii in the range of 100–200 nm. The probability of finding larger droplets is very low. Due to the lower mobility of larger droplets, their growth becomes slower. This calculation method will be validated by comparing the simulation results with experimental data.

References

1. J. H. Ferziger, H. G. Kaper, *Mathematical Theory of Transport Process in Gases* (North-Holland, Amsterdam, 1972).
2. E. M. Lifshitz, L. P. Pitaevskii, *Physical Kinetics* (Pergamon, Oxford, 1981).

STRUCTURE AND MAGNETIC PROPERTIES OF TEMPLATE SYNTHESIZED FeCo NANOTUBES

A. Shumskaya, E. Kaniukov
Division of Cryogenic Research, Scientific-Practical Materials Research Center NASB
P. Browka 19, 220072 Minsk, Belarus
Lunka7@mail.ru

A. Kozlovskiy, M. Zdorovets
Institute of Nuclear Physics, Ibragimov St., 050032 Almaty, Kazakhstan
L.N. Gumilyov Eurasian National University
Satpayev Str. 2, 010008 Astana, Kazakhstan
artem88@mail.ru

FeCo nanotubes with the length of 12 μm and diameters of 110±5 nm have been formed by template synthesis method. Comprehensive study of structural parameters has shown that the nanotubes have polycrystalline walls of 20 nm. Investigation of the magnetic properties of FeCo nanotubes in the temperature range of 50–300 K allowed us to establish the anisotropy of main magnetic parameters.

1. Introduction

Last years there is a great interest to metallic nanostructures (NSs) caused by both a demand to understand their fundamental properties and to apply them as a new material in biomedicine, chemistry, physics, electronics [1–3]. The template synthesis method is one of the most promising technique owing to its low cost and simplicity. It enables to synthesize nanotubes with predetermined parameters (size, form and composition) by control of deposition conditions and template parameters [4].

Nowadays, properties of magnetic NSs, composed of pure ferromagnetic metals Co, Ni, Fe and their alloys are well studied [5–9]. Iron-cobalt alloy (FeCo) is more attractive because of a huge saturation magnetization and coercivity in comparison with pure ferromagnetic metals [10]. It should be noted that the structural and magnetic characteristics of FeCo NSs were carefully analyzed in [10–11]. However, parameters of the FeCo nanotubes (NTs) are still not well studied, that is connected with the lack of a reliable synthesis method [11]. In this regard, in our work the simple method of FeCo NTs formation and results of investigations of their structural and magnetic parameters are presented.

2. Experimental

Track-etched membranes based on polyethylene terephthalate (PET) were used as templates with a thickness of 12 μm, the nominal pore diameter of 110±3 nm and the density of 10^9 cm^{-2}. Deposition of FeCo was carried out by electrochemical method in potentiostatic mode with the potential $U = 1.5$ V using the electrolyte $CoSO_4 \times 7H_2O$ (110 g/l), $FeSO_4 \times 7H_2O$ (110 g/l), H_3BO_3 (25 g/l), $C_6H_8O_6$ (3 g/l).

Characterization of structural features was made by scanning electron microscopy (SEM, Hitachi TM3030), energy dispersive analysis (EDA, Bruker XFlash MIN SVE), X-ray diffraction analysis (XRD, D8 ADVANCE) using Cu K_α radiation and selected area electron diffraction (SAED, JEM-100). Internal diameters and wall thicknesses were controlled by gas permeability (Sartocheck® 3 Plus 16290) and transmission electron microscopy (TEM, JEM-100). Magnetic characteristics of FeCo NTs were studied on universal measuring system (automated vibrating magnetometer) «Liquid Helium Free High Field Measurement System» (Cryogenic LTD) in the magnetic fields of ±3 T in the temperature range of 50–300 K.

3. Results and discussion

The SEM image presented in Fig. 1a indicates that the formed NSs have a hollow shape. TEM images of individual NTs allow us to determine external diameters of FeCo NTs which equal to 110 nm over the entire length with deviations within 5–8 % of the average values (Fig. 1b). Internal diameters of 70 nm were established by gas permeability method, accordingly walls thickness amount 20 nm. EDA spectra analysis reveal the ratio of cobalt and iron atoms in the alloy as $Fe_{60}Co_{40}$.

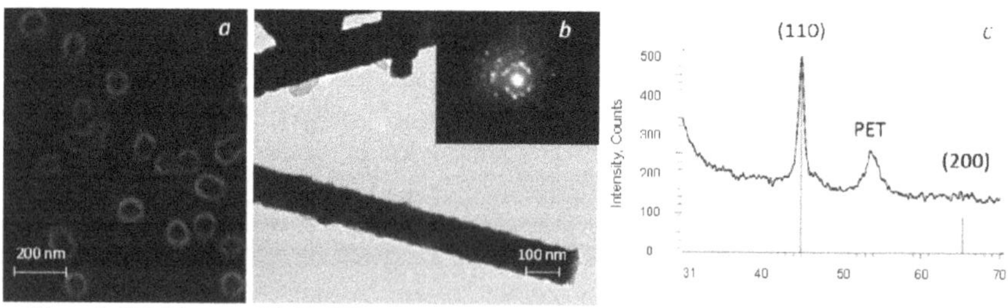

Figure 1. SEM image of FeCo NTs array (a), TEM image of individual NTs with SAED on insert (b), X-ray diagram of FeCo NTs array into PET template (c).

The crystal structure of FeCo NTs was studied on the basis of the analysis of SAED for individual NTs (insert in Fig. 1b) and the X-ray diffraction spectra of NTs arrays into PET template (Fig. 1c). The fabricated NTs have a preferred direction of the growth to be (110), which is also confirmed by the large ratio of the peaks (110) and (200) on the X-ray diffraction spectra. The analysis of SAED indicates a polycrystalline structure of NTs. Studies of the X-ray spectra show that NTs have a body centered cubic (BCC) structure. The lattice parameter a=2.8535 Å and the average crystallite size is 16.7 nm.

Measurements of the magnetization dependence on the magnetic field $M(H)$ were made for parallel and perpendicular directions of the field with respect to the orientation of the NTs axis. The main magnetic characteristics obtained (H_c – coercivity, M_s – magnetization of saturation, M_r / M_s – squareness ratio of hysteresis loop) are summarized in Table 1.

Table 1. Magnetic properties of FeCo NTs arrays.

T, K	Parallel to NTs				Perpendicular to NTs			
	H_c, Oe	M_r, emu/g	M_s, emu/g	M_r/M_s	H_c, Oe	M_r, emu/g	M_s, emu/g	M_r/M_s
50	516	26.4	80.7	0.33	243.5	11.2	76.6	0.15
100	501	28.1	79.3	0.35	201.7	10.1	72.6	0.14
150	510	30.5	77.5	0.39	195.3	9.8	70.3	0.14
200	587	29.5	76.1	0.385	197	9.5	71	0.13
250	607	36	74	0.48	197	9.5	69.5	0.14
300	641	40.1	75.5	0.53	212	13.1	69.1	0.19

The values of main magnetic characteristics of FeCo NTs at the perpendicular magnetic fields are twice less than at parallel ones. Taking into account the high aspect ratio of the investigated NTs (~100), the dependence on the shape anisotropy should be absent [9]. Their difference is caused by the crystalline anisotropy. The temperature dependence of H_c has unusual for ferromagnetic material character. It could be connected with a presence of several magnetic phases of FeCo in NTs.

4. Conclusion

Using PET track-etched membranes with the thickness of 12 μm as a template FeCo NTs with the outer diameter of 110 nm and the wall thickness of 20 nm were synthesized by electrochemical method. A study of their morphological features shows a polycrystalline nature of FeCo NTs walls with the composition of 60 % of Fe and 40 % of Co. The NTs material has a body centered cubic structure with a = 2.8535 Å and the average crystallite size of 16.7 nm. The investigation of the main magnetic characteristics in the temperature range of

50–300 K made it possible to determine the magnetic anisotropy of the FeCo NTs. Untypical changes in coercivity observed with rising temperature is due to the presence of several magnetic phases.

References

1. S. K. Yen, P. Padmanabhan, S. T. Selvan, *Theranostics* **3**, 986 (2013).
2. A. Kros, R. J. M. Nolte, N. A. J. M. Sommerdijk, *Adv. Mater.* **14**, 1779 (2002).
3. H. Dai, J. H. Hafner, A. G. Rinzler, *Nature*. **384**, 147 (1996).
4. C. R. Martin, *Science* **266,** 1961 (1994).
5. V. Haehnel, S. Fähler, P. Schaaf, *Acta Mater.* **58**, 2330 (2010).
6. A. Kozlovskiy, A. Zhanbotin, M. Zdorovets, *Nucl. Instruments Methods. Phys. Res. Sect. B Beam Interact. with Mater. Atoms.* **365**, 663 (2015).
7. H. Cao, L. Wang, Y. Qiu, Q. Wu, G. Wang, L. Zhang, X. Liu, *ChemPhysChem* 7, 1500 (2006).
8. A. Kozlovskiy, D. Shlimas, E. Shumskaya, E. Kaniukov, M. Zdorovets, K. Kadyrzhanov, *Phys. Met. Metallogr.* **118**, 174 (2017).
9. E. E. Shumskaya, E. Yu. Kaniukov, A. L. Kozlovskiy, M. V. Zdorovets, V. S. Rusakov, K. K. Kadyrzhanov, *Ph. Solid State* **4**, 766 (2017).
10. F. S. Li, D. Zhou, T. Wang, Y. Wang, L. J. Song, C. T. Xu, *J. Appl. Phys.* **101**, 014309 (2007).
11. Z. Dong, L. Zhi-Wei, Y. Xu, W. Fu-Sheng, L. Fa-Shen, *Chinese Phys. Lett.* **25**, 1865 (2008).

NANO- AND MICROPARTICLES OF Fe_3O_4 OBTAINED BY SPRAY PYROLYSIS FOR MAGNETORHEOLOGICAL APPLICATIONS

D. Kotsikau, A. Eroma, V. Natarov, V. Pankov

Belarusian State University, Leningradskaya 14, 220030 Minsk, Belarus
kotsikau@bsu.by

Ultrasonic spray pyrolysis approach was used to prepare Fe_3O_4 powders with particles of ~40 nm and ~1.5 μm in size as potential additives to magnetorheological suspensions. The magnetization of the magnetite powders was shown to depend on the particles size. The difference in the size, magnetic features and specific surface is expected to produce different effects on the magnetorheological behavior of suspensions fabricated on their base. The measured growth in shear stress (τ/τ_0) of the suspensions has reached 4 units for micro-sized Fe_3O_4 under an applied magnetic field.

1. Introduction

Ferrimagnetic oxides, including γ-Fe_2O_3, Fe_3O_4 and metal ferrites, have been a subject of interest due to their attractive electronic, magnetic and rheological properties [1–4]. Semiactive control devices for various applications have received a significant attention in recent years since they offer an adaptability of active control devices without requiring large power sources.

Magnetorheological (MR) dampers are semiactive control devices that use MR fluids to produce controllable dampers. When a magnetic field is applied to the MR fluids, particle chains are formed. The fluid becomes semisolid and exhibits a viscoplastic behavior. They potentially offer highly reliable operations and can be viewed as fail-safe passive dampers in the case of control hardware or software malfunctions [3].

A wide range of additives is used with MR fluids to enhance their stability, MR response, seal life. Powdered magnetic materials are promising additives to these fluids since they combine several functions. In particular, they enhance the sedimentation stability and increase the MR response of suspensions. A synergizing effect of the magnetic field on the response could be achieved. However, there is no strict background for the compositions of MR fluids to control their behavior. The particle size and magnetic properties are important parameters affecting their rheological features [1,3,4].

Herein, we represent a study on the preparation of Fe_3O_4 powders with particles ranging in size from nanometers to micrometers as potential modifying additives for MR suspensions of iron carbonyl in mineral oil.

2. Experimental

Aqueous solutions of $Fe(NO_3)_3$ and citric acid (CA) were taken in various proportions to synthesize magnetite (Fe_3O_4) by a spray pyrolysis approach. CA played a role of reducing agent. The solution was transformed into an aerosol state with an ultrasonic atomizer and directed to the reaction zone with a flow of nitrogen. The reaction zone was placed in a furnace and heated at 500–650 °C. In order to prepare nano-sized particles, an inert component (NaCl) was additionally introduced into the reactant mixture. The resulting powder was washed with distilled water to remove NaCl.

Powdered samples were characterized by scanning electron microscopy (SEM), X-ray diffraction (XRD) and infrared spectroscopy (IR). Particle size and morphology were estimated on a LEO 1420 microscope. The XRD analysis was carried out on a DRON-2 diffractometer using Co-Kα radiation. IR spectra were recorded from powdered samples with an AVATAR FTIR-330 spectrometer.

Magnetization measurements were performed with a cryogen free measurement system (CFMS) Cryogenic Ltd (T = 7–300 K, B_{max} = 18 T).

Rheological parameters of the suspensions made of the prepared powders in mineral oil (15 wt.%) were measured on a Rheotest RN-4.1 rheometer (T = 20 °C, shear rate is 200 s^{-1}). In order to determine the effect of magnetic field, a toroidal ferrite magnet with B = 0.5 T was used. The MR response was measured here as τ/τ_0, where τ and τ_0 are the shear stresses with and without magnetic field, respectively.

3. Results and discussion

In order to obtain powders with spinel-type Fe_3O_4 structure, different conditions of the synthesis were used. Proportions of the iron salt and CA, as well as the pyrolysis temperature were varied. XRD patterns of the products are given in Fig. 1. The single-phase powder based on Fe_3O_4 that does not contain admixtures of paramagnetic α-Fe_2O_3 phase has been achieved in the sample 3. In spite of the high temperature of the synthesis, the diffraction lines are broad, which reflects poorly crystallized structure of the sample. This is evidently caused by a short-time exposure of the reactant mixture to heat. Note that the presence of the inert additive (NaCl) did not influence noticeably the shape of XRD patterns. IR-spectroscopy examination of the samples revealed a consistence with the results of XRD analysis.

SEM picture of the powder synthesized without NaCl is given in Fig. 2a. It displays that the sample consists of spherical particles of the size ranging from

1 to 2 μm. The described morphology results from drying the droplets of initial solution dispersed with ultrasonic atomizer.

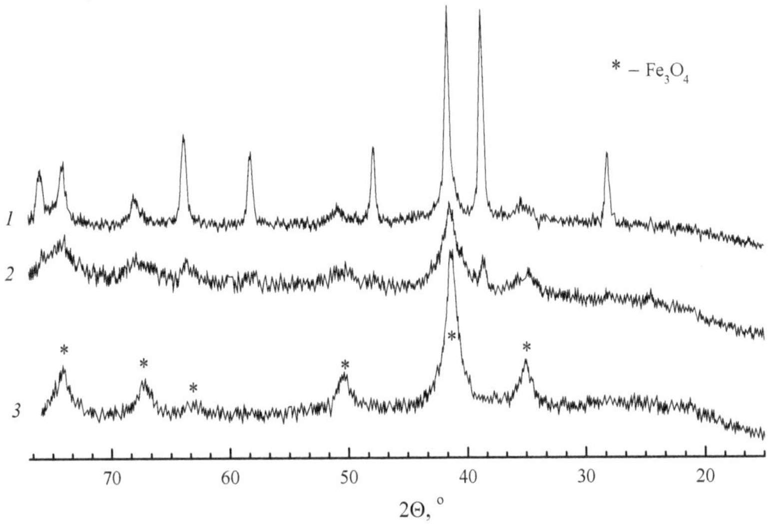

Figure 1. XRD patterns of the pyrolysis products at different conditions: *1* – 15 g of $Fe(NO_3)_3$, 5 g of CA, 650 °C; *2* – 15 g of $Fe(NO_3)_3$, 10 g of CA, 500 °C; *3* – 15 g of $Fe(NO_3)_3$, 10 g of CA, 650 °C.

When NaCl was added to the starting reactant mixture, the resulting powder had the same size and shape of particles. However, the spheres consisted of Fe_3O_4 nanoparticles distributed over NaCl matrix. The inert component could be washed out with distilled water to separate the oxide nanoparticles. A typical SEM image of the powder is given in Fig. 2b.

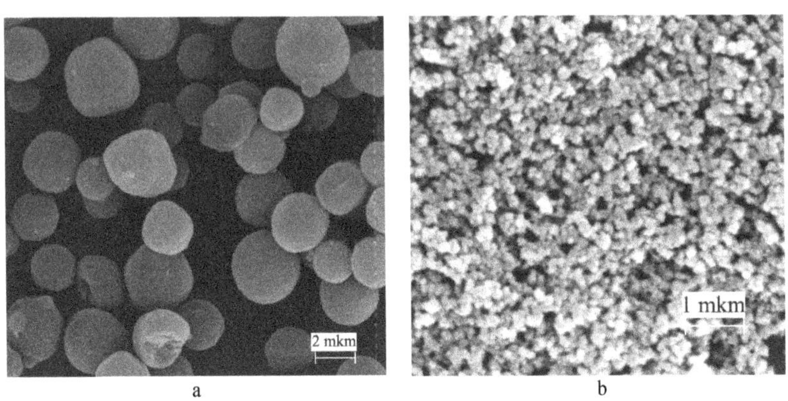

Figure 2. SEM images of the pyrolysis products without (a) and with (b) NaCl as an inert additive.

Note that the diameter of the particles could be controlled by varying the amount of the inert component. Under the conditions used, Fe_3O_4 particles have an average diameter of ~40 nm.

The magnetic measurements have revealed a noticeable difference between the nano- and micro-sized Fe_3O_4 samples. Thus, the oxide with the size of 1.5 μm is characterized by enhanced specific magnetization and coercivity. Rheological tests have shown that in the case of powder with the greater particle size, the magnetic field had more pronounced effect upon the viscosity (τ/τ_0) of the suspension. However, the smaller particles could possess a stronger stabilizing effect when used as a modifier of MR fluids based on iron carbonyl. The structural and functional features of the obtained materials are summarized in Table 1.

Table 1. The characteristics of the synthesized powders and MR fluids on their base.

Phase composition	Grain size	M_{sat}, $A \cdot m^2 \cdot kg^{-1}$	H_c, $kA\ m^{-1}$	Calculated S, $m^2 \cdot g^{-1}$	τ/τ_0, a.u.
Fe_3O_4	~40 nm	58±1	4.5±0.1	~20	~2.5
Fe_3O_4	~1.5 μm	65±1	6.5±0.1	~0.5	~4

4. Conclusion

A spray pyrolysis method has been employed to prepare Fe_3O_4 powders from inorganic iron precursor (iron nitrate) and organic reducing agent (citric acid). The conditions of the synthesis that allowed formation both nano- and micro-sized particles have been chosen. The effect of the particle size on magnetic properties of the powders and MR behavior of the suspensions on their base has been revealed. The synthesized material will be used as modifying and stabilizing additives to iron carbonyl in mineral oil MR fluids for semiactive dumping devices.

References

1. M. Sugimoto, *J. Am. Ceram. Soc.* **82**, 269 (1999).
2. Z. Pedzich *et al.*, *J. Europ. Ceram. Soc.* **24**, 1053 (2004).
3. B. F. Spencer *et al.*, *J. Eng. Mechan.* **123**, 230 (1997).
4. I. Bica, Y. D. Liu, H. J. Choi, *J. Ind. Eng. Chem.* **19**, 394 (2013).

EFFECT OF THE SURFACE STRUCTURE OF TITANIUM IMPLANTS ON THE RESPONSE TO THE BONE TISSUE

S. K. Lazarouk, O. V. Kupreeva
Belarusian State University of Informatics and Radioelectronics
P. Browka 6, 220013 Minsk, Belarus
serg@nano.bsuir.edu.by

O. N. Dudich, V. L. Krasilnikova
Belarusian Medical Academy of Postgraduate Education
P. Browka 3, 220013 Minsk, Belarus

S. I. Bahayeu
Physical-Technical Institute NASB
Kuprevich 10, 220141 Minsk, Belarus

The growth of bone tissue on titanium plates and mounting screws during surgical treatment of rabbit bones was studied. The growth of bone tissue was more intensive on the developed nanostructured titania surface in comparison with the planar titania surface. The methods of electrochemical treatment of a titanium surface in order to control the regeneration process of the bone tissue on the titanium constructions were proposed.

1. Introduction

An influence of the surface structure of titanium implants on the response to the bone tissue during surgical treatment has been actively studied for the last ten years [1,2]. It should be noted that in some cases surgical treatment needs maximal integration of bone tissue with titanium implants. While in the other cases the penetration of bone tissue inside the titanium construction is undesirable. In particular, when titanium plates and screws are used for temporal fixation of bones fragments, the penetration of bone tissue inside titanium details should be minimized. In our work the growth of the rabbit bone tissue on titanium plates and mounting screws after electrochemical treatment with different surface conditions has been studied.

2. Experimental

We have used plates and mounting screws made of Ti_6Al_4V alloy fabricated by "Medbiotech" (Republic of Belarus) as initial samples. The titanium plates did not undergo electrochemical treatment. The first group of titanium screws has been anodized in 10 % phosphoric acid electrolyte in order to grow up the barrier titanium oxide with the thickness of about 180 nm. The second group of titanium screws has been anodized in the 0.3 % ammonium fluoride solution in the ethylene glycol with 2 vol.% of water at the current densities of 2 mA/cm^2 in order to form porous anodic oxide with the thickness of about 1.5 μm. The titanium plates and all of the mounting titanium screws were surgically installed in to the rabbit bones. After one year of observation of rabbits all experimental parts were extracted. Their surface has been studied by scanning electron microscopy.

3. Results and discussion

Fig. 1 shows the SEM image of the porous titania film formed by electrochemical anodization of titanium in the NH_4F electrolyte. The porous titania thickness, the average pore diameter and the interpore distance are found to be 1.5 μm, 50 nm and 150 nm, respectively. The upper titania layer has a "honeycomb" structure while the lower layer has a tubular structure [3].

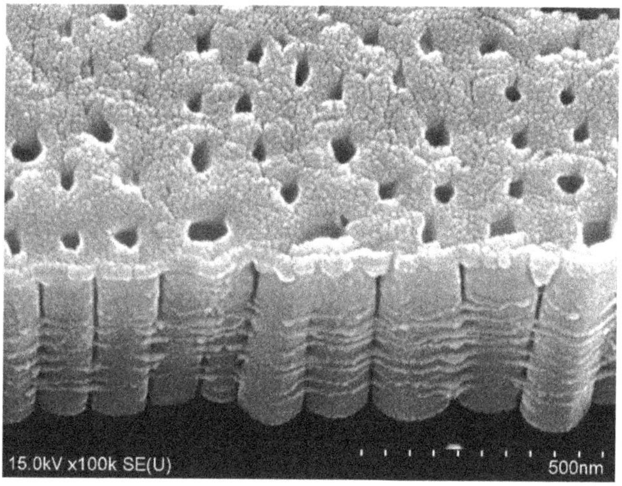

Figure 1. The SEM image of the cross-section of porous titania film formed by electrochemical anodization of titanium screw.

The electrochemical anodization of a titanium surface in the phosphoric acid electrolyte results in the formation of the titania film with the homogeneous barrier structure. The titania thickness is about 180 nm. It should be noted that the specific area of porous titania films is considerably higher in comparison with the titania films of the barrier type.

Fig. 2 shows SEM images of titanium plates and mounting screws after they have been used for the bone anchorage during a surgery operation of rabbit lower limbs. The titanium plates were not treated, while the mounting screws were anodized before use in order to form porous oxide (Fig. 2a,b,c) or barrier titanium oxide (Fig. 2d,e,f) on their surfaces. As we can see in the SEM images with porous titania the bone tissue covers the top spherical part of the screw as well as the lower cone part with the thread (Fig. 2a). The bone tissue covering the porous surface can be removed only by special mechanical treatment. The cone screw-thread surface after additional mechanical treatment is shown in Fig. 2b,c. The porous titania surface promotes the bone tissue growth [1,2,4].

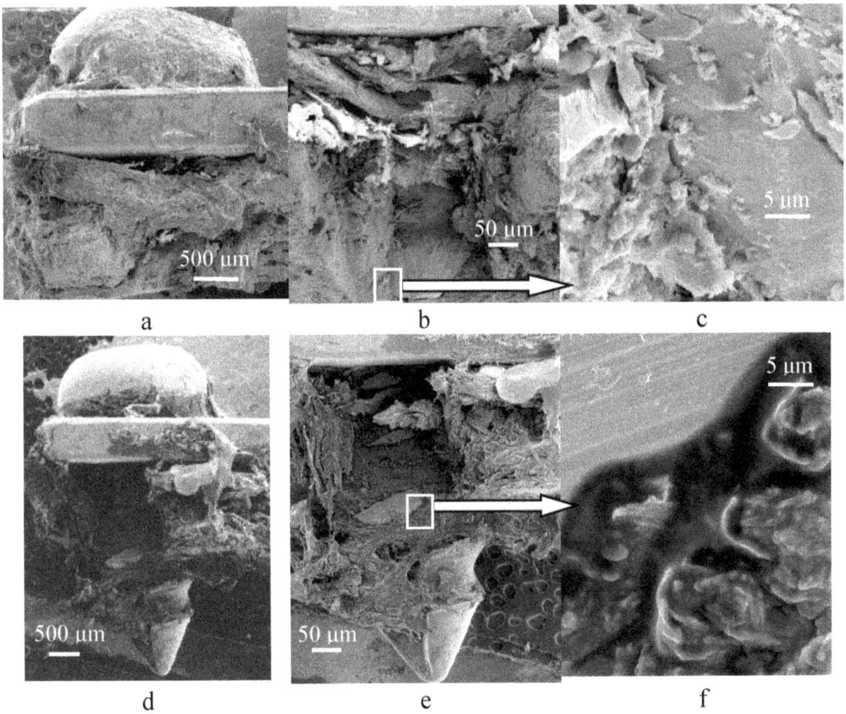

Figure 2. The SEM images of titanium plates and mounting screws used as bone anchoring during surgery operation of rabbit lower limbs: a-c – titanium mounting screw covered by porous titania, d-f – titanium mounting screw covered by barrier titania.

The bone tissue could be removed from the screw surface with the barrier titania without additional mechanical treatment (Fig. 2d,e,f). The difference between barrier and porous surface could be observed in the high resolution SEM images (Fig. 2c,f). The border between the bone tissue and porous titania is rough. Moreover, the remaining bone tissue can be seen on the porous titania surface. It can be explained by the penetration of the bone tissue into pores. Meantime the bone tissue is removed from the barrier titania completely, and the studied titania surface is smooth.

Thus, it is recommended to use the barrier titania coating for temporal fixation of bone fragments during surgery due to its bioinert properties [5,6]. When titanium constructions are used for permanent fixation of bone fragments it is recommended to use nanoporous titania coatings.

It should be noted that anodic titania has a negative electrical charge which could promote bone tissue regeneration and decrease a blood coagulation around this surface [7].

4. Conclusion

Anodic treatment of a titanium surface allows to form either the barrier or porous titania films. It can be used in surgery in order to control the regeneration process of the bone tissue on the titanium constructions. The barrier titania is a bioinert material, while the porous titania promotes the bone tissue growth with its penetration into pores.

References

1. Zvi Schwartz *et al.*, *J. Bone Joint Surg. Am.* **90**, 2485 (2008).
2. M. Jäger *et al.*, *J. Biomed. Biotechnol.* **2007**, 69036 (2007).
3. S. K. Lazarouk *et al.*, *Thin Solid Films* **526**, 41 (2012).
4. S. K. Lazarouk *et al.*, *Doklady BSUIR* **1**, 100 (2016).
5. S. I. Bagaev *et al.*, in: *Proc. "Modern Methods and Technologies of Creation and Processing of Materials"* (Minsk, Belarus, 2016).
6. E. A. Gvozdovskaya *et al.*, *Doklady BSUIR* **5**, 48 (2016).
7. M. Kulkarni *et al.*, *Nanotechnology* **26**, 062002 (2015).

NANOSTRUCTURING OF DIAMOND AND OPTICAL DIFFRACTION GRID FORMATION BY BORON ION IMPLANTATION

A. L. Stepanov, V. I. Nuzdin, V. F. Valeev, N. V. Kurbatova, M. F. Galyautdinov

Kazan Physical-Technical Institute, RAS
Sibirsky Trakt 10/7, 420029 Kazan, Russia
aanstep@gmail.com

V. V. Vorobev, Y. N. Osin

Kazan Federal University, Kremlyovskaya 18, 420008 Kazan, Russia

The present study relates to a new fabrication method of diffractive optical elements with diamond nanostructuring. The diffraction grating was obtained on diamond by implantation of boron ions through a mask. Formation of periodic diffraction microstructures on the diamond surface was monitored by optical, electron and atomic force microscopy techniques.

1. Introduction

Modern tasks of integrated optics require the use of special new materials and the development of novel technologies for production of components and devices based on them. One of the specific areas is diamond optics. The interest to diamond is explained by its radiation resistance and high thermal conductivity. Diamond optical elements possess a wide transparency window from 0.2 to 5 μm. Diamond is stable up to very high temperature and in aggressive environments. In practice, diamond is used for production of various diffractive optical elements (DOE) as gratings, focusers, equalizers, *etc.* Diamond DOE can be applied for the high-power beam of a CO_2-laser up to 20 kW/cm^2 power density [1,2] to create photonic crystal resonators, quantum information storage devices [1] and to control the radiation fluxes in X-ray optics, for example, using a diamond Bragg mirror with a reflectivity of 100 % [3]. For fabrication of periodic DOE on the diamond surface various technological methods were used, such as manufacturing with powerful pulses of excimer laser, etching in a gas stream of transport gas.

In the present paper, a new approach is suggested in which diamond periodic diffractive structures are formed by ion implantation through the mask surface. Previously, this technology was successfully tested for the DOE production on silica glass by implantation of noble metals [5]. The objective of this study is to analyze a possibility to fabricate DOE on diamond surface by boron ions implantation.

2. Experimental

Formation of a DOE on the diamond polished surface was done by B^+-ion implantation with energy of 40 keV, dose of 1.3×10^{18} ion/cm^2 and current density of 15 μA/cm^2 using an ion accelerator ILU-3 through the metal mask with the grid square cells of 40 μm. Local morphology and surface structure of the implanted diamond was analyzed by scanning electron microscopy (SEM, Merlin, Carl Zeiss), equipped with a detector of electron backscatter diffraction HKL NordLys (Oxford Instruments), and atomic force microscopy (AFM, FastScan, Brucker). Optical characterization of the fabricated DOE was performed by an optical microscope Polar-1 (Micromed) and Raman spectroscopy. Analysis of optical diffraction patterns from the DOE was carried out with a He-Ne laser at the wavelength of 632.8 nm. Modelling of concentration profiles of implanted boron in diamond by the SRIM-2013 algorithm showed that boron atoms were accumulated in the 100 nm thick surface layer of diamond.

3. Results and discussion

Fig. 1 shows the grid image formed on the diamond surface by B^+ implantation through the mask observed in optical microscope (Fig. 1a) and SEM (Fig. 1b). The periodic microstructure is composed of alternating dark square cells belonging to the implanted areas of the sample surface separated by walls (open area) of unimplanted diamond.

a b

Figure 1. Images of the diamond surface implanted with boron ions through the mask observed by optical microscopy (a) and SEM (b).

Structural characterization of the implanted area of diamond was carried out by electron backscatter diffraction probing the near-surface region of the sample. Unlike the Kikuchi diffraction in the form of strips parallel to the planes of the diamond crystal lattice for implanted areas, in our diffraction pattern only broad diffuse rings were observed. This is indicating on some destruction of the crystal lattice in the implanted diamond surface and formation of amorphous carbon nanostructures.

The AFM image of the diamond fragment DOE near the wall (dark area in Fig. 1a) between the implanted cells (light rough area in Fig. 1a) evidences that the implanted diamond surface (cells) are raised above the original surface by ~100 nm. This effect is attributed to swelling of irradiated regions (grids) of the sample characterized by the lower density (graphite = 2.09–2.23 g/cm^3) as compared with the density of diamond, 3.47–3.55 g/cm^3.

Raman spectra of the implanted samples under excitation with an argon laser at 522 nm confirms the destruction of diamond. Together with a well-known diamond intense line at 1336 cm^{-1} after the ion implantation weaker lines in the region near 1560 cm^{-1} appeared indicating the graphitization process. As a result of B^+-ion implantation of diamond a formation of boron carbide could be expected. However, boron carbide lines in the Raman spectrum in the range of 200–1200 cm^{-1} [5] were not detected. Meanwhile, some chemical bonds or small fragments consisting of boron and carbon atoms not recognized by Raman spectroscopy, may be present in the graphitized diamond implanted with boron ions [6].

It should be noticed, that diamond implanted with boron ions leads to both changes in chemical composition (accumulation of boron) and to modification of carbon substrate nanostructure, *i.e.* the formation of periodic domain with graphitic matter. The implantation through a mask has resulted in specific periodic microstructures with periodically variable distribution of optical properties in the surface layer, *i.e.*, between the walls of the diamond lattice with the refractive index $n = 2.42$ and graphitized cells, $n_{grafit} = 2.1–2.23$. Therefore, the periodic microstructure formed with the graphitized areas on diamond can be used as a two-dimensional photonic crystal, or DOE. As an illustration, Fig. 2 shows the diffraction image that was initiated by optical reflection from the diamond grid using a He-Ne laser at 632.8 nm. Manipulating with the ion implantation regimes, thereby changing the effective refractive index of individual areas in the DOE, one can control its optical and diffractive characteristics.

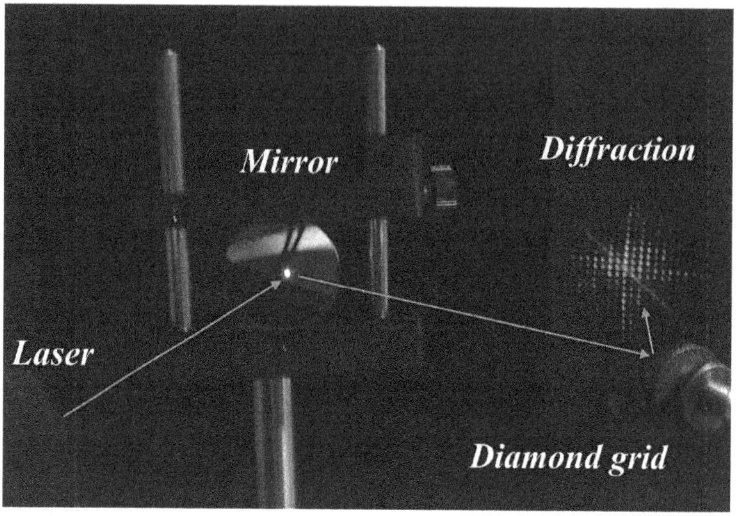

Figure 2. Image of reflected diffraction pattern on the screen from the diamond grid by probing with a He-Ne laser. A diamond sample is fixed to the metal holder.

4. Conclusion

We consider the process of low-energy high-dose implantation of diamond with boron ions through the surface mask and demonstrated successive realization of the new approach for creation of DOE. On the diamond surface periodic microstructures were fabricated and the phase contrast was provided by the graphitized diamond areas.

Acknowledgments

This work was supported by the Russian Foundation for Basic Research (grant 15-48-02525).

References

1. V. V. Konenko *et al.*, *Quant. Electr.* **29**, 9 (1999).
2. M. Karlsson, F. Nikolajeff, *Opt. Express.* **11**, 502 (2003).
3. Y. Shvydko, S. Stopin, V. Blank, S. Terentev, *Nature Photon.* **5**, 539 (2011).
4. A. L. Stepanov *et al.*, *Appl. Phys. A* **111**, 261 (2013).
5. A. Hushur *et al.*, *J. Phys.: Condens. Matter.* **28**, 045403 (2016).
6. V. Domnich *et al.*, *J. Am. Ceram. Soc.* **94**, 3605 (2011).

THE METHOD TO CREATE THE SELF-ORGANIZED NANOSTRUCTURES BASED ON INDOTRICARBOCYANINE DYE

N. V. Belko, M. P. Samtsov, G. A. Gusakov
A. N. Sevchenko Institute for Applied Physical Problems of the Belarusian State University
Kurchatov Str. 7, 220045 Minsk, Belarus
samtsov@ bsu.by

E. S. Voropay
Department of Physics, Belarusian State University, 220030 Minsk, Belarus
voropay@bsu.by

This paper presents the results obtained during studies of self-organization in water of the nanostructures based on the water-insoluble indotricarbocyanine dye when subjected to ultrasonic treatment without the use of additional substances. As distinct from the nanoobjects produced due to the normal reprecipitation, the ultrasonic treatment of the dye in water leads to the formation of the particles cylindrical in form, with the diameter of about 200 nm and the height of about 10 nm.

1. Introduction

Organic self-organized nanostructures are characterized by unique properties and may be used as elements for organic optoelectronic devices, components of solar batteries, biological markers [1].

Most often, the self-organization processes of organic molecules and the formation of nanostructures take place upon an introduction of additional substances into solutions of the compounds under study. In the case of water-insoluble compounds which are soluble in organic solvents the reprecipitation method is a usual practice [2]. A concentrated solution of the required compound in some organic solvent is added into water to initiate the molecular self-organization process with the formation of nanonstructures. Naturally, the reprecipitation method may be used for water-soluble substances as well. Still, in some cases the water-soluble organic dyes are forming nanostructures in water solutions without an addition of extra compounds [3].

The indotricarbocyanine dye (CD-1) studied in this work is insoluble in water in ordinary conditions. It is introduced from the crystalline state into a water solution by means of the ultrasonic effect. In time, in such a solution we can observe self-organization of the dye molecules and the nanostructure formation. Previously, the nanostructures of CD-1 have been formed by the reprecipitation method in water–ethanol mixtures.

2. Methods

The compound CD-1 has been developed at the A. N. Sevchenko Research Institute for Applied Physical Problems of the Belarusian State University [4]. CD-1 is a symmetric indotricarbocyanine dye with the molar weight 765.5 g/mol. The dye is insoluble in water in the normal conditions.

The experiments have been performed with the use of deionized water. Dissolution of CD-1 molecules in water was realized in an ultrasonic bath (at the frequency of 35 kHz). The treatment was performed for 30–60 min.

3. Results

The crystals of CD-1 that is ordinary insoluble in water were introduced into water for the ultrasonic treatment. The resultant green suspension was kept for 5–10 days at room temperature. In the process of storage the absorption spectra of the solution have revealed changes illustrated in Fig. 1.

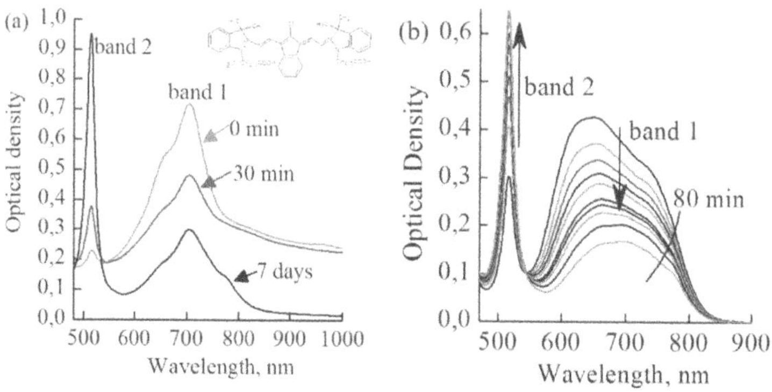

Figure 1. Absorption spectra of CD-1 in water as a function of time after the ultrasonic treatment during 40 min (a). Time evolution of the absorption spectra for CD-1 in water-ethanol mixture (b).

Immediately after the ultrasonic treatment, in the initial time of recording, the absorption spectra exhibited a wide absorption band with the maximum at 704 nm (band 1). The absorption over the spectral region of 600–800 nm is due to monomers and associates of CD-1 molecules. The long-wavelength wing of this band (800–1000 nm) seems to be due to scattering from large particles in the solution. As the particles are settling on the bottom, the intensity in the wing is decreased.

The optical density in the absorption band 1 for the solution is decreasing in the process of its storage. At the same time, the spectrum has revealed the short-wavelength absorption band 2 (with the maximum at 513 nm and the half-width of 22 nm), the intensity of which is increased monotonically. The spectral changes are terminated in 5–10 days. Based on the results of the studies using a chromate-mass spectrometer, it is shown that after the ultrasonic treatment the molecular structure of CD-1 in water is unaltered.

By the reprecipitation method, a concentrated solution of CD-1 in ethanol was added into water to give a homogeneous solution whose color was changing from green to pink. In the process, absorption spectra reveal the changes similar to those mentioned above. However, there is no long-wavelength wing (800–1000 nm) in these spectra. As distinct from water solutions, changes in the absorption spectra for such a solution are terminated 60–100 min after its preparation.

It should be noted that the nanostructures with the absorption band at 513 nm have no photoluminescence when prepared by any of the described methods.

The morphology of the formed nanostructures on silicon substrates has been studied by atomic force microscopy. The rod-like and cylindrical objects 10–20 nm in height, and also single particles with the dimensions of several nanometers have been revealed. They are shown in Fig. 2.

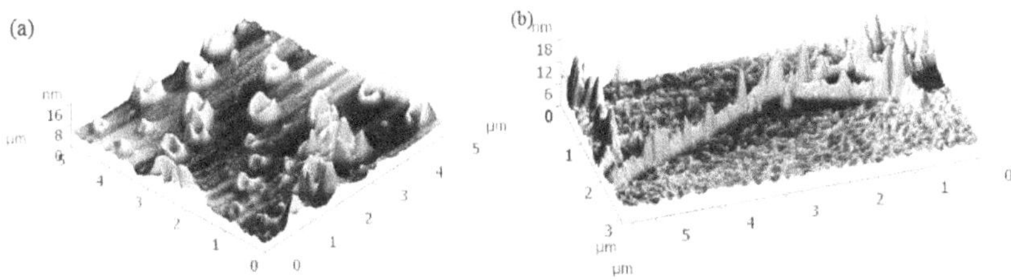

Figure 2. The CD-1 nanostructures on crystalline silicon imaged with atomic force microscope after the preparation with the ultrasonic treatment in water (a) and obtained by reprecipitation in water-ethanol mixtures (b).

Besides, the rod-like structures and the particles have been observed earlier when using the reprecipitation method in water-ethanol mixtures. The cylindrical formations are found only in water solutions with a new method for the preparation of nanostructures.

4. Conclusion

The method to produce nanostructures of water-insoluble organic dyes employing self-organization in a water solution subjected to ultrasonic treatment without a use of any additional substances is presented. The formed nanostructures are cylindrical in form and characterized by a narrow absorption band that is shifted to the short-wavelength region as compared to the spectra of the dye monomers.

References

1. Y. S. Zhao, H. Fu, A. Peng, Y. Ma, D. Xiao, J. Yao, *Adv. Mater.* **20**, 2859 (2008).
2. H. Kasai, H. S. Nalwa, H. Oikawa, S. Okada, H. Matsuda, N. Minami, A. Kakuta, K. Ono, A. Mukoh, H. Nakanishi, *Jpn. J. Appl. Phys.* **31**, L1132 (1992).
3. S. Kirstein, S. Daehne, *Int. J. Photoenergy* **2006**, 20363 (2006).
4. A. A. Lugovski, M. P. Samtsov, K. N. Kaplevsky, D. Tarasau, E. S. Voropay, P. T. Petrov, Yu. P. Istomin, *J. Photochem. and Photobiol. A* **316**, 31 (2016).

SIMS PROFILING OF ORGANIC-BASED MULTILAYER NANOSTRUCTURES WITH Al/LiF TOP ELECTRODE

M. N. Drozdov, G. L. Pakhomov, V. V. Travkin
Institute for Physics of Microstructures RAS, GSP-105 603950 Nizhny Novgorod, Russia
pakhomov@ipmras.ru

M. N. Bochkarev
Razuvaev Institute of Organometallic Chemistry RAS
Tropinina 49, 603950 Nizhny Novgorod, Russia
mboch@iomc.ras.ru

The archetype photovoltaic devices containing an organic photoabsorber were fabricated. The bottom transparent electrode was a sandwich-type structure composed of molybdenum oxide and silver nanolayers MoO_x/Ag/MoO_x, while the top nontransparent electrode was an Al/LiF bilayer. The devices were analyzed using the time-of-flight secondary ion mass spectrometry (SIMS) technique with depth profiling. It was found that the capping aluminum layer is severely oxidized. Lithium coming from the 1 nm thick LiF in the bilayer cathode is distributed within the underlying organic layer reaching the bottom electrode. Further diffusion of lithium is mostly restricted by the Ag interlayer in the MoO_x/Ag/MoO_x sandwich.

1. Introduction

Al/LiF bilayers are used as reflecting electrodes in organic photovoltaic or light-emitting devices [1–4]. To provide an efficient injection of the charge carriers, the thickness of LiF in these electrodes should be about 1 nm or less [1,2]. There are several, sometimes conflicting, explanations of the injection efficiency. Some of them rely on a thermodynamic forbiddance of the reaction in bilayers and hence existence of a tunneling barrier formed by insolating LiF [3]. Others fully admit this reaction and hence doping of underlying organics by released lithium species (atoms or ions) [1–4]. In this paper we show that spatial distribution of lithium species is wider than the nominal LiF underlayer, and release of lithium is not associated with deposition of aluminum on the LiF layer.

2. Experimental

Materials and methods of the device preparation are described elsewhere [5]. The devices were fabricated by depositing the bottom electrode (a three-layer MoO_x/Ag/MoO_x sandwich), organic photoabsorber boron chloride subphthalocyanine (SubPc), LiF and Al (top electrode), in that order, on the glass substrate.

The fabricated devices were analyzed with a TOF-SIMS5 (IONTOF GmbH) instrument. Depth profiles were obtained in the positive mode (sputtering with oxygen ions) and in the negative mode (sputtering with cesium ions). Accordingly, characteristic positive or negative secondary ions were selected from the raw data and their yields, after corresponding verification, were analyzed as a function of the sputtering time.

3. Results and discussion

SIMS depth profiles of the secondary ions are shown in Fig. 1 (after smoothing). Note that the top panels display the sequence of layers during the etching, *i.e.* from the top electrode down to the substrate. The time scale does not correlate with the actual thickness of the layers, since etching rates differ i) for different materials and ii) for two types of sputtering ions used. For instance, etching of top Al layer with Cs ions requires twice less time than with O ions, so the right and left panels do not coincide in length.

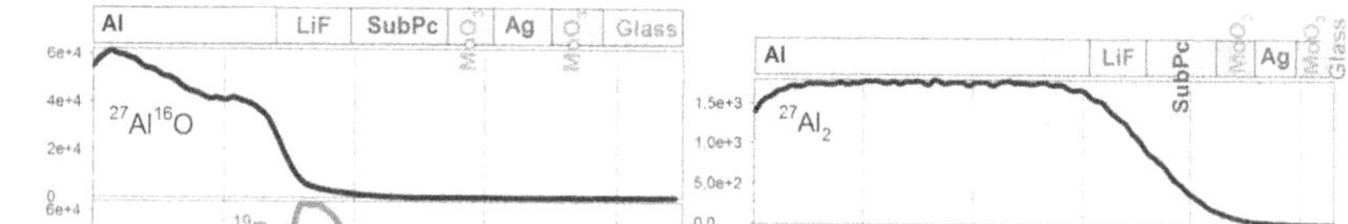

Figure 1. SIMS profiles of the characteristic secondary ions recorded in the negative mode (left graph) and in the positive mode (right graph). Top panels show the sequence of the layers in the device according to the time scale. The thickness of the layers (in the order of appearance): Al = 40 nm, LiF = 1 nm, SubPc = 17 nm, MoO_x = 5 nm, Ag = 10 nm, MoO_x = 5 nm, bulk substrate = glass.

It is evident that all components of the multilayer nanostructure can neatly be distinguished by the SIMS depth profiling. In general, the profiles recorded in the negative mode seem to suffer more from the side-effects (artefacts) of SIMS sputtering. Following points should be highlighted:

1) Top Al layer contains significant amount of oxygen (not shown), a trend also confirmed by the profile of AlO^- secondary ions in Fig. 1. The concentration of oxygen drops from the surface to the depth, suggesting that the oxidation (and hence degradation) of the electrode is caused by admixtures coming from the ambient.

2) Though LiF is only 1 nm thick, it is distinctly visualized due to the high yields of fluorine and lithium secondary ions in the negative and positive mode, respectively. However, there is an important difference between the profiles of fluorine- and lithium-related species. While the ^{19}F signal fades in the middle of the SubPc layer, the ^{6}Li signal becomes weaker twice only reaching the surface of the Ag interlayer in the bottom electrode (*cf.* [1,5]). Notably, there is no important difference in the profiles acquired when sputtering starts from Al or from LiF, *i.e.* in the samples without Al. This implies that knocking out lithium atoms from the LiF layer by hot depositing Al atoms, or any substitution reaction, is not necessary to observe the migration of lithium species towards the device bulk. Therefore, an intensive release of lithium from LiF into underlying layer is not specific for Al/LiF/SubPc nanostructures (*cf.* [1–4]).

3) SubPc molecule (chemical formula $C_{24}N_6H_{12}BCl$) can be traced by using C_8^- secondary ions, a fragment of molecule corresponding to the carbon skeleton of isoindole, or by using boron secondary ions in the positive mode. Typically for organics, sputtering of the SubPc layer with O ions takes less time than sputtering of the inorganic components of the device.

4) An occurrence of the substoichiometric molybdenum trioxide MoO_x, where $2.7<x<3.0$, is seen as a doublet peak of MoO_3^- and Mo^+ secondary ions at ~650 and ~950 s of sputtering, respectively (Fig. 1). Interestingly, the signal of oxygen-containing MoO^- (not shown) or MoO_3^- secondary ions tends to zero in the region between the two MoO_x layers in the bottom electrode, while the profile of Mo^+ displays a presence of molybdenum inside of the Ag interlayer. A replacement of molybdenum from its highest oxide by silver is hardly possible [5], we therefore assume that Mo appears due to mixing effects.

5) Silver is better seen as a polyatomic cluster ion, especially in the negative mode. The allocation of Ag-related peaks exactly in the middle of the MoO_x envelope suggests minimal upward or downward diffusion of silver through the neighboring oxide layers. This is an important feature when considering the three-layer MoO_x/Ag/MoO_x sandwich as a transparent electrode for

organic-based devices. The Ag interlayer contains some amount of oxygen (not shown). It is yet unclear whether this should be attributed to the partial oxidation of silver or to mixing with MoO_x.

4. Conclusion

We demonstrate usefulness of the SIMS depth profiling for analysis of hybrid (organic/inorganic) multilayer devices where the nominal thickness of layers lies in the range of 1–40 nm. The accuracy of analysis is limited by the energy of sputtering ions and by the natural roughness of the layers obtained by the thermal vacuum evaporation technique. The obtained results on the actual distribution of the components throughout the device structure are potentially interesting for understanding photoelectrical properties of the devices. Particularly, they might explain short-circuiting of the devices where organic layers are too thin, or unintentional doping of organic layers by lithium species released from the LiF cathode underlayer.

Acknowledgments

This work was supported by RFBR grants 15-43-02496, 15-08-06364.

References

1. H. Heil, J. Steiger, S. Karg, M. Gastel, H. Ortner, H. von Seggern, M. Stößel, *J. Appl. Phys.* **89**, 420 (2001).
2. S. K. M. Jonsson, E. Carlegrim, F. Zhang *et al.*, *Jpn. J. Appl. Phys.* **44**, 3695 (2005).
3. W. J. H. van Gennip, J. K. J. van Duren, P. C. Thüne *et al.*, *J. Chem. Phys.* **117**, 5031 (2002).
4. E. Ahlswede, J. Hanisch, M. Powalla, *Appl. Phys. Lett.* **90**, 163504 (2007).
5. V. V. Travkin, A. Yu. Luk'yanov, M. N. Drozdov *et al., Appl. Surf. Sci.* **390**, 703 (2016).

EFFECT OF Cr ON HYDROPHOBICITY OF RAPIDLY SOLIDIFIED Al

I. I. Tashlykova-Bushkevich
Belarusian State University of Informatics and Radioelectronics
P. Browka 6, 220013 Minsk, Belarus
iya.itb@bsuir.by

J. S. Yakovenko
Belarusian State Pedagogical University, Sovetskaja 18, 220050 Minsk, Belarus

The effect of Cr on hydrophobicity of Al has been investigated in rapidly solidified Al-1.0Cr and Al-3.0Cr (at.%) alloys by means of SPEM, AFM and distilled water CA measurements. The results demonstrate the crucial importance of both chemical composition and surface morphology of alloys for the control of their wettability.

1. Introduction

In hydrogen storage technologies, aluminum-based materials are a favorite candidate for hydrogen tanks because of an excellent gas barrier property coupled with their lightness [1]. The progress in commercialization of hydrogen fuel cell vehicles requires a development of high-strength aluminum alloys immune to hydrogen embrittlement (HE) for a use in compressed hydrogen storage systems where maximum filling pressure is increased up to 70–100 MPa. In response, remarkable international efforts have been deployed for an assessment of factors that account for hydrogen-assisted fracture behavior of Al materials. In spite of controversy on HE mechanisms in Al alloys in the literature, a consensus has emerged that H provides the dominant damage mechanism for high strength alloys stressed in moist environments. Particular attention is focused on corrosion effects which can occur at wetted surfaces when atomic H enter and diffuse through the metal.

In our recent work H desorption measurements were successfully accompanied by nanoscale microstructural evaluation of rapidly solidified (RS) Al and an Al-Cr alloy by means of synchrotron-based photoelectron microscopy [2] to gain insight into hydrogen/microstructure interactions developing H embrittlement mechanisms in Al-based alloys. The objective of this paper is to investigate how chemical composition and microstructure affect surface wettability of RS Al-Cr alloys.

2. Experimental

Foils of pure Al, Al-1.0Cr and Al-3.0Cr alloys (at.%) were produced by the centrifugal melt quenching method in air using a Cu wheel. The cooling rate was of the order of 10^6 K/s. The thickness of foil bulk parts was 50–100 μm.

The chemical composition was studied using scanning photoelectron microscopy (SPEM) at the ESCAmicroscopy beamline at the 3rd generation Elettra Synchrotron Laboratory. In prior experiments, samples were sputter cleaned in 3 steps of 10 min by Ar^+ bombardment with an ion energy of 2 keV to eliminate surface contamination by carbon. Air-facing surface was examined with a photon energy of 665 eV, energy resolution of 0.2 eV, and spatial resolution of less than 0.1 μm. The SPEM set-up has been described in details elsewhere [2]. The foil roughness was characterized by means of atomic force microscopy (AFM) by a NT-206 instrument. Average surface roughness characteristics were calculated using SurfaceXplorer software. The contact angles (CAs) for the air-facing foil side were determined using the sessile-drop technique. The measurements were conducted using a 5 μl droplet of distilled water at room temperature with the relative error of 1 %.

3. Results and discussion

Fig. 1 shows SPEM images of an air-side surface of RS Al-1.0Cr and Al-3.0 Cr alloys. The contrast of the raw images in Figs. 1a and c reflect both the foil surface elemental composition and morphology. Figs. 1b,d present a lateral distribution of elements mapped in SPEM images.

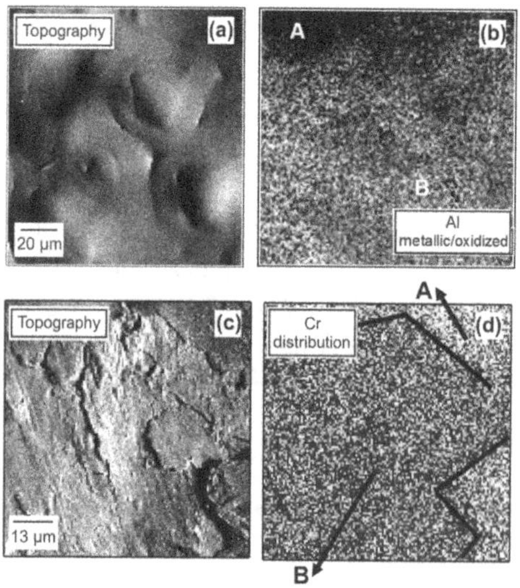

Figure 1. SPEM images for Al $2p$ (a) and chemical map (b) obtained as the ratio of Al $2p$ metallic and oxidized species maps from the same area (a) of Al-1.0Cr alloy foil, illustrating the relative local concentration of metallic Al. SPEM images for Cr $2p_{3/2}$ (c) and Cr/background (d) maps acquired for Al-3.0Cr alloy. The black solid lines in (d) indicate Cr-rich areas at upper and bottom right side corners.

The changes in the elemental composition can be observed as a difference in brightness in the image proportional to the relative concentration of indicated elements, see Figs. 1b and d. Comparing the Cr and Al metal/oxide maps one can see a clear correlation between the regions of higher/lower Cr content and lower/higher Al oxide content. Black solid lines in the map of Fig. 1d mark the boundary between the Cr-rich and Cr-depleted (areas A and B, respectively).

The foil morphology in AFM images shown in Fig. 2 demonstrates that RS Al-1.0Cr and Al-3.0 Cr alloys are rough on a sub-micron and nanometer scale and include ridge-rich surface regions and microvoids. The surface roughness is increased from 39.2 nm for Al-1.0Cr alloy to 42.7 nm for Al-3.0Cr alloy. The CA measured at the surface of Al-1.0Cr alloy (86.2 deg) was larger than that measured at the surface of Al-3.0Cr alloy (75.2 deg).

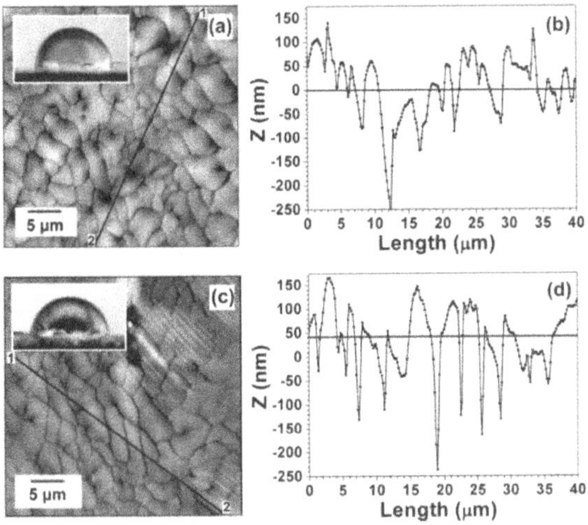

Figure 2. AFM surface images of Al-1.0Cr (a) and Al-3.0Cr (c) alloys and correspondent section analysis (b) and (d) performed along the lines shown in (a) and (c), respectively. The inserts in left upper corner of images (a) and (c) show the shape of distilled water droplets on foil surfaces.

The wettability behavior has been investigated for nano-rough surfaces of RS Al-Cr alloys concerning specific surface properties, such as morphology, roughness, and surface chemistry. The present SPEM results reflect the chemical composition of foil surface, rich in Al oxide species with some scarce contribution of Cr. The Cr-rich areas (regions A in Fig. 1) exhibit a high metallic component of Al at the foil surface, while Cr-free areas (regions B in Fig. 1) indicate Al almost entirely in the oxidized state. Recently, we reported that oxide inclusions of Al_2O_3 as well as $Al(OH)_3$ and AlOOH are found both on and below the Al and Al-Cr alloy surfaces [2]. As the RS Al materials can contain a high density of lattice defects, we can attribute the presence of aluminum hydroxides and oxyhydroxides to the formation of Al-H bonds during such non-equilibrium processing as rapid solidification processing.

It was found that Al-1.0Cr alloy exhibits hydrophobic behavior, as its large CA value is in a good agreement with the data which we obtained for RS pure Al (87.7 deg) [3]. One can argue that detected widespread regions with lack of Cr at foil surface of Al-1.0Cr alloy account for it. This fact attracts a vast interest because Al is a high surface-energy metallic material expected to be hydrophilic [4]. At the same time, the decrease of CA measured on Al-3.0Cr alloy, see Figs. 2c,d, shows a wetting transition from hydrophobic to hydrophilic state of the samples with the increase of Cr content when the surface roughness tends to be increased and according to the Wenzel model [4] the water wets the surface homogeneously and enters the cavities and microvoids.

4. Conclusion

The wettability of RS Al-*x*Cr alloys, $x = 1.0$; 3.0 at.%, has been observed to depend on chemical composition and surface morphology. The wetting transition from hydrophobic to hydrophilic state of the foils was found with the increase of Cr content when stable Wenzel state of water droplet is observed.

Acknowledgements

This work was partially supported by the Abdus Salam International Centre for Theoretical Physics (Italy). The authors would also like to thank Professor V. G. Shepelevich (BSU, Belarus) for the help with the sample preparation.

References

1. J. R. Scully, G. A. Young, S. W. Smith, in: *Gaseous Hydrogen Embrittlement of Materials in Energy Technologies*, eds. R. P. Gangloff and B. P. Somerday, Vol. 1 (Woodhead Publishing Ltd., 2012), p. 707–768.
2. I. I. Tashlykova-Bushkevich, M. Amati, B. Alemán, H. Sezen, L. Gregoratti, M. Kiskinova, *Int. J. Hydrogen Energy* **41**, 9100 (2016).
3. I. I. Tashlykova-Bushkevich, J. S. Yakovenko, I. A. Bushkevich, P. Zukowski, *Przegląd Elektrotechniczny* **92**, 151 (2016).
4. W. Liu, L. Sun, Y. Luo, R. Wu, H. Jiang, Y. Chen, G. Zeng Y. Liu, *Appl. Surf. Sci.* **280**, 193 (2013).

NANOSTRUCTURE-BASED DEVICES

INVITED

PHOTOCHROMIC ORGANIC FIELD-EFFECT TRANSISTORS: MOLECULES, DEVICE PERFORMANCE AND LASER PATTERNING

Y. Wakayama
International Center for Materials Nanoarchitectonics (WPI-MANA)
National Institute for Materials Science (NIMS)
1-1 Namiki, Tsukuba 305-0044, Japan
WAKAYAMA.Yutaka@nims.go.jp

We developed an optically controllable organic field transistor (OFET) by employing photochromic diarylethene (DAE) molecules as a transistor channel layer. DAE molecules undergo typical photochromic reaction, *i.e.* reversible conformational change between closed- and open-ring isomers by alternating ultraviolet (UV) and visible (VIS) light irradiation. We found that the drain current in the DAE-based OFET also showed a reversible change accompanied by this conformational change. The closed-ring isomer exhibited a transistor operation under appropriate gate and drain bias voltages, meanwhile the open-ring isomer showed no drain current. As a result, a remarkably high on/off ratio of 1,000 was achieved. The drain current modulation can be attributed to the drastic transformation in the π-conjugation system in association with the photoisomerization. These results present two important messages. The first one is that this compound can be regarded as a new material, which has dual properties: organic semiconductor and photochromism. The second is that a phase transition between semiconductor and insulator can be induced by light irradiation. Based on these achievements, we demonstrate laser drawing of one-dimensional (1D) channels on an OFET with a photochromic DAE layer. The main findings are: (i) a number of 1D channels can be written and erased repeatedly in the DAE layer by scanning UV and VIS focused laser spots alternately between the source and drain electrodes, (ii) the conductivity of the 1D channel can be controlled by the illumination condition such as the laser power density and the scan speed, and (iii) it is possible to draw an analogue adder circuit by optically written 1D channels so as to overlap a portion of the channels and perform optical summing operations by local laser illumination of the respective channels. These findings open new possibilities for realization of various optically reconfigurable low-dimensional organic transistor circuits, which are not possible with conventional thin film OFETs.

1. Introduction

In recent years, organic field-effect transistors (OFETs) have attracted much attention as key devices for developing printable and flexible electronic circuits [1–3]. Another important direction of the OFET research is in miniaturizing device size for logic and memory elements for future opto-electrical circuits. One of the most challenging studies in this field is to utilize OFETs in the development of opto-electrical switches [4] or optically switchable transistors [3], where drain current between source and drain electrodes is controllable by light irradiation with different wavelengths under certain bias voltages. For this purpose, photochromic molecules are key components to trigger optically induced variations in transport properties of transistors [6,7]. Photochromic molecules show a variety of reversible changes, such as a cis/trans conformational change in azobenzene [8,9], a non-polar/polar variation in spiropyran [10,11], and an open/closed ring isomerization in diarylethene (DAE), when the molecules are exposed to a light stimulus that has a wavelength corresponding to an absorption peak of the counter isomeric state. These photoisomerizations accompany variations in properties those are energy levels, molecular structures and π-electronic conjugation [12–14]. Therefore, these molecules are suited for controlling electrical charge transport upon photoinduced isomerization.

In many cases, DAEs are doped into organic semiconductor matrices or are inserted at the interfaces between organic semiconductor and gate insulator [15–17]. However, the light-induced current modulations have been very small, in the range of 10–80 %. This is because the doped photochromic molecules can cause only marginal variations in the drain current. To overcome this issue, we have employed a DAE thin film directly as a channel layer in an OFET [18]. Important parts of the DAE molecule are biphenyl substituents attached to both sides of the main framework, which realize a drastic change in the π-conjunction involved in photoisomerization. As a result, distinct semiconductor-insulator phase transition is induced. A DAE film with closed-ring isomer works as a *p*-type transistor channel, while the transistor channel is completely insulated when the DAE film is changed to the open-ring isomer. Alternate irradiation of VIS and UV light induced large modulations ($\sim 10^3$) in the drain current. We confirmed that the modulated drain current was very stable, which can be maintained even after a week in dark conditions. These results demonstrate that the direct use of DAE molecules for transistor channels is promising for future high-performance opto-electrical devices.

Other challenges of OFETs are patterning and downsizing of transistor channels, because conventional lithography and lift-off process are not suitable for organic materials. In this regard, various techniques have been proposed for fabricating low-dimensional organic structures those are laser interference patterning [19], maskless oxygen plasma etching [20], vapor-phase deposition [21,22], directed assembly in anodized aluminum oxide templates [23] and in lithographically patterned SiO_2 gate layer [24]. However, these approaches are difficult to apply to the fabrication of transistor structures and thus are not compatible with patterning, downscaling, and large-scale integration of OFETs.

In this paper, we present an optical patterning method. Focused laser beams were scanned between the source and drain electrodes on a DAE thin film to draw 1D transistor channels. Importantly, these channels can be written and erased repeatedly by tuning the wavelengths appropriately due to the reversible photoisomerization of DAE molecules. Moreover, we also demonstrate various channel networks and device operations, such as repeatable writing/erasing of multi-channels, laser-power-controlled conductivity of the 1D channels, and an analogue adder circuit patterning. These results open new possibilities for optically reconfigurable, low-dimensional OFETs.

2. Experimental

2.1. *Device fabrication*

A top-contact and bottom-gate transistor was fabricated on a highly doped p^{+}-Si(001) wafer with a 200-nm-thick SiO_2 layer. The SiO_2 layer was coated with a 10-nm-thick PMMA film for passivation of the SiO_2 surface. Then, a 25-nm-thick DAE thin film was thermally evaporated through a shadow mask at the substrate temperature of 40 °C in a vacuum deposition chamber with the base pressure below 5×10^{-7} Pa. Finally, 6-nm-thick α-sexithiophene (6T) and 150-nm-thick Au films were sequentially deposited on top of the DAE film using a shadow mask as source and drain electrodes, respectively. Here, the 6T layer promotes carrier injection to increase the drain current.

2.2. *Optical patterning and OFET measurements*

An experimental setup for laser patterning and transistor measurements, the molecular structures of the open- and closed-ring DAEs, together with their optical absorption are shown in Fig. 1. Ne-Cd laser (325.0 nm) and He-Ne laser (632.8 nm) lines were used as UV and VIS lights, respectively. The DAE films were irradiated by the focused beams with 5 μm diameter in an optical microscope with a 50× objective lens. The DAE transistor was put on a stage, the position of which can be moved by a stepping motor and actuator-driven XY-axes stage. The scan speed were controlled from 0.1 μm/s to 2 mm/s. The transistor properties were measured using a home-made prober system and a dual-channel source/measure unit.

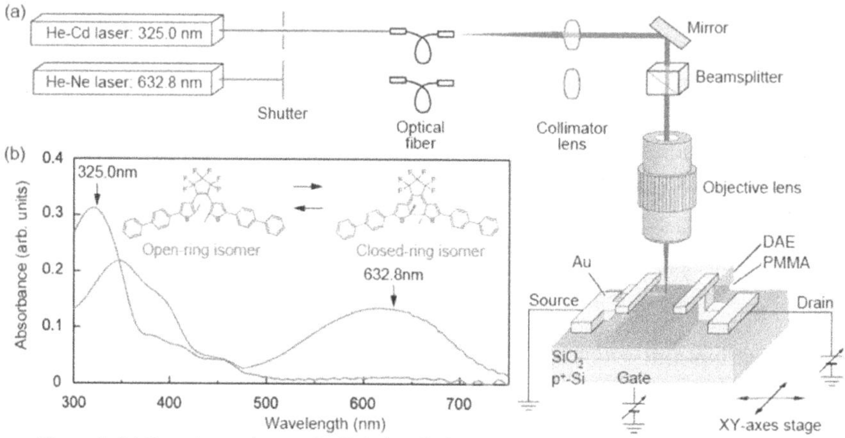

Figure 1. (a) Experimental setup for light irradiation, transistor measurements and XY-stage. (b) Absorption spectra of closed- and open-ring isomers.

3. Results and discussion

The setup in Fig. 1a makes various kinds of laser patterning possible to form networks of electrical flow paths, such as multi-channel transistor, laser-power-controlled drain current, branched transistor channels and optical valves. These basic techniques enable us to design a variety of electronic circuits. Adder circuit patterning is demonstrated as an example.

3.1. *Multi-channel transistor*

Fig. 2 demonstrates writing and erasing of 1D transistor channels. In Fig. 2a, the drain current was increased with the numbers of the channels. Here, UV spotlight was scanned between a pair of electrodes on the open-ring (insulator) DAE thin film.

The scanned area was turned to the closed-ring isomer (semiconductor) to work as a transistor channel. Up to triple channels were connected to common source-drain electrodes in parallel. Next, thus patterned transistor channels can be erased by scanning VIS spotlight on the same area. Drain current was decreased with decreasing connected channels. In this manner, repeatable writing/erasing 1D transistor channel can be easily performed.

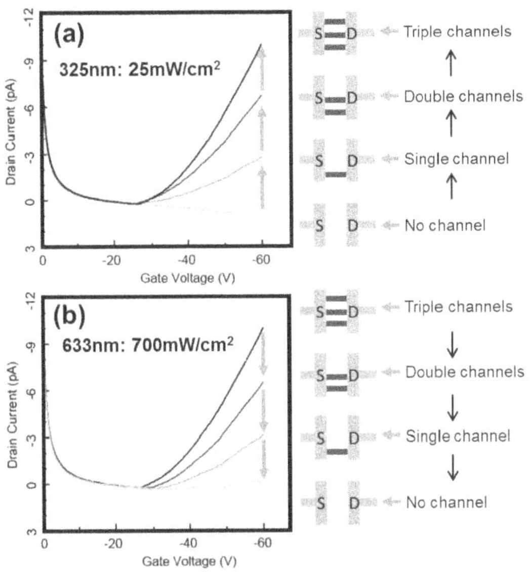

Figure 2. (a) 1D transistor patterning by UV light scanning. Multiple channels were connected with common electrodes in parallel. Drain current was increased with the number of channels. (b) Erasing transistor channel by VIS light scanning. Drain current was decreased with the number of channels.

3.2. *Laser-power-controlled drain current*

The ratio of the open/closed ring isomers is proportional to the laser power and the drain current is dependent on the ratio. Therefore, drain current was precisely controlled by tuning irradiated laser power. Fig. 3 shows stepwise drain current through a 1D transistor channel. Here, UV and VIS spotlights with different intensities were scanned sequentially on the DAE film. In this manner, multilevel drain current can be achieved by controlling the laser power.

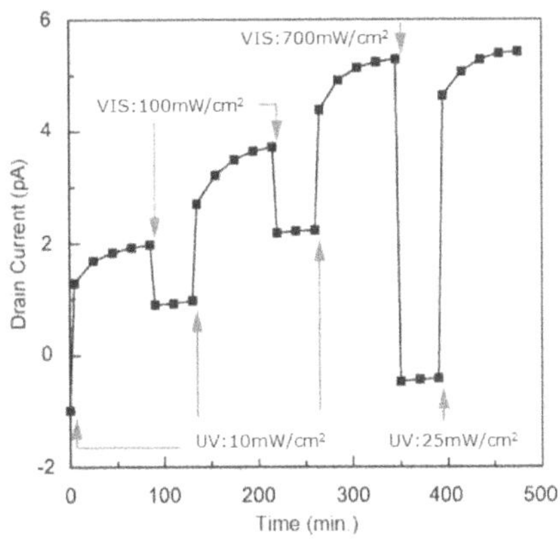

Figure 3. Drain current is proportional to the laser power, enabling stepwise control of drain current.

3.3. *Adder circuit drawing*

A strong point of our technique is that various kind of electrical flow paths can be patterned, which are zigzag channel, branched channel and optical valves. It is illustrated in Fig. 4.

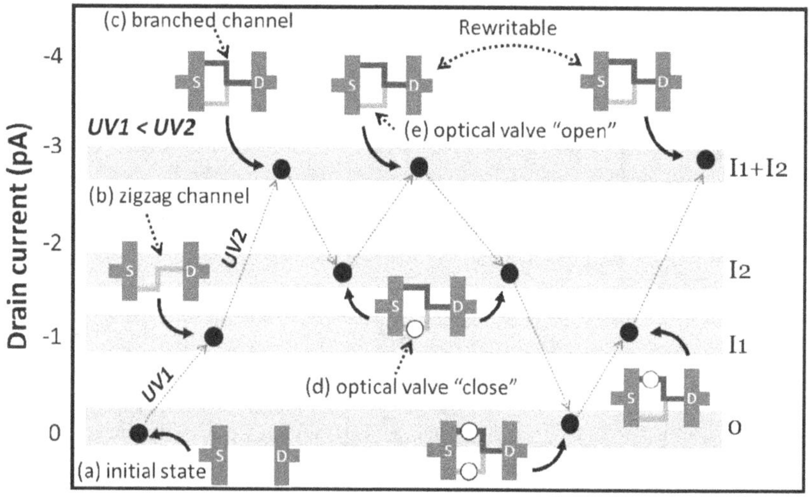

Figure 4. Adder circuit with branched-shaped channels. The electrical current through respective channels can be controlled by UV and VIS spotlight irradiation ("optical valves").

At the beginning, no current was observe at the initial state (Fig. 4a). The first channel with zigzag shape was draw as shown in Fig. 4b with certain UV intensity (UV1) to yield drain current level "I1". In similar manner, the second zigzag shape channel was patterned, where the first and the second channels were partially overlapped to form a branched shape channel (Fig. 4c). Importantly, the second channel was drawn with different intensities (UV2) to yield different drain current levels "I2". As a result, total current of "I1+I2" flowed to the drain electrode.

Next, the first channel was irradiated with a VIS spotlight to close the drain current through this channel (Fig. 4d). The drain was reduced from I1+I2 to I2, and returned to I1+I2 again by irradiating with VIS spotlight on the same position (Fig. 4e). In this manner, the electrical current through the channels can be opened and closed repeatedly. We call this function "optical valves". Consequently, drain current level can be controlled at I1, I2 and I1+I2 to be used in an adder circuit operation. Other examples of logic circuits are shown in Fig. 5.

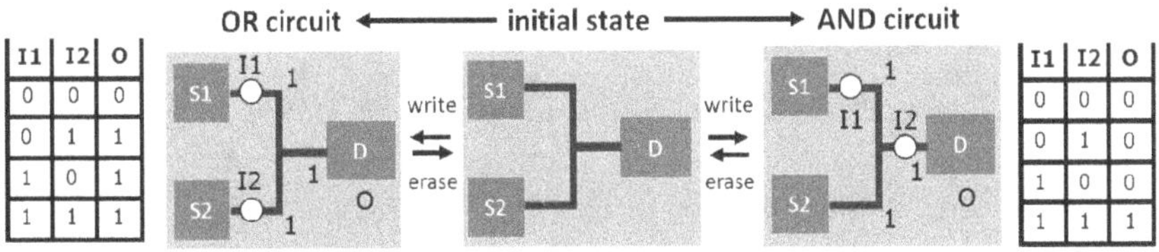

Figure 5. Logic circuit drawing. In the initial state, dual source electrodes are connected with a drain electrode by branched-shape transistor channel. According to the location of optical valve, the device can work as OR (left) or AND (right) circuit.

4. Conclusion and future perspective

We successfully developed laser patterning techniques for organic electronic devices. This device patterning is achievable owing to the unique property of the DAE molecule in which semiconductor-insulator phase transition can be induced by UV and VIS light irradiation. As a result, electrical path network can be drawn in an insulator thin film. In this study, the channel lengths were in the range of several tens micrometers. However, the device size can be reduced by utilizing a near-field optical microscope. Another merit of our technique is that patterning of a variety of logic circuits are available. Device architectures with multiple electrodes connected by multiple 1D channels can work as "OR" or "AND" circuit according to the positions of the optical valves. As demonstrated here, our technique has potential of various optically reconfigurable low-dimensional organic transistor circuits, which are not possible with conventional thin film OFETs.

References

1. C. D. Dimitrakopoulos, R. P. L. Malenfant, *Adv. Mater.* **14**, 99 (2002).
2. G. Gelinck, P. Heremans, K. Nomoto, T. D. Anthopoulos, *Adv. Mater.* **22**, 3778 (2010).
3. Y.-Y. Noh, N. Zhao, M. Caironi, H. Sirringhaus, *Nat. Nanotechnol.* **2**, 784 (2007).
4. A. J. Kronenmeijer, H. B. Akkernab, T. Kudernac, B. J. van Wees, B. L. Feringa, P. W. M. Blom, B. de Boer, *Adv. Mater.* **20**, 1467 (2008).
5. K.-J. Baeg, M. Binda, D. Natali, M. Caironi, Y.-Y. Noh, *Adv. Mater.* **25**, 4267 (2013).
6. M. Irie, T. Fukaminato, K. Matsuda, S. Kobatake, *Chem. Rev.* **114**, 12174 (2014).
7. M.-H. Russew, S. Hecht, *Adv. Mater.* **22**, 3348 (2010).

8. N. Crivillers, E. Orgiu, F. Reinders, M. Mayer, P. Samori, *Adv. Mater.* **23**, 1447 (2011).
9. C. W. Tseng, D. C. Huang, Y. T. Tao, *ACS Appl. Mater. Interfaces* **4**, 5483 (2012).
10. H. Zhang, X. Guo, J. Hui, S. Hu, W. Xu, D. Zhu, *Nano Lett.* **11**, 4939 (2011).
11. L. A. Frolova, P. A. Troshin, D. K. Susarova, A. V. Kulikov, N. A. Sanina, S. M. Aldoshin, *Chem. Comm.* **51**, 6130 (2015).
12. D. Dulic, S. J. van der Molen, T. Kudernac, H. T. Jonkman, J. J. D. de Jong, T. N. Bowden, J. van Esch, B. L. Feringa, B. J. Wees, *Phys. Rev. Lett.* **91**, 207102 (2003).
13. T. Tsujioka, M. Irie, *J. Photochem. Photobio.* C **11**, 1 (2010).
14. Y. Kim, T. J. Hellmuch, D. Sysoiev, F. Pauly, T. Pietsch, J. Wolf, A. Erbe, T. Huhn, U. Groth, U. E. Steiner, E. Scheer, *Nano Lett.* **12**, 3736 (2012).
15. E. Orgiu, N. Crivillers, M. Herder, L. Grubert, M. Pätzel, J. Frisch, E. Pavlica, D. T. Duong, G. Bratina, A. Salleo, N. Koch, S. Hecht, P. Samori, *Nat. Chem.* **4**, 675 (2012).
16. M. E. Gemayel, K. Börjesson, M. Herder, D. T. Doung, J. A. Hutchison, C. Ruzié, G. Schweicher, A. Salleo, Y. Geerts, S. Hecht, E. Orgiu, P. Samori, *Nat. Comm.* **6**, 1 (2015).
17. Q. Shen, L. Wang, S. Liu, Y. Gao, L. Gan, X. Guo, M. L. Steigerwald, Z. Shuai, Z. Liu, C. Nuckolls, *Adv. Mater.* **22**, 3282 (2010).
18. R. Hayakawa, K. Higashiguchi, K. Matsuda, T. Chikyo, Y. Wakayama, *ACS Appl. Mater. Interfaces* **5**, 3625 (2013).
19. H. Fang, D. Yuan, R. Guo, Su. Zhang, R. P. S. Han, S. Das, Z. L. Wang, *ACS Nano* **5**, 1476 (2011).
20. K. Du, I. Wathuthanthri, Y. Liu, Y. Tae Kang, C.-H. Choi, *Nanotechnology* **25**, 165301 (2014).
21. Y. S. Zhao, D. Xiao, W. Yang, A. Peng, J. Yao, *Chem. Mater.* **18**, 2302 (2006).
22. Y. Wakayama, R. Hayakawa, T. Chikyow, S. Machida, T. Nakayama, S. Egger, D. G. de Oteyza, H. Dosch, K. Kobayashi, *Nano Lett.* **8**, 3273 (2008).
23. J. Hu, K. W. Clark, R. Hayakawa, A.-P. Li, Y. Wakayama, *Langmuir* **29**, 7266 (2013).
24. S. Oh, R. Hayakawa, T. Chikyow, Y. Wakayama, *Appl. Phys. Lett.* **106**, 243301 (2015).
25. T. Tsuruoka, R. Hayakawa, K. Kobashi, K. Higashiguchi, K. Matsuda, Y. Wakayama, *Nano Lett.* **16**, 7474 (2016).

COMPUTATIONAL EXPLORATION ON THE APPLICATION OF CARBON NANOTUBES IN PLANAR HELIX SLOW-WAVE STRUCTURE TRAVELING-WAVE TUBE

Y. D. Lim, S. Wang, S. Aditya
School of Electrical and Electronics Engineering, Nanyang Technological University
50 Nanyang Avenue 639798, Singapore

B. K. Tay
School of Electrical and Electronics Engineering, and CINTRA CNRS/NTU/THALES
Nanyang Technological University
50 Nanyang Avenue 639798, Singapore

V. Labunov
Belarusian State University of Informatics and Radioelectronics
P. Browka 6, 220013 Minsk, Belarus

Application of carbon nanotubes field emitter array (CNT FEA) in planar helix-slow wave structure traveling wave tubes (PH-SWS TWTs) is explored. From the fabricated CNT FEA, electron beam current of 8.8 mA. From the obtained electron beam current, simulation studies on the operation of the PH-SWS TWT is carried out. Signal gain of 31.32 dB with the maximum output power of 3.2 W is obtained at the operation frequency of 105 GHz.

1. Introduction

Modern satellite communication systems require low-interruption signal transmissions. To relieve the data transmission congestion in the existing Ku-band (12–18 GHz), high frequency Ka-band (26–40 GHz) has been introduced [1]. In high frequency operation, traveling-wave tube (TWT) is usually preferred over silicon-based devices due to its high frequency capability [2]. In the effort of miniaturizing TWT, our research group has proposed a microfabricated planar helix-slow wave structure (PH-SWS) which can reduce the device size of the TWT [3]. From the reported simulation study, it has been found that 10 mA electron beam gives TWT 3.7 W peak power at 110 GHz with a 3-dB bandwidth of 60 % [4].

In this study, we propose to explore the possibility in using carbon nanotubes (CNTs) as the field emitter in the PH-SWS TWT. CNT field emitter array (FEA) is first microfabricated by growing CNTs in the pre-fabricated structure. After that, the field emission current obtained from the fabricated

CNT FEA is used as the current of the electron beam in the PH-SWS TWT simulation.

2. Experiment

In the fabrication of CNT FEA, titanium disilicide ($TiSi_2$) barrier layer is first deposited on the *n*-doped Si substrate, followed by the deposition of 2 μm SiO_2. The SiO_2 is then dry-etched selectively to form 500 nm diameter pits by means of a photolithography technique. After that, nickel catalyst is electroplated in the SiO_2 pits. CNTs are grown on the deposited nickel layer using volatile catalyst source (ferrocene/xylene) chemical vapor deposition (CVD) process. CNT bundles of ~2 μm height is then grown in the fabricated pits.

Field emission measurement of the CNT FEA is carried out in A.C. conditions, where the cathode-anode distance and the effective emission area are estimated to be 18 μm and 0.75 mm^2, respectively. From the obtained field emission current density, particle-in-cell simulation is carried out on the proposed traveling-wave tube with the planar helix slow-wave structure [4].

3. Results and discussion

Fig. 1 shows the scanning electron microscope (SEM) images which describes the fabrication process of the CNT FEA in this study. A series of SiO_2 pits with the aspect ratio of ~4 (2/0.5 μm) is first obtained from SiO_2 etching. $TiSi_2$ is anticipated at the bottom of the pits. After that, nickel catalyst dots are electroplated in the pits where CNTs grown selectively on the dots.

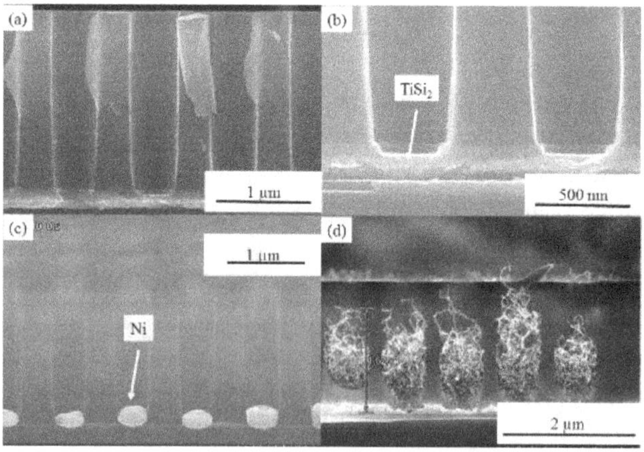

Figure 1. SEM images of pre-fabricated SiO_2 pits at different magnifications (a,b), electroplated Ni catalyst (c) and CNT bundles grown in SiO_2 the pits (d).

The field emission (FE) measurement and the Fowler-Nordheim (FN) plot of the CNT FEA are shown in Fig. 2. The FN plot shows a linear trend, which indicates that its FE characteristics can be resembled by the FN theory. The emission current of 8.8 mA is obtained. By using the obtained field emission current, we have performed a realistic simulation on the TWTs as proposed in our previous reported studies [3,4].

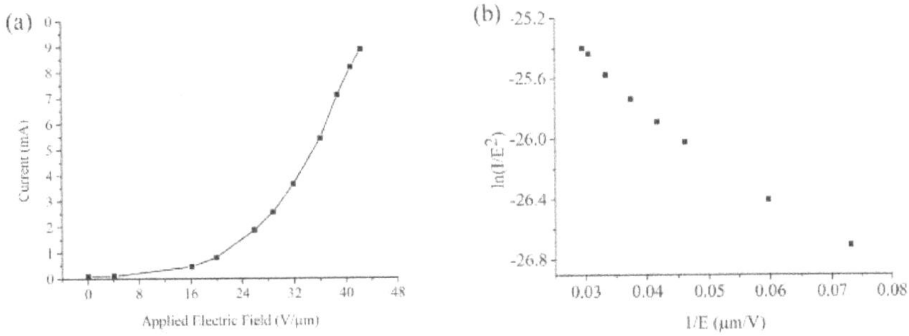

Figure 2. Field emission measurement (a) and Fowler-Nordheim plot (b) of CNT FEA.

Fig. 3 shows the simulation outcomes of the 8.8 mA field emission current obtained from CNT FEA in the operation of TWT. The simulation is carried out based on the 200 period PH-SWS TWT with the input power of 2.5 mW.

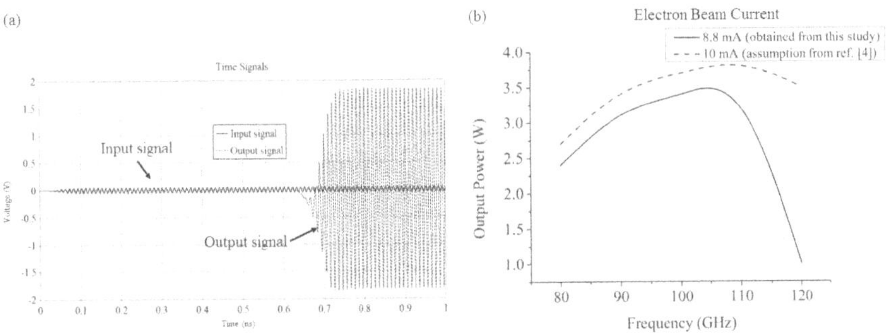

Figure 3. Simulated port signals at 100 GHz (a) and saturation power of the planar helix SWS-TWT at different electron beam currents (b) [4].

Fig. 3a shows that when the input signal is set to 50 mV, the power signal can reach 1.84 V. Effective gain of 31.32 dB is calculated. Fig. 3b shows the saturation output power gain for an input signal of 2.5 mW for different electron beam currents. For the electron beam current of 10 mA, the maximum output power is 3.7 W at the operation frequency of 110 GHz. By using the electron beam current of 8.8 mA emitted from the CNT FEA obtained in our current study, the maximum output power reduced to 3.2 W at 105 GHz. The reduction in the signal gain and maximum output power can be attribute to the reduction in the electron beam current [5]. Therefore, for further improvement of the PH-SWS TWT, the emission current density from CNT FEA needs to be improved for higher signal gain.

4. Conclusion

Possible application of carbon nanotubes field emitter array in planar helix-slow wave structure traveling-wave tube has been explored. The carbon nanotubes field emitter array is first fabricated and its field emission current is measured. By using the emission current of 8.8 mA from the fabricated field emitter array, the signal gain of 31.32 dB and maximum output power of 3.2 W at 105 GHz were obtained from the simulation. The simulation results have been presented to show that the signal gain and the maximum output power of the planar helix-slow wave structure traveling-wave tube reduces with the reduction of electron beam. These results together demonstrate the possibility of carbon nanotubes field emitter array as an electron beam source for the planar helix-slow wave structure traveling-wave tube.

Acknowledgments

This work has been supported by the Office for Space Technology and Industry (OSTIn), Singapore, under the project S14-1126-NRF OSTIn-SRP.

References

1. B. Lewis, *Ka-band vs. Ku-band: the Knowledge You Need to Make the Right Decision for Your Organization* (2013), http://www.intelsat.com/blog/ka-band-vs-ku-band-the-knowledge-you-need-to-make-the-right-decision-for-your-organization/.
2. J. H. Booske *et al.*, *IEEE Trans. Terahertz Sci. Technol.* **1**, 54 (2011).
3. C. Chua *et al.*, *IEEE Trans. Electron Devices* **58**, 4098 (2011).
4. S. Wang, S. Aditya, S. Member, *IEEE Trans. Electron Devices* **63**, 3725 (2016).
5. J. H. Booske *et al.*, *Travelling-Wave Tubes*, in: *Modern Microwave and Millimeter-Wave Power Electronics* (2003), pp. 171–245.

ANTIMONIDE-BASED SUPERLATTICE NANOSTRUCTURES TO DESIGN LONGWAVE INFRARED BARRIER DETECTORS

M. Delmas[1], P. Christol[2,3], R. Rossignol[2,3], J. B. Rodriguez[2,3]

[1]*School of Physics and Astronomy, Cardiff University, The Parade, Cardiff CF243AA, UK*

[2]*University of Montpellier, IES, UMR 5214, F- 34000 Montpellier, France*

[3]*CNRS, IES, UMR 5214, F- 34000 Montpellier, France*

Barrier structure is a new kind of infrared photodetector device while type-II InAs/GaSb superlattice (T2SL) is a new infrared material technology. This communication reports the way to combine both in order to design T2SL barrier photodetector for longwave infrared (LWIR) operation. The expected results in term of dark current at 77 K, achieved by TCAD simulation, are compared with those obtained by standard T2SL pin photodiode.

1. Introduction

The last past years, type-II superlattice (T2SL) made of InAs/GaSb nanostructures has emerged as a new material technology suitable for high performance cooled infrared detectors required for specific military and civilian applications. This was possible because T2SL is a particular quantum nanostructure with several advantages such as the possibility to address, thanks to SL bandgap engineering, the midwave and the longwave infrared (MWIR and LWIR, respectively) domains or to design, thanks to band offset engineering, barrier structures [1]. Recently, the T2SL technology has reached the point of being commercially available both in the MWIR [2] and the LWIR [3] but there is still a room for improvements to enhance performances.

The barrier photodetector is a minority carrier unipolar device, hybrid between a photovoltaic and a photoconductive system, with a potential barrier to block the majority carrier transport. The bulk generation-recombination (GR) current, which is the dominant current in a pin photodiode at low temperature, is suppressed by excluding the depleted electric field region from the photon absorbing layer. Consequently, the unipolar barrier detector device is diffusion-current limited whatever the temperature and compared to a pin diode configuration, the dark current is strongly reduced inducing an improvement of the electrical performance at iso-temperature or an enhancement of the operating temperature at iso-performance [4].

The objective of this paper is to present a design of InAs/GaSb T2SL barrier detector operating in the LWIR.

2. Design of the LWIR T2SL barrier detector

The barrier photodetector is composed of a contact layer (X), a barrier layer (B) and an active absorbing layer (n- or p-type). Depending on the minority carriers (electrons versus holes), several device designs (called XBp versus XBn) can be taken into account [5]. However, because high quantum efficiency is achieved with p-type T2SL absorbing zone [6], XBp configuration involving electron minority carrier is better.

The LWIR XBp architecture detector on GaSb substrate consists of a 4 μm thick 15/7 InAs/GaSb SL (dimension in monolayers (MLs)) slightly p-type active region, chosen to exhibit a cut-off wavelength around 12 μm at 77 K [5], an appropriate p-type barrier layer to block majority hole carriers and finally, a highly n-type doped contact layer composed of the 15/7 InAs/GaSb SL and an InAs cap layer. The main difficulty in designing a XBp barrier structure resides in the choice of the barrier layer: high and thick enough to block the hole majority carrier transport while allowing the electron minority to reach the contact. A solution to have a barrier in the valence band is to consider a InAs/GaSb/AlSb/GaSb SL to tailor the hole potential barrier (ΔE_v). A ΔE_v higher than 250 meV is adjusted by difference between the first heavy hole minibands of 15/7 InAs/GaSb SL and the one of InAs/GaSb/AlSb/GaSb SL for which a period 8/1/5/1 has been calculated.

The corresponding XBp band diagram structure calculated under no bias at 77 K is shown in Fig. 1a, where the doping of each layer is specified. These have been chosen in order to confine the electric field in the barrier layer and to generate flat band (no depleted region) condition in the absorbing zone. Consequently, the device should be diffusion current limited. A small electron potential barrier (ΔE_c) which can penalize the quantum efficiency can be noticed from Fig. 1a. This can be overcome by applying a small reverse bias. The T2SL XBp design is compared with an equivalent T2SL pin photodiode (same doping and thickness layers) which is displayed in Fig. 1b.

In the case of pin photodiode, the dark-current under moderate reverse bias is mainly limited by the GR current at low temperature and by the diffusion current at high temperature. In addition, under some conditions of reverse bias and residual doping level, the trap-assisted tunneling (TAT) and the band-to-band tunneling (BTB) can affect the dark current. Same currents are present in a XBp barrier detector but, compared to a pin diode at a given temperature, GR current appears at higher reverse voltage. The dark current of the XBp T2SL structure was calculated using the ATLAS software from SILVACO. ATLAS is a commercially available physically based device TCAD simulator often used for the simulation of the dark current-voltage (*J-V*) characteristics of IR photodiodes [7].

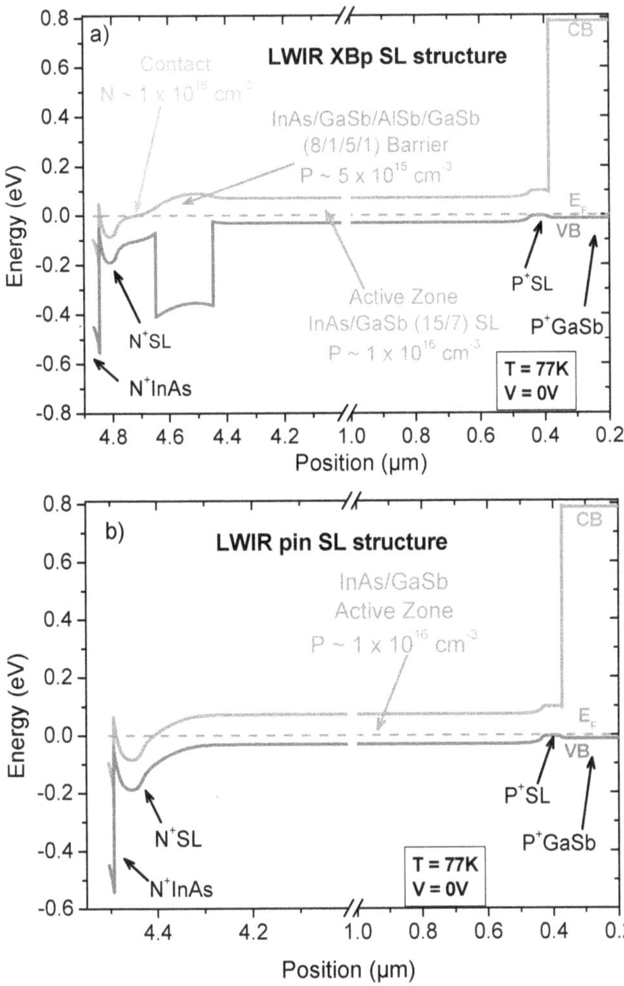

Figure 1. Calculated energy band diagram profile at $T = 77$ K under no bias of the LWIR XBp (a) and pin (b) SL structure.

The advantage of the barrier device is demonstrated in Fig. 2, which compares the calculated current-voltage characteristics at 77 K obtained for the XBp and standard pin LWIR SL diodes. At low reverse bias, the pin diode is GR current limited with the dark current level at –50 mV of 5.7×10^{-3} A/cm^2. It can be seen that the BTB contribution appears quickly started from 120 mV reverse bias. On the other hand, the XBp device is, as expected, diffusion current limited at low reverse bias until –0.3 V, showing a dark current density at –50 mV of 3.7×10^{-4} A/cm^2, which is more than one decade lower than that corresponding to pin diode. This result is at the state of the art compared to the well-known "rule 07" rule of thumb [8].

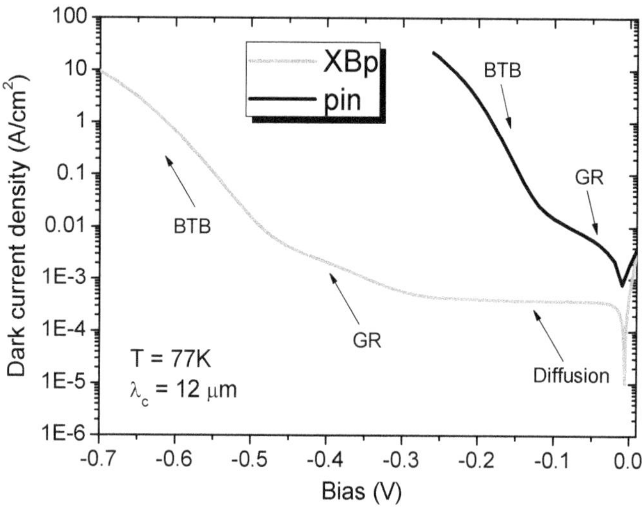

Figure 2. Simulated dark current density at 77K of LWIR SL detector: XBp and pin designs.

3. Conclusion

Using the Silvaco ATLAS simulator, current-voltage characteristic of LWIR InAs/GaSb XBp detector was modelled. Compared to standard equivalent pin diode, the result shows an advantage of barrier detector design with dark current density to be as low as the ones describing the "rule 07" in the LWIR. Such result will have to be confirmed by measurements performed on fabricated T2SL XBp device.

References

1. M. Razeghi, B-M. Nguyen, *Phys. Procedia* **3**, 1207 (2010).
2. L. Höglund *et al.*, *Proc. SPIE* **9819**, 98190Z (2016).
3. P. C. Klipstein *et al.*, *Proc. SPIE* **9819**, 98190T (2016).
4. S. Maimon, G. W. Wicks, *Appl. Phys. Lett.* **89**, 151109 (2006).
5. P. C. Klipstein, *J. Cryst. Growth* **425**, 351 (2015).
6. E. Giard *et al., J. Appl. Phys.* **116**, 043101 (2014).
7. M. Delmas *et al., J. Appl. Phys.* **116**, 113101 (2014).
8. D. R. Rhiger, *J. Electron Mater.* **40**, 1815 (2011).

GENERATION OF NEW SPATIAL AND TEMPORAL COHERENT LIGHT STATES USING III-V SEMICONDUCTOR METASURFACE BASED LASER TECHNOLOGY: VORTEX, CONTINUUM, DUAL FREQUENCY FOR THz

A. Garnache, M. Seghilani, M. Sellahi, R. Paquet, B. Chomet, M. Myara, S. Blin
IES - CNRS UMR5214, Université Montpellier, Montpellier, France
arnaud.garnache@ies.univ-montp2.fr

L. Legratiet, G. Beaudoin, I. Sagnes
LPN-CNRS - UPR20, Marcoussis, France

P. Lalanne
LP2N - IOGS-Bordeaux, Talence, France

We take an advantage of III-V semiconductor surface emitting laser technology integrating flat-photonics for the generation of new coherent states, thanks to insertion of intracavity functions based on the metasurface: vortex beam, dual-frequency spectrum for THz, modeless spectrum. These new kinds of coherent light states target many applications including optical tweezers, telecommunications, fundamental physics, sensors.

1. Introduction

Since years, III-V semiconductor based external-cavity VCSEL (VECSEL) concept is pointed out as a technology of choice for beyond-state-of-the-art laser light sources, demonstrating wavelength flexibility, high power, high spatial, temporal and polarization coherence, CW or fs ultra-short pulsed operation, compactness and functionalities. The targeted coherent state is typically a common circular low divergence fundamental gaussian TEM00 mode, linearly polarized state, single frequency state or modelocked comb. Such high-Q laser cavity exhibits a class-A dynamics with a low intensity noise at a shot noise level in MHz-RF range, as well as a quantum limit optical frequency noise at the Hz level. Integration and packaging of such high performances sources is in progress (by Coherent, Innoptics, Fraunhofer ILT...).

In this work, we take an advantage of diode-pumped VECSEL based on III-V nanotechnologies integrating flat-photonics (quantum-well, Bragg mirror, metasurface, metallic mask), and fundamental physical features (axial symmetry of the laser design, orthogonality of the cavity modes, high finesse/low losses operation, homogeneous gain, birefringence, dichroism, spatial hole burning) for the generation of non-conventional coherent states, thanks to the insertion of new intracavity functions: vortex beam, dual-frequency spectrum for THz, modeless spectrum.

2. Controlling the transverse mode emission: vortex beam and dual-mode/frequency laser

A first part of this work aims at demonstrating new spatial domain coherent state. We developed a semiconductor flat photonics technology for transverse phase and intensity control, as well as spectral and polarization control, inside the cavity. The intensity control is obtained thanks to sub-wavelength metallic masks deposited on the quantum well gain structure. For a phase-amplitude and polarization control, we developed a high-Q metasurface based Bragg mirror technology [2]. This concept allows generation of highly coherent high order Laguerre-Gauss modes, including single frequency VORTEX beam exhibiting controlled orbital angular momentum (Figs. 1,2) [1], or dual-mode/frequency laser for THz source (Fig. 3) [3]. It preserves the full coherence properties (diffraction limited low divergence beam, highly polarized E-field, ultra low noise) of usual TEM00 VeCSEL source. It also paves the way for the generation of other coherent states (Bessel beams, light with spin orbital momentum control) or new functionalities (wavelength filtering, chirality).

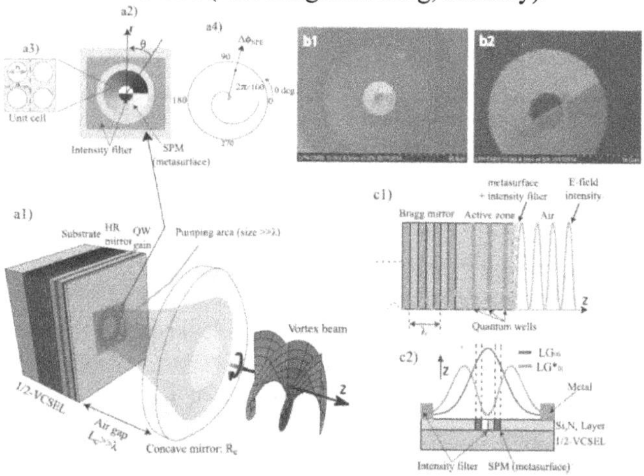

Figure 1. (a1) Vortex laser based on III-V semiconductor technology. Stable cavity with a millimeter air gap *Lc*. Pump diameter is 90 mm. (a2) SPM and intensity filter (loss masks). (a3) Elementary cell of the 2D sub-wavelength grating forming the metasurface. (a4) Perturbative azimuthal phase distribution induced by the SPM. (b1) Scanning electron microscope image of right-handed SPM surrounded by a Cr metallic mask. (b2) Zoom on SPM (20 mm external diameter). (c1) Longitudinal E-field distribution in the 1/2 VCSEL structure and SPM position. (c2) Longitudinal (XZ) cross section showing the overlap of the LG00 and LG01 modes with the SPM and the intensity filter, the peak of unwanted LG00 mode coincides with the high central loss [1,2].

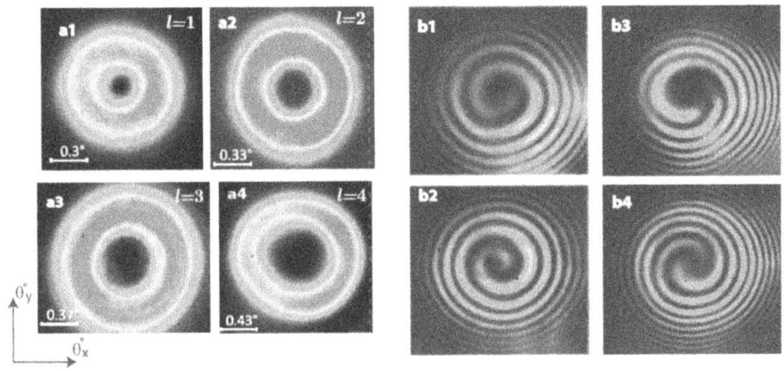

Figure 2. Intensity profiles of obtained vortex LG modes having helical wavefront: (a1) LG01, (a2) LG02, (a3) LG03 and (a4) LG04. Mach-Zehnder interferogram: (b1 (b2)) One-start spiral fringes corresponding to right (left) handed LG01 with OAM charge l = 1(-1). (b3 (b4)) Two-starts spiral fringes corresponding to right (left) handed LG02 with OAM charge l = 2(-2) [1].

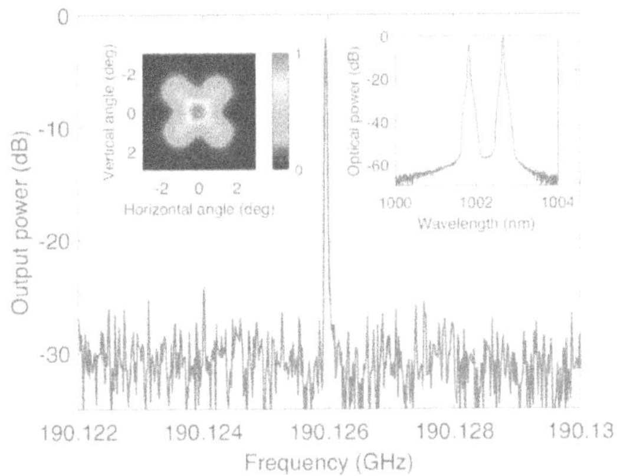

Figure 3. THz emission using UTC-PD and the dual-mode laser. The inset shows the far-field intensity map of the laser (left) and the dual-frequency spectrum (right) [3].

3. Coherent continuum generation: modeless laser

A modeless laser is a device which provides intense coherent broadband light fields. It finds applications in a high resolution spectroscopy, medicine, radar-lidar, metrology where a coherent broadband radiation is needed. In a modeless laser, the constructive interference of the optical wave, which would lead to a spectral mode structure, is prevented by the insertion of an intra-cavity frequency shift on each round trip. In order to eliminate the mode structure, the frequency shift should occur by a mechanism which does not simultaneously change the cavity length. In this work, we demonstrate for the first time a 300 GHz bandwidth (FWHM) continuous wave operation of a modeless

VECSEL [4]. The laser cavity design (Fig. 4) is based on a frequency-shifted-feedback laser design using an intracavity acousto optic frequency shifter (110 MHz shift, 92 % efficiency).

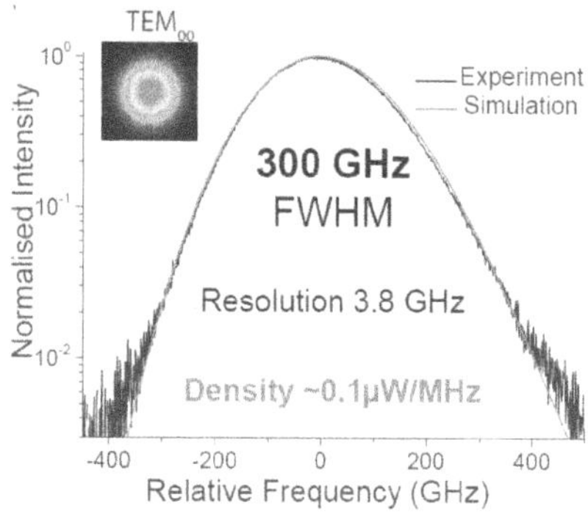

Figure 4. Modeless VECSEL optical spectrum. (inset) Far field beam profile.

The gain is provided by high gain GaAs based multiple quantum well semiconductor nanostructure emitting at 1070 nm. This semiconductor nanostructure includes a HR Bragg mirror. It is pumped with the single-mode pump, focused on a 50 μm spot. The emitted power is 30 mW, with TEM00 emission and strongly linear polarization (>20 dB). The measured coherence time is 19 μs, which is greater than that of a single frequency DFB laser.

References

1. M. S. Seghilani, M. Sellahi, M. Myara, I. Sagnes, G. Beaudoin, L. Legratiet, P. Lalanne, A. Garnache, *Scientific Reports* **6**, 38156 (2016).
2. M. S. Seghilani, M. Sellahi, M. Devautour, P. Lalanne, I. Sagnes, G. Beaudoin, X. Lafosse, L. Legratiet, M. Myara, A. Garnache, *Optics Express* **22**, 5962 (2014).
3. R. Paquet, S. Blin, M. Myara, L. Legratiet, M. Sellahi, B. Chomet, G. Beaudoin, I. Sagnes, A. Garnache, *Optics Letters* **41**, 3751 (2016).
4. M. Sellahi, M. Myara, G. Beaudoin, I. Sagnes, A. Garnache, *Optics Letters* **40**, 4301 (2015).

TEMPERATURE AND DEFORMATION DEPENDENCE OF OUTPUT WAVELENGTH AND POLARIZATION IN LASER DIODES BASED ON *p*-AlGaAs/GaAsP/*n*-AlGaAs NANOSTRUCTURES

E. V. Bogdanov, K. I. Kolokolov, N. Ya. Minina

M. V. Lomonosov Moscow State University, 119991 Moscow, Russia

bev@mig.phys.msu.ru

Numerical calculations of the optical energy gap and the optical gains g_{TE}, g_{TM} of TE and TM polarization modes in p-$Al_xGa_{1-x}As/GaAs_{0.84}P_{0.16}/n$-$Al_xGa_{1-x}As$ laser diode nanostructure are carried out for uniaxial compression up to 10 kbar along in-plane and normal to the heterostructure directions at 77–300 K.

1. Introduction

The results of this paper mainly concern the packaging-induced strain that might be caused in AlGaAs based laser devices by their soldering to a heat sink and, as well, the strain influence on the performance properties. Previously, strong effects of anisotropic stress on wavelength and polarization of the $A^{III}B^{V}$ based diode lasers emission were detected and investigated, for example by producing a bending stress in InGaAsP semiconductor laser chips [1].

In our recent papers [2,3] systematic experimental study and numerical calculations of external uniaxial compression P influence on the energy spectrum, wave functions, and the optical gain g_{TE}, g_{TM} of TE and TM polarization modes in p-$Al_xGa_{1-x}As/GaAs_{1-y}P_y/n$-$Al_xGa_{1-x}As$ laser diode nanostructures of different quantum well (QW) thickness and composition were performed for 77 K and $P \| [110]$ direction. Very good agreement between numerical calculations and experimental results for the structure with $GaAs_{0.84}P_{0.16}$ quantum well of 14 nm width under uniaxial compression along [110] direction makes the results of numerical calculations for p-$Al_xGa_{1-x}As/GaAs_{1-y}P_y/n$-$Al_xGa_{1-x}As$ laser diode structures, performed with other QW compositions, thickness, and other directions of applied stress, quite reliable.

Taking into account that laser diodes operate at about room temperature and mechanical strains, introduced by their soldering on a heat sink, may arise in different directions, we performed analogous calculations for compressions up to 10 kbar along [100] and [001] directions (in-plane and normal to the heterostructure) at 77 K and 300 K, shown together with the previous results [2,3].

2. Numerical calculations

Numerical calculations of uniaxial stress influence on the energy spectrum, wave functions, matrix elements of electron-photon interaction operator as well as the optical gain of TE and TM polarization modes in strained laser diode nanostructures have been performed for the experimentally investigated in Refs. [2,3] structure with $GaAs_{0.84}P_{0.16}$ QW of 14 nm width under uniaxial compression up to 10 kbar along [110], [100], and [001] directions at 77 K and 300 K. The Luttinger-Kohn Hamiltonian with strain terms was self-consistently solved together with Poisson's equation for the electrostatic potential using the finite-difference **k·p** method [2]. The program "Heterostructure Design Studio 2.1" wsa used. All necessary parameters were taken from [4,5].

3. Results and discussion

The results of calculations for $P\|[100]$ and $P\|[001]$ together with the previous data for $P\|[110]$ are represented in Fig. 1. The emitted light wavelength is determined by the optical gap E_{opt} that is equal to the difference of the energies of the electron and hole ground states in a QW. In the temperature interval under consideration E_{opt} increases and $\Delta E_{opt} \approx 77$ meV. In comparison with this value, its shift under compression is moderate and does not change between 77 K and 300 K being substantially anisotropic (Fig. 1a). This $\Delta E_{opt}(P)$ temperature behavior is explained by the absence of deformation potentials dependence on temperature and very low temperature dependence of elastic modulus [5]. From the other side, the results show that polarization of the emitted light can be extremely sensitive to the external uniaxial stress (Fig. 1b) due to the change of the wave functions and optical transitions between the subband levels in the quantum well [2,3]. If $P\|[100]$ (or $P\|[110]$) the increase of the applied uniaxial

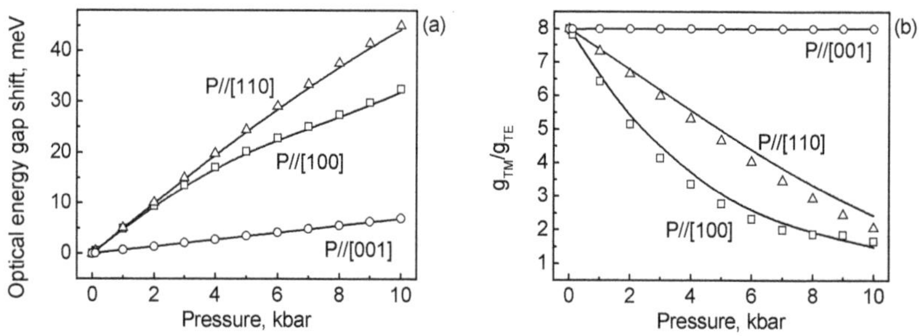

Figure 1. Pressure dependence of the optical energy gap (a) and g_{TM}/g_{TE} ratio (b) for the $GaAs_{0.84}P_{0.16}$ QW of 14 nm width under uniaxial compression along different directions at 77 K (lines) and 300 K (symbols).

compression leads to the strong change of the energy spectrum of light (LH) and heavy (HH) holes (Fig. 2). At $P = 0$ the uppermost level h1 and the next level h2 are pure LH states, at the same time the level h3 is a pure HH state. While the compression increases the HH state is admixed to LH ones. Finally the share of HH state becomes even dominant on the level h2 and it becomes almost of HH nature (Fig. 2c). These changes in the topmost states remove the prohibition on the interband transitions to occur only in the TM mode.

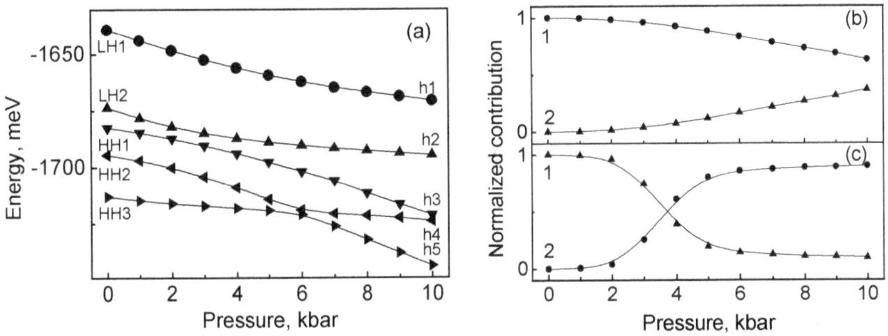

Figure 2. Calculated energy shifts of five hole levels at the zone center under uniaxial compression (a) and pressure dependence of light (1) and heavy (2) holes normalized contributions into wave functions of h1 (b) and h2 (c) levels for the $GaAs_{0.84}P_{0.16}$ QW of 14 nm width ($P \| [100]$, $T = 77$ K).

The results of calculations for the experimentally investigated laser diode structure demonstrate (Fig. 1b) that in the 14 nm QW with phosphorus fraction of 0.16 the g_{TM}/g_{TE} ratio at zero compression is equals to about 8. Under the compression of 5 kbar in [110] direction the g_{TM}/g_{TE} ratio strongly drops to approximately 5, *i.e.* in 1.6 times, that is in a good agreement with the experimental data on the decrease of the relative TM/TE intensity ratio with about 5 % of variance [3]. Even more strong decrease reveals g_{TM}/g_{TE} ratio in the case of compression along [100] direction (Fig. 1b). At the same time the g_{TM}/g_{TE} ratio does not show any change under compression along [001] because, being perpendicular to the heterostructure, uniaxial compression in this direction does not influence the wave function symmetry. The fact that under in-plane compression g_{TM}/g_{TE} ratio shows noticeable decrease even at moderate uniaxial stress and a small temperature dependence indicates, that p-$Al_xGa_{1-x}As/GaAs_{1-y}P_y/n$-$Al_xGa_{1-x}As$ nanostructures can be used in devices for TM-TE modes switching and, as well, for stress detection and characterization.

4. Conclusion

In this paper we have presented the results of numerical calculations of uniaxial stress influence on the optical gap and TM-TE polarization modes intensity in

p-$Al_xGa_{1-x}As/GaAs_{0.84}P_{0.16}/n$-$Al_xGa_{1-x}As$ laser diode nanostructure usually used in TM emitting 808 nm high-power semiconductor lasers. They indicate that the most effective blue shift of the emission spectrum is possible under compression in [110] direction. At the same time, the optical gain g_{TM}/g_{TE} ratio reveals strong several times decrease under in-plane [110] and [100] compression and, as it was shown previously [3], this effect may be specially increased by the proper choice of quantum well thickness and composition. In this aspect, some devices for effective TM and TE polarization modes tuning and switching under uniaxial stress based on p-$Al_xGa_{1-x}As/GaAs_{1-y}P_y/n$-$Al_xGa_{1-x}As$ laser diode nanostructures are possible. The calculations are also useful to anticipate and evaluate mounting and temperature induced strain effects on the diode laser performance and, quite the reverse, the detected shifts in parameters of laser emission may be applied for analysis of the arising strain. It should be mentioned that the demonstrated numerical calculations in the framework of the program "Heterostructure Design Studio 2.1" can be applied to other layered nanostructures of zinc blend $A^{III}B^{V}$ semiconductor materials.

References

1. C. S. Adams, D. T. Cassidy, *J. Appl. Phys.* **64**, 6631 (1988).
2. E. V. Bogdanov, N. Ya. Minina, J. W. Tomm, H. Kissel, *J. Appl. Phys.* **112**, 093113 (2012).
3. E. V. Bogdanov, H. Kissel, K. I. Kolokolov, N. Ya. Minina, *Semicond. Sci. Technol.* **31**, 035008 (2016).
4. I. Vurgaftman, J. R. Meyer, L. R. Ram-Mohan, *J. Appl. Phys.* **89**, 5815 (2001).
5. S. Adachi, *GaAs and Related Materials: Bulk Semiconducting and Superlattice Properties* (World Scientific, Singapore, 1999).

OPTICAL INTERCONNECTS BETWEEN SILICON CHIPS BASED ON LIGHT-EMITTING DIODES ON NANOSTRUCTURED SILICON

A. A. Leshok, A. V. Dolbik, Le Dinh Vi, A. I. Matskevich
Belarusian State University of Informatics and Radioelectronics
P. Browka 6, 220013 Minsk, Belarus

V. B. Vysotskii, S. V. Shvedov
OJSC "Integral" Kazintsa Sq. 1, 220108 Minsk, Belarus

Design and manufacturing technology of optical interconnects between silicon chips have been developed. Structures based on nanosize silicon clusters embedded in an alumina matrix are used as light-emitting diodes and photodetectors. The silicon microchannel plate vias are used as light-guiding channels. An experimental structure is fabricated and characterized with the current conversion efficiency of 0.3 %.

1. Introduction

The performance increase is currently one of the main tasks of the integrated electronics development. Replacement of electronic interconnects to optical ones will improve the performance of integrated circuits (ICs) by eliminating the resistive-capacitive delay of the metal wiring. We have developed and manufactured a system that provides optical interconnects between silicon chips based on silicon avalanche LEDs and silicon microchannel plate vias. Measurements showed a reproducible response of the optical signal of the avalanche LEDs registered by photodiodes on a silicon chip, which has no electrical commutation to another silicon chip on which the sources of the optical signal are located.

2. Experimental

Experimental structures were created by magnetron sputtering deposition of the aluminum/silicon composite film with 0.1 μm thickness and subsequent deposition of pure aluminium film with 1.0 μm thickness on n-type monocrystalline silicon wafer with a resistivity of 0.3 Ω·cm. An aluminum/silicon alloy with 30 at.% silicon content was used as a target material. The deposited films were subjected to anodic treatment in a 20 % aqueous solution of orthophosphoric acid via preformed photoresist masks at their surfaces. Various velocities of aluminum and silicon anodizing have resulted in the formation of the composite film of nanostructured silicon embedded in the alumina matrix. Not anodized areas protected by the photoresist mask formed metal electrodes between anodized regions. The photoresist mask has been removed after anodizing.

3. Results and discussion

The developed construction shown in Fig. 1 consists of two Schottky contacts and an anodic alumina layer separating aluminum electrodes. The anodic alumina layer contains silicon nanoparticles that emit light in the Schottky contact avalanche breakdown mode (LED). These Schottky contacts at a reverse bias below breakdown state (in this case up to 12 V) are sensitive to an external light signal, and therefore can be used as photodiodes (PD) [1,2].

Along with the silicon chips with diode structures that are able to generate and register the optical signal microchannel wafers of 100–150 μm thickness with through-holes (vias) with diameters of 5–6 μm are also produced. The distance between the centers of vias was 10 μm. These wafers are capable of transmitting an optical signal through the vias with the transmission ratio of 15–20 %. Such microchannel wafers called interposer, as they can be used to implement the commutation of "stackable" silicon ICs that has been demonstrated previously by filling the microchannels with copper [3]. The interposer formation was carried out using electrochemical anodization and subsequent thinning of silicon wafers by mechanical polishing [3].

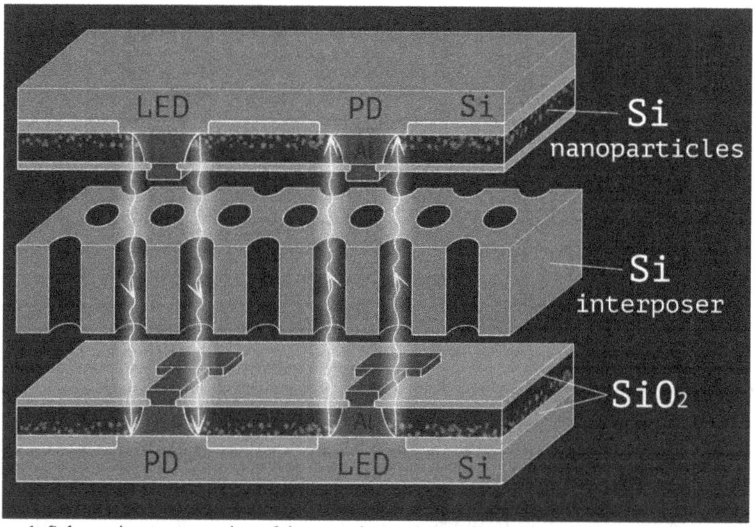

Figure 1. Schematic representation of the experimental structure of interchip optical interconnects.

Fig. 2 shows SEM images of the microchannel silicon wafer, providing an optical signal transmission in the vertical direction relative to the surface of silicon chips. The vias area occupies 20–25 % of the entire surface of the wafer.

a b

Figure 2. Cross-section (a) and bottom (b) images of the silicon microchannel wafer used for optical interconnects between silicon chips.

Fig. 3 shows the current-voltage characteristics of the avalanche diodes operating in the light signal detection mode. The optical signal is detected in the reverse bias range from 0 to 12 V as seen from the characteristics.

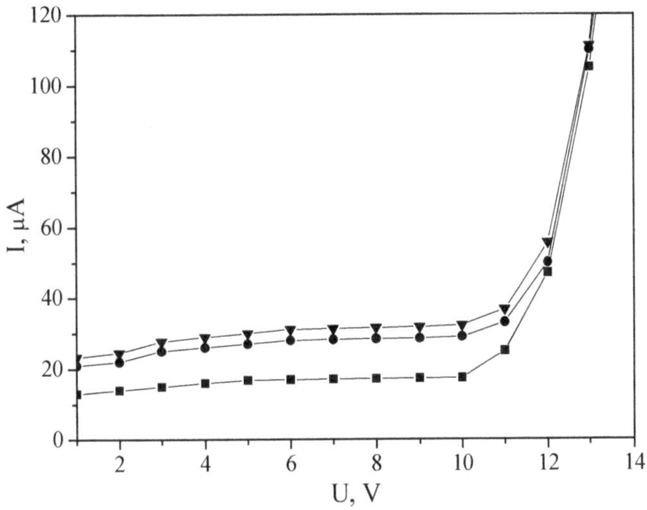

Figure 3. The current-voltage characteristics of the avalanche diodes with different methods of light exposure: squares – the dark characteristic, circles – light exposure with avalanche LEDs at a current of 30 mA; triangles – light exposure with a tungsten lamp.

The current-voltage characteristics with light exposure and without it do not differ after exceeding the reverse bias of 12 V, due to the appearance of self light emission in the investigated structures. Current-voltage characteristic of the diodes with exposure from an external light source (a tungsten lamp) that provides surface irradiance of 100 mW/cm^2 is also shown in Fig. 3. As can be seen from the characteristics, the response of the diodes to radiation from the tungsten lamp is comparable with the response to radiation from avalanche LED at the bias voltage of 15 V and current of 30 mA, which indicates the proportionality of optical flow of the both light sources.

4. Conclusion

The design and manufacturing technology of optical interconnects between silicon chips have been developed. Among the main characteristics of the developed system it should be noted that the ratio of the current conversion reaches 0.3 %, which opens up new possibilities for the development of silicon optoelectronics.

References

1. S. K. Lazarouk, A. A. Leshok, V. E. Borisenko, C. Mazzoleni, L. Pavesi, *Microelectron. Eng.* **50**, 81 (2000).
2. S. K. Lazarouk, P. V. Jaguiro, A. A. Leshok, V. E. Borisenko, *Physica E* **16**, 495 (2003).
3. S. K. Lazarouk, A. V. Dolbik, L. I. Stepanova, T. I. Bodryh, A. S. Turtsevich, S. V. Shvedov, V. A. Labunov in: *Mat. 22th Int. Conf. «Microwave & Telecommunication Technology»* (Sevastopol, Veber, 2012), p. 665.

PHOTOCURRENT IN THE DIODE STRUCTURE WITH STRONTIUM TITANATE XEROGEL FILMS ON P-TYPE SILICON

Xiang Wang[1], Zhiqun Cheng[1], H. Sohrabi Anaraki[2], N. V. Gaponenko[2], B. S. Kolosnitsin[2], P. A. Kholov[2], N. D. Manarkhovich[2], D. A. Golosov[2], V. A. Ivanov[3], N. I. Kargin[4]

[1]*Key Laboratory of RF Circuits and Systems, Ministry of Education Hangzhou Dianzi University, Hangzhou 310018, China*
[2]*Belarusian State University of Informatics and Radioelectronics P. Browka 6, 220013 Minsk, Belarus*
nik@nano.bsuir.edu.by
[3]*SSPA Scientific and Practical Materials Research Center NASB P. Browka 19, 220072 Minsk, Belarus*
[4]*National Research Nuclear University MEPhI Kashirskoe Shosse 31, 115409 Moscow, Russia*

A photosensitive thin film structure Ni/$SrTiO_3$/Si was fabricated using the sol-gel method. Current-voltage characteristics reveal a photocurrent at the forward and reverse bias under visible and UV illumination. The photocurrent effect was proposed to be originated due to non-equilibrium carriers generating at the defect level 1.27 eV below the bottom of the $SrTiO_3$ conduction band.

1. Introduction

Strontium titanate ($SrTiO_3$) is a wide-band gap semiconductor (> 3.2 eV) with interest for design of electronic devices due to its high dielectric constant, switching ability from high to low resistance states and photoconductivity. Strontium titanate films can be fabricated by the sol-gel method on silicon, mesoporous layers and quartz substrates.

Optical properties of the films strongly depend on synthesis conditions and thickness. The band gap of strontium titanate xerogels ranges from 3.4 eV to 3.9 eV for polycrystalline materials and attains 4.2 eV for amorphous state [1,2].

In this paper, we present the evaluation of current-voltage characteristics of the strontium titanate film formed by the sol-gel method and compare the data measured for Ni/$SrTiO_3$/Si structure without illumination and under illumination (white or UV light). Also, we have determined the band gap of strontium titanate films from optical transmission spectra and discussed the photocurrent nature in this structure.

2. Experimental

P-type silicon was used as a substrate. Strontium titanate single layer was deposited by spinning of the sol described in [3,4]. The main components of the sol were strontium acetate $Sr(CH_3COO)_2$ and titanium tetraisopropoxide $Ti(OCH(CH_3)_2)_4$. After spinning and drying, the film was subjected to final heat treatment at 750 °C for 60 min, revealing usually the polycrystalline structure [3,4]. To prepare the diode structure ($Ni/SrTiO_3/Si$) and measure its current-voltage characteristics, square shaped nickel electrodes (0.8×0.8 mm^2) were formed by magnetron sputtering.

The current-voltage characteristics were measured at room temperature in the voltage range of ± 3 V. The current-voltage characteristics were measured without illumination as well as under visible or UV illumination. A halogen lamp with intensity 57 mW/cm^2 was used as a visible white light source. To study the effect of UV irradiation on the photocurrent, the structure was illuminated with a 250 W mercury lamp irradiation passing through 310–390 nm filters.

3. Results and discussion

The fabricated diode structure $Ni/SrTiO_3/Si$ is shown in Fig. 1. The inset depicts the SEM image of the strontium titanate film in this construction.

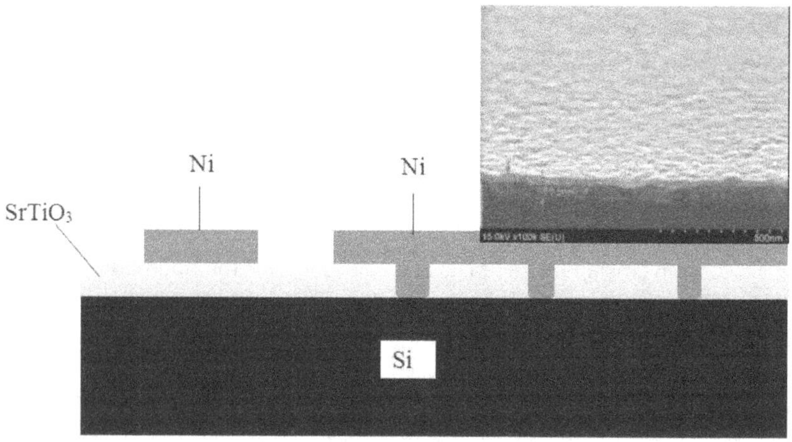

Figure 1. The metal/semiconductor/metal structure of the strontium titanate film.

According to the SEM data, the $SrTiO_3$ single layer thickness is 65 nm. The current-voltage characteristic of this structure without illumination corresponds to the typical Schottky barrier structure with a zero current at the reverse bias

(Fig. 2a). We assume that the bigger top electrode at the $SrTiO_3$ film provides an ohmic contact with silicon due to pinholes, while the smaller upper electrode provides the Schottky-type contact with $SrTiO_3$.

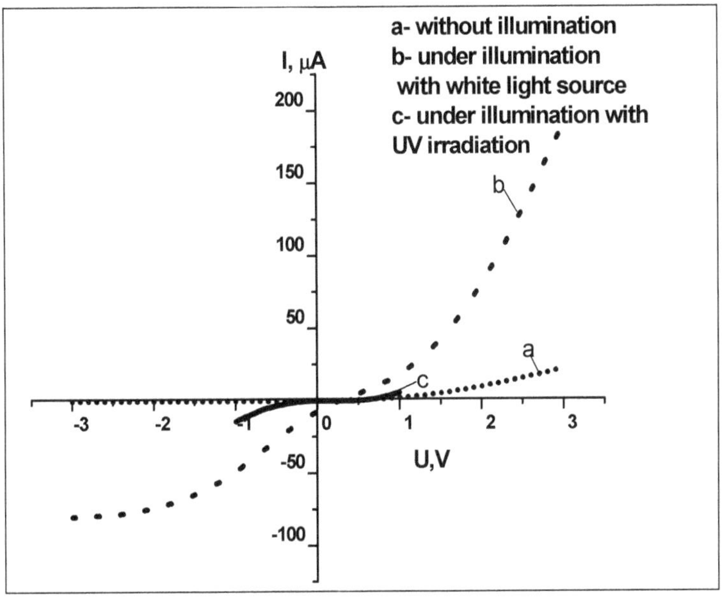

Figure 2. Current-voltage characteristics of Ni/$SrTiO_3$/Si structure with single-layer $SrTiO_3$ film: (a) without illumination; (b) under white light source illumination; (c) under UV illumination.

Current-voltage characteristics of the film illuminated with white light source (Fig. 2b) show that the current significantly increases both in forward and reverse bias. When the reverse bias voltage reaches −1 V, the photocurrent reaches 53 μA, whereas without illumination the reverse current is not registered. When the forward bias voltage is 1 V the photocurrent is 10 μA, however without illumination the current was much lower and did not exceed 1 μA (Fig. 2b).

The current-voltage characteristic of the diode structure under UV illumination is given in Fig. 2. Under UV illumination, the current increases basically at the reverse bias. When the reverse bias voltage is −1 V, the photocurrent is 14 μA, whereas the current at the forward bias voltage of 1 V is 6 μA.

The energy band diagram of the structure under study was depicted according to the band gap of strontium titanate (3.8 eV) determined from optical transmission measurements of similar $SrTiO_3$ film deposited on fused silica [2].

The presence of defect level located at 1.27 eV below the bottom of the conductive band of $SrTiO_3$ was confirmed in [5].

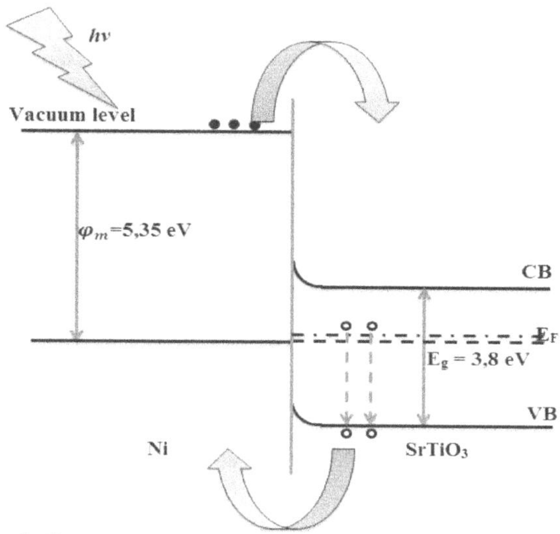

Figure 3. The energy band diagram of the $Ni/SrTiO_3/Si$ diode structure.

4. Conclusion

We have fabricated the diode $Ni/SrTiO_3/Si$ structure with the sol-gel derived $SrTiO_3$ film and investigated its current-voltage characteristics without illumination and under white light and UV irradiation. Based on the obtained data of photocurrent measurements, we can assume that in addition to the generation of non-equilibrium carriers in silicon, the carriers can be also generated due to ionization of defect levels in strontium titanate contributing to the observed photocurrent in the $Ni/SrTiO_3/Si$ structure under visible and UV irradiation.

References

1. G. Panomsuwan *et al.*, *J. Mater. Engin. Perf.* **22**, 863 (2012).
2. A. I. Kulak *et al.*, *J. Appl. Spectr.* **84** (2017) in press.
3. H. Sohrabi Anaraki *et al.*, *Semiconductors* **48**, 1685 (2014).
4. H. Sohrabi Anaraki *et al.*, *Phys. Sol. State* **57**, 2030 (2015).
5. Y. S. Kim *et al.*, *Appl. Phys. Lett.* **94**, 202906 (2009).

INTERRELATION OF PLANCK THERMAL GLOW WITH CURRENT SATURATION EFFECT OF CARBON NANOTUBE FIELD EMITTERS

J. Wu, Z. Cheng, L. Sun
Institute of Electron Device & Application, Hangzhou Dianzi University
Hangzhou, Zhejiang 310018, China
wujun@hdu.edu.cn

V. Labunov, B. Shulitski, I. Kashko, D. Grapov
Belarusian State University of Informatics and Radioelectronics
P. Browka 6, 220013 Minsk, Belarus
shulitski@bsuir.by

N. Kargin
National Research Nuclear University MEPhI, 115409 Moscow, Russia
NIKargin@mephi.ru

The interrelation of the emission current saturation effect with Planck glow in the visible spectral range of the matrix field emission carbon nanotube based cathodes was investigated. It was found that the beginning of the glow corresponds to the beginning of the transition into the emission current saturation mode. It was found that the main reason for the emission current saturation is high internal resistance of carbon nanotubes and its contact with the substrate.

1. Introduction

Since the discovery of carbon nanotubes (CNTs) by Iijima in 1991, continuous research of them reveals more and more unique properties and new areas of applications [1]. One of the promising and intensively studied fields of CNT applications is field-emission cathodes (FECs). FECs can be used in a series of devices, such as high brightness flat-panel displays, compact X-ray sources, powerful microwave amplifiers, high-performance luminescent lamps, compact high-frequency electron sources, sources of synchrotron radiation. The attractiveness of CNTs for these applications is due to their electrical, chemical, thermal and mechanical properties, that are reached owing to quasi one-dimensional structure and naturally high aspect ratio. Regardless of the field of FECs application, a number of requirements to their characteristics must be satisfied. Among them there are high emission efficiency, low threshold and operating voltages, high homogeneity, density and stability of the electron emission.

The theoretical limit of the current density of the field emission cathodes based on CNTs reaches 10^9 A/cm^2 [2]. For a number of important applications, particularly in the field of vacuum microwave electronics (transistors, klystrons, traveling wave tubes) the required current density is more than 1–10 A/cm^2 [3]. However, until recently the practical implementation of such cathodes is problematic. One of the reasons is the so-called emission current saturation effect [4]. This effect is manifested in the pronounced inflection, of the current-voltage characteristics and Fowler-Nordheim dependences at sufficiently large emission current densities.

A number of attempts to explain the origin of the saturation effect was taken. In [5] this effect was associated with the field amplification factor β, in the papers [6,7] with the presence of adsorbents at the CNTs apex. Authors of articles [4,8] demonstrated that it was due to a large voltage drop along the CNT emitter and emitter/substrate contact. Due to the complex nature of the field emission processes other explanations may be attracted. For example, in the case of silicon emitters the change of current-voltage characteristics slope and Fowler-Nordheim dependence have been attributed to the problem of the potential supply to the emitting part of FEC through a complex chain comprising silicon substrate, metal underlayer, carbon tubes and contact layers including semiconductor junctions and Schottky barriers creation in the chain of FEC feeding [9].

In this paper, we propose a mechanism of the emission current saturation by attracting well known experimental effect demonstrating an appearance of glow in the visible spectral range in the area of the field emission at sufficiently large emission current densities. The appearance of the glow in CNT FECs has been noted in [10–12], but its nature is still a controversial issue. In [11] the glow was associated with the phenomenon of fluorescence, in [12] with the effect of the filament, but without any proof. In [10], based on the results of emitted electrons energy distribution studies, it was shown that in the process of multi-walled CNTs emission the emitters could be heated up to 2000 K and above without substantial loss of emission properties. There has also been supposed that the glow in the CNT FECs may be related to the Plancks thermal radiation. In the case of sufficiently large current densities the temperature of emitters can be raised up to the temperature of the carbon material decomposition (4500–5000 K) [13]. It was shown by simulation [14,15] that at such temperatures there was a probability of transition from field electron emission to the thermo-field emission. However, as it is noted in [15], due to the considerable uncertainty of available data on the temperature dependences of transport coefficients for carbon nanotubes, it is not possible to construct by modeling the current-voltage characteristics of CNT FEC at high operating temperatures and high densities of the emission current. At the same time, in practice, just at these conditions a glow effect reveals under emissions, leading to the CNT FEC current saturation effect.

Thus, the experimental study of the relationship of the emission current saturation effect of the matrix CNT FEC with Planck glow of the emission area in the visible spectral range represents the doubtless interest. Such studies may provide useful information on the specificity of current saturation effect mechanism and the possibility to eliminate this effect and create as a result the CNT FECs with high density of the emission current (up to tens of A/cm^2 and more) that is required for the high frequency vacuum microelectronics.

2. Experimental

The arrays of CNTs were synthesized by CVD method realized by a high temperature pyrolysis of fluid hydrocarbon (*p*-xylene [C_8H_{10}]) mixed with the volatile source of catalyst (ferrocene [$Fe(C_5H_5)_2$]) using pre-structured localized catalyst on the *n*-type silicon wafers with high doping level KES-0, 01 (100) of the size of 5×5 mm^2. The content of ferrocene in this "feeding solution" was 0.1 wt.%. The process was conducted under atmospheric pressure by using Ar as a gas-carrier. The aerosol of the feeding solution is delivered into the synthesis zone by dosed injection at the rate of 0.15–0.5 cm$^3\cdot$min^{-1}, the speed of the flow of Ar was 100 cm$^3\cdot$min^{-1}. The synthesis process was realized in the tubular type quartz reactor of the specially constructed equipment at relatively low working temperature of 650 °C and synthesis time of 4 min. The speed of the cooling was determined by the process of natural cooling of reactor.

The localized catalyst is a metallic layer of Al\Ni with the thickness of 10 nm and 1 nm, respectively. To prevent interaction of Al with Si the Si-substrate was coated with a molybdenum film of 200 nm thickness. The topology of the test structure is obtained by lift-off photolithography.

As it is shown in our patent [16], the use of volatile catalyst Fe from $(C_5H_5)_2Fe$ source in combination with the multilayer localized catalyst Mo\Al\Ni, playing a role of the activator of metal-organic compound dissociation, insures obtaining the few-wall (2–6 walls) CNTs with a narrow diameter distribution of low defect density, high adhesion to the substrate and minimum micro and macro nonuniformity of CNTs height.

Study of the geometric features of the experimental samples of CNT FECs was carried out using a scanning (JEOL 6510) and transmission (JEOL 100CX) electron microscopes.

The field emission (FE) properties of the experimental samples of CNT FECs were investigated with the field emission scanning microscope (FESM) (Wuppertal University, Germany) at a residual gases pressure of 10^{-9} Torr in a cyclic measurement mode (multiple raising and dropping of the operating voltage). High-resolution video-camera built into the vacuum chamber, allowed to observe the emission region in real time.

SEM images of the matrix FECs emitting elements are presented in Fig. 1. The topology of emitting elements represents a set of simple elements of 2×4 μm^2 with 5 μm spacing. Active area of the samples is 0.25×0.5 mm^2 with an overall chip size of 1.2×3 mm^2.

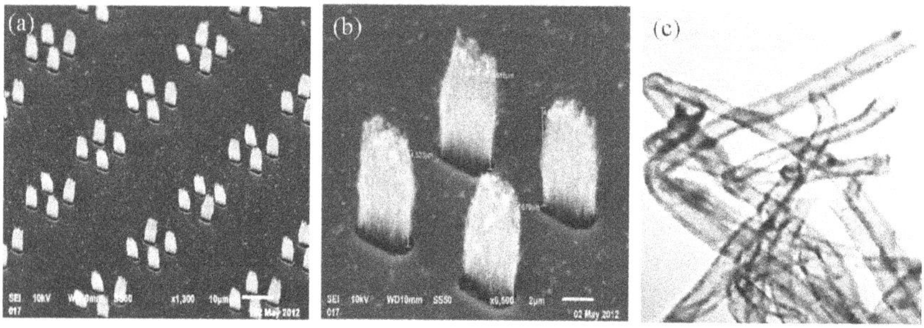

Figure 1. SEM images of the matrix FECs emitting elements: a, b – scanning electron microscopy; c – transmission electron microscopy.

Fig. 2 shows the current-voltage characteristics (CVC) and Fowler-Nordheim dependences of the CNT FEC samples in a cyclic measurement mode. Voltage sweep rate was 5 V/s. CVC is measured to a value of the emission current of 8×10^{-4} A, which corresponds to the emission current density of 0.64 A/cm^2 from cathode area of 1.25×10^{-3} cm^2 suitable for CNT FEC practical use. These parameters correspond to the best published results on CNT FECs.

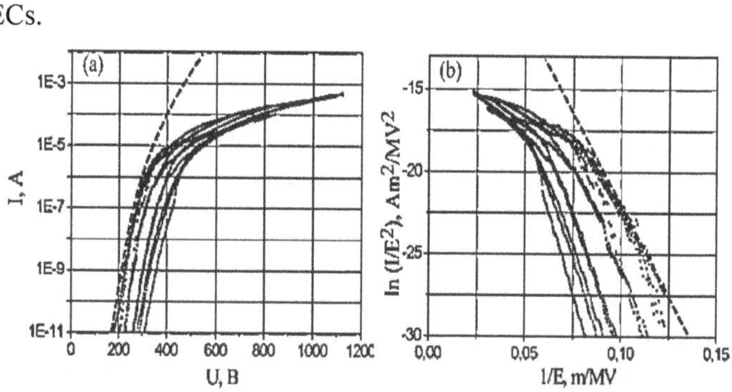

Figure 2. The current-voltage (a) and according Fowler-Nordheim (b) characteristics of CNT FEC sample with emission work area of 0.25–0.5 mm^2 obtained by the cyclic measurements.

As can be seen from the curves, cyclic loading leads to a shift of the emission threshold voltages towards higher values (from 170 V to 310 V), what may be associated with the processes of CNT emitters degradation at large emission current densities. Fig. 2 shows (dashed line) the theoretically expected Fowler-Nordheim dependences. But starting from the current of 10^{-6}–10^{-5} A the deviation of the experimental curves from the theoretical to the lower current values takes place (so-called "emission current saturation effect").

Fig. 3 shows the emission area of the same sample at different values of the emission current in the process of CVC measurements within the first cycle of measurements. The separate glowing centers in the emission area appear at the emission current of 10^{-5} A (Fig. 3a), what correlates with the appearing of the saturation effect of emission current on the current-voltage characteristics and the Fowler-Nordheim dependences (Figs. 2a,b). Nonhomogeneity of the glow along the emitting area is determined by the variation of geometrical parameters of the carbon nanotubes (height, diameter, distance).

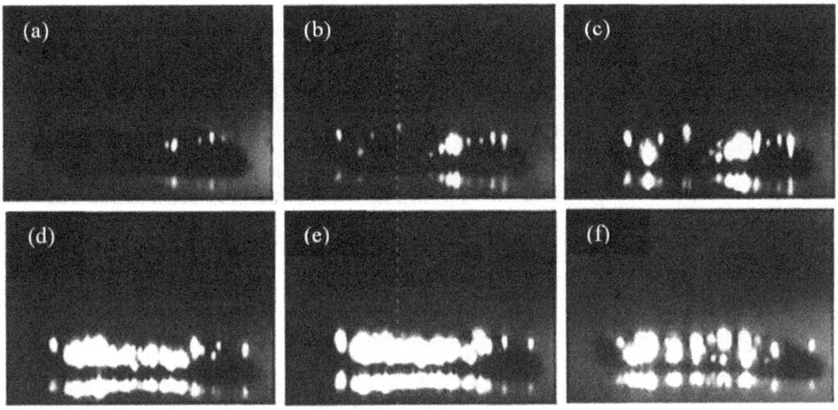

Figure 3. The micrographs of the CNT FEC emission area at different emission currents: (a) 10^{-5} A; (b) 1.5×10^{-5} A; (c) 4.0×10^{-5} A; (d) 10^{-4} A; (e) 5.4×10^{-4} A; (f) current failure to $I = 4.8\times10^{-4}$ A.

With the increasing voltage on electrodes the number, size and brightness of the emission centers increase, but the localization of the centers remains unchanged what is clearly seeing from Fig. 3b ($I = 1.5\times10^{-5}$ A) and Fig. 3c ($I = 4.0\times10^{-5}$ A).

With the further growth of the emission current the glowing centers conjugate into the continuous area of the glowing surface what is retained up to the top point on the CVC (8×10^{-4} A), which corresponds to Fig. 3e. With the further voltage increasing the emission current decreases sharply up to 4.8×10^{-4} A what indicates on the beginning of the process of overloaded emitters destruction (Fig. 3f). Thus, the experimental value of the emission current is many orders of magnitude smaller in the tested samples then theoretically expected values, which is shown by a dotted line in Figs. 2a,b.

3. Results and discussion

As it was already mentioned, many authors in a number of papers have tried to explain the origin of the emission current saturation effect [4–8]. One of the approaches is following. As it is known, the slope angle (tgα) in the F-N coordinates is proportional to the ratio of the material work function φ and the electric field enhancement factor β (tg$\alpha \approx \varphi/\beta$). The work function depends on the sorption-desorption processes of the residual gases on the emitter apex, and the emitter temperature [14], while the electric field enhancement factor β is mainly determined by the geometric parameters (say, aspect ratio) of the emitter. In [5] the effect of emitter current saturation is attributed to the change of the field enhancement factor β. The authors of [6,7] believe that the reason is the presence of adsorbates (adsorbed molecules or impurities) at the CNT apex. Another hypothesis is [8] that the presence of a resistance in series with the emitter can induce a saturation at high applied fields. The authors of [4] demonstrate both experimentally and through simulations that the emitter current saturation is due to a large voltage drop along the CNT emitter and/or at the CNT/substrate interface. Our experimental results show that the last hypothesis as the most convincing may be considered.

Indeed, in practice, the individual emitters, as a rule, are multi-walled CNTs. They are characterized by a significant defect structure of graphene walls, causing a low electrical conductivity of CNTs. Another problem is poor mechanical, electrical and thermal contact with the substrate. As follows from [17], the conductivity of multiwall CNTs in practice corresponds to the conductivity of graphite, therefore the resistance value of the CNT array is of the order of several MOhms. Moreover, one should keep in mind that due to the significant dispersion of CNTs geometric parameters in the cathode (height, diameter and distance) the main contribution to the emission current yield gives at best a few percent of the CNTs. Taking into account all this together the actual resistance of the CNT emitters can reach hundreds of MOhms.

Further, in accordance with an exponential nature of the Fowler-Nordheim dependence the resistance of the interelectrode gap is reduced with the increasing of the emission current, whereas the emitter resistance is constant. This situation leads to the fact that, beginning from the certain values of operating voltage only some part of the applied potential falls on the discharge gap, and a significant part of the potential falls on the high resistance emitting nanotubes. This in turn should lead to intensive heating of CNT emitters due to Joule heat, particularly on structural defects and in poor contact with the substrate. As a result, starting from a certain temperature the glow of the field emission area in the visible spectral range appears. Besides the further rise of the electrode voltage does not lead to a proportional increase in the voltage across the discharge gap, what eliminates a possibility of exponential increase of the emission current in the interelectrode gap.

Thus, it can be stated that the main reason for the emission current limitation (effect of current saturation) is the high internal resistance of CNTs and CNT-substrate contact junctions. This hypothesis is confirmed by the correlation of the beginning of current curves inflection and the intensive Planck glow of the field emission area in the visible spectral range. The appearance of the glow indicates CNT emitters heating to the temperatures over 1500 K. The heating in this case is due to the Joule heat generated by the current passing through the emitting nanotube array. At sufficiently large emission currents ($\geq 10^{-4}$ A) almost all supplied to the electrodes potential falls on the CNT array and its contact with the substrate, what leads to the further emitter heating and its irreversible thermal degradation and destruction.

4. Conclusion

The interrelation of the field emission current saturation effect with the field emission area glow in the visible spectral range was investigated. Field emission scanning microscope for spatially resolved CVC measurements and video camera built into the vacuum chamber of the measuring unit allowed to observe the emission region in real time.

It was found that the main reason for the emission current saturation can be associated with a high internal resistance of the CNT array. This is evidenced by the overlap of the beginning of the Fowler-Nordheim curves inflection and the intense glow of the electron emission area in the visible spectral range. For sufficiently large emission currents almost all supplied to the electrode potential falls on the CNT array and its contact with the substrate, what limits the possibility of further increase of the emission current.

Thus, to avoid the effect of the field emission current limitation of the CNT FECs a search of synthesis technology of carbon nanotubes with low electric resistance, minimal structural imperfections, unified geometric parameters (height, diameter, distance) on the entire working area of the cathode is required. Solving these problems will pave the way to obtaining high emission current densities for CNT FECs, required for creation a new generation of instruments and devices of vacuum microwave nanoelectronics.

Acknowledgments

We wish to thank professor G. Muller (University of Wuppertal, Germany) for his help in arranging the measurements on the field emission by scanning microscope.

References

1. L. A. Chernozatonskii *et al.*, *Chem. Phys. Lett.* **233**, 63 (1995).
2. G. N. Fursey *et al.*, *J. Vac. Sci. Techn. B* **16**, 232 (1998).
3. E. Minoux *et al.*, *NanoLetters* **5**, 2135 (2005).
4. W. I. Milne *et al.*, *J. Vac. Sci. Technol. B* **24**, 345 (2006).
5. H. M. Manohara *et al.*, *J. Vac. Sci. Techn. B* **23**, 157 (2005).
6. V. Semet, *Appl. Phys. Lett.* **81**, 343 (2002).
7. K. A. Dean, *Appl. Phys. Lett.* **76**, 375 (2000).
8. J.-M. Bonard, *Phys. Rev. B* **67**, 115406 (2003).
9. P. Serbun *et al.*, *J. Vac. Sci. Techn. B* **31**, 02B10 (2013).
10. S. Purcell *et al.*, *Phys. Rev. Lett.* **88**, 105502 (2002).
11. J.-M. Bonard *et al.*, *Appl. Phys. A* **69**, 245 (1999).
12. A. G. Rinzler *et al.*, *Science* **269**, 1550 (1995).
13. D. A. Usanov, R. K. Yafarov, *Study of Autoelectronic Emission from Nanocarbon Materials* (Saratov, 2006), 23 p.
14. A. B. Petrin, *JETP* **109**, 314 (2009).
15. G. S. Bocharov, A. V. Eletskii, *Technical Physics* **52**, 498 (2007).
16. A. N. Saurov *et al.*, *Eurasian Patent* EA016557. Date of issue 30.05.2012.
17. H. Kataura *et al.*, *Synthetic Metals* **103**, 2555 (1999).

TOP AND BOTTOM SPIN VALVES BASED ON NiFeMn ANTIFERROMAGNETIC TRIPLE ALLOYS

I. V. Blinov, T. P. Krinitsina, M. A. Milyaev, V. V. Popov, V. V. Ustinov
M. N. Miheev Institute of Metal Physics, 620990 Ekaterinburg, Russia
blinov@imp.uran.ru

Structure, magnetic and magnetoresistive properties of spin valves with NiFeMn antiferromagnet as a pinning layer have been studied. A technique of fabrication of spin valves with an enhanced thermal stability and improved hysteretic characteristics has been elaborated.

1. Introduction

Bilayers of permalloy/antiferromagnetic triple NiFeMn alloy have been studied, as well as magnetic and magnetoresistive properties of spin valves (SVs) with NiFeMn antiferromagnet (AF) as a pinning layer. A technique of fabrication of bottom spin valves with an enhanced thermal stability and improved hysteretic characteristics has been elaborated.

2. Experimental details

The samples were made by *DC* magnetron sputtering and by electron-beam evaporation on glass (Corning) substrates and single-crystalline sapphire (1012). To form a unidirectional anisotropy the magnetic field of 110 Oe was applied in the process of nanostructures growth, and the thermal-magnetic treatment was performed at pressure of 10^{-4} Pa in the permanent magnetic field of 2 kOe applied in the sample plane at 260 °C for 4 h. The exchange bias field temperature dependence was measured in the temperature range of 20–260 °C. Etching of samples for preparation of bottom SVs was carried out in a device for reactive ion-plasmic etching PlasmaPro NGP 80 RIE Oxford.

3. Results and discussion

According to the results of the studies carried out, the bilayers with antiferromagnetic triple alloy $Ni_{14}Fe_{6}Mn_{80}$ are characterized by moderate energy of exchange interaction $J_{ex} = 0.05$ erg/cm^2 and medium blocking temperature $T_b = 150$ °C.

Higher $T_b = 270$ °C (Fig. 1) and $J_{ex} = 0.27$ erg/cm^2 were obtained for the annealed manganese/permalloy bilayers. In this case, an ordered AF phase of NiFeMn is formed [1]. The ordered AF phase formation is testified by an appearance of super-structural Debye rings (100), (110), (210), (211) (indicated by arrows in Fig. 2a) in electron diffraction patterns of sample Al_2O_3/Mn(50 nm)/$Ni_{77}Fe_{23}$(30 nm)/Ta(5 nm) after its annealing in the magnetic field at 260 °C for 4 h. In this case in the electron-microscope images one can see the columnar structure, and an intermediate layer between manganese and permalloy layers is absent (Fig. 2b).

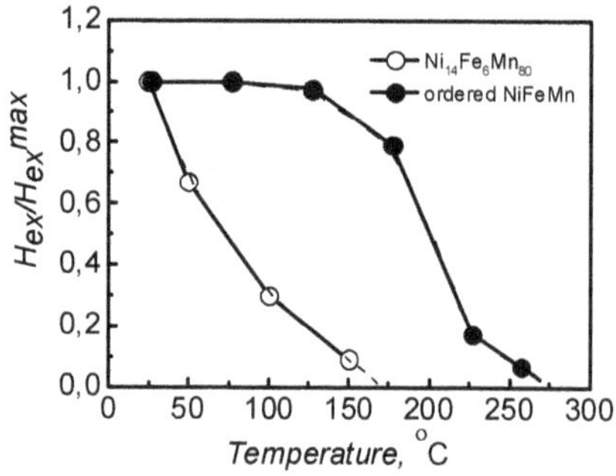

Figure 1. Temperature dependence of the bias field (H_{ex}) for two types of AF used.

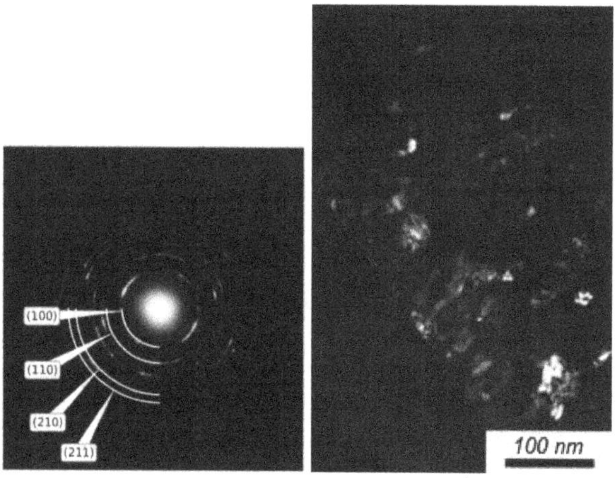

Figure 2. Electron diffraction pattern (a) and dark-field image in (100) reflection (b) of a cross-section TEM image of Al_2O_3/Mn(50 nm)/$Ni_{77}Fe_{23}$(30 nm)/Ta(5 nm) after thermal-magnetic treatment at 260 °C for 4 h. Super-structural reflections are shown by arrows.

Based on the ordered AF NiFeMn phase the spin valves with an enhanced thermal stability can be fabricated.

As shown by the studies of an effect of the AF layer $Ni_{14}Fe_6Mn_{80}$ thickness on the magnetoresistance and shift of the pinned layer $Co_{90}Fe_{10}$ loop of the top spin valve glass/Ta/$Ni_{80}Fe_{20}$/$Co_{90}Fe_{10}$/Cu/$Co_{90}Fe_{10}$/$Ni_{14}Fe_6Mn_{80}$/Ta, the combination of maximal magnetoresistive effect and pinned layer $Co_{90}Fe_{10}$ loop shift is observed at $t_{AF} = 25$ nm. At this thickness $(\Delta R/R_s) = 6.95$ %.

To study the magnetoresistance dependence of a spin valve on the copper layer thickness the samples with $t_{AF} = 25$ nm were prepared. With increasing copper layer thickness the magnetoresistance $\Delta R/R_s$ at first increases and then decreases [2]. The maximal value of $\Delta R/R_s = 7.30$ % corresponds to $t_{Cu} = 2.8$ nm (Fig. 3). The magnetoresistance sensitivity determined as an average value for the ascending and descending hysteresis loop of the free layer $Ni_{80}Fe_{20}/Co_{90}Fe_{10}$ is $\Delta(\Delta R/R_s)/\Delta H = 0.75$ %/Oe. The data obtained demonstrate possibility of using disordered alloy $Ni_{14}Fe_6Mn_{80}$ as a pinning layer in top spin valves.

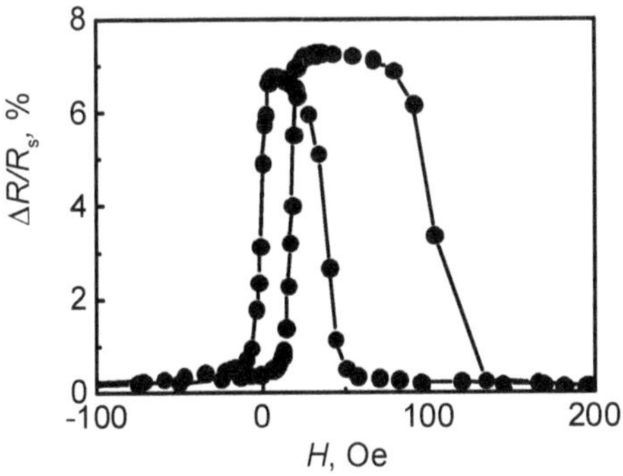

Figure 3. Field dependence of magnetoresistance of a top spin valve glass/Ta(5 nm)/$Ni_{80}Fe_{20}$(2 nm)/$Co_{90}Fe_{10}$(5.5 nm)/Cu(2.8 nm)/$Co_{90}Fe_{10}$(5.5 nm)/$Ni_{14}Fe_6Mn_{80}$ (25 nm)/Ta(3 nm).

To fabricate a bottom SV with the ordered AF phase NiFeMn a technological cycle has been worked out, including the following operations:

1) Formation of the ordered AF phase NiFeMn in a sample $Al_2O_3/Ni_{77}Fe_{23}$(5 nm)/Mn(50 nm)/$Ni_{77}Fe_{23}$(30 nm)/Ta(5 nm) by the thermal-magnetic treatment at 260 °C for 4 h.

2) Ion etching for 20 min of the annealed sample for the surface layer removing. The ion etching duration was chosen to assure the removal of the contaminated surface layer and retaining the ferromagnetic area together with the ordered AF phase required for the formation of the unidirectional anisotropy in the FM layer sputtered on the sample after etching.

3) Magnetron sputtering of the layered structure consisting of ferromagnetic $Co_{90}Fe_{10}$ layers separated by Cu on the as-prepared sample Al_2O_3/NiFeMn.

4) Annealing of the as-prepared structure in a magnetic field at $T_{ann} = 300$ °C for 15 min, the annealing temperature $T_{ann} = 300$ °C being higher

than $T_b = 270$ °C for $Al_2O_3/Ni_{77}Fe_{23}$(5 nm)/Mn(50 nm)/$Ni_{77}Fe_{23}$(30 nm)/Ta(5 nm).

The magnetoresistance of the as-fabricated spin valve Al_2O_3/NiFeMn/$Co_{90}Fe_{10}$(5.5 nm)/Cu(3.6 nm)/$Co_{90}Fe_{10}$(5.5 nm)/Ta(5 nm) is $\Delta R/R_s = 3.8$ % (Fig. 4).

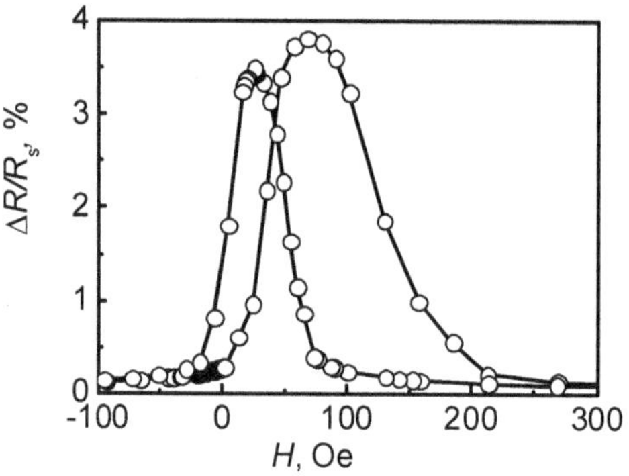

Figure 4. Field dependence of the magnetoresistance of the bottom spin valve Al_2O_3/NiFeMn/$Co_{90}Fe_{10}$(5.5 nm)/Cu(3.6 nm)/$Co_{90}Fe_{10}$(5.5 nm)/Ta(5 nm).

4. Conclusion

A possibility of application of the NiFeMn AFM for SV devices is demonstrated on a sample of composition Ta/$Ni_{80}Fe_{20}$ /$Co_{90}Fe_{10}$/Cu/$Co_{90}Fe_{10}$/$Ni_{14}Fe_6Mn_{80}$/Ta. A technology of fabrication of a spin valve based on the ordered AF phase NiFeMn has been worked out. A spin valve with giant magnetic resistance (GMR) exceeding the GMR of previously known structures has been created, and the possibility of using the ordered AF phase NiFeMn as a pinning layer in a spin valve has been proved.

Acknowledgments

The research has been done within the IMP state assignment “Spin”, with partial support by UB RAS (project 15-9-2-44) and RFBR (project 16-32-00128).

References

1. I. V. Blinov, T. P. Krinitsina, S. A. Matveev, M. A. Milyaev, P. N. Sedova, V. V. Popov, V. V. Ustinov, *J. Nanosci. Nanotech.* **12**, 7562 (2002).
2. I. V. Blinov, T. P. Krinitsina, M. A. Milyaev, V. V. Popov, V. V. Ustinov, *Sol. Stat. Phenomena* **233-234**, 517 (2015).

SnS_x-BASED NANOSTRUCTURED LAYERS FOR SOLAR CELLS

E. A. Outkina, A. I. Vorobyova
Belarusian State University of Informatics and Radioelectronics
P. Browka 6, 220013 Minsk, Belarus
outkina@bsuir.by

A. A. Khodin
"Optics, Optoelectronics and Laser Techniques" Association, NASB
220072 Minsk, Belarus
alexhodin@msn.com

SnS_x-based nanostructured thin films and photocells thereof fabricated by chemical processing were investigated. Optical absorption characteristics of single-layer SnS_2 and double-layer SnS_x/SnS_2 composites have been studied demonstrating $E_g \sim 2.1$ to 2.9 eV due to ~10 nm nanograins morphology. The test SnS_x/ZnO photocells have been fabricated and tested showing the conversion efficiency ~1.6 %.

1. Introduction

Tin sulfide based thin-film semiconductors such as SnS_x (and its derivative Cu_2SnZnS_4) are considered as promising materials for solar cells. They possess optical band gap close to the optimum for efficient light absorption in a thin film, exhibit mainly *p*-type electrical conductivity, holes mobility of $\sim 90\ cm^2V^{-1}s^{-1}$ [1] and high light absorption coefficient $\alpha > 10^4\ cm^{-1}$ [2]. A valuable advantage of SnS_x is that it could be produced of abundant elements by simple low-cost techniques. Due to weak Van der Waals forces between Sn-S double layers in the crystal lattice it could facilitate oriented growth of crystals, especially on template substrates.

In the paper, the results on SnS_x-based nanostructured thin films and photocells thereof fabricated by electrochemical and chemical processing are presented. The optical transmission and absorption characteristics of single-layer SnS_2 and double-layer SnS_x/SnS_2 composites have been investigated. The test SnS_x/ZnO photocells have been fabricated and tested.

2. Experimental

To deposit SnS films, the SILAR deposition process was performed using two types of substrates: glass-ceramic and glass of 60×48 mm with 30 nm thickness Ta sublayer and 1 μm thick Al layer.

The 0.1 M Na_2S water solution (pH=12.5) was used as an anion precursor followed by dilution to the concentration 0.05 M. The 0.05 M $SnCl_2$ water solution (pH=1.25) was used as a cation precursor. The amount of the prepared solution was 100 ml. Then, 5 ml of 0.05M NaCl and ~1 ml of triethanolamine were added to this solution.

The SILAR process included: substrate vertical dipping into the NaCl solution for 8 s at room temperature, dipping into water for 8 s (not rinsing), dipping into Na_2S solution for 8 s, dipping into water for 8 s. The water bath for the dippings was the same. The number of dipping cycles (for example, 30) depended on desired thickness of the deposited layer.

The test photocells using SnS_x were fabricated using porous alumina anodizing technology to prepare bottom nanostructured electric contact and windows for subsequent deposition of semiconductor layers. Wide bandgap ZnO was vacuum deposited through the contact mask followed by the frontal electric contact grid vacuum deposition.

The optical transmission and absorption characteristics of SnS_x thin films were measured with PV 1251 spectrophotometer (SOLAR). The fabricated experimental photocells were measured using laboratory simulator under ~AM1.5 conditions at 100 mW/cm^2 flux, 16 °C.

3. Results and discussion

To determine the semiconductors bandgap E_g, the known dependence of the absorption coefficient α on photons energy $h\nu$ was used [3]:

$$\alpha \cdot h\nu \sim \left(h\nu - E_g\right)^m, \qquad (1)$$

where m = 1/2, 2, 3/2, 3 for direct allowed, indirect allowed, direct forbidden and indirect forbidden inter-band transitions in semiconductor, respectively.

As a result of analysis of the characteristics obtained, the different behavior of $\alpha(h\nu)$ for the single-layer SnS_2 thin-film system and double-layer SnS_x/SnS_2 heterostructure has been established. The single-layer SnS_2 structure was characterized by spectral absorption edge corresponding to indirect forbidden transitions for $E_g = 2.92$ eV. For SnS_x/SnS_2 double-layer structure, the indirect allowed transitions were observed for $E_g = 2.11$ eV and 2.23 eV (Fig. 1).

The observed E_g values do not match to bulk stoichiometric SnS or SnS_2. At the same time, $E_g \sim 2.1$ eV could correspond to intermediate Sn_2S_3. For two-layer system, the E_g~2.9 eV is not observed apparently, because the less-bandgap layer with E_g~2.1 eV masks absorption in the wider bandgap SnS_2.

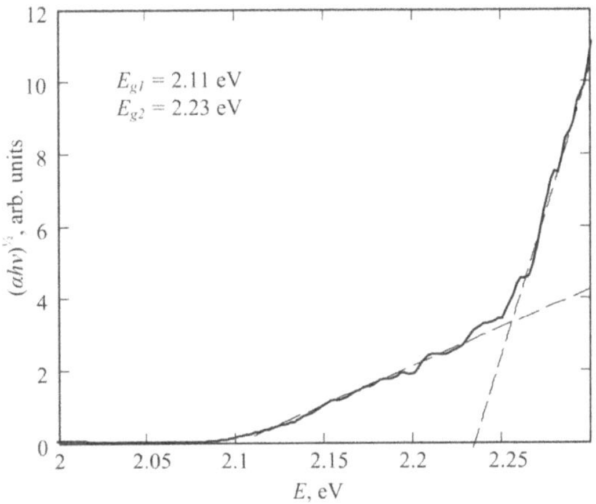

Figure 1. Optical transmission of double-layer SnS_x/SnS_2 thin-film system on glass substrate.

The $E_g \sim 2.9$ eV value for SnS_2 layer could indicate the grain size dependence of E_g [4]:

$$E_g^{nano} = E_g^{bulk} + \frac{\hbar^2 \pi^2}{2m^* r^2}, \quad (2)$$

where r is the grains size, m^* is the semiconductor charge carriers effective mass. As it was mentioned in [5], the E_g for SnS_2 films could vary in the range of 2.8–3.0 eV for the film thickness 350–580 nm and grains size of 26–38 nm. The effect is applicable to monosulphide SnS, too, where E_g could gain ~2.1 eV for the grains of 10–12 nm [6].

Thus, the results obtained evidence the nanograins (~10 nm) effects on the bandgap in SnS_x-based thin film systems. This is supported by preliminary theoretical analysis of nanometer-size effects in tin sulphide films [7,8]

The test cells topology is presented in Fig. 2a. The measured photocells show the conversion efficiency up to 1.59 % (Fig. 2b).

The rather low conversion efficiency is because of high resistive losses in ZnO layer (which resistivity was ~ 9 kOhm/sq.) and shunting losses R_s was from 1.1 to 1.6 Ohm, and R_{sh} was ~0.3 Ohm. ZnO doping is underway to improve the photocell performance.

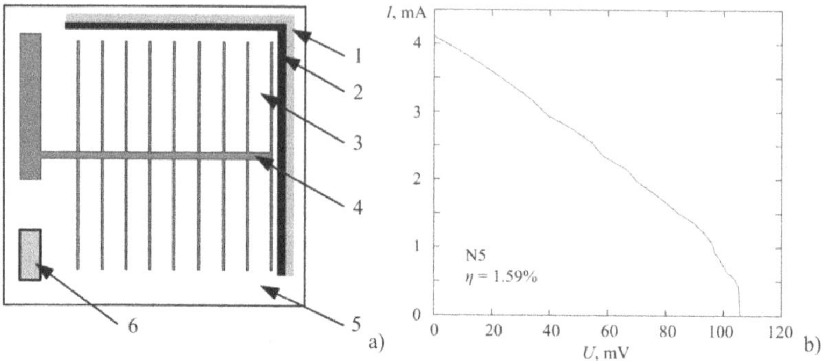

Figure 2. Topology schematic view (a) and current-voltage characteristic of test SnS/ZnO photocell (b): 1 – back nanostructured contact; 2 – SnS; 3 – ZnO; 4 – frontal electrode grid; 5 – barrier (dense) alumina; 6 – contact pad to back nanostructured electrode.

4. Conclusion

The results obtained show the impact of SnS_x layers nanoscale morphology on their optoelectronic characteristics. The SnS/ZnO test photocells demonstrated performance which could be improved by ZnO layer doping and shunting resistance optimization.

References

1. J. B. Johnson, H. Jones, B. S. Latham, J. D. Parker, R. D. Engelken, C. Barber, *Semicond. Sci. Technol.* **14**, 501 (1999).
2. N. K. Reddy, K. Ramesh, R. Ganesan, K. T. R. Reddy, K. R. Gunasekhar, E. S. R. Gopal, *Appl. Phys. A* **83**, 133 (2006).
3. G. Burns, *Solid State Physics* (Academic Press, Orlando, 1985).
4. L. E. Brus. *J. Chem. Phys*. **80**, 4403 (1984).
5. K. T. R. Reddy, G. Sreedevi, R. W. Miles, *J. Mater. Sci. Eng. A* **3**, 182 (2013).
6. Priyal Jain, P. Arun, arXiv: 1207.2830v2 [*cond-mat.mtrl-sci*] 6 Nov 2012.
7. V. L. Shaposhnikov, A. V. Krivosheeva, V. E. Borisenko, J.-L. Lazzari, *ScienceJet* **1**, 1 (2012).
8. V. L. Shaposhnikov, A. V. Krivosheeva, *in*: *Physics, Chemistry and Application of Nanostructures*, ed. by V. E. Borisenko, S. V. Gaponenko, V. S. Gurin, C. H. Kam (World Scientific, Singapore, 2015), pp. 39–42.

AUTHOR INDEX

www.ingramcontent.com/pod-product-compliance
Lightning Source LLC
LaVergne TN
LVHW010538300926
849162LV00002B/171
* 9 7 8 9 8 1 3 2 2 4 5 2 0 *